**교육의 힘으로
세상의 차이를 좁혀 갑니다**

차이가 차별로 이어지지 않는 미래를 위해
EBS가 가장 든든한 친구가 되겠습니다.

모든 교재 정보와 다양한 이벤트가 가득!
EBS 교재사이트 book.ebs.co.kr

본 교재는 EBS 교재사이트에서
eBook으로도 구입하실 수 있습니다.

수능특강

영어영역 | 영어독해연습

기획 및 개발	감수	책임 편집
김현경(EBS 교과위원)	한국교육과정평가원	이진옥
김 단(EBS 교과위원)		
양성심(EBS 교과위원)		

본 교재의 강의는 TV와 모바일 APP, EBS*i* 사이트(www.ebsi.co.kr)에서 무료로 제공됩니다.

발행일 2025. 1. 31. **1쇄 인쇄일** 2025. 1. 24. **신고번호** 제2017-000193호 **펴낸곳** 한국교육방송공사 경기도 고양시 일산동구 한류월드로 281
표지디자인 디자인싹 **내지디자인** ㈜글사랑 **내지조판** 다우 **인쇄** ㈜타라티피에스
인쇄 과정 중 잘못된 교재는 구입하신 곳에서 교환하여 드립니다. **신규 사업 및 교재 광고 문의** pub@ebs.co.kr

정답과 해설 PDF 파일은 EBS*i* 사이트(www.ebsi.co.kr)에서 내려받으실 수 있습니다.

교 재 내 용 문 의 교재 및 강의 내용 문의는 EBS*i* 사이트 (www.ebsi.co.kr)의 학습 Q&A 서비스를 활용하시기 바랍니다.	**교 재 정오표 공 지** 발행 이후 발견된 정오 사항을 EBS*i* 사이트 정오표 코너에서 알려 드립니다. 교재 → 교재 자료실 → 교재 정오표	**교 재 정 정 신 청** 공지된 정오 내용 외에 발견된 정오 사항이 있다면 EBS*i* 사이트를 통해 알려 주세요. 교재 → 교재 정정 신청

나를 가장 나답게!

덕성은 자유다

수도권 대학 최초 전면 자유전공제 실시,
2024학년도 가상현실융합학과와
데이터사이언스학과 신설에 이어
2025학년도 AI신약학과와 자유전공학부로
또 한 걸음 나아갑니다.

노위연 재학생
(심리학전공 23학번)

유전공학부로 자유로운 전공 탐색

동안 전공 탐색 후 2학년 진학 시 계열과 무관하게 전공·학부 선택 가능

학 내 모든 전공을 인원 제한 없이 100% 제1전공으로 선택 가능

단, 유아교육과, 약학과, Art & Design대학, 미래인재대학(가상현실융합학과, 데이터사이언스학과, AI신약학과)은 제외

2026학년도 신·편입학 안내 | 입학안내 **enter.duksung.ac.kr** 문의전화 **02-901-8189/8190**

덕성여자대학교
DUKSUNG WOMEN'S UNIVERSITY

서일에서 내일로

입학안내

서일에서 LEVEL UP

7호선
(서일대입구)
면목

**2026 학년도 서일대학교
신입생 모집일정**

[수시 1차] 2025. 09. 08(월) ~ 2025. 09. 30(화)
[수시 2차] 2025. 11. 07(금) ~ 2025. 11. 21(금)
[정　　시] 2025. 12. 29(월) ~ 2026. 01. 14(수)

서일대학교
SEOIL UNIVERSITY

수능특강

영어영역 | 영어독해연습

이 책의 특징, 활용법 및 구성

이 책의 **특징**

본 교재는 2026학년도 대학수학능력시험을 준비하는 데 도움을 주고자 제작되었으며, 2015 개정 영어과 교육과정을 반영하였다. 특히 학생들의 읽기 능력 신장을 목적으로 하여 교과 내용을 토대로 더욱 다양하고 참신한 주제·소재·분야의 글과 정보를 제시하였고, 교육과정상의 어휘 범주와 지문의 내용을 고려하여 제작되었다. 학교에서 영어 I, 영어 II 교과서를 바탕으로 영어 독해의 기본 개념과 어휘를 철저히 익힌 후, 본 교재를 활용한 독해 훈련으로 실제 응용력을 키우면, 교육과정 성취 목표 도달과 함께 대학수학능력시험 대비에 크게 도움이 될 것으로 기대된다.

이 책의 **효과적인 활용법**

본 교재를 효과적으로 활용하여 독해 능력을 향상하고 수능 영어 읽기 영역에 완벽하게 대비하기 위해서 학업 성취 기간을 정하여 '이 책의 차례'에 제시된 학습 내용을 차근차근 공부하되 다음과 같은 사항에 유의하도록 하자.

능동적이고 적극적인 학습 자세

수동적이고 소극적으로 강의를 듣기만 할 것이 아니라 강의하는 선생님과 토론하고 대화한다는 자세로 임하는 것이 중요하다. 이와 같은 자세는 핵심적인 부분의 이해와 학습 내용의 기억에 도움이 된다. 또한 중요한 사항을 교재 여백에 메모하여 강의를 듣는 것이 좋다. 이것은 학습 내용의 심층적인 이해와 효율적인 복습을 위해 꼭 필요하다.

예·복습과 정리를 통한 내재화

영어 학습은 유의미한 내재화가 중요하다. 따라서 학습 내용을 미리 예습하고 여러 번 복습하면서 주요 어휘와 문장 구조를 파악하고 글의 내용을 자신이 이미 알고 있는 지식과 연관하여 문제 풀이에 접목할 수 있어야 한다.

이 책의 **구성**

틀리기 쉬운 유형편 01 ~ 12강

수능 읽기 영역에서 틀리기 쉬운 11가지 유형을 집중 학습할 수 있도록 강과 문항을 유형별로 구성하였다. 다양한 소재와 주제의 지문을 통해 어려운 유형을 집중적으로 학습하여 해당 유형에 대한 문제 해결력을 높일 수 있도록 하였다.

Sentence Structure

각 지문 중에서 어법상 중요하거나 글의 내용 파악에 핵심이 되는 문장을 발췌하여 철저한 분석과 구문 설명을 제시함으로써 지문에 대한 정확한 이해를 돕고자 하였다.

Word Search

지문에서 알아두어야 할 어휘의 예문을 제시하고 문맥상 적절한 영어 단어를 지문에서 찾아 빈칸에 직접 써 보게 함으로써 어휘의 뜻을 정확히 파악할 수 있도록 하였다.

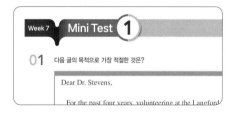

Mini Test 1 ~ 3회

Mini Test 3회분으로 구성되어 있으며, 2026학년도 대학수학능력시험 문제로 출제될 가능성이 있는 모든 독해 유형을 포함하였다. 1강에서 12강까지 모두 학습한 후 Mini Test를 통해 자신의 실력을 최종적으로 간단히 점검할 수 있다.

정답과 해설

[정답] – [소재] – [해석] – [문제 해설] – [구조 해설] – [어휘 및 어구] – [Tips]의 순서로 구성되어 있다. 먼저 각 문항의 소재를 제시하여 글의 전반적인 내용에 대한 이해를 유도하고자 하였고, 이어서 정확한 해석과 함께, 정답에 이르는 과정을 상세하게 해설하여 자기 주도 학습이 가능하도록 하였다. 그리고 구조 해설과 어휘 및 어구를 통해 정확하고 핵심적인 구문 이해에 도움이 되도록 구성하였으며, Tips에서는 지문의 내용에 대한 배경 지식이나 부가적 정보를 제공하였다.

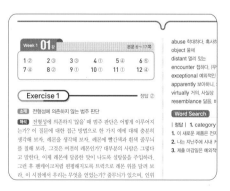

이 책의 차례

틀리기 쉬운 **유형편**

학업 성취 9주 계획	강	제목	페이지
Week 1	01	빈칸 추론 함축 의미 추론	6
	02		18
Week 2	03		30
	04		42
Week 3	05	어법 주제 추론 제목 추론 무관한 문장 글의 순서	54
	06		66
Week 4	07		78
	08		90
Week 5	09	어휘 문장 삽입 요약문 완성 1지문 2문항	102
	10		114
Week 6	11		126
	12		138

Contents

Mini Test

학업 성취 9주 계획	제목	페이지
Week 7	**Mini Test 1** 수능 독해 실전모의고사	150
Week 8	**Mini Test 2** 수능 독해 실전모의고사	176
Week 9	**Mini Test 3** 수능 독해 실전모의고사	202

학생

인공지능 DANCHOQ 푸리봇 문|제|검|색

EBS_i_ 사이트와 **EBS_i_ 고교강의 APP** 하단의 **AI 학습도우미 푸리봇**을 통해 문항코드를 검색하면 푸리봇이 해당 문제의 해설과 해설 강의를 찾아 줍니다. **사진 촬영으로도 검색**할 수 있습니다.

문제별 문항코드 확인

[25007-0001]

1. 아래 그래프를 이해한 내용으로 가장 적절한 것은?

문항코드 검색

25007-0001

사진 촬영 검색

선생님

EBS 교사지원센터 교재 관련 자|료|제|공

교재의 문항 한글(HWP) 파일과 교재이미지, 강의자료를 무료로 제공합니다.

⬇ 한글다운로드   🖼 교재이미지   📊 강의자료

• 교사지원센터(teacher.ebsi.co.kr)에서 '교사인증' 이후 이용하실 수 있습니다.
• 교사지원센터에서 제공하는 자료는 교재별로 다를 수 있습니다.

Exercise 1 　빈칸 추론

다음 빈칸에 들어갈 말로 가장 적절한 것은?

25007-0001

　　How are category judgments made when they *don't* rely on _____? As an approach to this question, let's think through an example. Consider a lemon. Paint the lemon with red and white stripes. Is it still a lemon? Most people say that it is. Now, inject the lemon with sugar water, so it has a sweet taste. Then, run over the lemon with a truck, so that it's flat as a pancake. What have we got at this point? Do we have a striped, artificially sweet, flattened lemon? Or do we have a non-lemon? Most people still accept this poor, abused fruit as a lemon, but consider what this judgment involves. We've taken steps to make this object more and more distant from the prototype and also very different from any specific lemon you've ever encountered. But this seems not to shake your faith that the object remains a lemon. To be sure, we have a not-easily-recognized lemon, an exceptional lemon, but it's still a lemon. Apparently, something can be a lemon with virtually no resemblance to other lemons.

* prototype: 원형

① tradition　　　　　　② typicality　　　　　　③ classification
④ rationality　　　　　　⑤ functionality

Sentence Structure

● We've taken steps [to make this object {more and more distant from the prototype} and also {very different from any specific lemon ⟨you've ever encountered⟩}].
　[ ]는 steps를 수식하는 to부정사구이다. 그 안의 두 개의 { }는 and로 연결되어 make의 목적격 보어 역할을 하는 형용사구이다.
　⟨ ⟩는 any specific lemon을 수식하는 관계절이다.

Word Search　주어진 문장에 들어갈 적절한 영어 단어를 본문에서 찾아 쓰시오. (필요시 형태를 바꿀 것)

1. This new product falls under the electronics c_____.
2. I e_____ an old friend at the town coffee shop last week.
3. The submission deadline will be extended only in e_____ circumstances.

Exercise 2 빈칸 추론

다음 빈칸에 들어갈 말로 가장 적절한 것은?   25007-0002

There's a reason so many of the studies that found that _____ were conducted with children. Children are busy figuring out their likes and dislikes. When I ask my eight-year-old if he likes a subject at school, he needs to think about it; he doesn't intuitively know the answer the way you might. Children are relatively new to a world that's largely controlled by adults, so many of the activities that occupy their days need explanation. They might ask themselves, "Am I drawing because I like to draw or because my teacher made me draw?" or "Does this food taste good to me or am I eating it because otherwise I won't get dessert?" Incentives give them the clues to start piecing together their likes and dislikes. And if you're a child, and an adult is willing to pay you to do something, that's a clue that you wouldn't otherwise enjoy doing it.

① it pays to trust our intuition
② we act in our own self-interests
③ incentives undermine motivation
④ old preferences can last a lifetime
⑤ rewards work better than punishment

Sentence Structure

● Children are relatively new to a world [that's largely controlled by adults], so [many of the activities {that occupy their days}] need explanation.
첫 번째 [ ]는 a world를 수식하는 관계절이다. 두 번째 [ ]는 so가 이끄는 절의 주어이고, 그 안의 { }는 the activities를 수식하는 관계절이다.

Word Search   주어진 문장에 들어갈 적절한 영어 단어를 본문에서 찾아 쓰시오. (필요시 형태를 바꿀 것)

1. He has a d_____ for people who are rude and impolite.
2. Problems at work continued to o_____ his mind for some time.
3. The cash bonus turned out to be a powerful i_____ for employees.

Exercise 3 　빈칸 추론

다음 빈칸에 들어갈 말로 가장 적절한 것은?　　　　　　　　25007-0003

　　Because both parties typically suffer costs when a competition escalates to violence, contests between members of the same species are typically a blend of truth and exaggeration by each party intended to convince the other party to back down. Exaggeration would disappear if there were no cost to testing one's abilities against those of one's opponent. If we're competing over the last slice of cake and I think I might be stronger than you, I'll just punch you and find out. But there is a notable cost to this test, as you are likely to punch me back — a bummer under the best of circumstances, but particularly so if you're stronger than I am. It is this _____ that allows deceptive individuals to exaggerate their strengths and play down their weaknesses without necessarily getting caught. This type of exaggeration can be seen throughout the animal kingdom, such as when moose or hyenas raise the hackles on their back to appear larger, or when crabs grow unnecessarily large claw shells that they do not fill with muscle.

* bummer: 불쾌한 경험　** hackles: 목덜미 털

① heavy price of inaction
② constant pursuit of harmony
③ guaranteed cost of competition
④ innate preference for physical conflict
⑤ excessive underestimation of the opponent

Sentence Structure

● This type of exaggeration can be seen throughout the animal kingdom, such as [when moose or hyenas raise the hackles on their back to appear larger], or [when crabs grow unnecessarily large claw shells {that they do not fill with muscle}].
두 개의 [ ]는 시간의 부사절로 or로 연결되어 such as에 이어진다. 두 번째 [ ] 안의 { }는 unnecessarily large claw shells를 수식하는 관계절이다.

Word Search　주어진 문장에 들어갈 적절한 영어 단어를 본문에서 찾아 쓰시오. (필요시 형태를 바꿀 것)

1. Tensions continued to e_____ between the two rival nations.
2. Her claim that the fish she caught was as big as a whale was a clear e_____.
3. The advertisement's d_____ claims misled many consumers.

Exercise 4 빈칸 추론

다음 빈칸에 들어갈 말로 가장 적절한 것은?   25007-0004

Other people's reactions can influence whether any one individual decides to help. No one wants to foolishly rush to help in a case that may not be an emergency after all. In fact, people sometimes fail to act because they fear appearing foolish in front of others. So we usually keep calm and check to see what others present are doing. Of course, if everyone else is also keeping calm while they check the reactions of others, everyone will conclude that help is not needed or that norms make helping inappropriate. In one series of studies, experimenters arranged for smoke to pour into a laboratory room in which students were sitting completing questionnaires. When the students were alone, their concern at the unusual situation soon led them to seek help. But when two confederates in the room showed no reaction to the smoke, participants also did nothing. When people notice that bystanders and passersby are unresponsive, that observation reduces the likelihood that they will help. Thus, one way that the presence of bystanders can influence helping is by _____.

* confederate: (실험) 공모자

① suggesting that helping is contrary to norms
② keeping others from taking unnecessary risks
③ following the norms that would secure safety
④ encouraging people to take notice of the rewards
⑤ emphasizing that norms should be applied to everyone

Sentence Structure

● Of course, [if everyone else is also keeping calm {while they check the reactions of others}], everyone will conclude [{that help is not needed} or {that norms make helping inappropriate}].
첫 번째 [ ]는 조건의 부사절이고, 그 안의 { }는 시간의 부사절이다. 두 번째 [ ]는 conclude의 목적어인데, 그 안에 명사절인 두 개의 { }가 or로 연결되어 있다.

Word Search   주어진 문장에 들어갈 적절한 영어 단어를 본문에서 찾아 쓰시오. (필요시 형태를 바꿀 것)

1. Each culture develops its own social n_____.
2. The p_____ quickly called for help after he witnessed the car accident on the busy street.
3. The customer service representative remained u_____ to my inquiries, leaving me dissatisfied.

Exercise 5 　빈칸 추론

다음 빈칸에 들어갈 말로 가장 적절한 것은?　　　　　　　　　　　　　　　　　25007-0005

> 　　There is a profound reason to start natural philosophy with the ancient Greeks rather than the older cultures (Egyptian, Babylonian, Indian, and Chinese), despite their many accomplishments. Although these older cultures had technical knowledge, keen observational skills, and vast resources of material and information, they failed to create natural philosophy because they _____. The religions of the old empires were predicated on the belief that the material world was controlled and inhabited by supernatural beings and forces, and that the reason for the behavior of these supernatural forces was largely unknowable. Although there were many technical developments in the societies of the four river cultures, the intellectual heritage was dominated by the priests, and their interest in the material world was an extension of their concepts of theology. Many ancient civilizations, such as the Egyptian, Babylonian, and Aztec empires, spent a large proportion of social capital (covering such things as the time, wealth, skill, and public space of the society) on religious activity.
>
> 　　　　　　　　　　　　　　* be predicated on: ~에 근거하다　** theology: 신학

① could not integrate technical knowledge into their religious frameworks
② regarded philosophical endeavors as secondary to practical pursuits
③ had no system through which to pass on their intellectual heritage
④ did not separate the natural world from the supernatural world
⑤ downplayed the importance of abstraction and spirituality

Sentence Structure

● The religions of the old empires were predicated on the belief [{that the material world was controlled and inhabited by supernatural beings and forces}, and {that ⟨the reason for the behavior of these supernatural forces⟩ **was** largely unknowable}].
　[ ]는 the belief와 동격 관계를 이루고, 그 안에 두 개의 { }가 and로 대등하게 연결되어 있다. 두 번째 { } 안에서 ⟨ ⟩는 절의 주어로, 주어의 핵심어가 the reason이므로 단수형 동사 was가 사용되었다.

Word Search　주어진 문장에 들어갈 적절한 영어 단어를 본문에서 찾아 쓰시오. (필요시 형태를 바꿀 것)

1. René Descartes is regarded as the founder of modern p_____.
2. The tribe's shared cultural h_____ fostered a deep sense of community and belonging.
3. The ancient temple held deep r_____ significance for the community.

Exercise 6　함축 의미 추론

밑줄 친 <u>the ruler sitting at his desk</u>가 다음 글에서 의미하는 바로 가장 적절한 것은?　　　25007-0006

The processes of state formation and the centralization of government in early modern Europe involved the use of increasing amounts of information. Historians have noted the rise of what the Canadian sociologist Dorothy Smith called 'textually mediated forms of ruling' such as writing letters, writing and annotating reports, issuing forms and questionnaires and so on, associated with what is variously known as the information state, archive state or paper state — now in the process of transforming itself into the digital state. This process may be described as the rise of 'bureaucracy' in the original sense of the term, the rule of the bureau, or office, and its officials. These officials both issued and followed written orders and recorded these orders in their files, together with the reports on the political situation at home and abroad that assisted decision-making. The ruler on horseback was gradually transformed into <u>the ruler sitting at his desk</u>, as in the famous cases of Philip II of Spain in the sixteenth century and Louis XIV of France in the seventeenth.

* mediate: (정보 등을) 매개하다　** annotate: 주석을 달다　*** bureaucracy: 관료제

① a governor who engages in extensive discussions before reaching a decision
② a dictator who controls the flow of information to suppress the opposition
③ a king without a centralized authority and a formal bureaucratic structure
④ a head of state who favors peaceful negotiations over armed conflict
⑤ a leader who relies on written communication to govern a state

Sentence Structure

● Historians have noted [the rise of {what the Canadian sociologist Dorothy Smith called 〈'textually mediated forms of ruling' such as writing letters, writing and annotating reports, issuing forms and questionnaires and so on〉}, {associated with what is variously known as the information state, archive state or paper state — now in the process of transforming itself into the digital state}].

[ ]는 noted의 목적어이다. 첫 번째 { }는 전치사 of의 목적어로 사용된 명사절이고, 그 안의 〈 〉는 called의 목적격 보어로 사용된 명사구이다. 두 번째 { }는 첫 번째 { }를 부가적으로 설명하는 분사구문이다.

Word Search　주어진 문장에 들어갈 적절한 영어 단어를 본문에서 찾아 쓰시오. (필요시 형태를 바꿀 것)

1. Excessive c＿＿＿＿＿＿ often hinders individual creativity and innovation.
2. New members will be i＿＿＿＿＿＿ with a temporary identity card.
3. The new software is designed to a＿＿＿＿＿＿ users in managing their finances.

Exercise 7 빈칸 추론

다음 빈칸에 들어갈 말로 가장 적절한 것은?   25007-0007

The primary goal of replication is to determine the extent to which an observed relationship generalizes across different tests of the research hypothesis. However, just because a finding does not generalize does not mean it is not interesting or important. Indeed, science proceeds by _____. Few relationships hold in all settings and for all people. Scientific theories are modified over time as more information about their limitations is discovered. As an example, one of the interesting questions in research investigating the effects of exposure to violent material on aggression concerns the fact that although it is well known that the viewing of violence tends to increase aggression on average, this does not happen for all people. So it is extremely important to conduct participant replications to determine which people will, and which will not, be influenced by exposure to violent material.

* replication: 반복 실험

① developing specialized knowledge designed for specific groups
② emphasizing the discovery of a hypothesis that can be generalized
③ generating numerous hypotheses across different fields of research
④ discovering limiting conditions for previously demonstrated relationships
⑤ helping the human population adapt to limits imposed by the environment

Sentence Structure

● As an example, one of the interesting questions in research investigating the effects of exposure to violent material on aggression concerns the fact [that {although **it** is well known ⟨that the viewing of violence tends to increase aggression on average⟩}, this does not happen for all people].
[ ]는 the fact의 구체적인 내용을 설명하는 동격절이고, 그 안의 { }는 양보의 의미를 나타내는 부사절이다. { } 안의 it은 형식상의 주어이고, ⟨ ⟩가 내용상의 주어이다.

Word Search 주어진 문장에 들어갈 적절한 영어 단어를 본문에서 찾아 쓰시오. (필요시 형태를 바꿀 것)

1. What is the p_____ cause of global warming?
2. Parents travel abroad with children to give them e_____ to other cultures.
3. In the wild, animals often show a_____ to protect their territory.

Exercise 8 빈칸 추론

다음 빈칸에 들어갈 말로 가장 적절한 것은?   25007-0008

One might ask why having a conversation on a mobile phone while driving is so much more disruptive than, for example, having a conversation with a passenger in the car. A likely reason is the _____ when having a mobile phone conversation. A passenger in the car will pick up from non-verbal cues that the driver needs to concentrate on the main task of driving at times when the latter becomes tricky. A remote interlocutor is much less likely to pick up these cues and therefore will continue to make cognitively demanding conversation at a time when the secondary task needs to be shut down to devote resources to the main driving task. A cognitively demanding conversation, especially one over which the driver has little or no control in terms of dynamically adjusting his or her allocation of cognitive resources, appears to interfere with computation of speeds, distances and widths as required by the driving task, probably as a result of diminished attention to sensory inputs. Use of a mobile phone also demands other secondary tasks, such as inputting of a telephone number on the keypad, which would also tend to interfere with the main driving task.

* interlocutor: 대화자

① influence of background noise
② loss of control over the situation
③ dependency on network reliability
④ inhibition of immediate interaction
⑤ difficulty in interpreting verbal cues

Sentence Structure

● A passenger in the car will pick up from non-verbal cues [that the driver needs to concentrate on the main task of driving] at times [when the latter becomes tricky].
  첫 번째 [ ]는 pick up의 목적어 역할을 하는 명사절이고, 두 번째 [ ]는 times를 수식하는 관계절이다.

Word Search   주어진 문장에 들어갈 적절한 영어 단어를 본문에서 찾아 쓰시오. (필요시 형태를 바꿀 것)

1. The loud music was a d_____ influence on the students taking the exam.
2. His raised eyebrow was a c_____ that he was uncertain about the plan.
3. Please try to c_____ fully on what the teacher is saying.

Exercise 9 | 빈칸 추론

다음 빈칸에 들어갈 말로 가장 적절한 것은?   25007-0009

Firms often bundle goods or services for convenience or marketing purposes. Shoe vendors could sell lefts and rights separately but nearly all consumers would rather buy the bundle. Bundling can _____ when consumers have imperfectly correlated preferences for related goods. For example, cable television services usually offer a wide range of programming, including channels that specialize in sports, food, drama, and news. Cable services could allow their customers to purchase channels "a la carte" — sports fans could purchase just the sports channels, and so forth. But cable services instead set a single bundled price that is not too much more than individual a la carte prices. (For example, the price for the "sports＋food＋drama＋news" bundle is not much more than the price the service would charge for the sports package alone.) Since the cable service has essentially zero marginal cost of selling the bundle, this practice helps increase its profits.

* vendor: 판매 회사 ** a la carte: 따로 골라
*** marginal cost: 한계 비용(생산물 한 단위를 추가로 생산할 때 필요한 총비용의 증가분)

① also help sellers extract higher profits
② lead to more efficient resource allocation
③ nonetheless increase confusion for customers
④ sometimes create dissatisfaction with bundled products
⑤ facilitate better matching of products to consumer needs

Sentence Structure

● But cable services instead set a single bundled price [that is not too much more than individual a la carte prices].
[ ]는 a single bundled price를 수식하는 관계절이다.

Word Search | 주어진 문장에 들어갈 적절한 영어 단어를 본문에서 찾아 쓰시오. (필요시 형태를 바꿀 것)

1. We extended online ordering options for the c_____ of our customers.
2. Investors are always seeking opportunities to maximize their p_____ in the stock market.
3. The lawyer chose to s_____ in human rights law to fight for freedom and justice.

Exercise 10 빈칸 추론

다음 빈칸에 들어갈 말로 가장 적절한 것은?

25007-0010

Many studies have shown that the brain cannot recognize the difference between a well-imagined experience and the real thing. Try this experiment. Imagine that you have a beautiful juicy yellow lemon in your hand. Imagine yourself slicing the lemon in half and looking at the juicy circle of the lemon. Now, imagine yourself biting into the lemon. If you are like many people, you begin to salivate. You may feel some tightness in your throat from the sourness. But you can see that since there is no real lemon, you are having a physiological reaction to an imagined experience. So, too, with organizing; the more vividly you can imagine arriving on time in a calm, relaxed fashion, the more your body receives signals from your brain that it is a true experience. Through visualizing, _____.

* salivate: 침이 나오다   ** physiological: 생리적인

① you are practicing for reality
② the brain absorbs new information better
③ you are challenging conventional thinking
④ the ability to discern falsehood is developed
⑤ you control your instinct to jump to conclusions

Sentence Structure

● So, too, with organizing; **the more vividly** you can imagine arriving on time in a calm, relaxed fashion, **the more** your body receives signals from your brain [that it is a true experience].

'~할수록 더 …하다'라는 의미의 「the+비교급 ~, the+비교급 …」 구문이 쓰이고 있다. [ ]는 signals의 내용을 설명하는 동격절이다.

Word Search   주어진 문장에 들어갈 적절한 영어 단어를 본문에서 찾아 쓰시오. (필요시 형태를 바꿀 것)

1. My mother is s_____ vegetables into small pieces to add them to the soup.
2. His r_____ to the news of his promotion was a mix of surprise and delight.
3. I v_____ remember the warmth of the sun as if it were yesterday.

Exercise 11   빈칸 추론

다음 빈칸에 들어갈 말로 가장 적절한 것은?   25007-0011

In addition to changing a hypothesis by being more specific about which amounts of one variable had what effect, you can change a hypothesis by being more specific about _____. Thus, if your hypothesis involves a general construct, you may be able to improve your hypothesis by breaking that multidimensional construct down into its individual dimensions and then making hypotheses involving those individual components. For example, rather than hypothesizing that love will increase over time, you might hypothesize that certain aspects of love (commitment, intimacy) will increase over time, whereas other parts (passionate love) will not. Similarly, rather than saying that stress will interfere with memory, you might try to find what part of memory is most affected by stress. Is it encoding, rehearsal, organization, or retrieval? The component strategy has paid off for social psychologists who have broken down prejudice into its conscious and unconscious dimensions and for personality psychologists who have broken down global (overall) self-esteem into different types (body self-esteem, academic self-esteem, social self-esteem, etc.).

* **retrieval:** (기억한 내용의) 인출, 상기 ** **self-esteem:** 자존감

---

① which aspect of a variable had what effect
② what other variables have the same impact
③ the role of sample size in identifying key variables
④ what methodology was used to measure a variable
⑤ the cultural and historical background of a variable

Sentence Structure

● The component strategy has paid off [for social psychologists {who have broken down prejudice into its conscious and unconscious dimensions}] and [for personality psychologists {who have broken down global (overall) self-esteem into different types (body self-esteem, academic self-esteem, social self-esteem, etc.)}]. 두 개의 [ ]는 and로 대등하게 연결되어 paid off에 이어지며, 첫 번째 { }는 social psychologists를, 두 번째 { }는 personality psychologists를 수식하는 관계절이다.

Word Search   주어진 문장에 들어갈 적절한 영어 단어를 본문에서 찾아 쓰시오. (필요시 형태를 바꿀 것)

1. He conducted a study to determine if diet could be a v_____ affecting mood.
2. Environmental issues have political d_____, which means they often involve government policies.
3. She realized her p_____ against people with disabilities was unfair.

Exercise 12) 함축 의미 추론

밑줄 친 a duck pond가 다음 글에서 의미하는 바로 가장 적절한 것은?

25007-0012

Most organizations and leaders get into trouble in the implementation phase of the leadership process. With self-serving leaders at the helm, the traditional hierarchical pyramid is kept alive and well. When that happens, who do people think they work for? The people above them. The minute you think you work for the person above you for implementation, you are assuming that person — your boss — is *responsible* and your job is being *responsive* to that boss and to his or her whims or wishes. Now "boss watching" becomes a popular sport and people get promoted on their upward-influencing skills. As a result, all the energy of the organization is moving up the hierarchy, away from customers and the frontline folks who are closest to the action. What you get is a duck pond. When there is a conflict between what the customers want and what the boss wants, the boss wins. You have people quacking like ducks: "It's our policy." "I just work here." "Would you like me to get my supervisor?"

* at the helm: 주도권을 잡은 ** whim: 변덕 *** quack: 꽥꽥거리다

① a company which appears peaceful but has hidden challenges posed by customers
② a company where the boss emphasizes a sense of belonging and tries to eliminate any conflict
③ an organization where the boss intensifies competition among employees for higher productivity
④ an organization filled with employees who prioritize pleasing their superiors over serving customers
⑤ an organization where the employees imitate their boss by doing repetitive tasks without questioning

Sentence Structure

● **The minute** you think you work for the person above you for implementation, you are assuming [{that person — your boss — is *responsible*} and {your job is being *responsive* to that boss and to his or her whims or wishes}].

The minute은 as soon as(~하는 순간, ~하자마자)의 의미를 지닌다. [ ]는 assuming의 목적어 역할을 하는 명사절이고, 그 안에서 두 개의 { }가 and로 대등하게 연결되어 있다.

Word Search 주어진 문장에 들어갈 적절한 영어 단어를 본문에서 찾아 쓰시오. (필요시 형태를 바꿀 것)

1. Never a_____ someone's intentions without clear communication.
2. A good leader is always r_____ to employees' needs and suggestions.
3. The c_____ between the two countries evolved into a full-scale war.

Exercise 1 빈칸 추론

다음 빈칸에 들어갈 말로 가장 적절한 것은?   25007-0013

> We are able to speak and comprehend language with great skill despite its quasiregularity — indeed, because of it. Communication requires shared knowledge, and so languages must be systematic rather than arbitrary. However, the demands of comprehending and producing language require additional _____ because speakers produce forms that deviate from standard patterns and listeners must be able to comprehend them. Many shortcuts that promote fluent speech eventually enter the language, such as "gonna," "hafta," and "tryna," which partially overlap with the source words. The product of these conflicting pressures is quasiregularity. These patterns can be mastered with extensive practice, which is easy to obtain if you've grown up speaking a language and become a fluent reader. Mastering stress patterns is much harder for people learning English as a second language, who often exhibit "stress deafness."
>
> * quasiregularity: 준규칙성 ** arbitrary: 자의적인 *** deviate from: ~에서 벗어나다

① flexibility   ② security   ③ integration
④ authorization   ⑤ documentation

Sentence Structure

- Many shortcuts [that promote fluent speech] eventually enter the language, such as "gonna," "hafta," and "tryna," [which partially overlap with the source words].
  첫 번째 [ ]는 Many shortcuts를 수식하는 관계절이고, 두 번째 [ ]는 "gonna," "hafta," and "tryna"를 부가적으로 설명하는 관계절이다.

Word Search 주어진 문장에 들어갈 적절한 영어 단어를 본문에서 찾아 쓰시오. (필요시 형태를 바꿀 것)

1. She stood staring at the accident, unable to c_____.
2. He studied five foreign languages but is f_____ in only three of them.
3. Football season o_____ the baseball season in September.

Exercise 2 　빈칸 추론

다음 빈칸에 들어갈 말로 가장 적절한 것은?　　　　　　　　　　　　　25007-0014

You know how people always tell you to "think outside the box"? Well, I hate that expression. I get the broader meaning of the phrase: to look for unexpected solutions that defy convention. Nothing wrong with that. But to me, advertising is all about thinking INSIDE the box. And advertising is full of boxes — or limitations, frameworks, and concrete realities. The budget is a box. The dimensions of the page are a box. The ingredients in the product are a box. The most important box of all is the strategy. If you can come up with a great creative idea that fits within the confines of the strategy, then you're a genius. Come up with a great idea that's wildly off the mark and NOT strategic, then you're an artist, not an advertiser. This is not to say that you can't wail against the box. Or try to change the dimensions of the box. But at its very essence, advertising can only truly be advertising when it is _____. The cleverest among us realize that the greatest fun of advertising is seeing how far we can go with an idea, an execution, a new media placement and *still be in the box*.

* defy: ~에 도전하다　** confines: 제한, 한계　*** wail: 투덜대다

① breaking new ground
② experience of an audience
③ a clear outgrowth of the box
④ inspired by other types of boxes
⑤ a challenge to existing conventions

Sentence Structure

● The cleverest among us realize [that the greatest fun of advertising is {seeing ⟨how far we can go with an idea, an execution, a new media placement and *still be in the box*⟩}].
[ ]는 realize의 목적어 역할을 하는 명사절이고, { }는 주격 보어이다. ⟨ ⟩는 seeing의 목적어 역할을 하는 명사절이다.

Word Search　　주어진 문장에 들어갈 적절한 영어 단어를 본문에서 찾아 쓰시오. (필요시 형태를 바꿀 것)

1. "It's only a suspicion," the police officer said, "nothing c_____."
2. The family struggle to balance their household b_____.
3. Einstein was a mathematical g_____.

Exercise 3 빈칸 추론

다음 빈칸에 들어갈 말로 가장 적절한 것은?   25007-0015

Life on our planet can be arranged, more or less, into autotrophs and heterotrophs, organisms that exploit energy from the sun or chemical reactions, and organisms that take energy from those who've already captured it. What is unusual about our species is that we've been able to use more and more energy without having to evolve into a different species. We've achieved this through a combination of social learning, complex culture, and technologies. We don't have to speciate to gain the claws of an allosaurus; we can share information to design a warhead or a power station. In other words, we _____.
Fire and spears did the trick for hundreds of thousands of years, until we devised the domestication of our food sources. The next big shift came in the mechanisation of processes that gave us the Industrial Revolution. This enabled us to draw ancient deposits of organic energy out of the Earth and burn them.

* **speciate**: 새로운 종으로 분화하다   ** **allosaurus**: 알로사우루스(육식 공룡)   *** **warhead**: 탄두

---

① spoil every frontier we encounter
② evolve to deal with long-term threats
③ are unlikely to use our tools optimally
④ change our tools rather than our bodies
⑤ leave genetic information everywhere we go

Sentence Structure

● Life on our planet can be arranged, more or less, into autotrophs and heterotrophs, [organisms {that exploit energy from the sun or chemical reactions}, and organisms {that take energy from those ⟨who've already captured it⟩}].

[ ]는 autotrophs and heterotrophs와 동격 관계에 있는 명사구이다. 두 개의 { }는 각각 바로 앞에 있는 organisms를 수식하는 관계절이다. ⟨ ⟩는 those를 수식하는 관계절이다.

Word Search   주어진 문장에 들어갈 적절한 영어 단어를 본문에서 찾아 쓰시오. (필요시 형태를 바꿀 것)

1. Many companies are coming to e_____ the natural resources.
2. A new system has been d_____ to control the air pollution in the city.
3. The region has many d_____ of valuable oil.

Exercise 4 빈칸 추론

다음 빈칸에 들어갈 말로 가장 적절한 것은?

25007-0016

Media institutions across the globe are facing multiple crises: of funding, trust, representation, accountability and legitimacy. In many of the countries that make up capitalism's core, the newspaper and magazine industry is in serious decline as large digital intermediaries take over the majority of advertising revenue. Much of the debate about the sustainability of the news industry circulates around debates relating to this 'broken business model'. Local news in particular is increasingly under threat. In the UK, the majority of the population (57.9%) is no longer served by a local daily newspaper. To retain high levels of profitability, media corporations have closed or merged titles and cut jobs, often moving journalists long distances away from the communities they serve and no longer being able to provide content of relevance to them. In short, a profit-driven response means _____.

* legitimacy: 합법성 ** intermediary: 중개업체, 중개자 *** revenue: 수익

① media become ever more unsustainable
② media have plenty of mistakes to point to
③ strengthening the journalists' social responsibility
④ improved sales revenue leads to editorial independence
⑤ improving the quality and availability of news and information

Sentence Structure

● In many of the countries [that make up capitalism's core], the newspaper and magazine industry is in serious decline [as large digital intermediaries take over the majority of advertising revenue].
첫 번째 [ ]는 the countries를 수식하는 관계절이다. 두 번째 [ ]는 접속사 as가 이끄는 부사절이다.

Word Search 주어진 문장에 들어갈 적절한 영어 단어를 본문에서 찾아 쓰시오. (필요시 형태를 바꿀 것)

1. She suffered m_____ injuries in the crash.
2. This issue has no r_____ to the current situation.
3. Efficient processes need to be introduced to increase p_____.

Exercise 5 빈칸 추론

다음 빈칸에 들어갈 말로 가장 적절한 것은?   25007-0017

The complexity of human intervention in nature means that the ecosystems have had to adapt — or in many cases die out. Ancient woodland exists only in small pieces in Britain now. Many of these remnants are enclosed in nature reserves and national parks. They need specific protection. New habitats have been created with their own ecosystems. The urbanization of the landscape and the creation of road and railway corridors have given us the garden habitats that many species thrive in. Motorway roadsides with higher salt deposits support salt-loving plants otherwise found along the coast. Roads provide abundant road-kill for scavengers. These may be poor substitutes for what they replace, but they are habitats that can add more if properly managed. The natural environment we may seek to conserve is the natural environment _____.

* remnant: (보통 복수로) 잔존물 ** corridor: 회랑, 좁고 긴 통로 *** scavenger: 죽은 동물을 먹는 동물

① we have in part created
② surviving would be a test
③ that have not yet been negatively affected
④ we are destroyed by planting exotic species
⑤ that attracts attention without requiring effort

Sentence Structure

● [The urbanization of the landscape] and [the creation of road and railway corridors] have given us the garden habitats [that many species thrive in].
첫 번째와 두 번째 [ ]가 and로 연결되어 문장의 주어를 구성한다. 세 번째 [ ]는 the garden habitats를 수식하는 관계절이다.

Word Search   주어진 문장에 들어갈 적절한 영어 단어를 본문에서 찾아 쓰시오. (필요시 형태를 바꿀 것)

1. The rate of u_____ is the challenge facing cities of the future.
2. His yard is e_____ with iron railings.
3. Many countries have to find s_____s for coal to protect the earth.

Exercise 6 함축 의미 추론

25007-0018

밑줄 친 <u>the code that has been programmed into your mind</u>가 다음 글에서 의미하는 바로 가장 적절한 것은?

For most of us, our minds have been programmed by a combination of factors — our friends, our parents, the mass media, and advertisers. Some of these agents of programming truly know you and have your best interests in mind as they reinforce your special strengths and help you overcome your troublesome weaknesses; they are trying to make you happier and make your life better. Other agents of programming are trying to use you as a tool to achieve their goals, which are often very different from your own goals. When this occurs, the programming makes you less and less happy as they "help" you solve problems you don't have and make worse the problems you do have. When you allow others to dominate the programming of your mind, then when your mind runs on automatic pilot, you end up behaving in ways that achieve the goals of those programmers rather than behaving in ways that would make you happier. Therefore, it is important that you periodically examine <u>the code that has been programmed into your mind</u>.

① the prejudices you hold against others
② the core means to achieve your objectives
③ the methods to enhance your brain activity
④ the qualities that tell true friends from pretenders
⑤ the influence others have on your thinking and acting

Sentence Structure

● When this occurs, the programming makes you less and less happy [as they "help" you solve problems {you don't have} and make worse the problems {you do have}].
[ ]는 이유를 나타내는 부사절이다. 두 개의 { }는 각각 바로 앞에 있는 problems와 the problems를 수식하는 관계절이다.

Word Search 주어진 문장에 들어갈 적절한 영어 단어를 본문에서 찾아 쓰시오. (필요시 형태를 바꿀 것)

1. Such jokes tend to r_____ harmful stereotypes.
2. She needed surgery to cure a t_____ foot injury.
3. It's important to know your own strengths and w_____.

Exercise 7 빈칸 추론

다음 빈칸에 들어갈 말로 가장 적절한 것은?

25007-0019

We tend to underestimate the _____ of oral cultures. We're all familiar with the children's game in which a message is whispered from one person to another until it goes around a room. The message invariably gets distorted — sometimes with hilarious results — when the original message and the final message are compared. But this is misleading. When it is important, oral cultures can accurately transmit information across long distances and through generations. For example, American author Alex Haley was able to discover an oral record of his ancestors in Africa, and his search is described in the 1976 book, *Roots: The Saga of an American Family*. Similarly, the *Odyssey* and *Iliad* were originally heroic oral histories of Greek culture that were only written down many centuries after they were composed.

* hilarious: 아주 재미있는

① creativity ② subtleness ③ immediacy
④ complexity ⑤ effectiveness

Sentence Structure

- [When it is important], oral cultures can accurately transmit information [{across long distances} and {through generations}].
  첫 번째 [ ]는 시간의 부사절이고, 두 번째 [ ]에서 전치사구인 두 개의 { }가 and로 대등하게 연결되어 있다.

Word Search 주어진 문장에 들어갈 적절한 영어 단어를 본문에서 찾아 쓰시오. (필요시 형태를 바꿀 것)

1. Many generations ago, my a_____ arrived in these lands, leaving behind a legacy of courage and adventure.
2. The pianist c_____ a famous piece of music that is still celebrated today.
3. The headline was m_____, giving a false impression of what the article was about.

Exercise 8 · 빈칸 추론

다음 빈칸에 들어갈 말로 가장 적절한 것은?   25007-0020

Emotions meet the criteria of being _____. Take, for example, two animals squaring off in a fight over food. As they prepare to lock horns, literally or figuratively, their intense feelings prompt a repertoire of bodily reactions. When an animal's back arches and its hair stands on end, it appears larger and stronger. When it bares its teeth, frowns its brows, makes fierce noises, or displays its horns, it signals to the other animal that fighting such a strong adversary may not be worth it. These signals — displays of aggression — directly improve the chances that the other animal will withdraw, thus preventing violence and avoiding potential injury or death. Sending these signals benefits the species, as does the ability to interpret these messages. It's a win-win.

* square off: 싸울 자세를 취하다   ** lock horns: 뿔을 맞대다, 다투다

① selfish tendencies
② injurious responses
③ aggressive instincts
④ information sources
⑤ advantageous adaptations

Sentence Structure

● These signals — displays of aggression — directly improve the chances [that the other animal will withdraw], thus [{preventing violence} and {avoiding potential injury or death}].
첫 번째 [ ]는 the chances의 내용을 구체적으로 설명하는 동격절이다. 두 번째 [ ]는 these signals를 의미상의 주어로 하는 분사구문인데, 그 안에서 두 개의 분사구문인 { }는 and로 연결되어 앞 절의 결과적인 상황을 나타낸다.

Word Search   주어진 문장에 들어갈 적절한 영어 단어를 본문에서 찾아 쓰시오. (필요시 형태를 바꿀 것)

1. The job applicants were evaluated based on strict c_____.
2. They began to p_____ for the storm by securing their homes.
3. The news about the company p_____ him to return to the office immediately.

Exercise 9 · 빈칸 추론

다음 빈칸에 들어갈 말로 가장 적절한 것은?   25007-0021

> With the construction and furnishing of interior space from the fifteenth to the seventeenth century, Italians created a world in which they could develop a different style of life and in which a new culture came to be defined. This is why so much was spent on objects, why so many new kinds of objects came into existence, why the arts flourished now in the domestic world as they had earlier in the ecclesiastical world. Consumption was a creative force to construct a cultural identity. In inventing all kinds of new furnishings ranging from pottery to paintings, in elaborating their forms, in refining their production, and in organizing them into new spatial arrangements within their homes, Italians discovered new values and pleasures for themselves, reordered their lives with new standards of comportment, communicated something about themselves to others — in short, generated culture, and in the process created identities for themselves. In this cultural development there was a dynamic for change that resulted from _____.
>
> * ecclesiastical: 종교의   ** comportment: 품위, 처신

① the pursuit of innovation and modernity
② the establishment of social classes by wealth
③ the intentional and systematic education of culture
④ the interaction between people and physical objects
⑤ the increasing recognition of the importance of arts

Sentence Structure

● With the construction and furnishing of interior space from the fifteenth to the seventeenth century, Italians created a world [in which they could develop a different style of life] and [in which a new culture came to be defined].

두 개의 [ ]는 and로 연결되어 선행하는 a world를 수식하는 관계절이다.

Word Search   주어진 문장에 들어갈 적절한 영어 단어를 본문에서 찾아 쓰시오. (필요시 형태를 바꿀 것)

1. In the Renaissance era, art f_____, giving rise to some celebrated masterpieces.
2. Engineers worked tirelessly to c_____ the new bridge.
3. The policies aimed at reducing energy c_____ have led to environmental benefits.

Exercise 10 빈칸 추론

다음 빈칸에 들어갈 말로 가장 적절한 것은?

25007-0022

The distribution of US farm homesteads granted to pioneer settlers is a classic case of _____. Properties were allocated based on surveys with regular north-south and east-west boundaries, regardless of the lay of the land. This meant boundaries of homesteads were unrelated to boundaries of watersheds. Some homesteads were high and dry. Others were lower and wetter, but subject to flooding. Downstream landowners could not control erosion and runoff from upstream properties. Thus, an uphill landowner's effects on the environment could harm another landowner downhill, but the lower landowner had no options. These nineteenth-century decisions have consequences to this day. For example, one of the most challenging problems for Austin College's Sneed grassland restoration is erosive flash-flood runoff from poorly managed upstream properties.

* homestead: (개척 시대에 미국 정부가 정착민에게 지급한) 자영 농지   ** watershed: (강의) 유역
*** runoff: (땅속으로 흡수되지 않고 흐르는) 유수(流水)

① shifting boundaries depending on the interests of pioneers
② land boundaries reflecting the boundaries along the waterways
③ conflicts between the natives and the pioneers over the boundaries
④ ineffective setting of boundaries according to natural boundaries
⑤ artificial boundaries not matching natural boundaries

Sentence Structure

● This meant [boundaries of homesteads were unrelated to boundaries of watersheds].
  [ ]는 that이 생략된 명사절로 meant의 목적어 역할을 한다.

Word Search 주어진 문장에 들어갈 적절한 영어 단어를 본문에서 찾아 쓰시오. (필요시 형태를 바꿀 것)

1. The charity decided to g_____ money for disaster relief efforts.
2. The p_____ faced numerous challenges while settling the new territory.
3. The manager must a_____ tasks efficiently to meet the deadline.

Exercise 11 — 빈칸 추론

다음 빈칸에 들어갈 말로 가장 적절한 것은?

25007-0023

Publication bias means that the size of an effect could be overstated for many behavioral phenomena reported in the peer-reviewed literature. For example, suppose you read a few studies showing that a new behavioral therapy for depression significantly reduces symptoms of depression in patients. If a researcher tests the effectiveness of this same behavioral therapy and finds no effect, it is likely that no peer-reviewed journal will accept the manuscript, so you will never find it or read about it. It is therefore possible that the effectiveness of this therapy is overstated because studies failing to show an effect are not included in the published peer-reviewed literature. Researchers stated that "scientific progress is made by trusting the bulk of current knowledge," and the publication bias compromises this trust. Keep in mind that while positive results reported in the peer-reviewed literature can certainly be trusted, also take caution in knowing that _____.

* **publication bias**: 출판 편향(실험이나 연구 결과가 출판이나 배포 여부에 영향을 미치는 것)

① the literature can be based just on a single experiment
② many negative results may not be included in your search
③ prejudice can lead to any outcome being interpreted as the opposite
④ the conditions of the experiment may have been intentionally altered
⑤ the next edition of the literature could reveal entirely different results

Sentence Structure

● Publication bias means [that the size of an effect could be overstated for many behavioral phenomena {reported in the peer-reviewed literature}].

[ ]는 that이 이끄는 명사절로 means의 목적어 역할을 하고, 그 안의 { }는 many behavioral phenomena를 수식하는 분사구이다.

Word Search 주어진 문장에 들어갈 적절한 영어 단어를 본문에서 찾아 쓰시오. (필요시 형태를 바꿀 것)

1. After his injury, he attended physical t_____ sessions twice a week.
2. She decided to a_____ the job offer, excited about the new opportunities.
3. The new advertising campaign had an immediate e_____ on sales.

Exercise 12 함축 의미 추론

25007-0024

밑줄 친 employed in a prison industry — consuming the world가 다음 글에서 의미하는 바로 가장 적절한 것은?

Relying on economic growth to overcome economic problems suffers a positive feedback pitfall. Governments encourage increased economic production as a means of lifting the poor out of poverty and satisfying the demands of the rich, but the appeal of the resulting new products creates new wants that get satisfied with new income generated from yet more production. In other words, people work to earn money to satisfy wants (and of course needs), but their effort results in production of goods, including new innovations, that, when marketed, increase others' desires. Those others then work to satisfy their new wants, producing yet more goods marketed to others, and so on in a positive feedback that grinds away at the planet's stock of resources and generates more waste and new types of wastes whose consequences we only partially understand. Reflecting on this circumstance, the wise gorilla Ishmael in Daniel Quinn's novel of the same name describes modern humans as prisoners of a mother culture, employed in a prison industry — consuming the world.

* pitfall: (숨겨진) 함정, 위험

① stuck in overproduction that exceeds consumer demand
② caught in a scenario where the law of the jungle applies
③ engaged in corporate competition for market dominance
④ trapped in an endless cycle of consumption and resource exhaustion
⑤ oppressed in a government-controlled economy where freedom is deprived

Sentence Structure

● Governments encourage increased economic production as a means of [lifting the poor out of poverty] and [satisfying the demands of the rich], but the appeal of the resulting new products creates new wants [that get satisfied with new income generated from yet more production].
첫 번째와 두 번째 [ ]는 동명사구로 and로 연결되어 전치사 of의 목적어 역할을 하고, 세 번째 [ ]는 선행하는 new wants를 수식하는 관계절이다.

Word Search 주어진 문장에 들어갈 적절한 영어 단어를 본문에서 찾아 쓰시오. (필요시 형태를 바꿀 것)

1. By delivering high-quality work, he managed to e_____ a promotion.
2. To s_____ the safety regulations, the manufacturer upgraded their equipment.
3. I_____ drives industries to evolve and adapt to new challenges.

Exercise 1 빈칸 추론

다음 빈칸에 들어갈 말로 가장 적절한 것은?   25007-0025

> What is said is never a _____. Even if one agrees word for word with something that has been said before, everything — the world, the speaker, the circumstances, the addressee, and the meaning of what was said — has changed. What one says is therefore in each case unique. Although most of our statements are unoriginal, they show through their particular, even unique, style (which can be dull, ugly, or trivial) that someone has appropriated and therewith personalized them. Every sentence proves that the author has changed a received gift into a thought of her own. An analysis of what is one's own might then reveal much of what the speaker has borrowed from parents, guides, friends, books, fashions, etc., while at the same time disclosing how all these influences have converged into the unique results of what the author said.
>
> * trivial: 진부한   ** therewith: 그와 함께   *** converge: 한데 모이다

① hidden truth
② mere repetition
③ biased statement
④ genuine discovery
⑤ precise representation

Sentence Structure

- [Even if one agrees word for word with something {that has been said before}], everything — the world, the speaker, the circumstances, the addressee, and the meaning of [what was said] — has changed.
  첫 번째 [ ]는 Even if가 이끄는 양보의 의미를 지닌 부사절이며, 그 안의 { }는 something을 수식하는 관계절이다. 두 번째 [ ]는 전치사 of의 목적어 역할을 하는 명사절이다.

Word Search 주어진 문장에 들어갈 적절한 영어 단어를 본문에서 찾아 쓰시오. (필요시 형태를 바꿀 것)

1. Under the current c_____, the company needs to ensure the safety of all employees.
2. This species has a u_____ song that is distinct from all other species.
3. His facial expression had r_____ his true feelings, despite his attempt to conceal them.

Exercise 2   빈칸 추론

다음 빈칸에 들어갈 말로 가장 적절한 것은?   25007-0026

Worry is often _____. In a world marred by uncertainty, doubt continuously swirling around you, a question rises to the surface of your awareness. And in that moment of recognition, you might choose to solve the mystery. You engage in worrying, hoping against hope that you'll finally nail down the unsolvable questions troubling your life. Uncertainty, or rather, the impatience with uncertainty, is a common thread running through many aspects of worrying. The quest for a neat and satisfying resolution frequently drives this behavior. It's understandable — it's likely born out of a tireless commitment to do better, driven by a willingness to pursue the important things in your life — but worrying begins to deviate from that path. Despite your good intentions, you get trapped in thought rather than called to action.

* mar: 손상시키다   ** swirl: 소용돌이치다   *** deviate: 벗어나다

① the realization of one's intentions
② the recognition of incompleteness
③ hesitation in acknowledging others' perfection
④ a reluctance to take action in the face of uncertainty
⑤ prioritizing important matters over controlling uncertainty

---

Sentence Structure

● [Uncertainty, **or rather**, the impatience with uncertainty], is a common thread [running through many aspects of worrying].
첫 번째 [ ]는 문장의 주어부로, 그 안의 or rather는 '더 정확하게는'이라는 의미이다. 두 번째 [ ]는 a common thread를 수식하는 분사구이다.

Word Search   주어진 문장에 들어갈 적절한 영어 단어를 본문에서 찾아 쓰시오. (필요시 형태를 바꿀 것)

1. He expressed significant d_____ about their claims.
2. The child's i_____ was clearly shown as he kept asking when they would arrive.
3. Her devoted c_____ to volunteering has made a notable difference in the community.

Exercise 3 빈칸 추론

다음 빈칸에 들어갈 말로 가장 적절한 것은? 25007-0027

In the early days of the commercial internet, scholars discovered that, in cyberspace, computer code operated as a kind of 'law'. Not law as we know it — public rules decided by legislators and judges — but a different kind of law, embedded in the tech itself. Whenever we use an app, platform, smartphone or computer, we have no choice but to follow the strict rules that are coded into these technologies. Some rules are commonplace, like the rule that *you cannot access this system without the correct password*. Hence the young man who lost more than $200 million because he couldn't remember the password to his virtual currency wallet. Other rules are more controversial. In late 2020, one social media platform made it impossible for users to share a controversial article containing allegations of corruption about a public figure's son, on the basis that it violated the platform's rules against sharing hacked material. As more and more of our actions, interactions and transactions are mediated through digital technology, those who write code increasingly _____. Software engineers are becoming social engineers.

* embedded: 내장된 ** allegation: 혐의, 주장

① need to make a conscious effort to advance their skills
② focus exclusively on improving the website platform
③ depend on legal principles to regulate coding rules
④ write the rules by which the rest of us live
⑤ exercise less power over societal norms

Sentence Structure

● Whenever we use an app, platform, smartphone or computer, we **have no choice but to follow** the strict rules [that are coded into these technologies].
「have no choice but to+동사원형」은 '~할 수밖에 없다'라는 의미이다. [ ]는 the strict rules를 수식하는 관계절이다.

Word Search 주어진 문장에 들어갈 적절한 영어 단어를 본문에서 찾아 쓰시오. (필요시 형태를 바꿀 것)

1. The l_____ proposed a new plan aimed at reforming healthcare.
2. The new policy proved to be c_____, sparking heated debates across social media platforms.
3. When you v_____ the law, you will face severe penalties, including fines.

Exercise 4) 빈칸 추론

다음 빈칸에 들어갈 말로 가장 적절한 것은?

25007-0028

It wasn't really until the late '70s and early '80s that the concept of brands began to extend into all areas of business. As entities were privatized, markets opened up and competition became fiercer, so the need to differentiate your business became greater. Utilities, telecoms providers, banks, insurance companies and airlines were all now enthusiastically embracing the power of branding. As brand owners fought for space in your mind, advertising became incredibly influential. The focus was on imbuing products and services with meaning so that you as the customer could surround yourself with the brands that best represented you — the very notion of shopping as a form of personal expression. Even businesses serving other businesses began to realize that having a brand was an important business support. Brands were now seen as more than just a logo or a tagline; they were _____.

* entity: 기업　** imbue: 불어넣다, 가득 채우다
*** tagline: 태그라인(기업, 브랜드 등에 꼬리표처럼 따라붙는 함축적 단어나 짧은 문구)

① seen as opportunities to create meaning
② becoming key to justifying a premium price
③ starting to focus on creating sustainable products
④ recognized as effective ways to boost employee spirits
⑤ reflected in customer satisfaction tied to functional benefits

Sentence Structure

● **It wasn't** really **until** the late '70s and early '80s **that** the concept of brands began to extend into all areas of business.
「It was not until ~ that ...」은 '~이 되어서야 비로소 …했다'의 의미이다.

Word Search　　주어진 문장에 들어갈 적절한 영어 단어를 본문에서 찾아 쓰시오. (필요시 형태를 바꿀 것)

1. The intense c_____ between the two companies pushed them to innovate rapidly.
2. The philosopher's i_____ writings have shaped the way people think about modern philosophy.
3. The n_____ of equality is fundamental to the principles of democracy.

Exercise 5 — 빈칸 추론

다음 빈칸에 들어갈 말로 가장 적절한 것은?

25007-0029

There are many ordinary, daily-life situations like the following example. A straight stick put in water looks bent; yet we do not believe it has become bent just because it was immersed in water, which is an easily penetrable liquid. Railroad tracks seem to converge in the distance, and yet when we walk to the spot where they apparently merged we find them to be parallel. The wheels of automobiles seen on television seem to be going backward when the automobile is seen to be moving forward. Yet this is impossible. Such examples of distorted perception could be multiplied endlessly. Each of these sense phenomena is thus misleading in some way. If human beings were to accept the world as being exactly how it looks, _____. They would think the stick in water really to be bent, the writing on pages really to be reversed, and the wheels really to be going backward.

* immerse: 담그다　** penetrable: 뚫고 들어갈 수 있는　*** converge: 수렴하다

① they would be deceived as to how things really are
② their senses would be seen as a gateway to imaginative worlds
③ understanding the physical world would become effortless and simple
④ they would realize the world is without illusions and full of objective truths
⑤ their distorted perceptions would quickly fade away

Sentence Structure

- A straight stick put in water looks bent; yet we do not believe [**it** has become bent just because it was immersed in water, {which is an easily penetrable liquid}].

  [ ]는 believe의 목적어 역할을 하는 명사절로, 그 안의 대명사 it은 a straight stick을 가리키며, { }는 앞에 나온 water를 부가적으로 설명하는 관계절이다.

Word Search   주어진 문장에 들어갈 적절한 영어 단어를 본문에서 찾아 쓰시오. (필요시 형태를 바꿀 것)

1. The roads run p_____ to each other for several miles, always maintaining the same distance apart.
2. The d_____ reflections in the funhouse mirrors amused the children.
3. Her p_____ of the situation was influenced by misinformation.

Exercise 6 함축 의미 추론

25007-0030

밑줄 친 the welcoming shelter of shade trees planted by others가 다음 글에서 의미하는 바로 가장 적절한 것은?

According to an ancient Greek proverb, "A society grows great when old men plant trees in whose shade they shall never sit." Likewise, an academic culture grows great when senior scholars perform acts of generosity for junior academics who may never know their names. Literary scholar and poet Lesley Wheeler remains "endlessly grateful" to the two anonymous readers whose thoughtful responses to her first book manuscript set her on the path to becoming a successful scholar: "They told me bluntly what was wrong with the book, but they also found the time to praise it; and that was enough encouragement." Having benefited from the welcoming shelter of shade trees planted by others, Wheeler has little patience for "cranky" referees who poison the air with mean-spirited reviews. She takes care to ensure that her own feedback to colleagues and students is always gracious and constructive: "The conscientiousness and generosity that I've seen directed at my work is something that I want to pay back."

* anonymous: 익명의   ** bluntly: 직설적으로   *** cranky: (성미가) 까다로운, 심기가 뒤틀린

① the implicit drawbacks and influence of an academic environment
② the support and guidance from senior scholars to junior academics
③ the satisfaction and fulfillment derived from academic achievement
④ the acknowledgment and validation of one's academic contributions
⑤ the advantage gained by outperforming others in academia through publishing

Sentence Structure

● Likewise, an academic culture grows great [when senior scholars perform acts of generosity for junior academics {who may never know their names}].
[ ]는 시간의 부사절이고, 그 안의 { }는 junior academics를 수식하는 관계절이다.

Word Search   주어진 문장에 들어갈 적절한 영어 단어를 본문에서 찾아 쓰시오. (필요시 형태를 바꿀 것)

1. His g_____ was evident when he donated a large sum to help rebuild the hospital.

2. The letter was a_____, so we couldn't figure out who sent it.

3. Her c_____ feedback helped improve the project and make it more effective.

Exercise 7 빈칸 추론

다음 빈칸에 들어갈 말로 가장 적절한 것은?

25007-0031

Working in small groups may foster the creativity of students, but just putting students together to work in small groups does not mean that creativity will automatically flourish. As with other aspects of productive teamwork, this process requires learning. According to Meissner, a notable researcher in educational psychology, to further creative thinking in mathematics education, we need to further _____. Students need to learn how to avoid the negative factors that affect creativity: *cognitive interference*, which includes production blocking, task-irrelevant behavior, and cognitive overload; and *social inhibition*, which includes social anxiety, free riding, and illusion of productivity. They also need to learn to recognize factors that can strengthen the potential of groups to generate ideas; for example, *social stimulation*, which includes both increased individual accountability and the development of shared standards for team performance; and *cognitive stimulation*, which includes stimulation of associations, attention to others' contributions, and opportunities to incubate ideas.

* incubate: 생각해 내다

① various educational technologies
② both individual and social abilities
③ a supportive and inclusive classroom culture
④ tailored instructions, especially in after-school activities
⑤ collaborative projects and structured teamwork exercises

Sentence Structure

● Working in small groups may foster the creativity of students, but [just putting students together to work in small groups] does not mean [that creativity will automatically flourish].

첫 번째 [ ]는 동명사구로 문장의 주어 역할을 하며, 두 번째 [ ]는 mean의 목적어 역할을 하는 명사절이다.

Word Search 주어진 문장에 들어갈 적절한 영어 단어를 본문에서 찾아 쓰시오. (필요시 형태를 바꿀 것)

1. The news report cast a n_____ light on the company's environmental practices, producing more protests from environmentalists.

2. Factory p_____ has increased due to improved automation.

3. She showed great p_____ as a young athlete.

Exercise 8) 빈칸 추론

다음 빈칸에 들어갈 말로 가장 적절한 것은?   25007-0032

Although it has been demonstrated that spaced learning sessions are usually more effective than massed learning sessions, in real-life settings this advantage may sometimes be _____. For example, there are some occasions when we only have a limited period of time available for study (for example, when we have only one hour left to revise before an exam); in such cases, it may be better to use the entire period rather than to take breaks, which will waste some of our available time. Again, a very busy person might have difficulty fitting a large number of separate learning sessions into their daily schedule. A further problem is that spaced learning obviously requires more time overall (i.e., total time including rest breaks) than massed learning, and therefore may not represent the most efficient use of that time unless the rest breaks can be used for something worthwhile. Because spaced learning can create practical problems of this kind, there is no clear agreement about its value in a real-life learning setting such as a school classroom.

* spaced learning: 분산 학습 ** massed learning: 집중 학습

① challenged by time constraints
② limited when preparation is rushed
③ compromised by practical considerations
④ weakened by inconsistent learning methods
⑤ ignored due to the perceived low value of breaks

Sentence Structure

● A further problem is [that spaced learning {obviously requires more time overall (i.e., total time including rest breaks) than massed learning}, and therefore {may not represent the most efficient use of that time unless the rest breaks can be used for something worthwhile}].

[ ]는 문장의 주격 보어 역할을 하는 명사절이며, 그 안에서 두 개의 { }가 and로 연결되어 명사절의 주어 spaced learning에 대등하게 이어져 있다.

Word Search   주어진 문장에 들어갈 적절한 영어 단어를 본문에서 찾아 쓰시오. (필요시 형태를 바꿀 것)

1. The library is a_____ for students to use from 8 AM to 8 PM every day.
2. Attending the conference was a w_____ experience because of the valuable insights gained.
3. The course includes both theoretical knowledge and p_____ training.

Exercise 9 빈칸 추론

다음 빈칸에 들어갈 말로 가장 적절한 것은?   25007-0033

> In recent years, more H_2O has been flowing from low-value crops (cotton and alfalfa) to high-value ones (nuts and berries). Ailing farms are selling their water rights to productive industry and rapidly growing cities. Food grown in wet, green climates (the northeastern United States, Brazil) is increasingly being exported to dry, brown ones (Arizona, India), allowing their water to be conserved for drinking supplies, for maintaining aquifer levels, or for other high-priority uses. Yet people have tended to dance around the question of pricing water in a way that reflects scarcity. Because water is an essential resource, it has no "market value," as, say, oil does. But with no price incentive to use it efficiently, people often waste water by using vast quantities for energy and mineral projects, and polluting it. In many places water is free, or priced so low that the revenue it generates is not enough to maintain, or upgrade, reservoirs, distribution pipes, and treatment plants. While citizens have good reason to be cautious of water privatizers, _____.
>
> * alfalfa: 알팔파, 자주개자리(사료 작물인 콩과(科) 식물) ** ailing: 침체된, 병든 *** aquifer: 대수층(지하수를 품고 있는 지층)

① cheap water invites waste
② the rush to price water could slow down
③ public control keeps water flowing for all
④ climate change will restore water resources
⑤ water conservation is less essential for urban residents

Sentence Structure

● Food grown in wet, green climates (the northeastern United States, Brazil) **is** increasingly **being exported** to dry, brown ones (Arizona, India), [allowing their water to be conserved for drinking supplies, for maintaining aquifer levels, or for other high-priority uses].
문장의 주어부 Food grown in wet, green climates가 export라는 동작을 하는 주체가 아니라 대상이므로 진행형 수동태 is being exported가 사용되었으며, [ ]는 주절이 기술하는 내용의 부수적 상황을 나타내는 분사구문이다.

Word Search   주어진 문장에 들어갈 적절한 영어 단어를 본문에서 찾아 쓰시오. (필요시 형태를 바꿀 것)

1. The economic policies should r_____ the needs and priorities of all citizens.
2. Good communication skills are e_____ for effective teamwork.
3. The company's total r_____ increased significantly after launching the new product line.

Exercise 10 빈칸 추론

다음 빈칸에 들어갈 말로 가장 적절한 것은? 25007-0034

At their most basic level, monsters represent fears held by society, fears associated with dangers perceived in the surrounding world. These fears have a powerful evolutionary history by encouraging people to flee instead of fighting suicidal battles. When ancient hunters encountered a saber-toothed tiger by accident, they ran. When the human ancestor *Homo erectus* caught angry cave bears by surprise, it ran. When chimpanzees and bonobos, the nearest genetic relatives to modern humans, encounter large predators in the wild, they run. While Hollywood heroes have made running away distinctly unpopular on the silver screen, every single actor who has ever portrayed a hero who stood his or her ground against some abominable terror comes from a long genetic lineage of cowards who fled in the face of danger. That is why they are here to act today. If their ancestors had fought against monsters far more powerful than themselves, as Hollywood heroes do all the time, their lineage would have been destroyed by predators long ago. Fear, in short, _____.

* saber-toothed tiger: 검치호랑이 ** abominable: 끔찍한 *** lineage: 혈통

① keeps people alive
② inspires heroic acts
③ drives people to fight
④ creates a sense of power
⑤ makes individuals stronger

Sentence Structure

● [While Hollywood heroes have made running away distinctly unpopular on the silver screen], every single actor [who has ever portrayed a hero {who stood his or her ground against some abominable terror}] comes from a long genetic lineage of cowards [who fled in the face of danger].
첫 번째 [ ]는 접속사 While이 이끄는 양보의 의미를 나타내는 부사절이다. 두 번째 [ ]는 every single actor를 수식하는 관계절이고, 그 안의 { }는 a hero를 수식하는 관계절이다. 세 번째 [ ]는 cowards를 수식하는 관계절이다.

Word Search 주어진 문장에 들어갈 적절한 영어 단어를 본문에서 찾아 쓰시오. (필요시 형태를 바꿀 것)

1. A_____ myths and legends often explain natural phenomena in imaginative ways.

2. When hiking in the forest, it's not uncommon to e_____ various wildlife unexpectedly.

3. The cheetah is a skilled p_____ known for its incredible speed when hunting prey.

Exercise 11 빈칸 추론

다음 빈칸에 들어갈 말로 가장 적절한 것은?   25007-0035

In the 1950s, hardly anyone was interested in yeast. To most, it didn't seem we could learn much about our complex selves by studying a tiny fungus. It was a struggle to convince the scientific community that yeast could be useful for something more than baking bread, brewing beer, and vinting wine. What Mortimer and Johnston recognized, and what many others began to realize in the years to come, was that _____. For their size, their genetic and biochemical makeup is extraordinarily complex, making them an exceptionally good model for understanding the biological processes that sustain life and control lifespans in large complex organisms such as ourselves. If you are doubtful that a yeast cell can tell us anything about cancer, Alzheimer's disease, rare diseases, or aging, consider that there have been five Nobel Prizes in Physiology or Medicine awarded for genetic studies in yeast, including the 2009 prize for discovering how cells counteract telomere shortening, one of the characteristics of aging.

* fungus: 균류   ** vint: (과실주를) 빚다, 만들다   *** telomere: 텔로미어, 말단 소립

① new discoveries often rediscover what has long existed
② those tiny yeast cells are not so different from ourselves
③ science must combine fields to grasp complex phenomena
④ we can improve the quality of our foods by including yeast
⑤ understanding cells is critical for discovering new treatments

Sentence Structure

● **It** was a struggle [to convince the scientific community {that yeast could be useful for something more than baking bread, brewing beer, and vinting wine}].
It은 형식상의 주어이고, [ ]는 내용상의 주어이며, 그 안의 { }는 convince의 직접목적어 역할을 하는 명사절이다.

Word Search 주어진 문장에 들어갈 적절한 영어 단어를 본문에서 찾아 쓰시오. (필요시 형태를 바꿀 것)

1. The ecosystem in the rainforest is highly c_____, with numerous species interacting in delicate balance.
2. Using g_____ engineering, scientists hope to increase crop resilience to diseases and pests.
3. The museum has a r_____ collection of artifacts from ancient civilizations.

Exercise 12 함축 의미 추론

25007-0036

밑줄 친 to arrive safely at your destination with money in your pocket이 다음 글에서 의미하는 바로 가장 적절한 것은?

Rules govern our daily lives. Some of these rules are explicit, imposed by government: "obey the speed limit," "no parking," "April 15 is tax day." But most are informal, often unspoken cultural norms — rules of politeness, rules of conduct in the business world, rules of interaction between people. Most are commonly understood traditions that have built up over time, habits so ordinary that we usually don't even think about them. Unfortunately, not all such involuntary habits and subconscious conventions are positive or productive. American business and political communication is rife with bad habits and unhelpful tendencies that can do serious damage to the companies and causes they seek to promote. Just as in every other field, there are rules to good, effective communication. They may not be as inflexible and absolute as the rules against speeding or avoiding your taxes, but they're just as important if you wish to arrive safely at your destination with money in your pocket.

* rife with: ~이 만연한, ~로 가득 찬

① to keep up with the rapidly changing trends of the industry
② to foster long-term relationships with community members
③ to capitalize on international opportunities without regard to risk
④ to maintain a minimum amount of taxes within the regulatory range
⑤ to effectively communicate to avoid suffering negative consequences

Sentence Structure

- Most are [commonly understood traditions {that have built up over time}], [habits {so ordinary that we usually don't even think about them}].

두 번째 [ ]는 첫 번째 [ ]의 내용을 구체적으로 재진술하는 동격어구이다. 첫 번째 [ ] 안의 { }는 commonly understood traditions를 수식하는 관계절이고, 두 번째 [ ] 안의 { }는 habits를 수식하며, 「so ~ that ...」 구문은 '···할 만큼 ~한'이라는 의미이다.

Word Search   주어진 문장에 들어갈 적절한 영어 단어를 본문에서 찾아 쓰시오. (필요시 형태를 바꿀 것)

1. O_____ people can make a big difference through small acts of kindness.
2. Blinking is an i_____ reaction that protects our vision.
3. There is a t_____ for people to resist change.

Exercise 1   빈칸 추론

다음 빈칸에 들어갈 말로 가장 적절한 것은?   25007-0037

Let's start with the statement "Humans are causing climate change by burning fossil fuels." It is the basis upon which people all over the world, including me, are calling for the rapid end to fossil fuel use and the transition to carbon-emission-free energy sources. It's a pretty bold statement, and it is very different from saying *that* the climate is changing — what scientists call "detection." If we're going to argue for a massive change in human society, which is what will be required to end our use of fossil fuels, it seems reasonable to ask that we move beyond detection. After all, fossil fuels, despite their problems, have provided tremendous benefits to society over the twentieth century. If we (the climate-concerned public) are going to insist that we stop using fossil fuels, it is incumbent upon us to prove that the downside is greater than the very real upside that fossil fuels have offered. We need to prove, beyond a reasonable doubt, that climate is changing *and* that human use of fossil fuels, not something else, is responsible for the climate change we are observing. _____.

* incumbent upon: ~의[에게] 의무인

① We need attribution in addition to detection
② It all starts and ends with saying that climate is changing
③ We must disengage from subjectivity to fight climate change
④ Detecting existing problems can lead to cooperative solutions
⑤ Every single change should be detected with scientific precision

Sentence Structure

● It is the basis [upon which people all over the world, including me, are calling for {the rapid end to fossil fuel use} and {the transition to carbon-emission-free energy sources}].

[ ]는 the basis를 수식하는 관계절이고, 두 개의 { }가 and로 대등하게 연결되어 calling for의 목적어 역할을 하고 있다.

Word Search   주어진 문장에 들어갈 적절한 영어 단어를 본문에서 찾아 쓰시오. (필요시 형태를 바꿀 것)

1. The patient made a r_____ recovery.
2. The country underwent a peaceful t_____ from dictatorship to democracy.
3. The team made the b_____ move of trading its star player.

Exercise 2 빈칸 추론

다음 빈칸에 들어갈 말로 가장 적절한 것은?   25007-0038

The grammatical structure of a language is a 'social fact' in Durkheim's sense of being external to and constraining for individual speakers. It is independent of their subjective preferences and they must follow the rules if they are to be understood. However, the grammatical rules of gender, agreement, number, subject and object, possessive, and so on are not, in general, consciously followed and applied by the individuals who speak to each other. Speakers typically have only a very limited and partial awareness of the rules of their own grammar, and speaking grammatically is a matter of unreflective habit rather than conscious rule following. The grammatical rules of a language, then, _____. They may be formulated in a book of grammar, but such a book *records* the grammar — more or less imperfectly — and does not *comprise* the grammar. The rules that are followed in forming a 'correct' utterance and a well-organised discourse exist only in the minds of the individual speakers as learned dispositions held in the neurophysiological memory traces of their brains.

* neurophysiological: 신경생리학적

① shape each speaker's understanding of the other person
② serve as the key framework for effective communication
③ do not exist apart from the minds of the individual speakers
④ influence the way in which subjective feelings are expressed
⑤ exert a tremendous constraint on recalling past conversations

Sentence Structure

● However, [the grammatical rules of gender, agreement, number, subject and object, possessive, and so on] **are** not, in general, consciously followed and applied by the individuals [who speak to each other].
첫 번째 [ ]는 문장의 주어 역할을 하고, 주어의 핵심어구가 the grammatical rules로 복수여서 복수형 동사 are가 이어졌다. 두 번째 [ ]는 the individuals를 수식하는 관계절이다.

Word Search   주어진 문장에 들어갈 적절한 영어 단어를 본문에서 찾아 쓰시오. (필요시 형태를 바꿀 것)

1. This lotion is for e_____ use only.
2. The capacity of those roads will c_____ the amount of car travel.
3. There was a complete lack of a_____ of the issues involved.

Exercise 3 · 빈칸 추론

다음 빈칸에 들어갈 말로 가장 적절한 것은?   25007-0039

When environment involves human interests, it must necessarily be understood in relation to humans and not as an assemblage of independent objects. We can find support for this in the work of social psychologists such as Kurt Lewin and J. J. Gibson. Lewin envisioned a social world comprised of vectors of force between participants and the things and conditions with which they interact. These vectors invite particular behaviors, and this led Lewin to call them by the German term *Affördungsqualitäten*, translated into English as "invitational qualities." More recently, the perceptual psychologist J. J. Gibson studied the ways in which the design and appearance of environmental configurations and objects encourage particular responses in human behavior. He called these connections "affordances" for behavior, clearly influenced by Lewin's terminology and resembling his observations. The work of Lewin and Gibson is important and instructive, for it suggests that environment is not just open space filled with arrangements of independent objects but rather is a field of forces in compelling relationships of attraction, repulsion, and neutrality or indifference. Environment is, then, _____.

* configuration: 구성, 배치   ** affordance: (행동) 유도성   *** repulsion: 반발, 혐오감

① a field that includes the human participant
② a reason behind the conflicts within society
③ an area that suppresses our emotions through objects
④ a trigger for negative emotional reactions in individuals
⑤ an invisible force that disconnects humans from objects

Sentence Structure

● Lewin envisioned a social world [comprised of vectors of force between participants and the things and conditions {with which they interact}].

[ ]는 a social world를 수식하는 분사구이고, { }는 the things and conditions를 수식하는 관계절이다.

Word Search 주어진 문장에 들어갈 적절한 영어 단어를 본문에서 찾아 쓰시오. (필요시 형태를 바꿀 것)

1. Humans are not just an a_____ of chemicals or a mindless machine.
2. It was very i_____ to watch the doctors work.
3. He was a sad man with a c_____ need to talk about his unhappiness.

Exercise 4 빈칸 추론

다음 빈칸에 들어갈 말로 가장 적절한 것은? 25007-0040

As many journalism scholars have argued, print-era journalists rejected audience research because doing so was one of the only means to protect their always-unstable professional status. Sociologist Andrew Abbott has characterized professions as "somewhat exclusive groups of individuals applying somewhat abstract knowledge to particular cases." Although it is commonly categorized as a profession, journalism has long struggled to comfortably inhabit this definition. Even before the rise of the internet helped shift institutional gatekeeping power away from news organizations and toward technology platforms, journalists had difficulty establishing themselves as a "somewhat exclusive group of individuals." Indeed, while traditional professions such as medicine and law rely on strict licensing requirements to limit entry into the profession, the First Amendment prohibits U.S. journalism from establishing any such thing. Nor can journalists lay a strong claim to jurisdiction over a form of abstract knowledge. As journalism scholar Matt Carlson has argued, "abstraction makes for bad journalism. Clarity, especially in the explanation of complex topics, makes for good journalism." The accessibility of journalistic language is helpful for informing the public, but it also renders _____ potentially suspect.

* First Amendment: (미국 헌법) 수정 제1조(언론·종교·집회의 자유를 정한 조항)
** jurisdiction: 관할(권)   *** render: ~을 (어떤 상태가 되게) 만들다

① journalists' claims to specialized expertise
② diverse opinions of journalists on social issues
③ journalism as a medium of conveying the truth
④ moral duties that journalists assert they perform
⑤ journalists' intention of accepting audience research

Sentence Structure

● As many journalism scholars have argued, print-era journalists rejected audience research because **doing so** was one of the only means [to protect their always-unstable professional status].
doing so는 rejecting audience research를 대신하고, [ ]는 the only means를 수식하는 to부정사구이다.

Word Search 주어진 문장에 들어갈 적절한 영어 단어를 본문에서 찾아 쓰시오. (필요시 형태를 바꿀 것)

1. Doctors have traditionally enjoyed high social s_____.
2. Most p_____ in the medical field require years of training.
3. The room is for the e_____ use of guests.

Exercise 5 　빈칸 추론

다음 빈칸에 들어갈 말로 가장 적절한 것은?　　　　　　　　　　　　　　　　　　25007-0041

Management, especially of anything as complex as a transportation system, is very difficult. There are often many different organizations involved, each of which has multiple divisions, with multiple levels of authority, and often many lengthy, written procedure manuals. To make matters worse, the manuals are seldom kept up to date and, in any event, cannot possibly consider every combination of factors that might occur. During the incident, the people responsible for maintaining control (the pilots, in most commercial aviation incidents) waste valuable time studying the different manuals, trying to find the relevant case. Modern computer systems attempt to help by automatically diagnosing the situation and either responding autonomously or offering operators the instructions to be followed, but the diagnosis or recommended course of actions is not always appropriate (because in each complex system, most accidents involve different unique factors). Different organizations might be involved: police and firefighters, company safety representatives, multiple teams from different divisions of a company, and different companies or government and regulatory agencies who must coordinate their decisions and actions. It is rare that _____.

* aviation: 항공　** autonomously: 독자적으로

① the result is smooth, flawless management
② a vision is universally shared within a system
③ managers look into the past to predict future risks
④ technology causes conflicting management problems
⑤ people use unbiased reasoning in their decision-making

Sentence Structure

● There are often many different organizations involved, [each of which has {multiple divisions, with multiple levels of authority}, and {often many lengthy, written procedure manuals}].
[ ]는 many different organizations involved에 대해 부가적으로 설명하는 관계절이고, 그 안의 두 개의 { }는 has의 목적어 역할을 하며 and로 대등하게 연결되어 있다.

Word Search 　주어진 문장에 들어갈 적절한 영어 단어를 본문에서 찾아 쓰시오. (필요시 형태를 바꿀 것)

1. The report blames bad m_____.
2. Campuses are usually accessible by public t_____.
3. I work in the administration d_____ as an office manager.

Exercise 6 함축 의미 추론

밑줄 친 Life has a way of clipping our wings.가 다음 글에서 의미하는 바로 가장 적절한 것은?　　　25007-0042

Age is not the only determinant of spatial skills. While thirteen-year-old children have all the cognitive attributes they need to be proficient at wayfinding, some are better at it than others. By this point, parental attitudes, freedom of movement, cognitive differences and life experience have already begun to leave their imprint, and they never ease off. All of us may be explorers when we're born, but few of us stay that way. We end up suppressing our childish natures, slipping into routines and following the routes we always take. A recent study by Canadian psychologists found that 84 per cent of eight-year-olds navigate by scrutinizing their surroundings and building a mental map, a so-called 'spatial' strategy that is also used by almost all competent adult navigators. The alternative is a more closed, 'egocentric' strategy, which entails learning and following a sequence of turns. Only 46 per cent of us still use the spatial approach in our twenties, and 39 per cent in our sixties. It seems that we all start off wandering free, but most of us end up on the straight and narrow. Life has a way of clipping our wings.

* imprint: 각인　** scrutinize: 세심히 살피다　*** entail: 수반하다

① Life forces us to explore our inner world, not the external one.
② As we age, our spatial perception becomes distorted by cognitive bias.
③ The biggest obstacle for children to navigate is their own fear of failure.
④ Our tendencies to follow determined paths are suppressed by social pressure.
⑤ Over time, we lose our sense of adventure, restrained from navigating boldly.

Sentence Structure

● [**While** thirteen-year-old children have all the cognitive attributes {they need ⟨to be proficient at wayfinding⟩}], some are better at **it** than others.
[ ]는 '~이기는 하지만'이라는 뜻의 While이 이끄는 부사절이며, { }는 관계사가 생략된 관계절로 all the cognitive attributes 를 수식한다. ⟨ ⟩는 목적을 나타내는 to부정사구이다. it은 wayfinding을 가리킨다.

Word Search　　주어진 문장에 들어갈 적절한 영어 단어를 본문에서 찾아 쓰시오. (필요시 형태를 바꿀 것)

1. Soil and climate are the main d_____ of how land is used.
2. This task is designed to test children's s_____ awareness.
3. He possesses the essential a_____ of a journalist.

Exercise 7 · 빈칸 추론

다음 빈칸에 들어갈 말로 가장 적절한 것은? 25007-0043

> The bird songs we hear every day are more than beautiful. They serve a practical purpose. Birds employ their voices to call their mates, find their flock, claim territory, scare off intruders, warn others about predators, and for countless other functions. For instance, Japanese and Swiss researchers recently discovered that Japanese great tits, small birds with jet-black heads and necks with prominent white cheeks, use syntax in their songs, just as humans do in their speech. Syntax is crucial to language. For example, if you say, "I love that restaurant," the message is clear. But not even Star Wars' Master Yoda could understand, "Restaurant love that I." Until recently, scientists believed that only humans could string together such vocalizations. The Japanese great tit, it turns out, is the first animal apart from humans who can use phonological syntax — the ability to combine sounds that individually have no meaning into a collective sound — that does. To instruct other members of his flock to scan for predators, or to attract a mate, a great tit must _____ — if the notes are sung differently, the study found, other birds will not react.
>
> * great tit: 박새 ** syntax: 구문 *** phonological: 음운론적인

① produce notes in unexpected patterns
② vary the volume of his song dynamically
③ sing several distinct notes in the correct order
④ deliver each song with its own unique rhythm
⑤ copy the exact sounds of different bird species

Sentence Structure

- For instance, Japanese and Swiss researchers recently discovered [that Japanese great tits, {small birds with jet-black heads and necks with prominent white cheeks}, use syntax in their songs, just as humans do in their speech].
  [ ]는 discovered의 목적어 역할을 하는 명사절이고, { }는 Japanese great tits와 동격 관계를 이룬다.

Word Search 주어진 문장에 들어갈 적절한 영어 단어를 본문에서 찾아 쓰시오. (필요시 형태를 바꿀 것)

1. The smell of freshly baked bread from the bakery was enough to a_____ passersby.
2. Artists often c_____ different materials to create unique and innovative works.
3. The p_____ features of the mountain range were clearly visible from miles away.

Exercise 8   빈칸 추론

다음 빈칸에 들어갈 말로 가장 적절한 것은?   25007-0044

When an organism is confronted with some sort of threat, it typically becomes vigilant, searches to gain information about the nature of the threat, struggles to find an effective coping response. And once a signal indicates safety — the lion has been evaded, the traffic cop buys the explanation and doesn't issue a ticket — the organism can relax. But this is not what occurs in an anxious individual. Instead, there is a nervous scrambling among coping responses — abruptly shifting from one to another without checking whether anything has worked, an agitated attempt to cover all the bases and attempt a variety of responses simultaneously. Or there is an inability to detect when the safety signal occurs, and the restless vigilance keeps going. By definition, anxiety makes little sense outside the context of what the environment is doing to an individual. In that framework, the brain chemicals and, ultimately, the genes relevant to anxiety don't make you anxious. They make you more responsive to anxiety-provoking situations, make it harder to _____.

* vigilant: 경계하는 ** abruptly: 갑자기 *** agitated: 동요된

① respect others' personal boundaries
② maintain focus on established goals
③ come up with a possible escape plan
④ detect safety signals in the environment
⑤ look at the situation from the bright side

Sentence Structure

● By definition, anxiety makes little sense outside the context of [what the environment is doing to an individual].
　[ ]는 of의 목적어 역할을 하는 명사절로 「what + 주어 + 동사」의 구조가 쓰이고 있다.

Word Search   주어진 문장에 들어갈 적절한 영어 단어를 본문에서 찾아 쓰시오. (필요시 형태를 바꿀 것)

1. The thief managed to e_____ the police by hiding in an abandoned building.
2. He had to s_____ his career from sports to academics after his injury.
3. The fire alarm can d_____ sudden increases in heat and smoke.

Exercise 9 빈칸 추론

다음 빈칸에 들어갈 말로 가장 적절한 것은?   25007-0045

Babies use statistical learning to make predictions about the world, guiding their actions. Like little statisticians, they form hypotheses, assess probabilities based on their knowledge, integrate new evidence from the environment, and perform tests. In one creative study by the developmental psychologist Fei Xu, ten- to fourteen-month-old children first expressed a preference for pink or black lollipops, then were shown two candy jars: one containing more black lollipops than pink, and one with more pink than black. The experimenter then closed her eyes and drew one lollipop from each jar so infants could see only the stick, not the color. Each lollipop was placed into a separate, opaque cup with only the stick showing. Infants crawled to the cup that was statistically more likely to contain their preferred color, because it came from a jar where that color was in the majority. Experiments like this demonstrate that infants are not merely reactive to the world. Even from a very young age, they actively _____, to maximize the outcomes they desire.

* lollipop: 막대 사탕   ** opaque: 불투명한

① absorb information selectively from their surroundings
② use observed exceptions to understand creative processes
③ improve their decision-making process utilizing input from adults
④ predict potential outcomes relying on their current emotional state
⑤ estimate probabilities based on patterns that they observe and learn

Sentence Structure

● Infants crawled to the cup [that was statistically more likely to contain their preferred color], because it came from a jar [where that color was in the majority].
첫 번째 [ ]는 the cup을, 두 번째 [ ]는 a jar를 각각 수식하는 관계절이다.

Word Search 주어진 문장에 들어갈 적절한 영어 단어를 본문에서 찾아 쓰시오. (필요시 형태를 바꿀 것)

1. His p_____ about the election results turned out to be accurate.
2. The lawyer presented e_____ to prove his client's innocence.
3. The first aid kit should c_____ bandages, pain relievers, and gauze.

Exercise 10 빈칸 추론

다음 빈칸에 들어갈 말로 가장 적절한 것은?   25007-0046

David Howes, a professor of anthropology, notes the frequent association in different cultures between scents and rituals of transition, such as funerals or rites of passage. He suggests that scent is felt to be symbolically appropriate for moments of social transition because it so frequently accompanies and marks other types of physical transition, as when cooking smells signal the transformation of raw ingredients into food. While scents tend to escape spaces and spread out of human control, our experience of them is frequently liminal, as we notice scents far more strongly when first entering their range. You smell baking bread strongly as you enter a house, but after a few minutes inside, you may no longer be able to smell it even with deliberate effort, a physical process known as olfactory adaptation or exhaustion. It takes an overwhelming smell to retain our notice after a period of constant exposure. Smells signal _____, and are thus used to mark socially important moments of change.

* rite of passage: 통과 의례 ** liminal: 전환적인, 경계에 있는 *** olfactory: 후각의

① adaptation in sensory perception over time
② transitions through space as well as changes of state
③ adjustments of setting to create the mood for an event
④ the social status assigned to each individual by society
⑤ the invisible architecture of social spaces and collective memory

Sentence Structure

- [While scents tend to {escape spaces} and {spread out of human control}] our experience of them is frequently liminal, [as we notice scents far more strongly when first entering their range].
첫 번째 [ ]는 양보의 의미를 가진 부사절이고, 두 개의 { }가 and로 대등하게 연결되어 to에 이어진다. 두 번째 [ ]는 이유를 나타내는 부사절이다.

Word Search 주어진 문장에 들어갈 적절한 영어 단어를 본문에서 찾아 쓰시오. (필요시 형태를 바꿀 것)

1. The s_____ of fresh flowers filled the room, creating a pleasant atmosphere.
2. She made a d_____ effort to stay calm during the stressful situation.
3. The walls are thick enough to r_____ heat during the winter.

Exercise 11 | 빈칸 추론

다음 빈칸에 들어갈 말로 가장 적절한 것은?   25007-0047

Variability in judgments is expected and welcome in a competitive situation in which the best judgments will be rewarded. When several companies (or several teams in the same organization) compete to generate innovative solutions to the same customer problem, we don't want them to focus on the same approach. The same is true when multiple teams of researchers attack a scientific problem, such as the development of a vaccine: we very much want them to look at it from different angles. Even forecasters sometimes behave like competitive players. The analyst who correctly calls a recession that no one else has anticipated is sure to gain fame, whereas the one who never strays from the agreement remains unnoticed. In such settings, variability in ideas and judgments is again welcome, because variation is only the first step. In a second phase, the results of these judgments will be pitted against one another, and the best will triumph. In a market as in nature, selection _____.

* recession: 불경기 ** stray from: ～에서 벗어나다 *** pit against: ～과 맞붙이다

① reflects general trends
② can lead to cooperation
③ should not allow flexibility
④ cannot work without variation
⑤ requires a step-by-step process

Sentence Structure

● [The analyst {who correctly calls a recession ⟨that no one else has anticipated⟩}] is sure to gain fame, whereas the one [who never strays from the agreement] remains unnoticed.
첫 번째 [ ]는 문장의 주어이고, { }는 The analyst를 수식하는 관계절이며, ⟨ ⟩는 a recession을 수식하는 관계절이다. 두 번째 [ ]는 the one을 수식하는 관계절이다.

Word Search 주어진 문장에 들어갈 적절한 영어 단어를 본문에서 찾아 쓰시오. (필요시 형태를 바꿀 것)

1. His j_____ was clouded by his emotions, leading to an unfair decision.
2. The engineer proposed an innovative s_____ to improve the machine's efficiency.
3. The app is designed to g_____ random numbers for the game.

Exercise 12 함축 의미 추론

밑줄 친 Do your homework.가 다음 글에서 의미하는 바로 가장 적절한 것은?

25007-0048

Intelligent failures begin with preparation. No scientist wants to waste time or materials on experiments that have been run before and failed. <u>Do your homework.</u> The classic intelligent failure is hypothesis driven. You've taken the time to think through what might happen — why you have reason to believe that you could be right about what will happen. My Harvard colleague Thomas Eisenmann, an entrepreneurship expert, finds that many start-up failures are caused by the skipping of basic homework. For example, Triangulate, an online dating start-up, rushed to launch fully functional offerings that didn't fit any market needs. Eager to launch fast, founders skipped the research — customer interviews to probe for unmet needs. Paying no attention to that crucial preparation, the company paid the price. Thomas attributes this common failure, in part, to "the 'fail fast' mantra," which overemphasizes action, shortchanging preparation. Moreover, while this might seem self-evident, once you've done the homework, you must pay attention to what it's telling you.

* entrepreneurship: 기업가 정신 ** mantra: (특히 기도나 명상을 할 때 외는) 주문

① Move quickly to get a competitive edge.
② Ask for advice when you encounter an obstacle.
③ Reflect on your mistakes to gain valuable lessons.
④ Regularly monitor progress and make adjustments.
⑤ Take time to conduct thorough research before actual work.

Sentence Structure

● [Paying no attention to that crucial preparation], the company paid the price.

[ ]는 이유를 나타내는 분사구문으로, As the company paid no attention to that crucial preparation과 비슷한 의미이다.

Word Search 주어진 문장에 들어갈 적절한 영어 단어를 본문에서 찾아 쓰시오. (필요시 형태를 바꿀 것)

1. To test the h_____, they conducted several controlled experiments.
2. She decided to s_____ breakfast and have an early lunch instead.
3. It's easy to o_____ the importance of grades while neglecting the value of learning.

Exercise 1 어법

다음 글의 밑줄 친 부분 중, 어법상 틀린 것은?

25007-0049

Cooperation is a hallmark of human society. Not merely ① do people behave considerately toward complete strangers, they sometimes make substantial financial and even physical sacrifices for them. Such behavior seems impossible to explain with classic models of the evolution of altruism ② which cooperation between unrelated others can arise only when the same individuals interact repeatedly. However, new models of the evolution of human sociality and the results of experiments to test ③ them suggest that a sense of fairness and other foundations of morality have deep evolutionary roots. Such models consider processes at the level of groups of individuals, but without relying on the discredited notion of group selection in which individuals act "for the good of the group." Genes promoting prosocial tendencies in individuals can arise when groups compete in ways ④ argued to be characteristic of the early stages of human evolution. In particular, such conditions may have promoted the evolution of *strong reciprocity*, a tendency to cooperate with anonymous unrelated others and ⑤ to punish those who do not do the same even when doing so is costly to the punisher.

* altruism: 이타주의 ** anonymous: 익명의

Sentence Structure

• Such models consider processes at the level of groups of individuals, but without relying on the discredited notion of group selection [in which individuals act "for the good of the group"].

[ ]는 the discredited notion of group selection을 수식하는 관계절이다.

Word Search   주어진 문장에 들어갈 적절한 영어 단어를 본문에서 찾아 쓰시오. (필요시 형태를 바꿀 것)

1. C_____ between countries helps share food and supplies to fight world hunger.

2. His decision to return the lost wallet showed his strong sense of m_____.

3. The organization led a campaign to p_____ awareness of environmental issues.

Exercise 2 주제 추론

다음 글의 주제로 가장 적절한 것은?

25007-0050

Deforestation drivers are very diverse and vary by countries and states. A number of important studies have attempted to generalise and pull together large numbers of local studies. This work is hampered, however, by the fact that most information on drivers is not quantitative, so that few direct quantitative correlations can be made linking certain quantities of deforestation to particular activities. Angelsen and Kaimowitz reviewed 140 economic models analysing the causes of tropical deforestation. They found that, when looking at proximate causes, deforestation is often associated with the presence of more roads, higher agricultural prices, lower wages, and a shortage of off-farm employment. Also they considered it likely that policy reforms associated with economic liberalisation and related adjustment increase the pressure on forests. They pointed out, however, that many research studies have adopted poor methodology and low quality data, which makes the drawing of clear conclusions about the role of macroeconomic factors difficult.

* hamper: 어렵게 하다, 방해하다 ** proximate cause: 직접적인 원인

① difficulty of clearly attributing deforestation to specific causes
② ways of assessing the impact of deforestation on local economies
③ significance of interdisciplinary approaches in deforestation studies
④ necessity of preserving forest biodiversity for sustainable ecosystems
⑤ challenges in balancing economic development with environmental protection

Sentence Structure

- Also they considered **it** likely [that policy reforms {associated with economic liberalisation and related adjustment} increase the pressure on forests].
  it은 형식상의 목적어이고 [ ]가 내용상의 목적어이다. [ ] 안의 { }는 policy reforms를 수식한다.

Word Search 주어진 문장에 들어갈 적절한 영어 단어를 본문에서 찾아 쓰시오. (필요시 형태를 바꿀 것)

1. Do not g_____ one person's experience to everyone.
2. In engineering, q_____ data is used to make precise measurements and calculations.
3. An a_____ to the policy is required to deal with the issue better.

Exercise 3 · 제목 추론

다음 글의 제목으로 가장 적절한 것은?   25007-0051

On the one hand, the astonishing success of modern science in predicting and explaining the world seems to justify relying on it and bestowing extraordinary power to scientific institutions (especially in the form of public funding). On the other hand, the dominant societal position of science risks marginalizing alternative forms of knowledge acquisition. It's not clear that there still is a fair competition between science and its alternatives; and it's not straightforward that science has to prove its epistemic superiority in the 21st century in a way comparable to the period of the scientific revolution (16th-18th centuries). This is Paul Feyerabend's major argument in *Science in a Free Society*, which he casts as: "today science prevails not because of its comparative merits, but because the show has been rigged in its favour." So, Feyerabend insists, the fact that science has been the most appropriate means for addressing problems in the past doesn't warrant that it's also the most reliable form of knowledge acquisition for our current and future problems.

* bestow: 부여하다 ** epistemic: 인식론적인 *** rig: 조작하다

① Integration of Science and Other Fields: A Must for Efficient Education
② Utilize Accumulated Scientific Knowledge to Thrive in Changing Times
③ From Lab to Society: Addressing the Gap in Science Communication
④ How Does Public Funding Undermine the Reliability of Science?
⑤ Is Science Still the Best Path to Knowledge?

Sentence Structure

- So, Feyerabend insists, the fact [that science has been the most appropriate means for addressing problems in the past] doesn't warrant [that it's also the most reliable form of knowledge acquisition for our current and future problems].
  첫 번째 [ ]는 the fact의 구체적 내용을 설명하는 동격절이고, 두 번째 [ ]는 warrant의 목적어 역할을 하는 명사절이다.

Word Search 주어진 문장에 들어갈 적절한 영어 단어를 본문에서 찾아 쓰시오. (필요시 형태를 바꿀 것)

1. She tried to j_____ her actions by explaining the circumstances that led to her decision.
2. The lion is the d_____ animal in the African savanna.
3. During challenging times, it's important to believe that goodness will p_____.

Exercise 4 · 어법

(A), (B), (C)의 각 네모 안에서 어법에 맞는 표현으로 가장 적절한 것은?   25007-0052

A hallmark of immigrant parents that has been found across socioeconomic groups and countries of origin (A) is / are that first-generation immigrant parents place a high value on the educational achievement of their adolescent children. Immigrant parents tend to view earning good grades, finishing high school, and attending college as the primary means (B) through which / which their children can establish themselves in their country of destination. Compared to native-born parents with similar education levels, immigrant parents have been found to have higher educational aspirations for their children. Immigrant adolescents internalize their parents' values regarding the importance of education and tend to study more and devote more effort to their schoolwork than (C) do / are nonimmigrant adolescents. During adolescence, immigrants' peer groups often become more segregated from nonimmigrants' peer groups, with the immigrant adolescents supporting their peers' academic achievement, particularly for higher SES (socioeconomic status) adolescents.

* hallmark: 특징

	(A)		(B)		(C)
①	is	⋯⋯	through which	⋯⋯	do
②	is	⋯⋯	which	⋯⋯	do
③	is	⋯⋯	through which	⋯⋯	are
④	are	⋯⋯	which	⋯⋯	are
⑤	are	⋯⋯	through which	⋯⋯	do

Sentence Structure

- Compared to native-born parents with similar education levels, immigrant parents **have been found** to have higher educational aspirations for their children.
  완료형 수동태 have been found가 쓰여 주어인 immigrant parents에 대한 특정 사실이 밝혀졌음을 나타낸다.

Word Search   주어진 문장에 들어갈 적절한 영어 단어를 본문에서 찾아 쓰시오. (필요시 형태를 바꿀 것)

1. Many i_____ families face unique challenges as they adjust to life in a new country.
2. The young athlete has had a_____ of competing in the Olympics since she was a child.
3. Children often i_____ the values and behaviors they observe from their parents.

Exercise 5 무관한 문장

다음 글에서 전체 흐름과 관계 <u>없는</u> 문장은?   25007-0053

In public speaking, delivery with little or no preparation is called *impromptu speaking*. You engage in impromptu speaking every day as you communicate thoughts and ideas that spring up in the moment with no preparation or practice whatsoever. ① For example, when you answer a question in class or speak up during a meeting of a campus organization, you're using impromptu speaking. ② In this respect, impromptu speaking is simply another way to use the basic communication skills you already have and use regularly. ③ Learning how to express yourself on the spot without relying on research, extensive preparation, or notes will help you do well in your public speaking class and in less-structured speaking situations beyond the classroom. ④ This expertise stems from the fact that the speaker presents relevant information prepared with careful research that had been carried out specifically for the speech. ⑤ In addition, developing your impromptu speaking skills increases your confidence and decreases your speaking anxiety in any presentation context.

Sentence Structure

- You engage in impromptu speaking every day [as you communicate thoughts and ideas {that spring up in the moment} with no preparation or practice whatsoever].

  [ ]는 시간을 나타내는 부사절이고, 그 안의 { }는 thoughts and ideas를 수식하는 관계절이다.

Word Search   주어진 문장에 들어갈 적절한 영어 단어를 본문에서 찾아 쓰시오. (필요시 형태를 바꿀 것)

1. Her research was so e_____ that it covered many different areas.
2. She approached the job interview with c_____ because she had prepared thoroughly.
3. Regular exercise can help d_____ the risk of heart disease.

Exercise 6 글의 순서

주어진 글 다음에 이어질 글의 순서로 가장 적절한 것은? 25007-0054

In principle, there is an intuitively plausible relationship between profit and team quality. Fans like to watch exciting games where the outcome is uncertain. This is a result of competitive balance between the opponents.

(A) Thus, improved team quality should translate into more wins and therefore increased attendance. With a given set of prices, increased attendance turns into increased revenue from gate receipts, parking, and concessions.

(B) This increased revenue should lead to higher profit, but it costs more to field a higher-quality team than a lower-quality team. Consequently, there is a trade-off and therefore an optimal level of team quality that will vary from team to team depending on variation in the fan base.

(C) Tight, hard-fought games are more fun to watch than lopsided games in which the outcome is never in doubt. However, fans also prefer to see their team win. As a result, attendance should increase with improvement in the team's performance. Presumably, better teams win more than weaker teams.

* plausible: 그럴듯한 ** revenue: 수입 *** lopsided: 한쪽으로 기운

① (A) − (C) − (B) ② (B) − (A) − (C) ③ (B) − (C) − (A)
④ (C) − (A) − (B) ⑤ (C) − (B) − (A)

Sentence Structure

● Fans like [to watch exciting games {where the outcome is uncertain}].
  [ ]는 like의 목적어 역할을 하는 to부정사구이고, 그 안의 { }는 exciting games를 수식하는 관계절이다.

Word Search 주어진 문장에 들어갈 적절한 영어 단어를 본문에서 찾아 쓰시오. (필요시 형태를 바꿀 것)

1. The o_____ of today's election will be announced tomorrow morning.
2. The temperatures can v_____ greatly between day and night in the desert.
3. His performance on the test showed a clear i_____ from last time.

Exercise 7　어법

다음 글의 밑줄 친 부분 중, 어법상 <u>틀린</u> 것은?　　　　　25007-0055

> The impacts that volcanoes can have on the weather can extend well beyond the eruption site; they are one of the largest drivers of climate change over geological timescales. The vast majority of the 65,500 billion tons of carbon on Earth is ① <u>held</u> within rocks. The remainder resides in the oceans, atmosphere, plants, soil and fossil fuels. Carbon dioxide inside Earth is released ② <u>continually</u> during volcanic eruptions, and before the Industrial Revolution volcanoes were the largest source of carbon dioxide entering Earth's atmosphere. However, the release of carbon is largely regulated by the natural carbon cycle, which draws down as much carbon from the atmosphere as volcanoes release into it, ③ <u>acting</u> as a planetary scale thermostat. If temperatures increase due to a period of intense volcanism, more carbon will be drawn down from the atmosphere, ④ <u>it</u> will take temperatures back to their previous levels. However, given the slow rate of some of these chemical reactions it can take hundreds of thousands of years for the system ⑤ <u>to stabilize</u>.
>
> * eruption: 분출, 분화　** thermostat: 자동 온도 조절 장치

Sentence Structure

● However, the release of carbon is largely regulated by the natural carbon cycle, [which draws down {**as much** carbon from the atmosphere **as** volcanoes release into it}, acting as a planetary scale thermostat].
[ ]는 the natural carbon cycle을 부가적으로 설명하는 관계절이고, 그 안의 { }는 draws down의 목적어이다. { } 안에는 '…만큼 ~인'의 의미를 나타내는 「as much ~ as …」가 사용되었으며, it은 문맥상 the atmosphere를 가리킨다.

Word Search　　주어진 문장에 들어갈 적절한 영어 단어를 본문에서 찾아 쓰시오. (필요시 형태를 바꿀 것)

1. The g＿＿＿＿＿＿ formations in this region provide valuable insights into Earth's history.
2. The majority of the country's population r＿＿＿＿＿＿ in urban areas.
3. The i＿＿＿＿＿＿ heat wave caused widespread power outages across the nation.

Exercise 8 주제 추론

다음 글의 주제로 가장 적절한 것은?   25007-0056

Ignacio Varchausky from the Buenos Aires tango orchestra El Arranque says in the documentary *Si Sos Brujo* that he and others tried learning from records how the older orchestras did what they did, but it was difficult, almost impossible. Eventually, El Arranque had to find the surviving players from those ensembles and ask them how it was done. The older players had to physically show the younger players how to replicate the effects they got, and which notes and beats should be emphasized. So, to some extent, music is still an oral (and physical) tradition, handed down from one person to another. Records may do a lot to preserve music and disseminate it, but they can't do what direct transmission does. In that same documentary, Wynton Marsalis says that the learning, the baton passing, happens on the bandstand — one has to play with others, to learn by watching and imitating. For Varchausky, when those older players are gone, the traditions (and techniques) will be lost if their knowledge is not passed on directly. History and culture can't really be preserved by technology alone.

* ensemble: 합주단 ** disseminate: 전파하다

① the limitation of learning music indirectly through records
② efforts to preserve underappreciated music of artistic value
③ the tension between innovation and tradition in the music world
④ effects of musical education overly centered on direct transmission
⑤ conflicting views on the effectiveness of imitation in learning music

Sentence Structure

● [Ignacio Varchausky from the Buenos Aires tango orchestra El Arranque] says in the documentary *Si Sos Brujo* [that he and others tried learning from records {how the older orchestras did what they did}, but it was difficult, almost impossible].
첫 번째 [ ]는 문장의 주어 역할을 하는 명사구이고, 두 번째 [ ]는 says의 목적어 역할을 하는 명사절이다. { }는 learning의 목적어 역할을 하는 명사절이다.

Word Search 주어진 문장에 들어갈 적절한 영어 단어를 본문에서 찾아 쓰시오. (필요시 형태를 바꿀 것)

1. Scientists struggled to r_____ the experiment's results in a different laboratory.
2. During the conference, she gave an impressive o_____ presentation on the impact of climate change on marine ecosystems.
3. The t_____ of cultural traditions from one generation to the next ensures their preservation.

Exercise 9 제목 추론

다음 글의 제목으로 가장 적절한 것은?   25007-0057

Teams have many communication technologies at their disposal, ranging from email and chat platforms to web conferencing and videoconferencing. People often default to using the tool that is most convenient or familiar to them, but some technologies are better suited to certain tasks than others, and choosing the wrong one can lead to trouble. Communication tools differ along a number of dimensions, including information richness (or the capacity to transfer nonverbal and other cues that help people interpret meaning) and the level of real-time interaction that is possible. A team's communication tasks likewise vary in complexity, depending on the need to reconcile different viewpoints, give and receive feedback, or avoid the potential for misunderstanding. The purpose of the communication should determine the delivery mechanism. So carefully consider your goals. Use leaner, text-based media such as email, chat, and bulletin boards when pushing information in one direction — for instance, when circulating routine information and plans, sharing ideas, and collecting simple data. Web conferencing and videoconferencing are richer, more interactive tools better suited to complex tasks such as problem solving and negotiation, which require squaring different ideas and perspectives.

* **default to:** ~을 기본적으로 선택하다   ** **reconcile:** 조화시키다   *** **square:** 조율하다, 맞추다

① Discover the Power of Purpose-driven Teamwork!
② Do Virtual Meetings Really Undermine Team Chemistry?
③ The Future of Work: Collaboration Redefined in the Digital Age
④ Match the Right Technology to Your Team's Communication Goals
⑤ Fueling Innovation: The Magic of Teams with Diverse Personalities

Sentence Structure

● A team's communication tasks likewise vary in complexity, depending on the need [to {reconcile different viewpoints}, {give and receive feedback}, or {avoid the potential for misunderstanding}].
  [ ]는 the need의 구체적 내용을 설명하는 to부정사구이다. 세 개의 { }는 콤마(,)와 or로 연결되어 to에 이어진다.

Word Search 주어진 문장에 들어갈 적절한 영어 단어를 본문에서 찾아 쓰시오. (필요시 형태를 바꿀 것)

1. The new online p_____ facilitated discussions on various topics among users.
2. Further tests are needed to d_____ the exact cause of the patient's illness.
3. The central bank is going to c_____ more money to stimulate the economy.

Exercise 10 어법

다음 글의 밑줄 친 부분 중, 어법상 틀린 것은?   25007-0058

Animals regularly engage in future-oriented behaviors, from nest-building to hibernation. Clearly these behaviors are functionally prospective, but the extent to which they are controlled by cognitive processes ① remains an open question. Migrating birds, for example, travel long distances to avoid cold winters, without ever ② having experienced a winter, warmer climates, or the dangers of travel. These birds are unable to *know* ③ that it is they are avoiding and why this course of action benefits them. Indeed, when two bird populations with different migratory paths are cross-bred, the resulting offspring migrate in a direction halfway between that of their parents, ④ suggesting that direction of migration is genetically determined. From this example we can see that not all prospective behaviors can ⑤ be considered to involve awareness of the future, but may instead be automatic responses to natural (e.g., seasonal changes in day length or hormones), or learned, cues.

* hibernation: 동면 ** cross-breed: 이종 교배하다 *** offspring: (동물의) 새끼

Sentence Structure

- Indeed, [when two bird populations with different migratory paths are cross-bred], the resulting offspring migrate in a direction [halfway between **that** of their parents], ~.

  첫 번째 [ ]는 접속사 when이 이끄는 부사절이다. 두 번째 [ ]는 a direction을 수식하고, that은 the direction을 대신한다.

Word Search 주어진 문장에 들어갈 적절한 영어 단어를 본문에서 찾아 쓰시오. (필요시 형태를 바꿀 것)

1. The young athlete's training program is a p_____ indicator of her future success in the competition.
2. Regular exercise has been shown to enhance c_____ function and improve memory.
3. Every winter, monarch butterflies m_____ thousands of miles to warmer climates.

Exercise 11   무관한 문장

다음 글에서 전체 흐름과 관계 <u>없는</u> 문장은?   25007-0059

Soil and the plants that grow in it are continually interacting in many ways, both directly and through the effects of the many kinds of bacteria, fungi and other small organisms that are always present. ① Most notably, plants are continually dropping leaves and other parts which fall to the ground to rot in place where they release their nutrients, although in some drier climates, lightning-set fires are necessary to liberate the mineral nutrients present in the plant litter. ② In any case, most of the nutrients present in dead plant material end up locally in the soil, while the vast networks of plant roots and fungal threads join with bacteria, lichens and sticky humus to effectively hold the soil together even in the face of heavy rains and strong winds. ③ The result is effective and continuous recycling of plant nutrients which are capable of moving back and forth between the soil and the plants almost indefinitely. ④ Methods for improving the quality of nutrient-poor soil include crop rotation and adding organic matter. ⑤ Soil is also a vast storehouse for the seeds and spores produced by the local plants, a key factor in allowing vegetation to regenerate following disturbance.

* lichen: 지의류, 이끼   ** humus: 부식토   *** spore: 포자(식물이 무성 생식을 하기 위하여 형성하는 생식 세포)

Sentence Structure

● Most notably, plants are continually dropping leaves and other parts [which fall to the ground {to rot in place ⟨where they release their nutrients⟩}], [although in some drier climates, lightning-set fires are necessary to liberate the mineral nutrients {present in the plant litter}].

첫 번째 [ ]는 leaves and other parts를 수식하는 관계절이다. 그 안의 { }는 결과의 의미를 나타내는 to부정사구이고, ⟨ ⟩는 place를 수식하는 관계절이다. 두 번째 [ ]는 부사절이고, 그 안의 { }는 the mineral nutrients를 수식하는 형용사구이다.

Word Search   주어진 문장에 들어갈 적절한 영어 단어를 본문에서 찾아 쓰시오. (필요시 형태를 바꿀 것)

1. The soil lacked essential n_____, hindering the plant's growth.

2. The ocean was v_____, stretching far and wide with no end in sight.

3. Ecological d_____ can disrupt delicate ecosystems, leading to significant ecological imbalances.

Exercise 12 글의 순서

주어진 글 다음에 이어질 글의 순서로 가장 적절한 것은? 25007-0060

When Darwin developed his theory of natural selection, he imagined evolution unrolling on geological timescales: the way a glacier sculpts a valley or weather beats on a rock, so time molds one species into another.

(A) The birds with the smallest and pointiest beaks were suddenly, and distinctly, advantaged. Their genes rapidly spread. Selection for traits can happen over the course of years or decades, not just eons. And Darwin's finches are exemplary, not exceptional. They are a privileged case of the paradigm that the biologist John Thompson has called "relentless evolution."

(B) It turns out that evolution works far more quickly than Darwin imagined. The finches of the Galápagos evolve on timescales we can observe. In 1983, for instance, a year with superabundant rainfall, a species of vine with tiny seeds overran the flora of Daphne Island.

(C) Such was the conventional wisdom for over a century. But in the late twentieth century, our understanding of evolution started to change, thanks in no small part to the continuing study of the Galápagos finches that Darwin himself had observed on the voyage of the *Beagle*.

* eon: 10억 년 ** finch: 핀치(부리가 짧은 작은 새) *** flora: 식물군

① (A) – (C) – (B)　　　② (B) – (A) – (C)　　　③ (B) – (C) – (A)
④ (C) – (A) – (B)　　　⑤ (C) – (B) – (A)

Sentence Structure

● But in the late twentieth century, our understanding of evolution started to change, [**thanks** {in no small part} **to** the continuing study of the Galápagos finches {that Darwin himself had observed on the voyage of the *Beagle*}].

[ ]의 thanks to는 '~ 덕분에, ~ 때문에'라는 의미이며, 첫 번째 { }가 thanks와 to 사이에 놓였다. 두 번째 { }는 the Galápagos finches를 수식하는 관계절이다.

Word Search 주어진 문장에 들어갈 적절한 영어 단어를 본문에서 찾아 쓰시오. (필요시 형태를 바꿀 것)

1. The sculptor's skilled hands can m_____ the clay into a breathtakingly lifelike figure.
2. The company decided to stick with c_____ methods rather than experimenting with new approaches.
3. The little girl was o_____ a butterfly flying from flower to flower.

Exercise 1   어법

다음 글의 밑줄 친 부분 중, 어법상 <u>틀린</u> 것은?   25007-0061

Profits are managed not only by setting a pricing strategy to maximize revenues and control costs; they are also influenced by a commitment to measure marketing performance. An organization cannot achieve its objectives if it does not know ① <u>where</u> it stands in meeting those objectives. Measurement of marketing activities is needed to assess past performance as well as ② <u>set</u> a future marketing strategy. For example, it is important for a minor league baseball team to understand the effectiveness of Saturday night fireworks promotions in terms of the number of additional tickets ③ <u>sold</u> and related spending from those customers. Tracking performance in this case would enable a comparison of revenue ④ <u>generating</u> with costs accumulated to execute the promotion. Such return on investment analysis should be performed whenever possible ⑤ <u>to understand</u> the relationship between marketing investment and profits, and guide decision-making for future marketing activities.

Sentence Structure

● Profits are managed not only by setting a pricing strategy [to {maximize revenues} and {control costs}]; ~.
  [ ]는 선행하는 a pricing strategy를 수식하는 to부정사구이고, 그 안에서 두 개의 { }가 and로 연결되어 to에 이어진다.

Word Search   주어진 문장에 들어갈 적절한 영어 단어를 본문에서 찾아 쓰시오. (필요시 형태를 바꿀 것)

1. The nonprofit o_____ works to support underprivileged communities.
2. He finally a_____ his lifelong dream of becoming a published author.
3. His ability to m_____ the resources ensured the project was completed within the budget.

Exercise 2 주제 추론

다음 글의 주제로 가장 적절한 것은?   25007-0062

Consider how some people, including some young children, enjoy doing jigsaw puzzles for hours. There are no big surprises. (Unless the last piece has been hidden by a sly sibling.) The players know exactly what the end result will look like and are rarely startled midway. But there can still be great satisfaction upon completion. When listening to an interesting story, we can be deeply disappointed if we don't get to hear the ending. When exploring a new place such as a house or a park, we often want to get a clear sense of the overall layout and can be frustrated if we don't know how different locations fit together. Infants, like all of us, strive to achieve closure. They have a need to identify *all* the toys that are in a box, or explore *every* room of a new space. They have similar needs for closure in their causal models of how things happen. This constellation of needs for closure often helps drive wonder, especially when there is a need to close an explanatory gap.

* sly: 장난꾸러기의 ** closure: 완결감 *** constellation: (유사한 것의) 집합체

① effects of unpredictability in completing familiar tasks
② reasons why young children avoid facing new challenges
③ the absence of conclusion as a driving force behind education
④ the minor role of completion in the enjoyment of jigsaw puzzles
⑤ the inherent human drive for completion, prominent from childhood

Sentence Structure

● The players [know exactly {what the end result will look like}] and [are rarely startled midway].
두 개의 [ ]가 and로 연결되어 문장의 술부를 이루고 있고, 첫 번째 [ ]에서 { }는 what이 이끄는 명사절로 동사 know의 목적어 역할을 한다.

Word Search 주어진 문장에 들어갈 적절한 영어 단어를 본문에서 찾아 쓰시오. (필요시 형태를 바꿀 것)

1. He felt f_____ by his inability to solve the complex puzzle.
2. The l_____ of the resort made it a popular destination for tourists.
3. The parents were delighted as they watched their i_____ take his first steps.

Exercise 3 제목 추론

다음 글의 제목으로 가장 적절한 것은?   25007-0063

Unlike Europe's national and partisan press, U.S. daily newspapers competed predominantly at the city level. The circulation range for a daily newspaper was primarily bounded by its distance from other cities and by its time of publication. Streetcars at the end of the 19th century, and motor trucks in the 20th permitted timely delivery of newspapers over larger regions. Different papers competing in the same towns chose to limit or spread out their physical distribution and editorial coverage to different radial distances from the central city. Morning newspapers printed during the night were, of course, able to distribute over a wider range than afternoon papers. By and large, however, the geographical area within which happenings were regularly reported as local news coincided with the area from which the newspaper drew the bulk of its readers.

* partisan: 당파적인   ** radial distance: 반경 거리

① There Is No Commonality in European Newspapers
② Why Do All U.S. Newspapers Cover the Same Content?
③ Global Competition: U.S. Newspapers' Editorial Strategies
④ Regional Focus: Characteristics of U.S. Daily Newspapers
⑤ Transportation Revolution, What Gave Birth to International Newspapers

Sentence Structure

● The circulation range for a daily newspaper was primarily bounded [by its distance from other cities] and [by its time of publication].

by가 이끄는 전치사구인 두 개의 [ ]는 and로 연결되어 was primarily bounded에 이어진다.

Word Search   주어진 문장에 들어갈 적절한 영어 단어를 본문에서 찾아 쓰시오. (필요시 형태를 바꿀 것)

1. Regular exercise can improve blood c_____.
2. Talented athletes c_____ in the international tournament.
3. The festival dates c_____ with the national holiday.

Exercise 4   어법

(A), (B), (C)의 각 네모 안에서 어법에 맞는 표현으로 가장 적절한 것은?   25007-0064

Wind plays an important role in Earth's water cycle. In recent years, the once-accepted view of the water cycle has changed, and wind's role in atmospheric circulation has found greater appreciation. Traditionally, it was thought that the planet's water cycle mostly involved evaporated moisture from large bodies of water (A) | being / was | distributed by wind across the Earth's surface. In the last fifty years, that view has changed — it is now known that forests play a significant role in generating the transpiration that drives rainfall. The various plants that live within forests capture water in their roots, and then release (B) | it / them | as vapour through transpiration. Prevailing winds transport the water through the air, delivering rainfall to other locations. When this water moves through the power of wind in large volumes, it becomes (C) | what / that | meteorologist José Marengo called a 'flying river'.

* transpiration: 증산 작용   ** prevailing wind: 탁월풍   *** meteorologist: 기상학자

	(A)	(B)	(C)
①	being ······	it ······	what
②	being ······	them ······	what
③	being ······	it ······	that
④	was ······	them ······	that
⑤	was ······	it ······	that

Sentence Structure

● ~ **it** is now known [that forests play a significant role in generating the transpiration {that drives rainfall}].
it은 형식상의 주어이고, that이 이끄는 명사절인 [ ]가 내용상의 주어이다. { }는 the transpiration을 수식하는 관계절이다.

Word Search   주어진 문장에 들어갈 적절한 영어 단어를 본문에서 찾아 쓰시오. (필요시 형태를 바꿀 것)

1. The team received widespread a＿＿＿＿＿＿ for their hard work.
2. The air was thick with m＿＿＿＿＿＿ after the rain.
3. The s＿＿＿＿＿＿ of the lake was so clear that it perfectly reflected the surroundings.

Exercise 5 무관한 문장

다음 글에서 전체 흐름과 관계 없는 문장은?   25007-0065

Caregivers and others in a child's immediate surroundings affect the language being acquired by that child. Later, social groups affect the child's language. ① This results in social varieties of language with differences for class, ethnicity, and age as well as gender and gender expression. ② However, when a social group undergoes significant changes across generations leading to the need for a different type of communication, an earlier variety might cease to be used. ③ This occurred with the English initially spoken in Cosme and Nueva Australia, which were rural Paraguayan communities founded by Australian immigrants in the 1890s. ④ As a means of adapting to new cultural norms and enhancing its own communication and understanding, the English language makes extensive use of borrowings. ⑤ The only remaining language vestiges from those English-speaking immigrants are common Paraguayan last names like Kennedy, Smith, Stanley, and Wood.

* vestige: 흔적, 자취

Sentence Structure

- However, [when a social group undergoes significant changes across generations {leading to the need for a different type of communication}], an earlier variety might cease to be used.

  [ ]는 시간의 부사절이며, 그 안의 { }는 앞선 significant changes across generations를 수식하는 분사구이다.

Word Search   주어진 문장에 들어갈 적절한 영어 단어를 본문에서 찾아 쓰시오. (필요시 형태를 바꿀 것)

1. The design has to appeal to all ages and s_____ groups.
2. Due to the lack of use, some languages have gradually c_____ to be spoken eventually.
3. Various c_____ in the region worked together to preserve their traditions.

Exercise 6   글의 순서

주어진 글 다음에 이어질 글의 순서로 가장 적절한 것은?   25007-0066

Pain is a neurological response to an external stimulus.

(A) The excitement of the brain pain neurons in turn asserts a presence in our consciousness. We become aware of something being on our arm. If it is a bee sting, where the bee leaves poisonous liquids at the sting site, more neurons at the site of the sting are triggered than with a mosquito.

(B) A mosquito bite, triggering a very small number of pain receptors, will create a mild sensation of awareness in the brain. The signal is relayed by a chain of neurons up the spine of the individual to a special place in the brain.

(C) Therefore, a stronger signal is sent to the brain, with a stronger insertion into the individual's consciousness. If a falling tree crushes a leg then many, many pain receptors are fired off and the consciousness is overwhelmed with pain signals. Torn flesh will continuously fire the signal until medical treatment deals with the problem.

* receptor: 수용기(受容器) ** spine: 척추

① (A) − (C) − (B)   ② (B) − (A) − (C)   ③ (B) − (C) − (A)
④ (C) − (A) − (B)   ⑤ (C) − (B) − (A)

Sentence Structure

● If it is a bee sting, [where the bee leaves poisonous liquids at the sting site], **more** neurons at the site of the sting are triggered **than** with a mosquito.

[ ]는 a bee sting을 추가적으로 설명하는 관계절이고, 주절에는 「more ~ than ...」의 비교 구문이 사용되었다.

Word Search   주어진 문장에 들어갈 적절한 영어 단어를 본문에서 찾아 쓰시오. (필요시 형태를 바꿀 것)

1. The teacher's r_____ to the student's question helped clarify the lesson.
2. The traffic light turned green, giving the s_____ for the cars to start moving.
3. The i_____ of normal genes into the diseased cells could cure the disease.

Exercise 7 어법

다음 글의 밑줄 친 부분 중, 어법상 틀린 것은?   25007-0067

Conflict is at the very heart of people-work, especially for professionals such as social workers, police and probation officers, who are often required to balance individual rights and freedoms against risks and harm to others. The very act of undertaking a risk assessment on an individual or family carries with it the likelihood of decisions being taken with which some of the people ① involved may violently disagree. To remove a child 'at risk' from a family to a place of safety, or ② to insist that someone in mental distress go into hospital for treatment, is to be involved in conflict. Within a multi-disciplinary hospital team, there could also be conflicting approaches to how to work most ③ effectively with, or treat, a particular individual or family. Most people-workers and healthcare workers can also give examples of times when they have had to deal with angry, distressed or disappointed people ④ whose response has often provoked high levels of conflict. To develop appropriate communication skills for dealing with conflict ⑤ being, therefore, essential.

* probation: 보호 관찰

Sentence Structure

● Conflict is at the very heart of people-work, especially for professionals such as social workers, police and probation officers, [who are often required to balance individual rights and freedoms against risks and harm to others].

[ ]는 professionals such as social workers, police and probation officers를 부가적으로 설명하는 관계절이다.

Word Search   주어진 문장에 들어갈 적절한 영어 단어를 본문에서 찾아 쓰시오. (필요시 형태를 바꿀 것)

1. Professors both teach and u_____ research.
2. You need to conduct the initial a_____ defining the scope of work.
3. We sensed his deep emotional d_____.

Exercise 8 주제 추론

다음 글의 주제로 가장 적절한 것은?   25007-0068

Drinking bypasses lots of our natural food control system. Just think how easy it is to drink a large glass of apple juice, yet you wouldn't sit down and eat eight apples in one sitting — after the second one, the control system would step in and say you have had enough. The whole apple contains fibre, which slows the release of the apple's sugar and, in doing so, balances the effect of eating it on our energy. Without the fibre in the juice, the sugar sends us on a rollercoaster of energy cycles, starting with a boost and ending in a crash. The crash may be just at the point when you are asked to stand up and present your ideas or you need to kick off that tricky task on your to-do list. It's just not worth it. Sugary caffeine drinks have a similar, if not more profound, effect, yet large coffee shop chains take great delight in selling us expensive, iced or cream-topped drinks that are equivalent to an extra meal, but will simply not register as one — starting a blood sugar crash cycle all over again.

* bypass: 무시하다   ** equivalent to: ~에 상당하는

① essential roles of sugar in human physiology
② common myths about how fruits help our health
③ the impact of drinking sugary beverages on our body
④ understanding the factors helping nutrient absorption
⑤ how to balance the energy we take in and the energy we burn

Sentence Structure

• The whole apple contains fibre, [which {slows the release of the apple's sugar} and, in doing so, {balances the effect of eating it on our energy}].
[ ]는 fibre를 부가적으로 설명하는 관계절이다. 두 개의 { }가 and로 연결되어 관계절의 술부를 구성한다.

Word Search   주어진 문장에 들어갈 적절한 영어 단어를 본문에서 찾아 쓰시오. (필요시 형태를 바꿀 것)

1. Eating enough f_____ can help waste to move smoothly through the body.
2. Passing the test was such a b_____ to my confidence.
3. This equipment was too t_____ to install.

Exercise 9   제목 추론

다음 글의 제목으로 가장 적절한 것은?   25007-0069

There are a lot of ways for an ad to go wrong. But its odds for success are increased by starting out with a smart, focused strategy. With such focus, the ad is able to deliver its singular message on the many levels in which advertising communicates. It bores down to make its point rather than making a big mess of itself. The chance for the reader to "get it" increases tremendously. The flip side is that an unfocused strategy leads to ads that must deliver many different points all at the same time. This is not effective. Here's an analogy: Try throwing a dozen balls at someone all at once. It's impossible to catch a single one. In fact, the impulse is to give up, shield your face, and try not to get hurt. Absolute chaos. But throw a single ball directly at the target and chances are, he or she will make the catch. That's the difference between an ineffective ad with a fuzzy strategy and an effective ad with a focused strategy.

* flip side: 이면   ** analogy: 비유, 유추   *** fuzzy: 모호한

① Navigating Responsibility in Advertising
② Ad Fusion: Blending Creativity with Data
③ Building Brand Loyalty Through Smart Ads
④ Understanding Consumer Behavior Through Ads
⑤ Importance of Crafting Focused Advertising Strategies

Sentence Structure

● With such focus, the ad is able to deliver **its** singular message on the many levels [in which advertising communicates].

its는 the ad를 가리키며, [ ]는 the many levels를 수식하는 관계절이다.

Word Search   주어진 문장에 들어갈 적절한 영어 단어를 본문에서 찾아 쓰시오. (필요시 형태를 바꿀 것)

1. He had a sudden i_____ to stand up and dance.
2. This video d_____ a clear message about water safety.
3. The o_____ of finding a parking space are very low.

Exercise 10   어법

다음 글의 밑줄 친 부분 중, 어법상 틀린 것은?   25007-0070

We are all able to think of words which have changed their meanings, or ① acquired additional meanings, in our own lifetimes, such as the changes in the use of words like chip, grass, joint and hardware. This demonstrates how words are mere labels for constructs of meanings. If these constructs have a different make-up from one person to another, the transitivity of communication between them ② is not going to be complete or accurate. For those who are learning English as an additional language, apprehension of English forms and structures, and their meanings may pose difficulties, or even may be tinged with aspects of meaning from the characteristics ③ owned by the native tongue. Gradually, with full fluency and the commerce of using the language, commonality of understanding will be ④ generally achieved. However, it is important for those teaching children at early stages of the acquisition of English in our schools ⑤ attend to building the understanding of meanings behind the labels of the words and phrases being taught.

* transitivity: 이행, 전이, 전달   ** tinge: 가미하다

Sentence Structure

- We are all able to think of words [which have {changed their meanings}, or {acquired additional meanings}, in our own lifetimes], such as the changes in the use of words like chip, grass, joint and hardware.
  [ ]는 words를 수식하는 관계절이고, 두 개의 { }가 or로 연결되어 have에 이어진다.

Word Search   주어진 문장에 들어갈 적절한 영어 단어를 본문에서 찾아 쓰시오. (필요시 형태를 바꿀 것)

1. A_____ measurements are essential for maintaining safety.
2. A full a_____ of all that is involved is utterly beyond the scientists.
3. Language a_____ means the process of learning and understanding a language by a child or adult.

Exercise 11 무관한 문장

다음 글에서 전체 흐름과 관계 <u>없는</u> 문장은? 25007-0071

Many of us are prone to wanting instant gratification: constantly buying new clothes to be 'trendy', wanting the latest thing to 'keep up' or seeking immediate results or pleasure. Even though choosing something now may feel good in the moment, taking a longer-term view and developing self-discipline can result in bigger and better rewards in the future. ① Delayed gratification is our ability to resist these temptations of instant pleasure, allowing us to stay focused on our long-term goals. ② It's important to have a big-picture goal, but it's the significant short-term goals that get you to the end result. ③ It involves budgeting and spending mindfully, as well as developing discipline and patience to avoid impulse spending and decision making. ④ Think about instant vs delayed gratification when you are developing your own financial goals. ⑤ Consider whether your goals are focused on short-term gains or designed to look forward to a bigger picture.

* gratification: 만족

Sentence Structure

● Even though [choosing something now] may feel good in the moment, [taking a longer-term view] and [developing self-discipline] can result in bigger and better rewards in the future.
첫 번째 [ ]는 부사절의 주어 역할을 하는 동명사구이다. 두 번째와 세 번째 [ ]는 동명사구로 and로 연결되어 주절의 주어 역할을 한다.

Word Search 주어진 문장에 들어갈 적절한 영어 단어를 본문에서 찾아 쓰시오. (필요시 형태를 바꿀 것)

1. His new novel was an i_____ hit.
2. An expensive jewelry is a t_____ to thieves.
3. It takes a lot of d_____ to stay away from snacks.

Exercise 12 글의 순서

주어진 글 다음에 이어질 글의 순서로 가장 적절한 것은?

25007-0072

Uncertainty about what tool or procedure to use, and the risk that results are not what they appear to be, are problems common to all the scientific disciplines.

(A) Waffling is annoying when you are trying to make decisions on the basis of the scientific information that comes your way. However, if a new technique is the source of the uncertainty, time and future experiments will confirm or disconfirm its usefulness and clear up uncertainty.

(B) The development of new tools allows scientists to answer questions they could not answer in the past, and the answers to those questions will lead to new questions, and so on. Therefore, new technologies and procedures are crucial to the progress of science.

(C) At the same time, other scientists unfamiliar with a new tool may express skepticism and call for others to replicate the experiments. Because this skepticism often comes to us in the form of sound bites, and because uncertainty about experimental tools is an aspect of science that is not familiar to most people, even people with a bachelor's degree in science, the skepticism may seem like waffling.

* waffle: (결정을 못 내리고) 미적거리다   ** skepticism: 회의적인 태도, 회의론
*** sound bites: 짧고 기억에 남기 쉬운 한두 문장으로 요약된 말

① (A) – (C) – (B)　　　② (B) – (A) – (C)　　　③ (B) – (C) – (A)
④ (C) – (A) – (B)　　　⑤ (C) – (B) – (A)

Sentence Structure

- [Uncertainty about what tool or procedure to use, and the risk {that results **are** not ⟨what they appear to be⟩}], **are** problems common to all the scientific disciplines.

  [ ]는 문장의 주어이고, 두 번째 are가 문장의 동사이다. { }는 the risk와 동격 관계에 있고, ⟨ ⟩는 주격 보어이다.

Word Search 주어진 문장에 들어갈 적절한 영어 단어를 본문에서 찾아 쓰시오. (필요시 형태를 바꿀 것)

1. There is a good deal of u_____ about the exact path of the typhoon.
2. His most a_____ habit was eating with his mouth open.
3. X-rays have c_____ that she has not broken any bones.

Exercise 1 어법

다음 글의 밑줄 친 부분 중, 어법상 틀린 것은?   25007-0073

It is not uncommon for children to report overconfidence about accomplishing difficult tasks. Even when they are given feedback indicating that they have performed ① poorly, their self-efficacy may not decline. The incongruence between children's self-efficacy and their actual performance can arise when children lack task familiarity and do not fully understand ② what is required to execute a task successfully. As they gain experience, their accuracy improves. Children may also be excessively swayed by certain task features and ③ decide based on these that they can or cannot perform the task. In subtraction, for example, children may focus on how many numbers or columns the problems contain and judge problems with fewer columns as less difficult than ④ those with more columns, even when the former are conceptually more difficult. This is an instance where higher self-efficacy is problematic because students feel unrealistically overconfident and not motivated to seek help and improve their skills. As children's capability to focus on multiple features improves, so ⑤ is their accuracy.

* incongruence: 불일치   ** sway: 영향을 주다   *** subtraction: 뺄셈

Sentence Structure

● **It** is not uncommon [{for children} to report overconfidence about {accomplishing difficult tasks}].
It은 형식상의 주어이고, [ ]는 내용상의 주어이다. 그 안에서 첫 번째 { }는 to부정사구의 의미상의 주어를 나타내고, 두 번째 { }는 전치사 about의 목적어 역할을 하는 동명사구이다.

Word Search   주어진 문장에 들어갈 적절한 영어 단어를 본문에서 찾아 쓰시오. (필요시 형태를 바꿀 것)

1. Even with repeated failures, his enthusiasm did not d_____.
2. Young children often l_____ the experience needed to assess their abilities.
3. Her c_____ to solve complex problems makes her an asset to the team.

Exercise 2 주제 추론

다음 글의 주제로 가장 적절한 것은?   25007-0074

Using generative AI in the hiring process still has its challenges and limitations. Generative AI may, for instance, produce inaccurate or misleading content that could harm the reputation or credibility of recruiters or candidates. It may also raise ethical or legal issues regarding data privacy, consent, or ownership. Thus, complete automation seems unlikely in the current field as hiring managers, legal departments, etc., still have to review and sign off on generated job ads and candidate communication. Therefore, generative AI should not replace human judgment or interaction in hiring but rather complement it. Generative AI should be used as a tool to augment human capabilities and creativity, not to automate them entirely. Recruiters should always verify the information and edit the content generated by generative AI before using it. They have to monitor the performance and impact of generative AI on their hiring outcomes and candidate experience.

* augment: 증강하다

① the crucial role AI plays in shaping the future job market
② the evolving legal frameworks surrounding AI use in candidate selection
③ the inevitable shift toward fully automating recruitment using generative AI
④ the technological advancements leading to the replacement of human roles with AI
⑤ the need for human participation in integrating generative AI into the hiring process

Sentence Structure

● Therefore, generative AI should **not** [replace human judgment or interaction in hiring] **but rather** [complement **it**].
두 개의 [ ]가 '~이 아니라 오히려 …인'이라는 의미의 「not ~ but rather ...」 구문으로 연결되었으며, it은 human judgment or interaction in hiring을 가리킨다.

Word Search   주어진 문장에 들어갈 적절한 영어 단어를 본문에서 찾아 쓰시오. (필요시 형태를 바꿀 것)

1. The company faced an e_____ dilemma when deciding whether to use animal testing for their new product.

2. Clear and concise c_____ is essential during a crisis to prevent misinformation.

3. The new regulations had a significant i_____ on the company's operations.

Exercise 3 제목 추론

다음 글의 제목으로 가장 적절한 것은?

25007-0075

Having asked the class a question, the average American teacher typically waits less than a second before picking a child to provide an answer — sending out a strong message that speed is valued over complex thinking. But a study from the University of Florida has found that something magical happens when the teacher takes a little more time — just three seconds — to wait to pick a child, and then for the child to think about the response. The most immediate benefit was seen in the length of the children's answers. The small amount of thinking time meant that the children spent between three and seven times as long elaborating their thoughts, including more evidence for their viewpoint and a greater consideration of alternative theories. The increased waiting time also encouraged the children to listen to each other's opinions and develop their ideas. Encouragingly, their more sophisticated thinking also translated to their writing, which became more nuanced and complex. That's an astonishing improvement from the simple act of exercising teacherly patience. As the researcher Mary Budd Rowe put it in her original paper: 'slowing down may be a way of speeding up'.

* nuanced: 미묘한 차이가 있는

① How Do Classroom Environments Affect Learning Skills?
② Why Active Participation Enhances Student Comprehension
③ Questions: The Key to Unlocking Deeper Thinking in Students
④ The Power of Pause: How Waiting Can Deepen Student Thinking
⑤ Fast-paced Learning: Encouraging Instant Understanding in Students

Sentence Structure

• The small amount of thinking time meant [that the children spent between three and seven times as long elaborating their thoughts, {including more evidence for their viewpoint and a greater consideration of alternative theories}].

[ ]는 meant의 목적어 역할을 하는 명사절이며, 그 안의 { }는 elaborating their thoughts에 대한 부가적인 정보를 더해 주는 전치사구이다.

Word Search 주어진 문장에 들어갈 적절한 영어 단어를 본문에서 찾아 쓰시오. (필요시 형태를 바꿀 것)

1. The a＿＿＿＿ temperature this month has been unusually high for the season.
2. Her i＿＿＿＿ reaction to the news was one of surprise and excitement.
3. The coach's motivational speech was meant to e＿＿＿＿ the team before the big game.

Exercise 4 어법

(A), (B), (C)의 각 네모 안에서 어법에 맞는 표현으로 가장 적절한 것은?

25007-0076

There is little hope of finding high grade sources of minerals other than those we know already. The planet's crust has been thoroughly (A) exploring / explored and digging deeper is not likely to help, since ores form mainly because of hydrothermal processes that operate near the surface. The oceanic floor is geologically too recent for containing ores; only the sea floor near the continents could be a useful source of minerals. The oceans themselves contain metal ions, but extracting rare metals from seawater is out of the question (B) because / because of their minute concentrations that make the process highly expensive in energy terms. In addition, the amounts dissolved are not very large. For instance, considering the concentration of copper in the oceans we can calculate (C) that / what the total amount dissolved corresponds to 10 years of the present mine production. Some suggest outer space as a source of minerals but the energy cost of leaving earth is a major barrier. Then, most bodies of the solar systems — e.g., the Moon and the asteroids — are geochemically "dead" and contain no ores.

* ore: 광물 ** hydrothermal: 열수(고온의 물) 작용의 *** asteroid: 소행성

	(A)		(B)		(C)
①	exploring	⋯⋯	because	⋯⋯	that
②	explored	⋯⋯	because	⋯⋯	that
③	explored	⋯⋯	because of	⋯⋯	that
④	explored	⋯⋯	because of	⋯⋯	what
⑤	exploring	⋯⋯	because of	⋯⋯	what

Sentence Structure

● There is little hope of [finding high grade sources of minerals other than **those** we know already].

[ ]는 hope와 동격 관계이며, those는 high grade sources of minerals를 대신한다.

Word Search 주어진 문장에 들어갈 적절한 영어 단어를 본문에서 찾아 쓰시오. (필요시 형태를 바꿀 것)

1. The ocean f＿＿＿＿＿ is a territory that hasn't been fully explored yet.

2. Before being turned in, the package must c＿＿＿＿＿ every necessary paper.

3. Sugar will d＿＿＿＿＿ quickly in hot water, creating a sweet solution.

Exercise 5　무관한 문장

다음 글에서 전체 흐름과 관계 <u>없는</u> 문장은?　　　　　　　　　　　　　　　　　25007-0077

When we apply a simple technological solution to a problem that is actually nested within a complex biological system, we cannot easily predict how the system will adapt to the technology. ① In fact, if we don't understand how the "problem" is linked to or supported by the ecosystem, then we are defining the problem in human terms without understanding its biological or ecological foundation. ② For example, we have used synthetic pesticides to control unwanted insect pests in agriculture since the late 1940s. ③ Each and every pesticide that has been in common and widespread use has resulted in the evolution of resistant pest species, and this has resulted in a constant search for replacement pesticides as a countermeasure. ④ As the development of new pesticides has progressed, crop yields have significantly increased, improving global food supply stability. ⑤ This back and forth battle to control a biological problem with a technological solution has no ending point and is the result of an unwillingness on the human side to recognize the underlying issue.

* synthetic pesticide: 합성 살충제

Sentence Structure

- When we apply a simple technological solution to a problem [that is actually nested within a complex biological system], we cannot easily predict [how the system will adapt to the technology].
  첫 번째 [ ]는 a problem을 수식하는 관계절이고, 두 번째 [ ]는 predict의 목적어 역할을 하는 명사절이다.

Word Search　주어진 문장에 들어갈 적절한 영어 단어를 본문에서 찾아 쓰시오. (필요시 형태를 바꿀 것)

1. The discovery of a new fossil could l_____ modern birds to their ancestors.
2. The f_____ of trust is essential for building strong relationships.
3. The e_____ of fashion reflects changing cultural values.

Exercise 6 · 글의 순서

주어진 글 다음에 이어질 글의 순서로 가장 적절한 것은?   25007-0078

> Psychiatry is often viewed as medicine's 'poor cousin', and there are good reasons for this.

(A) However, our understanding of how the human brain works as a whole still remains far from clear, and as a consequence, drug treatments for mental disorders have often been far from inspiring, with even the mechanisms of action of those drugs that do seem to have a positive impact on mental health remaining obscure.

(B) This lack of clarity is due to the brain being so much more complex than any other organ in the body, but it also reflects the fact that human conscious awareness is as much a social as a biological entity, and therefore human mental disorders have a major social input.

(C) During the past century, our understanding of bodily mechanisms has advanced through scientific understanding on a whole number of different fronts and led to important new drug treatments in areas as diverse as cardiology, cancer therapy, and treatment of bacterial and infectious diseases.

* psychiatry: 정신의학 ** obscure: 모호한 *** cardiology: 심장학

① (A) – (C) – (B) ② (B) – (A) – (C) ③ (B) – (C) – (A)
④ (C) – (A) – (B) ⑤ (C) – (B) – (A)

Sentence Structure

● However, our understanding of [how the human brain works as a whole] still remains far from clear, and as a consequence, drug treatments for mental disorders have often been far from inspiring, [**with even the mechanisms of action of those drugs** {that do seem to have a positive impact on mental health} **remaining obscure**].

첫 번째 [ ]는 전치사 of의 목적어 역할을 하는 명사절이다. 두 번째 [ ]는 「with＋명사구＋분사구」의 형태로 앞선 절의 부수적 상황을 표현하며, 그 안의 { }는 those drugs를 수식하는 관계절이다.

Word Search   주어진 문장에 들어갈 적절한 영어 단어를 본문에서 찾아 쓰시오. (필요시 형태를 바꿀 것)

1. Despite extensive research, many aspects of the disease still r_____ poorly understood.
2. The brain is the most complex o_____ in the human body.
3. Early t_____ of the disease can significantly improve outcomes for patients.

Exercise 7 　어법

다음 글의 밑줄 친 부분 중, 어법상 틀린 것은? 　　　　　　　　　　　　　　25007-0079

> When our brains encounter something unfamiliar or potentially dangerous, we quickly shift to *fight-or-flight response*, in which all of our mental (and sometimes physical) energy is devoted to ① addressing the perceived threat. Researchers at Harvard Medical School have done extensive research into this response and found that when ② faced with information that feels unknown or threatening, our brains send out a distress signal to the rest of the body through the autonomic nervous system, "which controls such involuntary body functions as breathing, blood pressure, heartbeat, and the dilation or constriction of key blood vessels and small airways." These physical responses can be uncomfortable or even ③ painful, but more importantly they then trigger emotions that end up guiding our decision-making. We all know what it's like to shake our heads and think, "I'd be so much smarter than that," while watching the main character in a horror movie ④ makes a bafflingly terrible decision in the face of extreme danger. Chances are, however, we wouldn't be smarter. In fight-or-flight mode, the part of our brain that takes over is only horror-movie-level smart, ⑤ which isn't very smart at all.
>
> * dilation: 확장(증)　** constriction: 수축(증)　*** bafflingly: 당황스럽게

Sentence Structure

● ~, but more importantly they then trigger emotions [that **end up guiding** our decision-making].
　[ ]는 emotions를 수식하는 관계절이다. 「end up+-ing」는 '결국 ~하게 되다'라는 의미이다.

Word Search　주어진 문장에 들어갈 적절한 영어 단어를 본문에서 찾아 쓰시오. (필요시 형태를 바꿀 것)

1. The growing t_____ of cybersecurity attacks has made companies spend a lot more on better security measures.
2. Exposure to bright flashing lights can t_____ severe headaches and discomfort in some people.
3. He was unable to walk any further due to e_____ fatigue.

Exercise 8 주제 추론

다음 글의 주제로 가장 적절한 것은?

25007-0080

In today's digitally dynamic landscape, consumer behavior and experience play crucial roles in shaping the success of businesses across industries. The advent of artificial intelligence (AI) has revolutionized the way companies understand, analyze, and respond to consumer needs. Through the amalgamation of advanced algorithms, machine learning, and data analytics, AI offers a profound opportunity to delve deeper into consumer behaviors, providing invaluable insights that facilitate personalized experiences and foster long-term relationships. Understanding consumer behavior lies at the heart of any successful business strategy. AI, with its ability to process vast amounts of data in real time, allows for a comprehensive understanding of consumer preferences, habits, and tendencies. By analyzing past behaviors and interactions, AI identifies patterns that might escape traditional analysis, enabling businesses to anticipate future needs and tailor offerings accordingly. Whether it's predicting purchase patterns, understanding browsing behaviors, or understanding sentiment through social media interactions, AI-driven insights empower businesses to fine-tune their strategies for maximum impact.

* amalgamation: 융합, 합병 ** delve into: ~을 철저하게 조사하다 *** sentiment: 감정, 정서

① challenges and constraints in applying AI-driven analytics
② ways of meeting the evolving demands of social media users
③ the need for ethical regulations surrounding the use of AI in businesses
④ the overestimated importance of data in developing effective marketing strategies
⑤ the critical role of AI in enhancing consumer understanding and driving profitability

Sentence Structure

● [Through the amalgamation of advanced algorithms, machine learning, and data analytics], AI offers a profound opportunity [to delve deeper into consumer behaviors], [providing invaluable insights {that facilitate personalized experiences and foster long-term relationships}].

첫 번째 [ ]는 전치사구이며, 두 번째 [ ]는 a profound opportunity의 내용을 구체적으로 설명하는 to부정사구이다. 세 번째 [ ]는 분사구문으로 앞선 절의 내용을 부가적으로 설명하며, 그 안의 { }는 invaluable insights를 수식하는 관계절이다.

Word Search 주어진 문장에 들어갈 적절한 영어 단어를 본문에서 찾아 쓰시오. (필요시 형태를 바꿀 것)

1. The scientist developed an a_____ heart to help patients with heart failure.
2. The Internet has r_____ global communication, shrinking distances and connecting cultures.
3. His speech had a p_____ impact on the audience.

Exercise 9 　제목 추론

다음 글의 제목으로 가장 적절한 것은? 　　　　　　　　　　　　　　　25007-0081

One of the key constraints on the development of sports tourism prior to the nineteenth century was the lack of suitable transport. There were incremental improvements from the fifteenth century onwards involving more comfortable coaches and, in the eighteenth century, greatly improved roads, at least in Britain if not everywhere in Europe. But transport was primarily slow and costly. For example, the journey time from London to Bath in 1680, a distance of 107 miles, was around sixty hours. Vastly improved roads had cut this time to ten hours by 1800 but the time and cost still meant that only the wealthy in society could travel substantial distances. It was not until the development of the railways in the nineteenth century that a relatively cheap and efficient form of transport was afforded to the population at large, enabling sports tourism to develop beyond the small and exclusive upper class activity that had existed hitherto. As Wray Vamplew points out, the railways 'revolutionized sport in England by widening the catchment area for spectators and by enabling participants to compete nationally'.

* incremental: 점진적인 　** hitherto: 그때까지 　*** spectator: 관중

① Regional Differences in Early Sports Tourism Practices
② Transport Breakthroughs: Decreasing the Cost of Traveling
③ The Influence of Local Rivalries on the Actions of Sports Fans
④ How Transport Advances Brought Change to Sports Tourism in England
⑤ Shift from Expensive Upper-class Travel Options to More Affordable Ones

Sentence Structure

● **It was not until** [the development of the railways in the nineteenth century] **that** [a relatively cheap and efficient form of transport was afforded to the population at large, {enabling sports tourism to develop beyond the small and exclusive upper class activity ⟨that had existed hitherto⟩}].

「It was not until A that B」 구문이 사용되어 'A하고 나서야 비로소 B했다'라는 의미를 나타내는데, 두 개의 [ ]가 각각 A와 B에 해당한다. 두 번째 [ ] 안에서 { }는 분사구문으로 앞의 내용을 추가적으로 설명하고 ⟨ ⟩는 the small and exclusive upper class activity를 수식하는 관계절이다.

Word Search 　주어진 문장에 들어갈 적절한 영어 단어를 본문에서 찾아 쓰시오. (필요시 형태를 바꿀 것)

1. Due to a financial c＿＿＿＿＿＿, the non-profit had to cut back on several planned activities.
2. She received s＿＿＿＿＿ recognition for her pioneering research in neuroscience.
3. The new scheduling system is more e＿＿＿＿＿, reducing the time it takes to coordinate meetings.

Exercise 10   어법

다음 글의 밑줄 친 부분 중, 어법상 **틀린** 것은?   25007-0082

Cambridge biologist William Hamilton studied the social behaviour of ants and wondered to himself not only why they often sacrificed ① themselves for their nestmates, but also how it was that sterile worker ants might have arisen, given these females don't breed but rather leave this to the queen. Hamilton solved this problem when he realised ② that, due to their particular method of breeding, these sterile worker ants share 75 per cent of their genes with their sisters. Hamilton then began to consider a 'gene's eye view' of natural selection, and immediately realised such non-queen worker ants could actually pass on a larger proportion of their genes indirectly by, instead of breeding themselves, ③ helped to raise their younger sisters. For each sister raised, they are passing on 75 per cent of their genes rather than the 50 per cent that would be ④ passed on via normal sexual reproduction. To Hamilton, ant colonies should really be perceived as a form of extended family ⑤ where the genetic interests of all are served by the apparent altruistic behaviour of some.

* sterile: 불임의   ** altruistic: 이타적인

Sentence Structure

● For each sister [raised], they are passing on 75 per cent of their genes rather than the 50 per cent [that would be passed on via normal sexual reproduction].
첫 번째 [ ]는 each sister를 수식하는 분사이고, 두 번째 [ ]는 the 50 per cent를 수식하는 관계절이다. the 50 per cent 뒤에는 of their genes가 생략된 것으로 이해할 수 있다.

Word Search   주어진 문장에 들어갈 적절한 영어 단어를 본문에서 찾아 쓰시오. (필요시 형태를 바꿀 것)

1. She's had to s_____ a lot for her family.
2. Many animals b_____ only at certain times of the year.
3. A large p_____ of income is spent on housing in urban areas.

Exercise 11 무관한 문장

다음 글에서 전체 흐름과 관계 <u>없는</u> 문장은?   25007-0083

The most common justification for a move to autonomous vehicles could well be called the argument from safety: the implementation of driving automation systems will reduce the risk of death or injury from automobile accidents. ① The basic premise seems reasonable enough: humans behind the wheel can be unreliable and, all too often, distracted, tired, or otherwise affected. ② This fallibility of human drivers no doubt contributes to a rate of injury and fatality that, in other contexts, would be considered a public health crisis. ③ Driving automation systems, by contrast, are supposed to be more reliable and, depending on the overall prevalence of autonomous vehicles and of the infrastructure to support them, may be able to coordinate with other vehicles to avoid accidents and other problems. ④ Despite the promise of safety, ethical concerns also arise regarding the use of autonomous vehicles, resulting in increased protests against their use. ⑤ On these premises, and assuming the new technology functions as advertised, it only makes sense to replace an unstable and unreliable human control system for something demonstrably better.

* autonomous: 자율 주행의, 자율의 ** premise: 전제 *** fallibility: 불완전성

Sentence Structure

● The most common justification [for a move to autonomous vehicles] could well be called the argument from safety: [the implementation of driving automation systems will reduce the risk of death or injury from automobile accidents].

첫 번째 [ ]는 The most common justification을 수식하는 전치사구이다. 두 번째 [ ]는 the argument from safety의 구체적인 내용을 나타낸다.

Word Search 주어진 문장에 들어갈 적절한 영어 단어를 본문에서 찾아 쓰시오. (필요시 형태를 바꿀 것)

1. His j_____ for the new policy was based on detailed data analysis and expert recommendations.
2. The i_____ of the new educational curriculum was a complex process that involved several parties.
3. The rise in the f_____ rate is caused by a number of factors.

Exercise 12 글의 순서

주어진 글 다음에 이어질 글의 순서로 가장 적절한 것은?

25007-0084

Most students can provide numerous examples of common sense, such as never putting a fork into an electrical outlet, not sticking your tongue on a pole covered with ice or placing your hands in boiling water, avoiding looking directly into the sun, and not attempting to attend a state dinner at the White House without an invitation.

(A) Experience teaches us to act routinely in most social situations. Before long, such expectations come to be viewed as common sense.

(B) And yet, none of these examples of common sense represents knowledge that we were born with; instead, we have learned all these things. The sociological perspective emphasizes that knowledge is gained through trial and error, experience, and the influences of others in the social environment.

(C) However, because each of us has unique social experiences, we come to view common sense differently — we see it from our unique perspective. Thus, if one has not been exposed to a particular behavior — labeled by some as commonsense knowledge — one is not capable of acting in an obvious, or routine, manner.

* outlet: 콘센트

① (A) – (C) – (B) ② (B) – (A) – (C) ③ (B) – (C) – (A)
④ (C) – (A) – (B) ⑤ (C) – (B) – (A)

Sentence Structure

● The sociological perspective emphasizes [that knowledge is gained through {trial and error}, {experience}, and {the influences of others in the social environment}].
[ ]는 emphasizes의 목적어 역할을 하는 명사절이고, 세 개의 { }는 콤마와 and에 의해 대등하게 연결되어 through에 공통으로 이어진다.

Word Search 주어진 문장에 들어갈 적절한 영어 단어를 본문에서 찾아 쓰시오. (필요시 형태를 바꿀 것)

1. The e_____ among fans was high as the team had been performing well all season.
2. The program is designed to e_____ students to various scientific fields.
3. It was o_____ to everyone in the room that the speaker was unprepared, as evidenced by his frequent pauses and nervous attitude.

Exercise 1　어법

다음 글의 밑줄 친 부분 중, 어법상 <u>틀린</u> 것은?

25007-0085

　　Emotion categories, in my view, ① <u>are made</u> real through collective intentionality. To communicate to someone else that you feel angry, both of you need a shared understanding of "Anger." If people agree that a particular combination of facial actions and cardiovascular changes ② <u>is</u> anger in a given context, then it is so. You needn't be explicitly aware of this agreement. You don't even have to agree ③ <u>whether</u> a particular instance is anger or not. You just have to agree in principle that anger exists with certain functions. At that point, people can transmit information about that concept among themselves so ④ <u>efficient</u> that anger seems inborn. If you and I agree that a furrowed brow indicates anger in a given context, and I furrow my brow, I am efficiently sharing information with you. My movement itself does not carry anger to you, any more than vibrations in the air carry sound. By virtue of the fact ⑤ <u>that</u> we share a concept, my movement initiates a prediction in your brain ... a uniquely human brand of magic. It is categorization as a cooperative act.

* cardiovascular: 심혈관의　** furrow: 찌푸리다　*** vibration: 진동

Sentence Structure

● You just have to agree in principle [that anger exists with certain functions].
　[ ]는 agree의 목적어 역할을 하는 명사절이다.

Word Search　주어진 문장에 들어갈 적절한 영어 단어를 본문에서 찾아 쓰시오. (필요시 형태를 바꿀 것)

1. The instructions e＿＿＿＿＿ stated that the deadline for submission was Friday.
2. The news spread quickly as people began to t＿＿＿＿＿ information through social media.
3. Some believe that musical talent is i＿＿＿＿＿, while others think it can be developed with practice.

Exercise 2 주제 추론

다음 글의 주제로 가장 적절한 것은?   25007-0086

In order to better understand the dynamics of online interactions and the psychology behind group decision-making processes, consider a question that bears directly on group decisions in general: how people judge comments on websites. Lev Muchnik, a professor at the Hebrew University of Jerusalem, and his colleagues carried out an experiment on a website that displays diverse stories and allows people to post comments, which can in turn be voted up or down. The researchers automatically and artificially gave certain comments on stories an immediate upvote — the first vote that a comment would receive. You might well think that after hundreds or thousands of visitors and ratings, a single initial vote on a comment could not possibly matter. That is a sensible thought, but it is wrong. After seeing an initial upvote (and recall that it was entirely artificial), the next viewer became 32% more likely to give an upvote.

* bear on: ~과 관련되다   ** upvote: 좋아요 투표

① effects of ratings and comments on self-esteem
② psychological reasons behind negative comments
③ common reactions to personal stories on websites
④ influence of initial votes on shaping online opinion
⑤ disadvantages of sharing opinions on social media

Sentence Structure

● Lev Muchnik, a professor at the Hebrew University of Jerusalem, and his colleagues carried out an experiment on a website [that displays diverse stories and allows people to post comments, {which can in turn be voted up or down}].

[ ]는 a website를 수식하는 관계절이다. 그 안에 있는 { }는 comments를 추가적으로 설명하는 관계절이다.

Word Search   주어진 문장에 들어갈 적절한 영어 단어를 본문에서 찾아 쓰시오. (필요시 형태를 바꿀 것)

1. It's inappropriate to make rude or offensive c_____ online.
2. The university is known for its d_____ student population from all over the world.
3. She made a s_____ suggestion to address the problem without causing unnecessary conflict.

Exercise 3 제목 추론

다음 글의 제목으로 가장 적절한 것은?   25007-0087

> We explain our world through stories. In our stories, we are both the storyteller and the principal character. Being able to tell our life story helps us to make sense of it. Perhaps we tell it to ourselves, quietly pondering. Perhaps we write it down, recognising in reading back something that we did not recognise at the time. But for most of us, chatting to a friend or reflecting with a confidant is the way we tell our stories. And as we tell them, we hear them anew. Telling helps us to notice and to interpret details, to become aware of the bigger picture or to discern aspects we previously overlooked or denied. Finding a listener who will give our story their full attention, who is prepared to become completely caught up in the tale, is an opportunity to meet ourselves, complete with our noble hopes and miserable failings, and to understand ourselves and the world around us in a truer, more helpful way.
>
> * ponder: 숙고하다   ** confidant: 절친한 친구

① Daydreaming: A Journey to Self-discovery
② Discovering Truths Through Others' Perspectives
③ Sharing Personal Stories, Building Stronger Relationships
④ The Art of Storytelling: A Necessary Skill to Master Success
⑤ Sharing Your Stories Opens a Door to Deeper Understanding of Yourself

---

Sentence Structure

● Telling helps us [{to notice} and {to interpret} details], [to become aware of the bigger picture] or [to discern aspects {we previously overlooked or denied}].
세 개의 [ ]가 콤마와 or로 연결되어 있으며 helps us에 이어진다. and로 연결된 첫 번째와 두 번째 { }에 details가 공통으로 이어진다. 세 번째 { }는 aspects를 수식하는 관계절이다.

Word Search   주어진 문장에 들어갈 적절한 영어 단어를 본문에서 찾아 쓰시오. (필요시 형태를 바꿀 것)

1. The company hired an expert to help i_____ the legal language in the contract.
2. It's easy to o_____ the importance of small gestures of kindness.
3. Being stuck at home with the flu made him feel m_____.

Exercise 4 어법

(A), (B), (C)의 각 네모 안에서 어법에 맞는 표현으로 가장 적절한 것은?

25007-0088

In one study, products to the left of center in a horizontal display were chosen only 24 percent of the time. Why? Bruce D. Sanders, author of *Retailer's Edge*, notes that our eyes move to the right when the left hemisphere of the brain, where we do the math, (A) get / gets active. Therefore, when we think we've gotten a good deal, our happy left side of the brain dominates, and our eyes shift right. It's an explanation for (B) what / why right-of-center products are more frequently purchased than left-of-center products. This preference for right over left has nothing to do with conscious or rational decision making. When asked to rate four *identical* pairs of stockings, the pair (C) positioned / was positioned to the far right was chosen four to one over the pair on the far left. Even when experimenters suggested that the position of the stockings might have influenced their selection, participants' responses ranged from confused to dismissive.

	(A)		(B)		(C)
①	get	·····	what	·····	positioned
②	get	·····	why	·····	was positioned
③	gets	·····	what	·····	was positioned
④	gets	·····	why	·····	was positioned
⑤	gets	·····	why	·····	positioned

Sentence Structure

● Even when experimenters suggested [that {the position of the stockings} might have influenced their selection], participants' responses ranged from confused to dismissive.

[ ]는 suggested의 목적어 역할을 하는 명사절이고, 그 안에서 { }가 주어의 역할을 한다.

Word Search   주어진 문장에 들어갈 적절한 영어 단어를 본문에서 찾아 쓰시오. (필요시 형태를 바꿀 것)

1. In the brain, the left h_____ is typically associated with language and analytical thinking.
2. The large corporation continues to d_____ the global market.
3. She made a c_____ effort to improve her communication skills, attending public speaking classes.

Exercise 5   무관한 문장

다음 글에서 전체 흐름과 관계 <u>없는</u> 문장은?   25007-0089

> With increased access to the internet across the world, consumers of healthcare (patients) are better educated. ① They are able to stay better informed about happenings around the world, timely news, and healthcare's latest trends than at any previous time in history. ② Today's consumers are demanding healthcare systems and experiences that accommodate their busy schedules, provide useful information that can be obtained quickly and easily, and that allow them to be involved in decision-making. ③ Even with the best of training, the highest of dedication, and the greatest of motivation, errors in medical procedure can still be made, but the number of errors can be diminished through the design of medical devices. ④ These consumers are different from those in the past because they also demand better access to care and better communication and a new level of participation with their healthcare providers. ⑤ Plans of treatment are often devised jointly and executed via creative methods.

Sentence Structure

- Today's consumers are demanding healthcare systems and experiences [that {accommodate their busy schedules}, {provide useful information ⟨that can be obtained quickly and easily⟩}], and [that allow them to be involved in decision-making].

  두 개의 [ ]는 healthcare systems and experiences를 수식하는 관계절로 and로 연결되어 있다. 첫 번째 [ ] 안에서 두 개의 { }가 콤마로 연결되어 있고, 두 번째 { } 안의 ⟨ ⟩는 useful information을 수식하는 관계절이다.

Word Search   주어진 문장에 들어갈 적절한 영어 단어를 본문에서 찾아 쓰시오. (필요시 형태를 바꿀 것)

1. Improving a_____ to healthcare is a priority for the government.
2. You can o_____ more information about the event from the official website.
3. C_____ are becoming more conscious of the environmental impact of their purchases.

Exercise 6 글의 순서

주어진 글 다음에 이어질 글의 순서로 가장 적절한 것은?   25007-0090

> While our gut may usually give us a good general sense of how the world works, it is frequently not precise. We need data to sharpen the picture.

(A) Just how significant is this effect? An optimistic read of the effectiveness of antidepressants would find that the most effective drugs decrease the incidence of depression by only about 20 percent. To judge from the search result numbers, a Chicago-to-Honolulu move would be at least twice as effective as medication for your winter blues.

(B) But you might not guess how big an impact this temperature difference can make. I looked for correlations between an area's Internet searches for depression and a wide range of factors, including economic conditions, education levels, and church attendance. Winter climate swamped all the rest. In winter months, warm climates, such as that of Honolulu, have 40 percent fewer depression searches than cold climates, such as that of Chicago.

(C) Consider, for example, the effects of weather on mood. You would probably guess that people are more likely to feel more gloomy on a 10-degree day than on a 70-degree day. Indeed, this is correct.

* gut: 직감 ** swamp: 압도하다

① (A) – (C) – (B)   ② (B) – (A) – (C)   ③ (B) – (C) – (A)
④ (C) – (A) – (B)   ⑤ (C) – (B) – (A)

Sentence Structure

- But you might not guess [**how big an impact** this temperature difference can make].

  [ ]는 guess의 목적어 역할을 하는 명사절이다. 간접의문문인 「how+형용사+a(n)+명사+주어+동사」의 어순이 쓰이고 있다.

Word Search 주어진 문장에 들어갈 적절한 영어 단어를 본문에서 찾아 쓰시오. (필요시 형태를 바꿀 것)

1. The architect sketched out a p_____ blueprint of the building's layout.
2. She lives each day with an o_____ mindset, seeing challenges as opportunities.
3. Access to education is a key f_____ in reducing inequality and poverty.

Exercise 7　어법

다음 글의 밑줄 친 부분 중, 어법상 틀린 것은?　　　　　　　　　　　　　25007-0091

Designers use diagrams, pictures, and simple illustrations to explain their work, ① whether it is how to turn the power on or off in a machine or how to use a DNA synthesizer or even a washing machine. Designers understand the need to test their efforts on the people ② for whom they are intended. Invariably, the tests reveal problems, places ③ where people do not understand, are confused, or, worse, could make serious errors in their use of the device under consideration: perhaps losing all of their work and, in the case of machinery or medical devices, perhaps leading to accidents or serious injuries. The reason for this early testing is ④ to find design flaws before the product is released, allowing the design to be corrected. In design, especially the set of procedures that goes under the general label *human-centered design*, multiple iterations of the design ⑤ is constructed and tested. Each new iteration — that is, each modification of the design — is guided by the results of tests.

* iteration: 수정판, 새로운 버전

Sentence Structure

● The reason for this early testing is to find design flaws before the product is released, [**allowing the design to be corrected**].

[ ]는 결과적인 상황을 설명하는 분사구문이다. 「allow+목적어+to부정사」는 '~이 …할 수 있게 하다'라는 의미이다.

Word Search　주어진 문장에 들어갈 적절한 영어 단어를 본문에서 찾아 쓰시오. (필요시 형태를 바꿀 것)

1. I didn't i_____ any disrespect.

2. He was momentarily c_____ by the foreign road signs.

3. The argument is full of fundamental f_____.

Exercise 8 주제 추론

다음 글의 주제로 가장 적절한 것은?   25007-0092

It is interesting to note that all inquiries were once a part of philosophy, that great mother of the sciences (*mater scientiarum*), and philosophy embraced them all in an undifferentiated and amorphous fashion. However, as Western civilisation developed, various sciences began to pursue separate and independent courses. Astronomy and physics were among the first to break away, and were followed thereafter by chemistry, biology and geology. In the nineteenth century, two new sciences appeared: psychology (the science of human behaviour) and sociology (the science of human society). Thus, what had once been natural philosophy became the science of physics; what had been mental philosophy, or the philosophy of mind, became the science of psychology; and what had once been social philosophy, or the philosophy of history, became the science of sociology. To the ancient mother, philosophy, still belong several important kinds of enquiries — notably metaphysics, logic, ethics and aesthetics — but the sciences themselves are no longer studied as subdivisions of philosophy.

* amorphous: 무정형(無定形)의   ** astronomy: 천문학   *** metaphysics: 형이상학

① reasons for people's avoidance of studying philosophy
② merits of balancing scientific and philosophical thinking
③ separation of the sciences from their philosophical roots
④ influence of scientific thinking on understanding philosophy
⑤ contribution of the sciences to forming a new branch of philosophy

Sentence Structure

● Astronomy and physics [were among the first {to break away}], and [were followed thereafter by chemistry, biology and geology].
두 개의 [ ]가 and로 연결되어 주어 Astronomy and physics의 술어 역할을 하고, { }는 the first를 수식하는 to부정사구이다.

Word Search   주어진 문장에 들어갈 적절한 영어 단어를 본문에서 찾아 쓰시오. (필요시 형태를 바꿀 것)

1. Scientific i_____ seeks to uncover truths about the universe.
2. This idea has been widely e_____ by the scientific community.
3. The treatment was applied in an u_____ manner to all patients.

Exercise 9 제목 추론

다음 글의 제목으로 가장 적절한 것은?   25007-0093

Suppose that people's aspirations rise as they get more money. When you live in a student dorm, you may aspire to have your own apartment, even if small. When you live in a small apartment, you may aspire to live in a bigger one. When you live in a bigger apartment, you may aspire to live in your own property. And so on: the more you have, the more you expect — and maybe the more you think you deserve. If so, the beneficial happiness effect of getting more money will be offset (at least partially) by the harmful effect on happiness of rising aspirations. Behavioural economists call this phenomenon the *aspiration treadmill*. The basic idea is ancient. The Stoic philosopher Seneca diagnosed the problem some 2,000 years ago. He wrote: 'Excessive prosperity does indeed create greed in men, and never are desires so well controlled that they vanish once satisfied.'

* offset: 상쇄하다

① Greed: The Cause of Internal Divisions Within Society
② Moral Dilemmas in the Pursuit of Accumulating Wealth
③ Why Does Achieving Happiness Matter in Modern Society?
④ For Financial Freedom, Shift Your Focus from How to When
⑤ Wanting More and More: How It Can Harm Your Happiness

Sentence Structure

● He wrote: 'Excessive prosperity **does** indeed create greed in men, and [**never are desires** so well controlled that **they** vanish {once satisfied}].'
조동사 does는 create를 강조하는 역할로 사용되었고, [ ]에서는 부정어 never가 앞에 위치하여 주어 desires와 동사 are가 도치되었다. they는 desires를 가리키고, { }에서는 접속사 once 뒤에 they are가 생략된 것으로 볼 수 있다.

Word Search   주어진 문장에 들어갈 적절한 영어 단어를 본문에서 찾아 쓰시오. (필요시 형태를 바꿀 것)

1. He said he never a_____ to become famous.
2. The price of p_____ has risen enormously.
3. I felt I d_____ better than that.

Exercise 10 어법

다음 글의 밑줄 친 부분 중, 어법상 틀린 것은?

25007-0094

Humanity has entered a new era. We are now living in a world that is increasingly wired by billions of predictive algorithms, a world ① in which *almost everything* can be predicted and risk and uncertainty appear to be diminishing in almost all areas of life. We are living ② longer, thanks to advances in health care and precision medicine. We have a greater mastery of the physical world that allows us to dream of, and build, new technologies that allow us to explore other planets and ③ visualizes billions of galaxies. We can model markets, disease, and traffic with increasingly greater precision, and we're getting very close to handing over the keys to the car so it can drive ④ itself. Even more striking, our tools may be revealing the genesis of some elusive and stubborn complexities of human behavior, and algorithms are even being used to ⑤ alter people's behavior. Predictive algorithms have changed the world, and all the worlds to come, and there is no going back.

* genesis: 기원 ** elusive: 이해하기 어려운 *** stubborn: 다루기 힘든

Sentence Structure

● We are now living in a world [that is increasingly wired by billions of predictive algorithms], ~.
  [ ]는 a world를 수식하는 관계절이다.

Word Search 주어진 문장에 들어갈 적절한 영어 단어를 본문에서 찾아 쓰시오. (필요시 형태를 바꿀 것)

1. Using p_____ modeling, weather experts can better forecast severe weather events.
2. The company's profits began to d_____ after the economic downturn.
3. The doctor performed the complex brain surgery with p_____.

Exercise 11   무관한 문장

다음 글에서 전체 흐름과 관계 <u>없는</u> 문장은?   25007-0095

It is worth noting that the progress in man's understanding about the physical aspects of his environment occurred much earlier in comparison to the social aspects. ① The main reason for this discrepancy in the development of man's understanding of the two distinct environments — the physical and the social — probably lies in the greater observability and control of physical phenomena, as well as in the impersonal approach that they afford. ② Physical phenomena are usually more concrete than social phenomena and hence are more observable. ③ Both the physical environment and social environment play crucial roles in shaping individual well-being and overall community health. ④ Samuel Koenig states that in his attempt to observe physical phenomena man was able quite early to develop a measure of detachment, but he found it very difficult to do so regarding social phenomena. ⑤ In the latter case he found himself too close to the object of investigation, too involved in it, to achieve the objectivity which is indispensable to all science.

* discrepancy: 차이 ** indispensable: 필수적인

Sentence Structure

● It **is worth noting** [that {the progress in man's understanding about the physical aspects of his environment} occurred much earlier in comparison to the social aspects].
「be worth+-ing」는 '~할 만한 가치가 있다'라는 표현이다. { }는 [ ] 안에서 주어로 쓰였다.

Word Search   주어진 문장에 들어갈 적절한 영어 단어를 본문에서 찾아 쓰시오. (필요시 형태를 바꿀 것)

1. The project showed slow but steady p_____.
2. They made a c_____ of different countries' eating habits.
3. This was one a_____ of her character he hadn't seen before.

Exercise 12 글의 순서

주어진 글 다음에 이어질 글의 순서로 가장 적절한 것은? 25007-0096

When computers process in time, we typically want them to also work in real time. For example, if you set up a microphone with a reverb effect for a performance, you would like it to work both in time and in real time.

(A) If the computer can process a three-minute sound file in seven seconds, it is unnecessary to wait three minutes to finish its processing. You may also happen to have a computationally heavy reverb model that runs "slower than real time." For a producer working in a studio, it is not crucial whether a process runs in real time.

(B) However, if you just want to add reverb to a prerecorded song, it does not matter when or how fast it happens. You may probably prefer that the computer does the processing "faster than real time."

(C) For a performer on stage, however, there is no other option than running in real-time mode. The concert is happening in the "now," hence the tools used need to run both in time and in real time.

* reverb: (음향 장치로 소리의 울림을 좋게 하기 위한) 에코

① (A) – (C) – (B) ② (B) – (A) – (C) ③ (B) – (C) – (A)
④ (C) – (A) – (B) ⑤ (C) – (B) – (A)

Sentence Structure

● [If the computer can process a three-minute sound file in seven seconds], **it** is unnecessary [to wait three minutes to finish its processing].

첫 번째 [ ]는 조건을 나타내는 부사절이다. it은 형식상의 주어이고, 두 번째 [ ]가 내용상의 주어에 해당하는 to부정사구이다.

Word Search 주어진 문장에 들어갈 적절한 영어 단어를 본문에서 찾아 쓰시오. (필요시 형태를 바꿀 것)

1. This evening's p_____ will start at 7 o'clock.
2. The information gathered by the telescopes will be p_____ by computers.
3. Some people like chocolate ice cream, but I p_____ vanilla.

Exercise 1   어휘

다음 글의 밑줄 친 부분 중, 문맥상 낱말의 쓰임이 적절하지 <u>않은</u> 것은?   25007-0097

At its most fundamental, relative brightness identifies the visual contrasts by which we live. Without contrast, the essentials of vision fail to operate; the brain will not identify edges and forms nor perceive depth or distance. The ① <u>fewer</u> the contrasts, the less able we are to identify details, and the less able we are to navigate the environment. Ideal visual conditions are those where contrasts of relative brightness are most ② <u>easily</u> identified, where the light source is strong and casts sharply defined shadows. It follows that ideal visual conditions are most likely to generate positive feeling-responses. At the opposite end of the safe/dangerous spectrum, the most ③ <u>secure</u> visual conditions are those where the contrasts given by a range of relative brightness no longer exist. Complete darkness (as in an underground cave) and complete brightness (as in a blizzard whiteout) are equally filled with danger. In these ④ <u>extreme</u> conditions of brightness, details are erased, and navigation becomes hazardous. Emotion-feelings aroused by conditions such as these are most likely to be ⑤ <u>negative</u>.

* blizzard: 눈보라   ** whiteout: 화이트아웃(천지가 모두 백색이 되어 방향 감각을 잃어버리는 상태)

Sentence Structure

- Without contrast, the essentials of vision fail to operate; the brain [will **not** identify edges and forms **nor** perceive depth or distance].

  [ ]는 절의 술부로 '~이 아니고 …도 아닌'이라는 의미의 「not ~ nor …」이 사용되었다.

Word Search   주어진 문장에 들어갈 적절한 영어 단어를 본문에서 찾아 쓰시오. (필요시 형태를 바꿀 것)

1. The remote control allows you to o_____ all the features of your TV.
2. Scientists work to i_____ new species in remote regions of the rainforest.
3. Refugees demanded c_____ freedom to do whatever they wanted.

Exercise 2) 문장 삽입

글의 흐름으로 보아, 주어진 문장이 들어가기에 가장 적절한 곳은?   25007-0098

Such a discovery may have enormous commercial implications, perhaps leading to the invention of a device that filters out the harmful effect of the waves.

The most important inherent limitation on patentable subject matter revolves around the sometimes blurred distinction between an *invention* and a *discovery*. (①) Laws of nature or scientific principles that a researcher may 'find' are not patentable. (②) For example, imagine that a scientist discovers that a certain previously unknown form of magnetic wave strikes the earth's atmosphere from deep space, and that this kind of wave has an adverse effect on the transmission of data between the earth and communications satellites. (③) The inventor of the filtering device may be entitled to a patent. (④) The discoverer of the magnetic wave cannot patent his or her discovery. (⑤) Today, the distinction between discovery and invention is of increasing importance in the fields of genetic engineering and biotechnology.

* blurred: 모호한

Sentence Structure

- Such a discovery may have enormous commercial implications, [perhaps leading to the invention of a device {that filters out the harmful effect of the waves}].

  [ ]는 주절의 내용에 부가적으로 이어지는 결과를 나타내는 분사구문이고, 그 안의 { }는 a device를 수식하는 관계절이다.

Word Search 주어진 문장에 들어갈 적절한 영어 단어를 본문에서 찾아 쓰시오. (필요시 형태를 바꿀 것)

1. The artist believed that creativity was i_____ in all individuals.
2. The p_____ of equality is fundamental to the legal system.
3. Every citizen is e_____ to basic human rights, regardless of their social status.

Exercise 3 요약문 완성

다음 글의 내용을 한 문장으로 요약하고자 한다. 빈칸 (A), (B)에 들어갈 말로 가장 적절한 것은?   25007-0099

"Farm animals, lab animals, sport horses...": our language when we refer to some animals often focuses on their exploitation and on the benefits humans get from them, for example, animals as a source of food, used in biomedical research or for recreational activities. When we refer to some animals as "pets," are we implying their use for any purpose? A pet can be defined, in a neutral manner, as an animal that lives with humans. Another largely used term is "companion animal," which, again, implies that we keep them in our home with the final aim of getting companionship. Some may argue that a fish in the fish tank is not really anyone's companion. However, this term well adapts to domestic species, such as dogs and cats, which we treat kindly, consider as members of the family and our "best friends," and with whom we share our time and financial resources. Differently from other species, companion animals thus appear to represent an emotional — rather than an economical — resource, providing humans with support, comfort and companionship.

When it comes to animal discourse, our language typically emphasizes the ___(A)___ of some animals, whereas the terms for those kept in homes suggest their ___(B)___ significance.

	(A)	(B)		(A)	(B)
①	behavior	cognitive	②	profitability	utilitarian
③	emotion	recreational	④	utility	relational
⑤	intelligence	cultural			

Sentence Structure

- Another largely used term is "companion animal," [which, again, implies {that we keep them in our home with the final aim of getting companionship}].
  [ ]는 "companion animal"을 부수적으로 설명하는 관계절이고, 그 안의 { }는 implies의 목적어 역할을 하는 명사절이다.

Word Search   주어진 문장에 들어갈 적절한 영어 단어를 본문에서 찾아 쓰시오. (필요시 형태를 바꿀 것)

1. Regular exercise offers a multitude of b_____, including better mood regulation.
2. Scientists conduct r_____ to advance knowledge and solve problems.
3. Pets provide valuable c_____ to their owners.

Exercise 4 · 어휘

(A), (B), (C)의 각 네모 안에서 문맥에 맞는 낱말로 가장 적절한 것은?   25007-0100

Whilst some stakeholders may try to protect the integrity of arts and cultural festivals from their reconfiguration as urban policy tools, we should recognise that policy-oriented festivals can still have very positive social and cultural effects. And we cannot (A) ignore / acknowledge the fact that some festivals were established to strategically assist urban areas. In other words, they have always been strategic interventions rather than artistic, social or cultural phenomena. Film festivals are a good example: many of these events were established for (B) economic / recreational reasons: for example, the Cannes Film Festival was launched to prolong the tourist season. The Brighton Festival was created for similar reasons. The re-establishment of the Venice Carnival in 1979, following a long break, was also a(n) (C) deliberate / accidental attempt to address some of the issues the city was facing at that time, including the lack of provision for young people. These festivities have not been appropriated as urban policy tools: they have always been staged with wider objectives in mind.

* **stakeholder**: 이해 당사자   ** **reconfiguration**: 재구성   *** **appropriate**: 전용[도용]하다, 사용(私用)하다

	(A)		(B)		(C)
①	ignore	······	economic	······	deliberate
②	ignore	······	recreational	······	deliberate
③	ignore	······	economic	······	accidental
④	acknowledge	······	recreational	······	accidental
⑤	acknowledge	······	economic	······	accidental

Sentence Structure

● [Whilst some stakeholders may try to protect the integrity of arts and cultural festivals from their reconfiguration as urban policy tools], we should recognise [that policy-oriented festivals can still have very positive social and cultural effects].

첫 번째 [ ]는 Whilst가 이끄는 양보의 부사절이고, 두 번째 [ ]는 recognise의 목적어 역할을 하는 명사절이다.

Word Search   주어진 문장에 들어갈 적절한 영어 단어를 본문에서 찾아 쓰시오. (필요시 형태를 바꿀 것)

1. The p_____ requires all employees to undergo annual training on data privacy.
2. P_____ thinking can lead to improved mental health and overall well-being.
3. Scientists study various weather p_____, such as tornadoes and hurricanes.

Exercise 5~6 1지문 2문항

[5~6] 다음 글을 읽고, 물음에 답하시오.

You may have heard "a calorie is a calorie is a calorie." While it is true that each calorie yields the same amount of energy, foods can be metabolized and used differently, some more efficiently than others. For example, carbohydrates, protein and fat, the macronutrients that provide calories, can each have (a) different effects on the hormones and brain centers that control hunger and eating behavior. Macronutrients have (b) standard amounts of calories. One gram of protein has 4 calories. One gram of carbohydrates has 4 calories. One gram of fat has 9 calories.

However, the source is (c) irrelevant. There are many compounds in foods, besides the macronutrients, that can influence your body's process and response mechanisms. Even though each gram of carbohydrates provides four calories per gram, not all carbohydrates are created (d) equal. For example, white bread has been refined and stripped of its fiber. Because of this, white bread may cause a spike in blood sugar, whereas whole grain bread is less processed and has its fiber intact. This (e) allows for a slower release of sugar into the blood, meaning that while white and whole grain bread may have the same total *amount* of carbohydrates, the way they *affect* your blood sugar is very different.

* metabolize: 대사시키다 ** macronutrient: 다량 영양소(생물체가 다량으로 필요로 하는 영양소) *** intact: 온전한

5 윗글의 제목으로 가장 적절한 것은?   25007-0101

① Do Your Foods Contain Enough Calories?
② Hidden Facts: Sugar Intake Makes You Healthy
③ Do Not Look at the Calorie Label, Just Enjoy Your Food
④ Beyond Calories: Unveiling the Impact of Nutrient Sources
⑤ Secret to Cooking: Different Tastes Depending on the Recipe

6 밑줄 친 (a)~(e) 중에서 문맥상 낱말의 쓰임이 적절하지 <u>않은</u> 것은?   25007-0102

① (a)   ② (b)   ③ (c)   ④ (d)   ⑤ (e)

Sentence Structure

● [While **it** is true {that each calorie yields the same amount of energy}], foods can be metabolized and used differently, some more efficiently than others.
[ ]는 While이 이끄는 양보의 부사절인데, 그 안에서 it은 형식상의 주어, that이 이끄는 명사절인 { }가 내용상의 주어이다.

Word Search 주어진 문장에 들어갈 적절한 영어 단어를 본문에서 찾아 쓰시오. (필요시 형태를 바꿀 것)

1. The decisions we take now may i_____ the course of events in the future.
2. The chef continued to r_____ the recipe to achieve the perfect flavor balance.
3. The company announced the r_____ of its latest product, generating excitement among consumers.

Exercise 7 어휘

다음 글의 밑줄 친 부분 중, 문맥상 낱말의 쓰임이 적절하지 <u>않은</u> 것은?   25007-0103

Sentiment is very close to emotion in meaning, but when social psychologists use the term sentiment, they emphasize the social aspect of emotion. Early social psychologists used sentiment to refer to the components of human responses that ① <u>separate</u> them from analogous responses that animals would have. For example, Cooley contrasts love and lust. Although lust is instinctive, we ② <u>learn</u> what love is through social interaction. In other words, sentiment relies not just on the responses of the individual to the stimulus but also on how other human beings understand that stimulus. In later years, as social psychologists have come to increasingly accept that social elements are a key piece of emotions, the idea of sentiment has become ③ <u>more</u> distinguishable from that of emotion. In contemporary work, social psychologists often use the term sentiment to distinguish ④ <u>immediate</u> emotional responses from longer-term emotional states such as love, grief, and jealousy. These sentiments can endure for days, weeks, and even years after the ⑤ <u>initial</u> event that triggered them.

* analogous: 유사한   ** lust: 욕망, 욕정

Sentence Structure

● In other words, sentiment relies **not just** [on the responses of the individual to the stimulus] **but also** [on how other human beings understand that stimulus].
두 개의 [ ]가 '~뿐만 아니라 …도'라는 의미의 「not just ~ but also ...」 구문으로 연결되어 relies에 이어진다.

Word Search   주어진 문장에 들어갈 적절한 영어 단어를 본문에서 찾아 쓰시오. (필요시 형태를 바꿀 것)

1. The study intends to c_____ urban and rural living conditions.
2. Each i_____ reacts differently to stress based on their unique circumstances.
3. The c_____ art exhibit featured works by artists from around the world.

Exercise 8 문장 삽입

글의 흐름으로 보아, 주어진 문장이 들어가기에 가장 적절한 곳은?

25007-0104

> On the other hand, if the organism dies a natural death, although the soft parts of the body will almost certainly rot away, neither shell is likely to be damaged.

　To become a fossil is a long and involved process, and is an unlikely outcome for the vast majority of individuals. The most obvious requirement for an organism to end up as a fossil is that it must become entombed within rock, usually by burial. (①) There are many chances that can prevent this happening. (②) If a mollusc with two shells has the misfortune to be killed by a predator, then the soft parts will be eaten. (③) The predator will probably have to break at least one of the shells to gain access to the flesh, so it is unlikely that both shells will survive intact. (④) Even if they do, the two shells may become separated. (⑤) The shells are held together at the hinge by fibrous tissues, which are among the last to rot, so there is a reasonable chance that the two shells will remain held together until they become buried.

* mollusc: 연체동물　** intact: 온전한　*** hinge: (껍질의) 이음매

Sentence Structure

● The shells are held together at the hinge by fibrous tissues, [which are among the last to rot], so there is a reasonable chance [that the two shells will remain held together {until they become buried}].
첫 번째 [ ]는 fibrous tissues를 부수적으로 설명하는 관계절이며, 두 번째 [ ]는 a reasonable chance와 동격 관계를 이루는 명사절이고, { }는 시간의 부사절이다.

Word Search　주어진 문장에 들어갈 적절한 영어 단어를 본문에서 찾아 쓰시오. (필요시 형태를 바꿀 것)

1. It is u_____ that the delicate flower will survive the harsh winter without protection.
2. The o_____ of the experiment will determine the next steps in the research process.
3. She had a r_____ amount of time to complete the project before the deadline.

Exercise 9 · 요약문 완성

다음 글의 내용을 한 문장으로 요약하고자 한다. 빈칸 (A), (B)에 들어갈 말로 가장 적절한 것은?   25007-0105

Sometimes, a monkey grooms another monkey and gets nothing in return. Sometimes an ape courts a mate and gets completely ignored. Cortisol prompts a mammal to try something different, but after a few disappointments it can be hard for the mammal to predict where to invest its energy. This is why we often fall back on the neural superhighways we myelinated in youth. Your electricity flows effortlessly down the pathways built by behaviors that were reliably rewarded in your past. Maybe it was scoring a touchdown, or joining friends to watch your favorite quarterback score. Of course, carrying a ball across a line does not meet real survival needs, but dopamine surges when you expect a social reward. Each brain predicts social rewards from its own life experience. Maybe you lived in a world where social rewards went to someone who cooked a big meal or solved a big equation or found a bar open after hours. There are limitless ways to get social rewards, but the ones we observe and enjoy in youth build expectations that last.

* court: 구애하다   ** cortisol: 코르티솔(부신 피질에서 생기는 스테로이드 호르몬의 일종)
*** myelinate: 미엘린(신경 세포를 싸고 있는 보호막)을 형성하다

Just as repeated social disappointments can lead individuals to return to ___(A)___ behaviors reinforced in their youth, the brain relies on past experiences and neural pathways to predict and seek out ___(B)___ experiences.

	(A)		(B)		(A)		(B)
①	impulsive	⋯⋯	rewarding	②	familiar	⋯⋯	rewarding
③	innovative	⋯⋯	exhausting	④	destructive	⋯⋯	frustrating
⑤	habitual	⋯⋯	frustrating				

Sentence Structure

- Your electricity flows effortlessly down the pathways [built by behaviors {that were reliably rewarded in your past}].

  [ ]는 the pathways를 수식하는 분사구이고, 그 안의 { }는 behaviors를 수식하는 관계절이다.

Word Search   주어진 문장에 들어갈 적절한 영어 단어를 본문에서 찾아 쓰시오. (필요시 형태를 바꿀 것)

1. The team's loss in the final game was a major d_____ to their fans.
2. The machine r_____ produces high-quality results.
3. Access to clean water is crucial for the s_____ of all living organisms.

Exercise 10   어휘

다음 글의 밑줄 친 부분 중, 문맥상 낱말의 쓰임이 적절하지 <u>않은</u> 것은?   25007-0106

To determine whether a person is lying, we tend to rely heavily on intuition. Unless we catch a lie on ① <u>factual</u> grounds, the only indicators of a person's dishonesty are tone of voice, body language, and facial expressions — signs that may be too subtle for us to consciously recognize but that can still evoke a strong gut feeling. The problem is that while we can practice our skills at evaluating others' truthfulness in social interactions, ② <u>without</u> clear feedback on whether our judgments are correct we don't know if we're erring on the side of gullibility or of distrustfulness. This means we're ③ <u>able</u> to improve over time. Though many people believe they're quite good at distinguishing truths from lies, almost no one in the general population performs with ④ <u>higher</u> than chance accuracy. On average, even police officers, lawyers, judges, psychiatrists, and members of other groups that encounter more frequent and serious lies than ⑤ <u>ordinary</u> people perform no better.

* gut feeling: 직감   ** err on the side of: 지나치다 싶을 정도로 ~하다   *** gullibility: 쉽게 속음

Sentence Structure

● [Unless we catch a lie on factual grounds], the only indicators of a person's dishonesty are tone of voice, body language, and facial expressions — signs [that may be too subtle for us to consciously recognize] but [that can still evoke a strong gut feeling].

첫 번째 [ ]는 조건을 나타내는 부사절이며, 두 번째와 세 번째 [ ]는 but으로 대등하게 연결되어 signs를 수식하는 관계절이다.

Word Search   주어진 문장에 들어갈 적절한 영어 단어를 본문에서 찾아 쓰시오. (필요시 형태를 바꿀 것)

1. It can be difficult to d_____ the cause of a problem without thorough investigation.
2. There was a s_____ change in her tone that indicated she was upset.
3. It can be hard to d_____ between genuine and fake products.

Exercise 11~12   1지문 2문항

[11~12] 다음 글을 읽고, 물음에 답하시오.

In ancient times, not each and every kind of suffering warranted pity. No one was supposed to feel pity for captured, tortured, and killed enemies. Neither slaves nor Christian martyrs deserved pity. Later on, religious communities that preached the gospel of brotherly love and love of neighbor found no difficulty in (a) <u>denying</u> it to those who did not belong and believe. This attitude only started to change during the 18th century. Playwrights reviving the Aristotelian concept of catharsis discovered pity as the most natural and most moral human faculty that should be (b) <u>cultivated</u> by theater, literature, and music. Philosophers who sought a moral foundation for modern civil society (c) <u>praised</u> pity and sympathy as counter-forces of self-love and egoism. Novelists were (d) <u>eager</u> to devise plots and stories that would elicit the readers' pity, inspiring them to become sensitive and sensible citizens. At the same time, hundreds of thousands of European and North American men and women campaigned for the ending of slavery and the liberation of slaves. Fueled by an "imagined empathy" and using a language of love for those "brothers" and "sisters" whose freedom and human dignity were (e) <u>respected</u>, they engaged in an unprecedented — and ultimately successful — struggle against the slave trade and the institution of slavery.

* **martyr**: 순교자 ** **gospel**: 복음, 기독교의 교의 *** **egoism**: 이기심

11 윗글의 제목으로 가장 적절한 것은?

25007-0107

① Tips for Nurturing Empathy and Kindness
② Pity's Paradox: Can Feeling Sorry Lead to Real Help?
③ Why Compassion Matters More Than Ever Nowadays
④ The Evolution of Pity: From Ancient Attitudes to Modern Ones
⑤ A Double-Edged Sword: Examining the Power and Limits of Pity

12 밑줄 친 (a)~(e) 중에서 문맥상 낱말의 쓰임이 적절하지 <u>않은</u> 것은?

25007-0108

① (a)   ② (b)   ③ (c)   ④ (d)   ⑤ (e)

Sentence Structure

• Playwrights [reviving the Aristotelian concept of catharsis] discovered pity as the most natural and most moral human faculty [that should be cultivated by theater, literature, and music].
첫 번째 [ ]는 Playwrights를 수식하는 분사구이며, 두 번째 [ ]는 the most natural and most moral human faculty를 수식하는 관계절이다.

Word Search 주어진 문장에 들어갈 적절한 영어 단어를 본문에서 찾아 쓰시오. (필요시 형태를 바꿀 것)

1. It is important to feel p_____ for those who are suffering and to offer help whenever possible.
2. Even if your r_____ faith is strong, it's normal to feel doubt.
3. The company's positive a_____ towards innovation has led to many successful products.

Exercise 1 어휘

다음 글의 밑줄 친 부분 중, 문맥상 낱말의 쓰임이 적절하지 <u>않은</u> 것은?

25007-0109

One of the striking features of many invasions is a massive, early population outbreak that makes the alien form conspicuous and often highly destructive. In some species this outbreak may occur ① <u>immediately</u> after initial introduction; in others it occurs only after a delay of years, decades, or even a century. Following the outbreak, however, the population of the alien, as well as its ecological impact, may ② <u>decline</u> substantially. In many cases, native species also make ecological and evolutionary adjustments that shield them from the impacts of aliens. In addition, the population outbreak of the alien ③ <u>creates</u> massive ecological and evolutionary opportunities for exploitation by members of the native community. Ecological responses may be rapid, as members of the native community ④ <u>learn</u> to exploit the new community member as a food source and to avoid its direct detrimental influences. Later, evolutionary adjustments ⑤ <u>reduce</u> the ability of members of the native community to use the alien as a resource or to avoid the negative impacts of the alien.

* conspicuous: 눈에 띄는 ** exploitation: 이용 *** detrimental: 해로운

Sentence Structure

● In many cases, native species also make ecological and evolutionary adjustments [that shield **them** from the impacts of aliens].

[ ]는 ecological and evolutionary adjustments를 수식하는 관계절이고, 그 안의 them은 native species를 가리킨다.

Word Search 주어진 문장에 들어갈 적절한 영어 단어를 본문에서 찾아 쓰시오. (필요시 형태를 바꿀 것)

1. The sudden i_____ of the village by the enemies surprised everyone.
2. The city hosted a m_____ festival that attracted thousands of people from all over the country.
3. The trees can s_____ the animals from harsh winds and heavy rain.

Exercise 2) 문장 삽입

글의 흐름으로 보아, 주어진 문장이 들어가기에 가장 적절한 곳은?

25007-0110

> In contrast, marketing through sport happens when a non-sport product is marketed through an association to sport.

It is useful to note that there are two angles to sport marketing. The first is that sport products and services can be marketed directly to the consumer. (①) The second is that other, non-sport products and services can be marketed through the use of sport. (②) In other words, sport marketing involves the marketing of sport and marketing through sport. (③) For example, the marketing of sport products and services directly to sport consumers could include sporting equipment, professional competitions, sport events and local clubs. (④) Other simple examples include team advertising, designing a publicity stunt to promote an athlete, selling season tickets, and developing licensed clothing for sale. (⑤) Some examples could include a professional athlete endorsing a breakfast cereal, a corporation sponsoring a sport event, or even a beer company arranging to have exclusive rights to provide beer at a sport venue or event.

* publicity stunt: 떠들썩한 선전   ** endorse: 홍보하다   *** venue: 행사장

Sentence Structure

● Some examples could include [{**a professional athlete** endorsing a breakfast cereal}, {**a corporation** sponsoring a sport event}, or even {**a beer company** arranging to have exclusive rights to provide beer at a sport venue or event}].

[ ] 안에서 세 개의 { }가 콤마와 or로 대등하게 연결되어 include의 목적어를 이룬다. 세 개의 { }에서 a professional athlete, a corporation, a beer company는 각각 이어지는 동명사의 의미상의 주어이다.

Word Search   주어진 문장에 들어갈 적절한 영어 단어를 본문에서 찾아 쓰시오. (필요시 형태를 바꿀 것)

1. The logo's design has a strong a_____ to the company's brand identity.

2. This cycling program provides necessary e_____ such as bikes and helmets.

3. The a_____ trains every morning to prepare for the big race.

Exercise 3 요약문 완성

다음 글의 내용을 한 문장으로 요약하고자 한다. 빈칸 (A), (B)에 들어갈 말로 가장 적절한 것은? 25007-0111

Determining the probability that a particular event will occur in a particular location within a particular time span is an essential goal of hazard evaluation. For many large rivers we have sufficient records of flow to develop probability models that can reasonably predict the average number of floods of a given magnitude that will occur in a given time period. Likewise, droughts may be assigned a probability on the basis of past occurrence of rainfall in the region. However, these probabilities are similar to the chances of throwing a particular number on a die or drawing an inside straight in poker; the element of chance is always present. For example, the 10-year flood may occur on the average of every 10 years, but it is possible for several floods of this magnitude to occur in any one year, just as it is possible to throw two straight sixes with a die.

* magnitude: 규모 ** die: 주사위
*** inside straight: 인사이드 스트레이트(4장의 숫자가 있을 때 5개의 연속된 숫자를 만들기 위해 특정 카드 한 장이 필요한 패)

Hazard evaluation involves estimating the ____(A)____ of events like floods or droughts in a specific time and space based on records, while always carrying an element of ____(B)____ with it.

	(A)		(B)			(A)		(B)
①	likelihood	……	randomness		②	likelihood	……	locality
③	intensity	……	locality		④	duration	……	randomness
⑤	duration	……	standardization					

Sentence Structure

• [Determining the probability {that a particular event will occur in a particular location within a particular time span}] is an essential goal of hazard evaluation.
[ ]는 문장의 주어 역할을 하는 동명사구이고, 그 안의 { }는 the probability의 구체적 내용을 설명하는 동격절이다.

Word Search 주어진 문장에 들어갈 적절한 영어 단어를 본문에서 찾아 쓰시오. (필요시 형태를 바꿀 것)

1. Simply put, p_____ is the possibility of an event happening expressed in numbers.
2. The family r_____ discussed the vacation plans, considering everyone's preferences.
3. Team collaboration is an important e_____ of the job.

Exercise 4 어휘

(A), (B), (C)의 각 네모 안에서 문맥에 맞는 낱말로 가장 적절한 것은? 25007-0112

Among the first studies to suggest the (A) use / avoidance of specifically a visual in contrast to verbal or phonologically based temporary memory for verbal stimuli were those conducted by psychologist Posner and his colleagues, who developed a visual letter-matching task in which pairs of letters were shown in their upper case (e.g., AB) or lower case (e.g., aa) versions or a mixture of both (e.g., Bb). The subjects' task was to respond on the basis of whether the letters in the pair had the same name (e.g., Aa) or had different names (e.g., Ab). When the letters were both in the same letter case and were physically identical (e.g., AA), subjects responded much more (B) quickly / slowly than if the upper and lower case versions were different. The advantage for physically identical letters remained when letters in each pair were shown one after another with interletter delays of up to 2 seconds. This suggested that subjects were (C) passing / relying on the visual code for the letters during the delay, after which a name code was being used for the decision.

* phonologically: 음운론적으로

	(A)		(B)		(C)
①	use	.....	quickly	.....	passing
②	use	.....	slowly	.....	relying
③	use	.....	quickly	.....	relying
④	avoidance	.....	slowly	.....	passing
⑤	avoidance	.....	quickly	.....	relying

Sentence Structure

● The subjects' task was [to respond on the basis of {whether the letters in the pair had the same name (e.g., Aa) or had different names (e.g., Ab)}].

[ ]는 주격 보어 역할을 하는 to부정사구이고, 그 안의 { }는 전치사 of의 목적어 역할을 하는 명사절이다.

Word Search 주어진 문장에 들어갈 적절한 영어 단어를 본문에서 찾아 쓰시오. (필요시 형태를 바꿀 것)

1. Plants respond to environmental s_____ such as light and temperature changes.
2. Because her handwriting was nearly i_____ to her mother's, it was hard to distinguish between the two.
3. Being fluent in multiple languages can be a considerable a_____ when traveling.

Exercise 5~6   1지문 2문항

[5~6] 다음 글을 읽고, 물음에 답하시오.

As more and more brands and organizations started working with influences, the return on influencer (yes, pun intended) has decreased. People realize that their heroes are being (a) <u>paid</u> to promote brands on social media. The amount celebrities can get for promoting brands goes up to half a million dollars for a mention in one single Instagram post. So, in the past few years, brands have slowly extended their influencer marketing strategies to so-called micro-influencers. These are people with 10,000 to 50,000 followers who are seen as experts or heroes to a smaller group of people. Because of the (b) <u>size</u> of their following, their messages on social media are perceived to be authentic and therefore more trustworthy. As this strategy is also gaining (too much) popularity, several experts have predicted that the 2020s will be the era of the nano influences. These are people with 1,000 – 10,000 followers and who are just like your own most popular friends and family members; their lack of any real fame will be a sign of their (c) <u>untrustworthiness</u>.

The expectation is that this nano category will also eventually (d) <u>lose</u> its authenticity and trust will move to the last category, the pico influences. This is the category of people with less than 1,000 connections: the level of you, me and my neighbor with her dog. These are the people that we know in (e) <u>real</u> life and with whom we have a relationship. The interesting thing is that we have then come full circle, back to where we were historically, when we trusted people who were actually very well known to us.

5 윗글의 제목으로 가장 적절한 것은?   25007-0113

① Trends and Predictions in Influencer Marketing
② Impact of Digital Marketing on Business Growth
③ Improving Brand Image Using Customer Reviews
④ Utilizing Social Media to Appeal to the Younger Generation
⑤ Enhancing Customer Loyalty with Positive Brand Experiences

6 밑줄 친 (a)~(e) 중에서 문맥상 낱말의 쓰임이 적절하지 <u>않은</u> 것은?   25007-0114

① (a)   ② (b)   ③ (c)   ④ (d)   ⑤ (e)

Sentence Structure

● The interesting thing is [that we have then come full circle, back to {where we were historically}, {when we trusted people who were actually very well known to us}].

[ ]는 주격 보어 역할을 하는 명사절이다. 첫 번째 { }는 의문사 where가 이끄는 명사절로서 전치사 to의 목적어 역할을 하고, 두 번째 { }는 앞선 { }를 구체적으로 설명하고 있다.

Word Search   주어진 문장에 들어갈 적절한 영어 단어를 본문에서 찾아 쓰시오. (필요시 형태를 바꿀 것)

1. Artists often p_____ the world in unique and imaginative ways.
2. Our team's winning s_____ is a combination of strong defense and quick counterattacks.
3. The website is considered t_____ because it provides accurate and reliable information.

Exercise 7 　어휘

다음 글의 밑줄 친 부분 중, 문맥상 낱말의 쓰임이 적절하지 <u>않은</u> 것은?　　　25007-0115

From an epistemic perspective, a central question concerns the extent to which advertising is expected to be truthful. Advertisements often use widely exaggerated or metaphorical claims, a practice called "puffery," which, in contrast to misleading advertising, is legally ① <u>permitted</u> in some countries, for example, the United States. Its legality is based on the assumption that individuals will not take such claims literally or act on them. This assumption, however, is ② <u>confirmed</u> by empirical evidence that shows that consumers *do* in fact react to puffed statements. Even though some consumers may indeed recognize puffery as what it is, others are more vulnerable and take it at face value. This practice thus also raises issues of ③ <u>fairness</u>: is it legitimate for companies to make false statements that better-informed or more reflective consumers will not believe, but others will fall prey to? Even sophisticated consumers ④ <u>suffer</u> from such strategies, however, because they have to double-check which advertisements to take seriously. In an analysis of the treatment of puffery in US law, legal scholar David Hoffman argues that it should be understood as ⑤ <u>causing</u> a negative externality: it creates "informational burdens currently borne by buyers, without compensation from sellers."

* **epistemic**: 인식론적인　** **puffery**: 과대광고
*** **externality**: 외부 효과(어떤 경제 주체의 행위가 제삼자에게 기대하지 않은 편익이나 비용을 발생시키는 현상)

Sentence Structure

● From an epistemic perspective, a central question concerns [the extent {to which advertising is expected to be truthful}].

[ ]는 concerns의 목적어이고, { }는 the extent를 수식하는 관계절이다.

Word Search　주어진 문장에 들어갈 적절한 영어 단어를 본문에서 찾아 쓰시오. (필요시 형태를 바꿀 것)

1. The lawyer questioned the l_____ of the contract due to its unclear terms.
2. The young birds in the nest were v_____ without their mother's protection.
3. The s_____ gentleman impressed everyone with his extensive knowledge of art.

Exercise 8　문장 삽입

글의 흐름으로 보아, 주어진 문장이 들어가기에 가장 적절한 곳은?　25007-0116

Not all ecological systems and spaces have the same ecosystem services and natural resource value.

Environmental economists often aim to put an economic valuation on ecosystem service. The reason for this is that ecosystem services historically have been underconsidered and underappreciated in environmental decision-making and policy. (①) Placing a monetary value on them increases their salience. (②) The economic value of global ecosystem services is estimated to be $125 trillion per year. (③) This puts into perspective just how crucial well-functioning ecological systems are to human well-being as well as the extent to which human systems are intertwined with and dependent upon them. (④) However, site- or system-specific economic valuations are often more important to environmental decision-making than is global or macro evaluation. (⑤) For example, an average hectare of open ocean provides fewer services than does an average hectare of reef, and an average hectare of desert provides fewer than does an average hectare of tropical rainforest.

* salience: 중요성　** intertwine: 밀접하게 관련짓다　*** reef: 암초

Sentence Structure

● This **puts into perspective** [{just how crucial well-functioning ecological systems are to human well-being} **as well as** {the extent ⟨to which human systems are intertwined with and dependent upon them⟩}].
[ ]는 '~을 분명하게 알 수 있게 하다'의 의미를 나타내는 「put ~ into perspective」의 동사 put의 목적어 역할을 한다. [ ] 안의 두 개의 { }는 as well as로 연결되어 있고, ⟨ ⟩는 the extent를 수식하는 관계절이다.

Word Search　주어진 문장에 들어갈 적절한 영어 단어를 본문에서 찾아 쓰시오. (필요시 형태를 바꿀 것)

1. A young businessperson received m_____ support from investors to start his new business.
2. Eating a balanced diet is c_____ for maintaining good health and preventing illness.
3. The small island's economy is heavily d_____ on tourism for its survival.

Exercise 9 요약문 완성

다음 글의 내용을 한 문장으로 요약하고자 한다. 빈칸 (A), (B)에 들어갈 말로 가장 적절한 것은?   25007-0117

Natural laws are not things that we can see or touch or hear; they are — according to Scottish philosopher David Hume — only descriptions of past groupings of events. For instance, consider the series of events: I let go of the pen; the pen falls; I let go of the pen; the pen falls; I let go of the pen; and so on. This series *may* be explained by a set of laws, such as gravity and inertia. But this series — by itself — simply *describes* events that have happened in the past; it does not tell us *why* pens behave that way, and it does not tell us that pens will always behave that way. And even if those past events happened because of laws, why believe those laws *will* continue to govern events in the future? If we say laws existed in the past, we need some additional law that tells us those laws will exist in the future. But, it seems we can only describe the way those laws behaved in the past.

* inertia: 관성

⬇

Natural laws, which David Hume claims merely ____(A)____ past grouping of events, do not guarantee future behavior since they lack an explanation for their ____(B)____ validity.

	(A)		(B)			(A)		(B)
①	distort	……	universal		②	represent	……	continuing
③	track	……	statistical		④	oversimplify	……	continuing
⑤	glorify	……	statistical					

Sentence Structure

● But this series — by itself — simply *describes* events [that have happened in the past]; it does not tell us [*why* pens behave that way], and it does not tell us [that pens will always behave that way].

첫 번째 [ ]는 events를 수식하는 관계절이다. 두 번째와 세 번째 [ ]는 각각 tell의 직접목적어 역할을 하는 명사절이다.

Word Search   주어진 문장에 들어갈 적절한 영어 단어를 본문에서 찾아 쓰시오. (필요시 형태를 바꿀 것)

1. The ball fell to the ground because of g_____, the force that pulls things down.
2. The new laws will g_____ how businesses operate in this country.
3. If you need a_____ help with your homework, please ask your teacher after class.

Exercise 10 어휘

다음 글의 밑줄 친 부분 중, 문맥상 낱말의 쓰임이 적절하지 <u>않은</u> 것은?

25007-0118

Even though the universe is consistently inflating, it is still orderly. There are consistent rules that can be observed and measured. There are ① dependable physical principles that govern the relationship of one particle to another at varying scales. Three hundred years ago, the brilliant mathematician Sir Isaac Newton formulated the laws of motion that define the movement of large objects in physical space. These are the reliable and reproducible physical rules that ② allow us to calculate the accurate trajectory of an artillery shell or a rocket ship. These apply to the scale of objects that we can observe ③ directly and are not microscopic in size; however, in the last century, a new set of quantum rules has been identified that govern how the atomic level operates. Some of these quantum principles contradict our ordinary senses. At this microscopic scale, the actions among particles ④ confirm our established expectations of clear-cut relationships between physical causes and effects. Instead, quantum measurements reveal inherent uncertainties about the relationship between one object and another. Yet, these ⑤ built-in uncertainties are just as relevant to our lives as those Newtonian forces that govern what happens when two cars crash into each other.

* trajectory: 궤도, 탄도 ** artillery shell: 포탄 *** quantum: 양자

Sentence Structure

● Three hundred years ago, [the brilliant mathematician Sir Isaac Newton] formulated [the laws of motion {that define the movement of large objects in physical space}].

첫 번째 [ ]는 문장의 주어이고, 두 번째 [ ]는 formulated의 목적어이다. { }는 the laws of motion을 수식하는 관계절이다.

Word Search 주어진 문장에 들어갈 적절한 영어 단어를 본문에서 찾아 쓰시오. (필요시 형태를 바꿀 것)

1. Kate's c_____ effort in practicing English resulted in significant improvement in her speaking skills.

2. The experiment must have clear instructions to ensure the results are r_____ by other researchers.

3. The politician's promises to lower taxes c_____ his voting record of supporting tax increases.

[11~12] 다음 글을 읽고, 물음에 답하시오.

Humans have an extremely extended period of brain development. This is especially true for circuits that mediate behavioral control, such as the prefrontal cortex, which do not fully mature until a person's early twenties. Until that time, synapses are still being modified on a (a) <u>massive</u> scale. This provides plentiful opportunity for these circuits to be shaped by experience and is likely one of the key factors in our ability to successfully populate what has been called the "cognitive niche." Rather than being adapted for specific environments, with a (b) <u>limited</u> set of hardwired, instinctive behaviors, we have evolved cognitive flexibility and responsiveness, allowing us to adapt ourselves to our individual environments. Repeated patterns are reinforced and habitual modes of behavior emerge. We gradually become ourselves.

But at some point we have to stop constantly becoming and just get on with things — important things like building a career or finding a mate. That means we have to consolidate the adaptations we have made and (c) <u>restrict</u> further changes. We can't have runaway positive feedback loops forever — we have to maintain these neural configurations to (d) <u>remain</u> ourselves. The periods of wholesale plasticity last considerably longer in behavioral and cognitive circuits than in sensory ones, but they still close as we reach adulthood. The plasticity processes themselves will have progressively (e) <u>expanded</u> the "degrees of freedom" of the developing brain, magnifying initial biases by both positive reinforcement and progressive elimination of connections mediating less-favored states. But the biochemistry of the brain also changes with maturation, so that mechanisms of plasticity and flexibility get replaced by mechanisms of stability and maintenance.

* cognitive niche: 인지적 적소(인간의 인지 능력이 특화된 생태학적 위치)
** consolidate: 공고히 하다   *** configuration: 구성, 배열

11 윗글의 제목으로 가장 적절한 것은?   25007-0119

① Ways to Trick Our Brain into Forming Lasting Habits
② Why a Flexible Mindset Contributes to Brain Development
③ The Aging Brain: Myths and Realities of Cognitive Decline
④ How Our Brain Responds to Changes in the Physical Environment
⑤ From Flexible to Stable: The Gradual Maturing of the Human Brain

12 밑줄 친 (a)~(e) 중에서 문맥상 낱말의 쓰임이 적절하지 <u>않은</u> 것은?   25007-0120

① (a)   ② (b)   ③ (c)   ④ (d)   ⑤ (e)

Sentence Structure

● This [provides plentiful opportunity {for these circuits to be shaped by experience}] and [is likely one of the key factors in our ability {to successfully populate 〈what has been called the "cognitive niche."〉}]
두 개의 [ ]는 and로 연결된 문장의 술어이다. 첫 번째 { }는 these circuits를 의미상의 주어로 하여 plentiful opportunity를 수식하는 to부정사구이다. 두 번째 { }는 our ability의 구체적 내용을 설명하는 to부정사구이고, 그 안의 〈 〉는 populate의 목적어 역할을 하는 명사절이다.

Word Search 주어진 문장에 들어갈 적절한 영어 단어를 본문에서 찾아 쓰시오. (필요시 형태를 바꿀 것)

1. The old man had an i_____ distrust of strangers.

2. The chameleon's ability to change its skin color is an a_____ that helps it hide from predators.

3. Our s_____ organs, like our eyes and ears, help us perceive the world around us.

Exercise 1 어휘

다음 글의 밑줄 친 부분 중, 문맥상 낱말의 쓰임이 적절하지 <u>않은</u> 것은?   25007-0121

Inspired by the pioneering work of the psychologists Daniel Kahneman and the late Amos Tversky, the field of psychology has cataloged a large inventory of behavioral anomalies in which people clearly ① <u>violate</u> the predictions and prescriptions of standard economic models. It is common, for example, for someone to be ② <u>willing</u> to drive across town to save $10 on a $20 clock radio, but unwilling to do so to save $10 on a $1,000 television set. Yet the benefit of making the drive is $10 in each case. So if the implicit cost of the drive were ③ <u>less</u> than $10, a rational person would drive across town in both cases. People often explain their ④ <u>reluctance</u> to make the drive for the television by saying the $10 savings is such a small percentage of its price. But a rational person reckons benefits and costs in ⑤ <u>relative</u> terms, not as percentages.

* behavioral anomaly: 이상 행동 ** reckon: 생각하다

Sentence Structure

- **It** is common, for example, [**for someone** to be willing to drive across town to save $10 on a $20 clock radio, but unwilling to do so to save $10 on a $1,000 television set].

It은 형식상의 주어이고, [ ]는 내용상의 주어이다. for someone은 부정사구 to be ~의 의미상의 주어를 나타낸다.

Word Search   주어진 문장에 들어갈 적절한 영어 단어를 본문에서 찾아 쓰시오. (필요시 형태를 바꿀 것)

1. The designer's latest collection was i_____ by vintage fashion trends.
2. There was a noticeable r_____ among the employees to adopt the new policy.
3. In order to solve the complex problem, we need a r_____ and systematic approach.

Exercise 2 문장 삽입

글의 흐름으로 보아, 주어진 문장이 들어가기에 가장 적절한 곳은?

25007-0122

> They walked to the offices of their Baby Boomer peers, emailed Gen X colleagues, and texted Millennials.

Regardless of their generation, everyone wants to be valued, respected, and heard. But each one of us might express our thoughts and receive information differently. (①) One of my clients was struggling with how some of their Baby Boomer employees were trying to build personal relationships with Millennial employees. (②) A few of these employees read articles that recommended specific methods for communicating with each generation. (③) The Boomer employees then went on to use this information to inform how they communicated with all of their coworkers of various generations. (④) On the surface, this seemed fine, but it soon started to stir up some controversy because people did not understand why they were getting an email, when someone else was getting a face-to-face conversation. (⑤) By limiting in-person conversation to only certain coworkers, these employees unintentionally made others feel left out and undervalued.

Sentence Structure

- The Boomer employees then went on to use this information [to inform {how they communicated with all of their coworkers of various generations}].

[ ]는 목적을 나타내는 to부정사구이고, 그 안의 { }는 inform의 목적어 역할을 하는 명사절이다.

Word Search 주어진 문장에 들어갈 적절한 영어 단어를 본문에서 찾아 쓰시오. (필요시 형태를 바꿀 것)

1. Many in the younger g_____ are pushing for social change.
2. She u_____ deleted the important file from her computer.
3. The decision to ban certain books from the library sparked a c_____ among students.

Exercise 3 요약문 완성

다음 글의 내용을 한 문장으로 요약하고자 한다. 빈칸 (A), (B)에 들어갈 말로 가장 적절한 것은?   25007-0123

Our species has been making music most likely for as long as we've been human. It seems to be a permanent part of us. The oldest known musical instruments date back to some 40,000 years ago. Among the most intriguing of these are delicate bone flutes, found in what is now southern Germany. These discoveries testify to the advanced technology that our ancestors applied to create music: the finger holes are carefully bevelled to allow the musician's fingers to make a tight seal; and the distances between the holes appear to have been precisely measured, perhaps to correspond to a specific musical scale. This time period corresponds to the last glaciation episode in the Northern Hemisphere — life could not have been easy for people living at that time. Yet time, energy, and the skills of craftworkers were used for making abstract sounds "of the least use ... to daily habits of life." So, music must have been very meaningful and important for them.

* **bevel**: 비스듬히 자르다   ** **glaciation**: 빙하

The fact that humans living 40,000 years ago had attentively crafted musical instruments employing ___(A)___ methods despite their harsh living conditions reflects the ___(B)___ they placed on music.

	(A)		(B)
①	sophisticated	·····	significance
②	sophisticated	·····	complexity
③	flexible	·····	complexity
④	outdated	·····	essence
⑤	outdated	·····	significance

Sentence Structure

● [Among the most intriguing of these] are [delicate bone flutes, {found in what is now southern Germany}].
전치사구인 첫 번째 [ ]가 문두에 위치하면서, 문장의 주어인 두 번째 [ ]가 술어동사 are 다음에 놓였다. { }는 delicate bone flutes를 추가적으로 설명하는 분사구이다.

Word Search   주어진 문장에 들어갈 적절한 영어 단어를 본문에서 찾아 쓰시오. (필요시 형태를 바꿀 것)

1. Experts will t_____ to the safety and effectiveness of the new medication during the trial.
2. The art exhibit featured highly i_____ pieces that sparked a lot of conversation.
3. The professor's lecture covered complex a_____ concepts in mathematics.

Exercise 4 어휘

(A), (B), (C)의 각 네모 안에서 문맥에 맞는 낱말로 가장 적절한 것은?

25007-0124

Sometimes, even seemingly unlucky starts turn out to make long-run success more likely. Gladwell cites the experience of Jews who immigrated to New York in the early twentieth century and went on to prosper in the garment industry. Many raised children who graduated from law school, only to be (A) rejected / hired by leading New York law firms, which in those days hired mostly lawyers from wealthy Protestant families. Jewish graduates were often left with few better options than to start small firms of their own. Those firms often specialized in cases that the elite law firms felt were (B) above / beneath them, such as the litigation of hostile corporate takeovers. The lawyers raised by garment workers were thus often the only ones who had developed the expertise to capitalize on the explosive growth of hostile takeover litigation that occurred in the 1970s and 1980s. By (C) sharing / dominating that new market, they went on to earn vastly more than the lawyers in the firms that had earlier shunned them.

* litigation: 소송 ** shun: 기피하다

(A)	(B)	(C)
① rejected	…… above	…… sharing
② hired	…… above	…… sharing
③ rejected	…… beneath	…… dominating
④ hired	…… beneath	…… dominating
⑤ rejected	…… beneath	…… sharing

Sentence Structure

• The lawyers [raised by garment workers] were thus often the only ones [who had developed the expertise {to capitalize on the explosive growth of hostile takeover litigation ⟨that occurred in the 1970s and 1980s⟩}].
첫 번째 [ ]는 The lawyers를 수식하는 분사구이고, 두 번째 [ ]는 the only ones를 수식하는 관계절이다. { }는 the expertise 를 수식하는 to부정사구이고, ⟨ ⟩는 the explosive growth of hostile takeover litigation을 수식하는 관계절이다.

Word Search 주어진 문장에 들어갈 적절한 영어 단어를 본문에서 찾아 쓰시오. (필요시 형태를 바꿀 것)

1. The singer's latest album received an e_____ response from fans.

2. His e_____ in engineering made him the ideal candidate for the project.

3. The new technology has v_____ improved efficiency in the workplace.

Exercise 5~6   1지문 2문항

[5~6] 다음 글을 읽고, 물음에 답하시오.

Let's take a classical example from cognitive psychology: the effect of word processing depth. Imagine that I present a list of sixty words to three groups of students. I ask the first group to decide whether the words' letters are upper- or lowercase; the second group, whether the words rhyme with "chair"; and the third, whether they are animal names or not. Once the students are finished, I give them a (a) memory test. Which group remembers the words best? Memory turns out to be much (b) better in the third group, who processed the words in depth, at the meaning level (75 percent success), than in the other two groups, who processed the more superficial sensory aspects of the words, either at the letter level (33 percent success) or the rhyme level (52 percent success). We do find a weak implicit and unconscious trace of the words in all groups: learning (c) leaves its subliminal mark within the spelling and phonological systems. However, only in-depth semantic processing guarantees explicit, detailed memory of the words. The same phenomenon (d) disappears at the level of sentences: students who make the effort to understand sentences on their own, without teacher guidance, show much better retention of information. This is a general rule, which the American psychologist Henry Roediger states as follows: "Making learning conditions (e) more difficult, thus requiring students to engage more cognitive effort, often leads to enhanced retention."

* subliminal: 잠재의식적인   ** phonological: 음운의   *** semantic: 의미적인

5 윗글의 제목으로 가장 적절한 것은?   25007-0125

① Ways to Support Children with Learning Difficulties
② Motivation and Its Impact on Memory Retention Rate
③ How the Depth of Semantic Processing Influences Memory
④ Subconscious Language Learning: Reality or Science Fiction?
⑤ The Language We Speak Can Influence Our Way of Thinking

6 밑줄 친 (a)~(e) 중에서 문맥상 낱말의 쓰임이 적절하지 <u>않은</u> 것은?   25007-0126

① (a)   ② (b)   ③ (c)   ④ (d)   ⑤ (e)

Sentence Structure

● I ask the first group [to decide whether the words' letters are upper- or lowercase]; [the second group, whether the words rhyme with "chair"]; and [the third, whether they are animal names or not].
첫 번째 [ ]는 문장의 목적격 보어 역할을 하는 to부정사구이다. 두 번째와 세 번째 [ ]에서는 각각 the second group과 the third의 뒤에 to decide가 생략된 것으로 이해할 수 있으며, 두 개의 [ ]는 모두 I ask에 이어진다.

Word Search   주어진 문장에 들어갈 적절한 영어 단어를 본문에서 찾아 쓰시오. (필요시 형태를 바꿀 것)

1. His understanding of the topic was only s_____; he didn't get the deeper concepts.
2. The detective found a t_____ of evidence that led to the suspect.
3. The teacher's g_____ helped the students improve their writing skills.

Exercise 7   어휘

다음 글의 밑줄 친 부분 중, 문맥상 낱말의 쓰임이 적절하지 않은 것은?   25007-0127

A successful farming community could support in a small region a much greater population than the foragers could have. There were, however, very significant ① downsides to this new way of life. First of all, storms, droughts, floods, torrential rains, and other severe weather anomalies could be ② disastrous for a farming community while posing only a major nuisance to foragers. The latter, with their ③ minimal lightweight possessions, could fairly easily up and move to some other area where the damage was not so great. But, more importantly, the natural home they depended on was much ④ more likely to be severely damaged than were the crops and structures the agricultural community relied on. By replacing their natural home with an ⑤ artificial construction, the farmers, paradoxically enough, had made themselves more defenseless to natural disasters; in fact, natural events that were not at all disasters for the foragers became disasters for them.

* forager: 수렵채집인 ** anomaly: 이변 *** nuisance: 골칫거리, 성가신 것[사람]

Sentence Structure

● **The latter**, with their minimal lightweight possessions, could fairly easily up and move to some other area [where the damage was not so great].

The latter는 앞 문장에 나온 foragers를 의미하고, [ ]는 some other area를 수식하는 관계절이다.

Word Search   주어진 문장에 들어갈 적절한 영어 단어를 본문에서 찾아 쓰시오. (필요시 형태를 바꿀 것)

1. Nuclear weapons p_____ a threat to everyone.

2. The necklace is one of her most treasured p_____.

3. Farmers here still plant and harvest their c_____ by hand.

Exercise 8 문장 삽입

글의 흐름으로 보아, 주어진 문장이 들어가기에 가장 적절한 곳은?

25007-0128

But even allowing nonbasic terms and even allowing the highly specialized vocabulary of the paint industry, the distinctions that languages make in the color space are astronomically small in comparison to well over 2 *million* distinctions in color that the human eye can discern.

There are special situations in which people may know dozens, if not hundreds, of distinct words for colors. One famous brand of house paints has more than two thousand colors in its commercial palette. (①) Many of these are labeled with unique (and decidedly nonbasic) English names: *violet posy, wing commander, Aztec tan.* (②) Even in highly specialized, technical vocabularies, the total inventory of color terms in use makes less than one-tenth of 1 percent of the discernible distinctions in color that the human visual system can make. (③) The everyday vocabulary that most people use will be one hundred times less than that again. (④) Our perceptual experience is rich, but our language comes nowhere close to that richness. (⑤) If you think that language is good for capturing perceptual experience, think again.

* astronomically: 어마어마하게, 천문학적으로   ** discern: 식별하다   *** inventory: 목록

Sentence Structure

● There are special situations [in which people may know dozens, {**if not** hundreds}, of distinct words for colors].

[ ]는 special situations를 수식하는 관계절이고, { }는 삽입구로, 「if not ~」은 '~은 아닐지라도'라는 의미이다.

Word Search 주어진 문장에 들어갈 적절한 영어 단어를 본문에서 찾아 쓰시오. (필요시 형태를 바꿀 것)

1. There are no obvious d_____ between the two designs.
2. Each box in the warehouse was l_____ with its contents.
3. Maps are a v_____ tool for learning.

Exercise 9   요약문 완성

다음 글의 내용을 한 문장으로 요약하고자 한다. 빈칸 (A), (B)에 들어갈 말로 가장 적절한 것은?   25007-0129

In Britain, large open areas of land known as the *commons* were the property of land-owning nobility but were accessible for use by peasants. They used the commons to conduct small-scale agriculture, to collect wood for fuel and cooking, and to raise small amounts of livestock for their families. As global demand for wool to be used in clothing and other goods increased, landowners sought more grazing land to increase wool production, so they privatized the commons, fencing them off and thus cutting off access to the peasants. By this process of the *enclosure* of the commons, peasants were cut off from land that had been their primary resource for survival for centuries. Today, there are interesting correlations between the physical commons of English pasturelands and the digital commons of the internet. Both are rich with opportunities for self-subsistence, expression, flourishing; both are full of challenges from *enclosure* and *landlords* (e.g., platform owners who charge fees for access or use).

* grazing land: 목초지   ** self-subsistence: 자립

In Britain, as the demand for wool increased, peasants' access to the commons was
_____(A)_____ , and this _____(B)_____ contemporary situations we are facing in the digital world.

	(A)		(B)		(A)		(B)
①	disturbed	·····	contrasts	②	guaranteed	·····	mirrors
③	blocked	·····	resembles	④	expanded	·····	parallels
⑤	restricted	·····	resolves				

Sentence Structure

● [**As** {global demand for wool to be used in clothing and other goods} increased], landowners sought more grazing land [to increase wool production], so they privatized the commons, [fencing them off] and thus [cutting off access to the peasants].

첫 번째 [ ]는 '~하면서'라는 의미의 As가 이끄는 부사절이고, 그 안의 { }가 주어이다. 두 번째 [ ]는 목적을 나타내는 to부정사구이고, 세 번째와 네 번째 [ ]는 모두 they를 의미상의 주어로 하는 분사구문이다.

Word Search   주어진 문장에 들어갈 적절한 영어 단어를 본문에서 찾아 쓰시오. (필요시 형태를 바꿀 것)

1. The experiments were c_____ by scientists.
2. The farmer tended to his l_____ early each morning.
3. Air traffic control has been p_____ .

Exercise 10 어휘

다음 글의 밑줄 친 부분 중, 문맥상 낱말의 쓰임이 적절하지 <u>않은</u> 것은?

25007-0130

The reason listening can be so difficult appears to be our narcissistic disposition. Too often, we ① <u>pretend</u> to be listening while our mind is racing in trying to think of something clever. However, being clever is not being wise. In addition, to exacerbate our narcissistic tendencies, there is also the kind of listening with half an ear that ② <u>presumes</u> that we already know what the other person is going to say. I am referring to an inattentive listening, only waiting for a chance to speak, and even becoming ③ <u>impatient</u>, wishing to get rid of the other person. As the philosopher and poet Ralph Waldo Emerson once said, 'There is a difference between truly listening and waiting for your turn to talk.' This ④ <u>reluctance</u> to interrupt and get a word in can be quite powerful. Some people just want to hear themselves speak just to confirm and validate their ⑤ <u>existence</u>. It has been said that big egos have little ears.

* disposition: 성향 ** exacerbate: 악화시키다

Sentence Structure

- In addition, to exacerbate our narcissistic tendencies, there is also the kind of listening with half an ear [that presumes that we already know {what the other person is going to say}].

  [ ]는 the kind of listening with half an ear를 수식하는 관계절이고, { }는 know의 목적어 역할을 하는 명사절이다.

Word Search   주어진 문장에 들어갈 적절한 영어 단어를 본문에서 찾아 쓰시오. (필요시 형태를 바꿀 것)

1. The p_____ read her latest poem at the restaurant.

2. Please don't i_____ me while I'm talking.

3. She seemed to need his admiration to v_____ her as a person.

Exercise 11~12   1지문 2문항

[11~12] 다음 글을 읽고, 물음에 답하시오.

The bank account analogy helps express why increased agricultural production must be accompanied by fertilizer use. Every time you harvest a crop and eat it, the nitrogen (and other nutrients) in those plants, the very nutrients that make that crop good food, are taken out of the soil and moved to wherever you are. Some of those nutrients (a) <u>accumulate</u> in your body (if you are growing), but most pass through. Either way, unless you and your waste are returned to the farm, there is a net (b) <u>loss</u> of nutrients from the soil — a net withdrawal from the nutrient bank. When there were relatively few people, most of whom lived, went to the bathroom, and died on or near the farm, leaving fields uncultivated or planting legumes and plowing them under was a reasonably (c) <u>sustainable</u> way to produce food. But for eight billion people? Or for ten? The food required to feed all of us requires a lot of nitrogen to be removed from farms, and it needs to be (d) <u>replaced</u>, or the soil bank account of nitrogen will run out. This means feeding the world requires industrially produced nitrogen fertilizer, at least for the foreseeable future, until we figure out a safe and effective way to return the nutrients passing through humans to the farm soil where they came. In other words, we thrive because of our innovations in (e) <u>eliminating</u> nitrogen.

* analogy: 비유   ** legume: 콩과(科) 식물   *** plow under: (작물 등을) 갈아엎다

11 윗글의 제목으로 가장 적절한 것은?   25007-0131

① Is Nitrogen the Key Element of Biodiversity?
② Why Nitrogen Fertilizer Is Essential in Today's World
③ The Total Amount in a Bank Account: A Mirror of Greed
④ Increasing Demand to Protect the Soil Against Chemical Use
⑤ Which Is the Cause of the Climate Crisis, Nitrogen or People?

12 밑줄 친 (a)~(e) 중에서 문맥상 낱말의 쓰임이 적절하지 <u>않은</u> 것은?   25007-0132

① (a)   ② (b)   ③ (c)   ④ (d)   ⑤ (e)

Sentence Structure

● [**Every time** you harvest a crop and eat it], [the nitrogen (and other nutrients) in those plants], [the very nutrients that make that crop good food], are taken out of the soil and moved to [**wherever** you are].

첫 번째 [ ]는 '~할 때마다'라는 의미의 Every time이 이끄는 부사절이고, 두 번째와 세 번째 [ ]는 의미상 동격 관계를 이룬다. 네 번째 [ ]는 복합관계절로 여기에서 wherever는 '~하는 곳이 어디든지'라는 뜻이다.

Word Search   주어진 문장에 들어갈 적절한 영어 단어를 본문에서 찾아 쓰시오. (필요시 형태를 바꿀 것)

1. The thunderstorm was a_____ by high winds.

2. Farmers sort the vegetables when they h_____.

3. You can make w_____ of up to $500 a day.

Exercise 1   어휘

다음 글의 밑줄 친 부분 중, 문맥상 낱말의 쓰임이 적절하지 <u>않은</u> 것은?   25007-0133

Must societies 'reproduce' their territory? On the one hand, history is full of territorial ① <u>conflicts</u>; a social system that has its territory removed is doomed to fail. What role does the territory play in the metabolic reproduction of the social system's population? Of key significance is the fact that territory offers the population a legitimate ② <u>physical</u> 'common living space', that is, it serves as a 'repository for humans and their infrastructures'. This 'repository function' provides the opportunity to participate in the ③ <u>consumption</u> of the so-called 'free goods', the ecosystem services within the territory (for example, clean air and water). In most cases, however, it means more than this. The state is in some sense answerable to the 'common good' and thus to ensuring conditions supporting the reproduction of its human subjects. In any case, at least within its territory, it must ensure that their metabolic reproduction is ④ <u>unattainable</u>. The territory is therefore meaningful in ⑤ <u>containing</u> natural resources which economic processes can appropriate. It also provides an outlet for the depositing of waste products from these processes, and it is a source of various non-provisioning ecosystem services.

* metabolic: (신진) 대사의   ** repository: 저장소   *** infrastructure: 하부 구조

Sentence Structure

● The state is in some sense answerable [to the 'common good'] and thus [to ensuring conditions {supporting the reproduction of its human subjects}].

두 개의 [ ]가 and로 연결되어 answerable에 이어진다. 두 번째 [ ] 안의 { }는 conditions를 수식하는 분사구이다.

Word Search   주어진 문장에 들어갈 적절한 영어 단어를 본문에서 찾아 쓰시오. (필요시 형태를 바꿀 것)

1. They have refused to allow the army to be stationed in their t_____.
2. Her passport seemed l_____, but it was found to have been altered.
3. Remember that councils should be a_____ to the people who elect them.

Exercise 2　문장 삽입

글의 흐름으로 보아, 주어진 문장이 들어가기에 가장 적절한 곳은?　　　　25007-0134

> Some people find such screens to be so impersonal that they provoke a suspicious, even aggressive, response.

There are some situations where physical barriers are both sensible and necessary for the protection of the worker. There are classic examples of this within the prison service and of how arrangements can be made for the safety of visitors. (①) But many reception areas for public services have, from time to time, employed security screens in the hope of discouraging violence or abusive behaviour and for the protection of staff. (②) These have not always worked, however, and have sometimes given the message that violent behaviour is being expected. (③) They much prefer to create an open, warm and human environment that respects the individual and encourages them in turn to behave respectfully towards the worker. (④) It is also possible that screens can put confidentiality at risk if people feel that they have to speak more loudly in order to make themselves heard. (⑤) Screens in GP surgeries are good examples of this.

* confidentiality: 비밀　** GP surgery: 지역 보건 진료소

Sentence Structure

● They much prefer to create an open, warm and human environment [that {respects the individual} and {encourages them in turn to behave respectfully towards the worker}].

[ ]는 an open, warm and human environment를 수식하는 관계절이고, 그 안에 두 개의 { }가 and로 연결되어 관계절의 술부를 구성한다.

Word Search　주어진 문장에 들어갈 적절한 영어 단어를 본문에서 찾아 쓰시오. (필요시 형태를 바꿀 것)

1. Rows of identical beds in dull grey rooms make hospitals seem i_____.

2. The officers on patrol became s_____ of the man in a car.

3. If we criticize her, she gets a_____ and starts shouting.

Exercise 3 | 요약문 완성

다음 글의 내용을 한 문장으로 요약하고자 한다. 빈칸 (A), (B)에 들어갈 말로 가장 적절한 것은?   25007-0135

Decoding is a verbal activity that regards its object as a set of signals; each signal carries a specific meaning determined by a code. The object of decoding may be verbal or nonverbal. The nonverbal object, however, becomes in the process of decoding analogous to verbal objects, that is, it is viewed as a set of signals. Decoding an enemy's message begins with the belief that the apparently random signals conceal a coherent meaning. As is the case with explanation, decoding aims to unify disparate facts. While explanation includes the particular covered by a general law, decoding substitutes one element with another in accordance with a rule that may be individual to the particular case: the enemy's second message may or may not be written in the same code as the first. The decoder cannot rely on seemingly similar cases to provide the right code — the apparent similarity may be misleading. By contrast, similar cases must be similarly explained, because explanations assume general, coherent rules common to all relevant cases.

* analogous to: ~과 유사한   ** coherent: 일관된   *** disparate: 이질적인

Decoding is the process of interpreting signals, often requiring ____(A)____ rules to particular cases, in contrast to explanation which relies on consistent rules across ___(B)___ cases.

	(A)		(B)
①	generalized	⋯⋯	random
②	generalized	⋯⋯	unique
③	specific	⋯⋯	accidental
④	specific	⋯⋯	similar
⑤	specific	⋯⋯	complex

Sentence Structure

● Decoding is a verbal activity [that regards its object as a set of signals]; each signal carries a specific meaning [determined by a code].

첫 번째 [ ]는 a verbal activity를 수식하는 관계절이고, 두 번째 [ ]는 a specific meaning을 수식하는 분사구이다.

Word Search | 주어진 문장에 들어갈 적절한 영어 단어를 본문에서 찾아 쓰시오. (필요시 형태를 바꿀 것)

1. Mark could barely c_____ his disappointment.

2. A series of a_____ unconnected events led to her resignation.

3. They have a v_____ agreement with him.

Exercise 4 · 어휘

(A), (B), (C)의 각 네모 안에서 문맥에 맞는 낱말로 가장 적절한 것은?

25007-0136

Carbon dioxide is a powerful regulator of breathing (it acts on a different set of chemoreceptors, found in the brain) and if its concentration in the blood falls, breathing is (A) facilitated / inhibited . It is possible to demonstrate this for yourself. You will find you are able to hold your breath for longer if you breathe very rapidly for a short time beforehand. (Do not do this for more than a minute or you may become dizzy.) The reason is that breath-holding is terminated not by the demand for oxygen but rather by the (B) rising / falling carbon dioxide concentration in the blood. When this reaches a critical level it stimulates you to take a breath. Hyperventilation before holding your breath blows off carbon dioxide from the body and enables a longer period to go by before carbon dioxide builds up to a level sufficient to (C) stimulate / obstruct breathing. The opposing drives from oxygen and carbon dioxide explain why no change in breathing occurs at altitudes of less than 3000 metres.

* chemoreceptor: 화학 수용체 ** hyperventilation: 과호흡 *** altitude: 해발

	(A)	(B)	(C)
①	facilitated	rising	stimulate
②	facilitated	falling	obstruct
③	inhibited	rising	stimulate
④	inhibited	rising	obstruct
⑤	inhibited	falling	stimulate

Sentence Structure

● [The opposing drives from oxygen and carbon dioxide] **explain** [why no change in breathing occurs at altitudes of less than 3000 metres].

첫 번째 [ ]는 문장의 주어이고 이에 이어지는 문장의 술어동사는 explain이다. 두 번째 [ ]는 explain의 목적어 역할을 하는 명사절이다.

Word Search · 주어진 문장에 들어갈 적절한 영어 단어를 본문에서 찾아 쓰시오. (필요시 형태를 바꿀 것)

1. An independent r_____ will be appointed for fair competition.
2. They t_____ his contract in December.
3. The government took steps to s_____ business development.

[5~6] 다음 글을 읽고, 물음에 답하시오.

It is perhaps self-evident that the outworking of policy is seen in the exercise of power if it compels individuals to behave in a way which they would not have chosen to do. Policy can be used to highlight power relationships, but it can also allow for depersonalisation of power and decision-making: a leader could imply that while they don't really agree with a particular course of action, the policy (a) <u>dictates</u> it and therefore it must be done. In the same way policy can act as the source of authority such that hierarchy is unnecessary: a group of individuals may (b) <u>resort</u> to policy to guide their decision-making when there is no established authority figure to set their direction. Ultimately those who define policy exert power over those who (c) <u>follow</u> policy and, in education at least, it is rare that policy is determined collectively: at best, leaders are elected democratically so that they may set policy for the electorate. There is therefore an (d) <u>inevitable</u> reliance on a political elite, and a consequent hierarchy of power. A frequently seen example of policy as power in schools is uniform. It is for the 'governing board' of a school to determine if it has a uniform, and if so what it is, but parents and children have a duty to comply, and school staff have a duty to enforce it. Countless children will have discovered that they cannot appeal to the logic that inadequate uniform does not affect their learning, and nor can they resort to any individual to plead their case, for the power rests within the policy itself: it is (e) <u>insufficient</u> to say that you have to wear school uniforms because that is what the policy says.

* hierarchy: 위계 ** electorate: 유권자 *** plead: 호소하다

5 윗글의 제목으로 가장 적절한 것은?   25007-0137

① Exploring Education Policy: Everyone Is the Victim
② Diversity Education: A Vital Inclusion in Educational Policy
③ Policy as Power: Understanding Its Implications in Education
④ Bridging Cultural Gaps Through Policy Engagement in Education
⑤ Resolving Negative Effects of Policy on Educational Decision-making

6 밑줄 친 (a)~(e) 중에서 문맥상 낱말의 쓰임이 적절하지 <u>않은</u> 것은?   25007-0138

① (a)   ② (b)   ③ (c)   ④ (d)   ⑤ (e)

Sentence Structure

- **It** is perhaps self-evident [that the outworking of policy is seen in the exercise of power {if it compels individuals to behave in a way ⟨which they would not have chosen to do⟩}].
  It은 형식상의 주어이고 [ ]가 내용상의 주어이다. { }는 if가 이끄는 부사절이고, ⟨ ⟩는 a way를 수식하는 관계절이다.

Word Search   주어진 문장에 들어갈 적절한 영어 단어를 본문에서 찾아 쓰시오. (필요시 형태를 바꿀 것)

1. The United Nations has used its a_____ to restore peace in the area.
2. He has e_____ considerable influence on the scientific community.
3. The accident was the i_____ consequence of carelessness.

Exercise 7 어휘

다음 글의 밑줄 친 부분 중, 문맥상 낱말의 쓰임이 적절하지 <u>않은</u> 것은?

25007-0139

Electric scooters are the latest vogue in urban transportation. They wait in clusters on the pavement, ready for hire by anyone with a smartphone and a credit card. Scooters are faster than walking, easier than cycling, and simpler than cars. They turn the urban landscape into a playground. Riding them is a carefree experience — but it is more ① <u>restrained</u> than it might seem. Every journey is tracked from start to finish. No matter how hard the throttle is pressed, the scooters will not go above a particular speed. They ② <u>refuse</u> to leave designated urban areas. And there is no haggling over the fare: an app charges a ③ <u>precise</u> sum depending on the length of the journey. None of this is inherently objectionable. But scooters do offer a helpful example of the paradox of digital technologies: they offer freedom, but only in exchange for some ④ <u>maintenance</u> of control. This is not a paradox that will ever be fully ⑤ <u>resolved</u>. The question will always be whether the balance between freedom and control is struck in the right place.

* **vogue**: 유행 ** **throttle**: (자동차 등의 연료) 조절판 *** **haggle**: 흥정하다

Sentence Structure

● [**No matter how** hard the throttle is pressed], the scooters will not go above a particular speed.

　[ ]는 양보의 부사절이고, No matter how는 However로 바꿔 쓸 수 있으며, '아무리 ~하더라도'의 의미이다.

Word Search 주어진 문장에 들어갈 적절한 영어 단어를 본문에서 찾아 쓰시오. (필요시 형태를 바꿀 것)

1. The city council is proposing a plan for a new u_____ park.

2. The parking lot is clearly d_____ for permit holders only.

3. The child possessed an i_____ curious nature, constantly asking questions.

Exercise 8 문장 삽입

글의 흐름으로 보아, 주어진 문장이 들어가기에 가장 적절한 곳은?

25007-0140

It is more likely that language evolved in several stages in the same way, for instance, as our large brains did or our tool-making ability.

It is possible, in principle, that language suddenly appeared fully formed during human evolution without any gradual or intermediate forms. That notion, a linguistic Big Bang, has been championed by linguists such as Noam Chomsky. (①) But it is extremely improbable, biologically speaking, that such a complex characteristic as our capacity for speech just popped up out of nowhere. (②) This may have been a process similar to the one children go through, or the process may have been entirely different and involved intermediate forms. (③) But some form of evolution must have occurred on the journey from non-speaking ape to speaking humans. (④) At some point there must have been linguistic precursors, simpler forms of language. (⑤) There must also have been a protolanguage, the first one that could be called a language.

* improbable: 가능성이 낮은, 일어날 듯하지 않은 ** precursor: 선행하는 형태
*** protolanguage: 조어(祖語), 공통 기어(基語)

Sentence Structure

- That notion, [a linguistic Big Bang], **has been championed** by linguists [such as Noam Chomsky].
  [ ]는 That notion과 동격 관계를 이루는 명사구이고, has been championed는 술부로서 현재완료형의 수동태이다. 두 번째
  [ ]는 linguists를 수식한다.

Word Search   주어진 문장에 들어갈 적절한 영어 단어를 본문에서 찾아 쓰시오. (필요시 형태를 바꿀 것)

1. Over time, the language has e_____ to include new words and phrases.
2. To understand the poem's full meaning, you need to consider the l_____ devices the author uses.
3. This is an i_____ step before the final approval.

Exercise 9   요약문 완성

다음 글의 내용을 한 문장으로 요약하고자 한다. 빈칸 (A), (B)에 들어갈 말로 가장 적절한 것은?   25007-0141

In prior ages to the Digital Age, the power to create information remained primarily in the hands of designated authorities, be they the King or the *New York Times*. Today however, advances in information technology — specifically, computers and telecommunicators — have resulted in a situation where information can be created by anyone and can flow from *anyone to everyone else*. Take for example these astounding statistics: at the time of writing this paper, the number of websites currently online is over 1.5 billion, the number of active users on some of the biggest social media platforms is over two billion, and the number of Internet users is over four billion. More specific to consumer-created content, 400 hours of videos are uploaded every minute and searches of "how to" videos are growing 70% year over year on popular video sites. These numbers will likely increase by the time this manuscript goes to print. People today are not simply innocent consumers of information, but also providers of information through their own content creation, which is "the material people contribute to the online world." Today, the production and distribution of information is a market in which nearly everyone can, and does, participate.

* astounding: 놀라운

In the Digital Age, advances in information technology have _____(A)_____ the power to create and share content from designated authorities to individuals, resulting in a _____(B)_____ information market where nearly everyone contributes.

(A)	(B)		(A)	(B)
① shifted	······ closed		② shifted	······ participatory
③ retained	······ competitive		④ maintained	······ privatized
⑤ maintained	······ consumer-created			

Sentence Structure

- In prior ages to the Digital Age, the power [to create information] remained primarily in the hands of designated authorities, [be they the King or the *New York Times*].

  첫 번째 [ ]는 the power를 수식하는 to부정사구이다. 두 번째 [ ]는 whether they are the King or the *New York Times*의 의미이다.

Word Search   주어진 문장에 들어갈 적절한 영어 단어를 본문에서 찾아 쓰시오. (필요시 형태를 바꿀 것)

1. The crime rate in this area is below the national s_____.
2. The design was i_____ yet charming, full of simplicity and purity.
3. The d_____ of the new software update will begin next week.

Exercise 10   어휘

다음 글의 밑줄 친 부분 중, 문맥상 낱말의 쓰임이 적절하지 <u>않은</u> 것은?

25007-0142

Scientific research is typically based on a cycle involving several distinct stages. The cycle typically starts with a collection of incidental observations and the formulation of a possible explanation for these observations. This explanation is known as a theory. Based on this theory, we can then make predictions about conditions that we have not yet ① observed. In other words, we can formulate specific hypotheses. Now, we can design an experiment that aims to disprove our theory: when the results of our experiment are consistent with the theory, we have no reason to suspect that anything is ② wrong with it. When the results are inconsistent, however, we will either have to ③ adjust our theory, or to reject it altogether. It is impossible to 'prove' a theory: as long as we find no conflicting evidence, the theory remains plausible. This does not exclude the possibility, however, that there might be future results that will ④ match. Although a theory becomes more plausible, the more results we find that are consistent with it, it remains the case that (at least in principle) a single experiment yielding conflicting results would be ⑤ sufficient to overthrow it.

* plausible: 타당한 것 같은   ** overthrow: 뒤집어엎다

Sentence Structure

- Scientific research is typically based on a cycle [involving several distinct stages].

  [ ]는 a cycle을 수식하는 분사구이다.

Word Search   주어진 문장에 들어갈 적절한 영어 단어를 본문에서 찾아 쓰시오. (필요시 형태를 바꿀 것)

1. Their paths crossed purely by i_____ chance at the coffee shop.
2. O_____ of the team's interactions showed the importance of patience and open-mindedness.
3. I s_____ she knows more than she's telling us.

Exercise 11~12 1지문 2문항

[11~12] 다음 글을 읽고, 물음에 답하시오.

One of the truths about writing in science is that it substantially limits the amount of insincere blather that can emanate from the mind of a student when he/she has absolutely no idea about a correct response. A student who does not have a basic understanding of scientific theories will have immense (a) difficulty in applying them. In responding to an authentic scientific problem through an essay, there is no way to "luck out" on a guess at the correct answer, there is no place to hide. In general, essays that require students to invoke relevant scientific principles and theories and to explore potential applications are more (b) genuine assessments of student knowledge than objective tests of short questions paired with possible answers already furnished.

Another (c) advantage of writing essays is that the teacher can get a sense of where student comprehension is strong and where understanding begins to break down. After all, writing is a form of concretized thought. Thus, when a student's grasp of a concept begins to get off-track, it is only through those moments when thought is made (d) visible that a teacher is able to identify the problem and provide help. Writing provides the opportunity for the kind of intervention that can (e) worsen conceptual misunderstandings that can haunt a student for his entire academic career. Indeed, Heddy & Sinatra (2013) and Francek (2013) have found that one of the most difficult areas of teaching is trying to get students to unlearn a misconception.

* blather: 허튼소리 ** emanate from: ~에서 나오다 *** invoke: (근거로 인물, 이론, 예 등을) 언급하다, 들다

11 윗글의 제목으로 가장 적절한 것은?

25007-0143

① The Benefits of Writing Essays in Science
② Essays in Science: Guesswork over Clarity
③ The Ugly Truth of Academic Scientific Writing
④ Unlearning Misconceptions Through Objective Tests
⑤ Enhancing Recall: How Writing Strengthens Knowledge

12 밑줄 친 (a)~(e) 중에서 문맥상 낱말의 쓰임이 적절하지 <u>않은</u> 것은?

25007-0144

① (a)   ② (b)   ③ (c)   ④ (d)   ⑤ (e)

Sentence Structure

● Indeed, Heddy & Sinatra (2013) and Francek (2013) have found [that **one of the most difficult areas of teaching** is {trying to get students to unlearn a misconception}].

[ ]는 found의 목적어 역할을 하는 명사절이다. 「one of the 최상급＋복수 명사」는 '가장 ~한 것 중 하나'의 의미이다. { }는 is의 주격 보어로 쓰인 명사구이다.

Word Search   주어진 문장에 들어갈 적절한 영어 단어를 본문에서 찾아 쓰시오. (필요시 형태를 바꿀 것)

1. The firefighters battled the i_____ forest fire for days, facing danger.
2. While interesting, the information about the celebrity's vacation was not entirely r_____ to the topic at hand.
3. The teacher's timely i_____ prevented a fight from breaking out between the two students.

01 다음 글의 목적으로 가장 적절한 것은?

25007-0145

Dear Dr. Stevens,

 For the past four years, volunteering at the Langford Science Museum has been a source of pride since my retirement as a university professor. However, recent changes in volunteer schedule management are proving increasingly challenging for me and many others to sustain our commitment. Previously, volunteers were asked to sign up for available slots on the schedule. This worked very well for all of us because it allowed the flexibility to fit our volunteer time with various other demands on our time. In the last six weeks, though, the new manager of volunteer services has begun assigning us to volunteer schedules without consulting us, causing a great deal of upset among the volunteers and resulting in numerous cancellations of shifts. As unpaid volunteers, we value our autonomy and should not be treated like salaried employees whose working hours can be assigned by management. I hope you can address this problem before it becomes too great to solve.

Sincerely,
Kate Greenbaum

* autonomy: 자율성

① 박물관 편의 시설의 확충을 요청하려고
② 박물관 직원의 친절한 안내에 감사하려고
③ 과학 박람회 자원봉사 활동 참여를 권유하려고
④ 자원봉사자 선정 기준에 대해 이의를 제기하려고
⑤ 변경된 자원봉사 일정 관리 방식에 대해 항의하려고

02 다음 글에 드러난 'I'의 심경 변화로 가장 적절한 것은?

25007-0146

A powerful earthquake struck our home at midnight, jolting me awake and nearly throwing me out of my bed. My instant response was to run and get my son, Dustin, who was in the room at the other end of our house. Everything was dark after the power went out. I ran across the house in total darkness, my hands outstretched, blindly feeling for walls and furniture to guide me towards Dustin's bedroom. A violent aftershock knocked me to my knees, but I couldn't stop. I had to get to my son, no matter what. Then, the shaking stopped, and everything became quiet as I peered into my son's bedroom. By now accustomed to the darkness, I could make out that Dustin's bookcases had fallen into the center of the room, narrowly missing his bed. Crawling over the debris, I finally reached Dustin. He was shaking with fear, but was huddling safely under the covers. I hugged Dustin tight and breathed a sigh of relief. In that moment, nothing else mattered but the fact that my son was safe.

* jolt: 갑자기 거칠게 흔들다   ** debris: 잔해   *** huddle: 웅크리고 있다

① surprised → curious
② ashamed → grateful
③ frustrated → confident
④ desperate → relieved
⑤ scared → confused

03 다음 글에서 필자가 주장하는 바로 가장 적절한 것은?

25007-0147

Good intuition is the first requirement for designing meaningful ecological studies. The best way to develop that intuition is by observing organisms in the field. Sadly, few of us "have the time" to just observe nature. Graduate committees and tenure reviewers are not likely to recommend investing precious time in this way. However, observations are absolutely essential for you to generate working hypotheses that are grounded in reality. So, carve out some time to get to know your organisms. If you are too busy with classes and other responsibilities, then reserve two days before you start your experiments to observe your ecological system with no manipulations (or preconceived notions). It's often fun to do this with a lab mate or colleague. The opposite can work well too: consider spending a whole day with no other people or distractions around, just looking at your system.

* tenure: (교수의) 종신(終身) 재직권   ** carve out: ~을 할애하다

① 생태계 현장 연구에 착수하기 전에 가설의 타당성을 사전에 점검하라.
② 생태계 연구에 필요한 영감을 얻기 위해 다양한 자료를 폭넓게 탐독하라.
③ 생물 관찰 여건이 조성되지 않을 경우를 대비하여 대안적인 계획을 마련하라.
④ 관찰 연구를 설계할 때는 연구 대상 생물에게 장기적으로 미칠 영향을 고려하라.
⑤ 생태 연구 설계에 필요한 직관 계발을 위해 현장에서 생태계를 있는 그대로 관찰하라.

04 밑줄 친 this is like saying you should carry an umbrella in Spain because it is going to rain in Vietnam이 다음 글에서 의미하는 바로 가장 적절한 것은?

25007-0148

We are getting better at understanding the influence of humans on the global climate system. As the planet gets warmer, high latitudes will warm faster than the tropics. The Mediterranean will become drier, and the tropics will be wetter. But this is like saying you should carry an umbrella in Spain because it is going to rain in Vietnam. To plan for and adapt effectively to climate change, we need information about the future climate at much finer scales than general circulation models (GCMs) can provide. To decide whether to put up a dike, move some houses, switch crops, or buy insurance, we need data at scales of less than 100 kilometers. One approach is to embed a finer-scale model of a particular area of interest into a larger-scale GCM. No region is isolated from the rest of the planet, so the GCM part of the model can keep track of what is going on globally and exchange information with the finer-scale regional climate model.

* general circulation model: (대기학) 대기 대순환 모델 ** dike: 제방 *** embed: 끼워 넣다

① It is hard to reach a consensus on how to tackle the climate crisis.
② What happens in the climate of one region could affect that of another.
③ Climate actions are sure to fail without the participation of major polluters.
④ We need to stop measuring the effects of climate change by using an outdated scale.
⑤ Relying solely on global-scale predictions ineffectively addresses local climate change issues.

05 다음 글의 요지로 가장 적절한 것은?

25007-0149

The bear not only knows where and when to find food, he also knows when to retreat to his den to ride out a challenging time. Like the bear, we sometimes feel a need to retreat from the world, particularly after periods of stress. When we feel this bear-like urge to carve out restorative time to "hibernate," we should think of it as the sensible impulse of our inner bear. We might consider withdrawing from some social activities to take stock of our lives, start a creative project, plan a trip, or plant seeds of thought that will hopefully spring up and come to fruition in the future. However, we also need to remember that bears come out of their dens once spring arrives. Spending too much time in isolation can deprive us of connection with and inspiration from the outside world. It's best to balance the urge to retreat for restoration with the opportunity to be revitalized by all that the world has to offer.

* den: (야생 동물이 사는) 굴   ** hibernate: 동면하다

① 인생의 겨울을 대비해 충분한 에너지를 비축하는 것이 필요하다.
② 변화에 대해 한결같은 자세로 대처하는 것이 스트레스를 방지한다.
③ 삶을 되돌아보는 시기에 이르러 새로운 인생을 계획하는 것은 어렵다.
④ 인생에서 인내와 노력의 시간이 반드시 결실을 가져오는 것은 아니다.
⑤ 휴식을 통한 재충전과 활력을 주는 사회 활동 참여 사이의 균형이 중요하다.

06 다음 글의 주제로 가장 적절한 것은?

25007-0150

Plants are not set pieces in our human drama, they are characters with distinct experiences and needs. Like us, they are capable of affection, care, and suffering. While we can (and do) dismiss the moral importance of plant experiences, to pretend that they don't exist is a very dangerous form of disregard. Our cold, utilitarian approach to reducing climate change has so far been largely ineffectual. To bring real change, we need to learn to *see* plants as independent, intrinsically valuable, *sentient* beings — not mere tools for our own human flourishing. Our welfare is deeply dependent upon the welfare of plants, and the best way for us to make *sense* of this is by reminding ourselves that they are *persons*. They have just as much of a right to clean water, healthy soil, and a liveable atmosphere as the rest of us, and this needs to be remembered when we consider the ethical dimensions of climate change.

* sentient: 지각력이 있는

① destructive consequences of non-native plants on regional ecosystems
② the reduction of plant species diversity stemming from climate change
③ the ecology of plants that provide a reference point to human ethical life
④ ethical lessons that humans can learn from plants for restoring human nature
⑤ the need to recognize plants as conscious beings in addressing climate change

07 다음 글의 제목으로 가장 적절한 것은?

25007-0151

Whereas contemporary cuisines represent a true blending of influences that developed all over the world, contemporary sports do not. In fact, global sports spread more like the bubonic plague than the burrito. They are global phenomena that emerged almost exclusively from Western civilization — from European nations and European settler societies — and spread to other parts of the world. Take, for instance, soccer (association football). Before European contact, ball games that limited or barred the use of the hands flourished in Mesoamerica. Today, *fútbol* ignites the passions of millions of Mexicans. Estadio Azteca has been filled to overflowing for two World Cup finals (1970 and 1986) and a multitude of other international matches. You might easily conclude that *fútbol* fused Mesoamerican and European sporting pastimes, much like pizza and pasta fused Old World and New World food. It did not.

* bubonic plague: 흑사병(黑死病) ** *fútbol*: football에 해당하는 스페인어의 단어

① New Taste, the Easternization of the West
② Is Europe Truly the Birthplace of Football?
③ The Similarities Between Cuisines and Sports
④ Why Sports Spread More Easily than Cuisines
⑤ Contemporary Sports: Western Roots and Global Expansion

08 다음 도표의 내용과 일치하지 <u>않는</u> 것은?

25007-0152

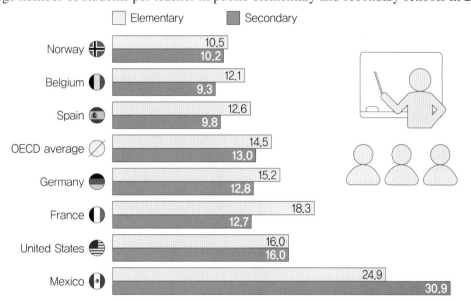

How Student-Teacher Ratios Vary Across the Globe

Average number of students per teacher in public elementary and secondary schools in 2019

Elementary ☐ Secondary ■

	Elementary	Secondary
Norway	10.5	10.2
Belgium	12.1	9.3
Spain	12.6	9.8
OECD average	14.5	13.0
Germany	15.2	12.8
France	18.3	12.7
United States	16.0	16.0
Mexico	24.9	30.9

* in selected OECD countries
Source: OECD

The chart above shows the differences among countries in the average number of students per teacher in public elementary and secondary schools for seven OECD countries in 2019. ① Norway had the lowest average number of students per teacher in elementary schools, while Belgium had the lowest in secondary schools. ② In Spain, the average number of students per teacher was 0.5 higher than in Belgium for both elementary and secondary schools. ③ France exhibited the second largest difference in the average number of students per teacher between elementary and secondary schools, with more than five additional students per teacher in secondary schools. ④ The United States uniquely recorded an equal student-to-teacher ratio between elementary and secondary schools. ⑤ Meanwhile, Mexico was the only country among those listed in the chart where the student-to-teacher ratio in secondary schools was higher than in elementary schools.

09 Helen Moore Sewell에 관한 다음 글의 내용과 일치하지 <u>않는</u> 것은?

25007-0153

Helen Moore Sewell was an American artist and author of children's books who was known for her illustrations. Sewell began drawing at an early age. At the age of 12, she became the youngest person ever to attend the Pratt Institute, which was especially renowned for art and design. She also studied under the Ukrainian American artist Alexander Archipenko, who dramatically influenced her style. Sewell's early work was as both an author and an illustrator. In 1924, she illustrated her first book, Susanne K. Langer's *The Cruise of the Little Dipper, and Other Fairy Tales*. She illustrated her own book, *ABC for Everyday*, in 1930 and a year later collaborated with her younger sister on *Building a House in Sweden*. She also illustrated classic works, including those by American poet Emily Dickinson and British authors like Jane Austen. During her career, Sewell illustrated more than 50 books. She won a Caldecott Honor in 1955 for her illustrations in *The Thanksgiving Story* by American author Alice Dalgliesh.

* renowned: 유명한, 명성 있는

① Pratt Institute를 다닌 사람 중 역대 최연소자였다.
② 우크라이나계 미국인 화가 Alexander Archipenko 밑에서 공부했다.
③ 1930년에 자신의 여동생과 함께 *Building a House in Sweden*을 공동으로 작업했다.
④ 미국 시인이나 영국 작가들의 고전 작품에 삽화를 그렸다.
⑤ 경력 기간 동안 50권이 넘는 책에 삽화를 그렸다.

10 Fun Unplugged Game Night에 관한 다음 안내문의 내용과 일치하지 <u>않는</u> 것은?

25007-0154

Fun Unplugged Game Night

The Town of Windsor Youth Service Center presents Fun Unplugged Game Night. Come out to play board games, card games, and dice games with your friends and family members!

When: Friday, April 18 from 6:00 P.M. to 8:00 P.M.
Where: Windsor Youth Service Center
Who: Open to all Windsor residents! Children under 12 must be accompanied by an adult.
Details:
- Free admission. (Drinks and snacks are available at no cost.)
- Games are provided, but you are also welcome to bring your own classic or non-electronic games.
- Seating is limited to 100 people.
- Pre-registration is required as we cannot accommodate walk-ins.
- To reserve your seat, contact Sarah Bunyan at 860-627-1482.

For more information, please email us at unpluggn@townwindsorgu.co.uk.

---

* dice: 주사위

① 금요일 저녁 2시간 동안 진행된다.
② 12세 미만은 반드시 성인과 동행해야 한다.
③ 음료와 간식이 무료로 제공된다.
④ 자신이 가지고 있는 고전 또는 비전자식 게임을 가져와도 된다.
⑤ 사전 등록 없이 현장에서 바로 참여가 가능하다.

11 GivingFriday Event에 관한 다음 안내문의 내용과 일치하는 것은? 25007-0155

GivingFriday Event

GivingFriday is an annual global event dedicated to promoting the spirit of giving. This event takes place every year on the Friday after American Thanksgiving. It lasts for 24 hours starting at midnight local time. It encourages nonprofit organizations to engage in activities that foster sharing and a sense of community.

Registration:
- There is no formal registration process to participate in this event.
- Any nonprofit organization is welcome to join.

Donations:
- The GivingFriday website does not accept or distribute donations directly.
- Nonprofit organizations must accept donations only through their own websites.

Cost: Participation in the event is free.

Ways to Participate: Organizations can join by hosting food drives, recruiting volunteers, or sharing their actions on social media using the hashtag #GivingFriday.

※ Free resources are available on the GivingFriday website.

① 미국 추수감사절 바로 직전 금요일에 진행된다.
② 현지 시간으로 정오부터 24시간 동안 진행된다.
③ 공식적인 등록 절차가 필요하다.
④ 기부하는 금요일 웹사이트를 통해 직접 기부금을 받거나 배분할 수 있다.
⑤ 자원봉사자를 모집하는 방식으로 참여할 수 있다.

12 다음 글의 밑줄 친 부분 중, 어법상 틀린 것은?

25007-0156

While new media allow us to do new things, make new kinds of meanings, and think, relate to others and enact our own identities in new ways, they also invariably introduce limitations on what we can do and mean, how we can think and relate, and who we can 'be' when we are using ① them. Television news, for example, ② allows for a vivid and dramatic presentation of a story, but may be less suitable than a newspaper or magazine for lengthy and probing analysis. Social networking sites make it easier for us to stay ③ connected to our friends, but make it more difficult to maintain our privacy (especially from advertisers). Caller identification, which is standard on most mobile phones, makes it easier for us to screen our calls, but it also makes it easier for calls that we make ④ is screened by others. Often the constraints of new technologies are less visible to us than their affordances. We tend to be so focused on the new things we *can* do with a new tool ⑤ that we don't pay attention to the things we *cannot* do with it.

* probing: 면밀히 살피는 ** affordance: 행동 유도성

13 다음 글의 밑줄 친 부분 중, 문맥상 낱말의 쓰임이 적절하지 <u>않은</u> 것은?

25007-0157

　　When given many chances, chance will tend to distribute random differences fairly equally. However, when given few chances, it may distribute random differences very unequally. Thus, if you assigned each individual to a group by flipping a coin and you had many participants, chance would do a ① <u>good</u> job of making your groups equivalent. Conversely, if you had few participants, chance would probably do a poor job of ② <u>balancing</u> the effects of individual differences between groups. Indeed, with too few participants, chance has no chance. For example, if you had four people in your study and only one of those was violent, flipping a coin could not give you ③ <u>equal</u> groups. Even if you had eight participants, four of whom were violent, flipping a coin might result in all four violent individuals ending up in the same group. Why? Because, in the short run, chance can be unpredictable. For instance, it is not that ④ <u>usual</u> to get four "heads" in a row. To appreciate that chance can be unpredictable in the short run but ⑤ <u>dependable</u> in the long run, realize that although a casino may lose several bets in a row, the casino always wins in the end.

* equivalent: 동등한

14 다음 빈칸에 들어갈 말로 가장 적절한 것은?

25007-0158

_____ of predator recognition makes functional sense because individuals that must experience predators for themselves to learn they are dangerous may not survive those experiences. The best-analyzed example involves monkeys' fear of snakes. Monkeys reared in captivity do not exhibit fear the first time they encounter live or toy snakes. If they watch another monkey behaving fearfully toward a snake, they later do the same themselves. During the learning trial the naive observer exhibits behavior like the model's (in this case responses such as withdrawal, vocalization, and piloerection). If naive monkeys observe a model behaving fearfully toward a snake and neutrally toward another object like a flower, they acquire the same discrimination. For example, if they are later offered raisins that are out of reach beyond a flower or a snake, they reach quickly over the flower but refuse to reach over the snake.

* piloerection: 털 세우기 ** raisin: 건포도

① The genetic basis
② Social transmission
③ Misleading guidance
④ Immediate comprehension
⑤ The unconscious inhibition

15

다음 빈칸에 들어갈 말로 가장 적절한 것은?

25007-0159

In the 1880s, a German psychologist named Hermann Ebbinghaus shut himself up in a room in Paris to test how memory works. He forced himself to learn, review, and recall nonsense words on a specific, timed schedule. What Ebbinghaus discovered was that the rate of forgetting was predictable. He discovered a pattern of exactly how long it took to forget. If he reminded himself of one of his nonsense words just before he knew he was about to forget it — but no sooner — he could save himself hours of studying but still recall the information correctly. The trick was knowing _____. Ebbinghaus's memorization technique became known as spaced repetition. Essentially, it was the most highly specific, scientifically based study schedule you could dream of. Over a hundred years later, specially designed computer programs made following a modified version of Ebbinghaus's schedules feasible.

* **feasible:** 실현 가능한

① how to make connections
② how he could ask meaningful questions
③ what the big picture was, not the details
④ which sources of motivation to rely on
⑤ when he was about to forget it

16 다음 빈칸에 들어갈 말로 가장 적절한 것은?

25007-0160

Linguist Philip Lieberman argues that human ancestors (e.g., *Homo erectus*) had the ability to speak, although their speech would not have been as refined as modern humans' speech. This conclusion is based on reasoning about why the human vocal tract has the shape it does. Lieberman notes that to produce vowel sounds such as /i/ (as in *meet*) and /u/ (as in *you*), the space above the larynx in the throat has to be about the same length as the horizontal space between the top of the throat and the mouth opening. For natural selection to produce and maintain this arrangement, Lieberman argues, some basic speech abilities must have been present beforehand. Natural selection could then have favored individuals who had physical characteristics that _____. Unless some basic speech abilities were present prior to the appearance of *Homo sapiens*, a lowered larynx, and the accompanying ability to produce more vowel sounds, would have to be the result of a massive and incredibly lucky mutation, rather than gradual evolution by natural selection.

* vocal tract: 성도(구강, 비강 등을 포함하여 말을 할 때 공기가 통과하는 후두 위의 통로)

** larynx: 후두   *** mutation: 돌연변이

① allowed them to produce a wider range of vowel sounds
② restricted the use of vowels to refine their speech abilities
③ improved their ability to reproduce and pass on their genes
④ enhanced their ability to communicate nonverbally for cooperation
⑤ made them more resistant to mutations affecting their speech abilities

17

다음 빈칸에 들어갈 말로 가장 적절한 것은?

25007-0161

What is the meaning of a concept and how does it contribute to the meaning of a sentence? Philosophers have been particularly vexed by this question and have developed a range of possible answers to it. On the one hand, the meaning of a concept seems to derive from the meaning of other concepts, as when a child is told the meaning of *sprint* by saying it is a kind of fast running. On the other hand, the meaning of a concept is connected to observations of things in the world, as when the child actually sees someone sprinting. A concept's meaning is normally not given by definition in terms of other concepts, since successful exact definitions are rare. Nor is meaning exhausted by a set of examples, as if one identified the concept of *dog* with a set of dogs. A theory of meaning of the concepts must therefore _____. Both aspects are necessary in order for us to understand how concepts underlie our ability to use language.

* vex: 고충을 느끼게 하다

① reflect both the linguistic ability of the language users and their cognitive process
② include an account of how concepts are related both to each other and to the world
③ account for the emotional part of the concepts as well as their practical applications
④ cover not only general terms but also diverse linguistic expressions from various cultures
⑤ address the inherent vagueness of concepts and the role of the interpretation by individuals

18 다음 글에서 전체 흐름과 관계 <u>없는</u> 문장은?

25007-0162

In local communities where they know each other well, speakers and listeners are able, for the most part, to draw on knowledge of overlapping language habits to converse or argue about moral and political issues. ① This may still be the case, to some extent, when communities of speakers who engage regularly with one another in practical activities do not all speak the same languages, or speak them equally fluently. ② Sometimes, however, potential parties to a verbal exchange find themselves sharing little more than physical proximity to one another. ③ The fact that physical contact has such strong cultural meanings shows that it's a vital element of non-verbal communication around the world. ④ Such situations arise when members of communities with radically different language traditions and no history of previous contact with one another come face to face and are forced to communicate. ⑤ There is no way to predict the outcome of such enforced contact on either speech community, yet from these new shared experiences, new forms of practice, including a new form of language — pidgin — may develop.

* proximity: 근접성 ** radically: 근본적으로 *** pidgin: 피진어(의사소통되지 않는 사람들 사이에 형성된 언어)

19 주어진 글 다음에 이어질 글의 순서로 가장 적절한 것은?   25007-0163

Machines have aided the scientific process for decades. Will they be able to take the next step, and help us automatically identify promising new discoveries and technologies?

(A) But, humans are still responsible for forming hypotheses, designing experiments, and drawing conclusions. What if a machine could be responsible for the entire scientific process — formulating a hypothesis, designing and running the experiment, analyzing data, and deciding which experiment to run next — all without human intervention?

(B) If so, it could drastically accelerate the advancement of science. Indeed, scientists have been relying on robot-driven laboratory instruments to screen for drugs and to sequence genomes.

(C) The idea may sound like a plot from a futuristic sci-fi novel, but that sci-fi scenario has, in fact, already happened. Indeed, back in 2009, a robotic system made a new scientific discovery with virtually no human intellectual input.

* genome: 게놈(세포나 생명체의 유전자 총체)

① (A) − (C) − (B)   ② (B) − (A) − (C)   ③ (B) − (C) − (A)
④ (C) − (A) − (B)   ⑤ (C) − (B) − (A)

20 주어진 글 다음에 이어질 글의 순서로 가장 적절한 것은?

25007-0164

Solutions we choose to solve environmental problems depend upon how we value people and the environment.

(A) This choice will reduce damage from flooding while providing habitat for a variety of animals including raccoons, foxes, beavers, and muskrats that use the stream environment; resident and migratory birds that nest, feed, and rest close to a river; and a variety of fish that live in the river system. We will also be more comfortable when interacting with the river. That is why river parks are so popular.

(B) For example, if we believe that human population growth is a problem, then conscious decisions to reduce human population growth reflect a value decision that we as a society choose to endorse and implement. As another example, consider flooding of small urban streams. Flooding is a hazard experienced by many communities.

(C) The study of rivers and their natural processes leads to a number of potential solutions for a given flood hazard. We may choose to place the stream in a concrete box — a remedy that can significantly reduce the flood hazard. Alternatively, we may choose to restore our urban streams and their floodplains, the flat land adjacent to the river that periodically floods, as greenbelts.

* muskrat: 사향쥐 ** endorse: 승인하다 *** adjacent to: ~에 인접한

① (A) - (C) - (B)　　　② (B) - (A) - (C)　　　③ (B) - (C) - (A)
④ (C) - (A) - (B)　　　⑤ (C) - (B) - (A)

21 글의 흐름으로 보아, 주어진 문장이 들어가기에 가장 적절한 곳은?

25007-0165

> But not all exercise in the ancient world was combat related.

The generalization, that adult exercise is modern, is kind of obvious. Early farmers had to toil as hard as if not harder than hunter-gatherers, and for the last few thousand years farmers primarily exercised, often through sports, to prepare for fighting. (①) Ancient texts like *The Iliad*, paintings from pharaonic Egypt, and Mesopotamian carvings testify that sports like wrestling, sprinting, and javelin throwing helped would-be warriors keep fit and hone combat skills. (②) If you were wealthy enough to attend one of the great Athenian schools of philosophy, you would have been advised to exercise as part of your physical education. (③) Philosophers like Plato, Socrates, and Zeno of Citium preached that to live the best possible life, one should exercise not only one's mind but also one's body. (④) This idea is not just Western. (⑤) Confucius and other prominent Chinese philosophers also taught that exercise was equally essential for physical and mental health and encouraged regular gymnastics and martial arts.

* toil: 힘들게 일하다　** javelin: 창　*** hone: 연마하다

22 글의 흐름으로 보아, 주어진 문장이 들어가기에 가장 적절한 곳은?

25007-0166

In much the same way, some scholars argue, the future somehow "grows" out of conditions in the past and the present.

There is considerable debate about whether or not the future really exists and, if it does, where one could find it. (①) Let us leave aside the realm of science fiction, where time travelers are able actually to experience the future "before it happens," and concentrate on those visions in which the future is somehow knowable because it is connected to the past and the present. (②) Some scholars see time in much the same way as biologists see living organisms. (③) If biologists know the genetic material of an organism and the conditions in which it lives (its nutrition, for example), and they know the laws that govern its development, they can predict how the mature organism will look and behave. (④) There is a direct and almost physical connection between past and future. (⑤) If one perceives the present and the past correctly, and if one also knows the laws that govern development or growth, then one can forecast fairly accurately what the future will be like.

* realm: 영역

23 다음 글의 내용을 한 문장으로 요약하고자 한다. 빈칸 (A), (B)에 들어갈 말로 가장 적절한 것은?　25007-0167

Emotions play powerful roles in health-related behaviors, motivating both current behavior and efforts to change future behavior, altering the way we process health-related information and shaping health-related judgments and decisions. After many years in which theoretical models of health behavior — and intervention efforts based on these models — emphasized knowledge, beliefs, attitudes, and self-efficacy, emotions are increasingly recognized as playing a crucial role. However, attempts to add an emotional component to behavior change interventions have focused almost exclusively on fear. Public service announcements (PSAs) promoting behavior change (e.g., smoking cessation, healthy diet) routinely present frightening facts and images in an effort to scare the viewer into adopting a healthier lifestyle. Although fear appeals are highly memorable, they are effective in promoting behavior change only if they also boost self-efficacy so viewers believe they are able to do whatever is needed to avoid the dangerous outcome; otherwise, they simply tune out the message. Because a typical viewer has already tried several times to make the change in question and failed, this is a substantial challenge.

* self-efficacy: 자기 효능감　** cessation: 중지

Emotions have played a vital role in ___(A)___ health-related behaviors, but ___(B)___ fear may not be effective without increasing self-efficacy.

	(A)		(B)
①	steering	······	evoking
②	steering	······	neutralizing
③	impacting	······	conquering
④	restraining	······	diminishing
⑤	restraining	······	addressing

[24~25] 다음 글을 읽고, 물음에 답하시오.

Education has always emphasized rules and logical thinking. Step-by-step learning proceeds in a predetermined direction (vertical thinking). Researchers have viewed creativity as a mysterious ability that can be fostered but not taught. Psychological studies have found that the natural creativity of children often declines after a couple of years in school.

According to a psychologist Edward de Bono, creativity requires the (a) restructuring of deep-rooted patterns. The ability, whether innate or developed, to go beyond the limitations of previous patterns is integral to that process. Lateral thinking is needed, the fundamental principle of which is that any particular point of view is but one of many possibilities. There are just as many descriptions of a phenomenon as there are perspectives. Lateral thinking seeks (b) alternative patterns instead of simply building on existing models.

Vertical thinking is a gradually increasing process, whereas lateral thinking permits significant leaps. The individual steps involved in lateral thinking need not be correct as long as the ultimate conclusion solves the problem. Instead of constantly evaluating and accepting that which appears to be correct at the moment, an assessment is (c) postponed. Once a certain point has been reached, it is often possible to retrace a logical path to the starting point. Once that has been accomplished, the direction and order of the various steps are (d) likely to matter very much. Lateral thinking is like building a vault or a bridge with scaffolding that is torn down once construction has been completed. The various parts of the bridge do not have to stand on their own during each phase of the project. But once the keystone has been laid, the entire structure must be (e) self-supporting.

* lateral: 수평적인, 측면의 ** vault: 아치형 천장 *** scaffolding: (공사장의) 비계, 발판

24 윗글의 제목으로 가장 적절한 것은?   25007-0168

① Embracing Chaos: The Path to Clearer Thinking
② Unveiling the Illusions of Creative Problem-solving
③ How to Improve Your Lateral Thinking Skills for Work
④ From Lateral to Vertical: Evolving Problem-solving Strategies
⑤ Lateral Thinking: Unlocking Creativity Through Significant Leaps

25 밑줄 친 (a)~(e) 중에서 문맥상 낱말의 쓰임이 적절하지 <u>않은</u> 것은?   25007-0169

① (a)   ② (b)   ③ (c)   ④ (d)   ⑤ (e)

[26~28] 다음 글을 읽고, 물음에 답하시오.

(A)

My son, David, participated in Scouts when he was younger and learned everything from how to build a fire to how to help those in our community. To say that I'm proud of him would be an understatement, but there was one specific time I remember that caused my chest to swell with pride. We had an elderly neighbor next door, Albert, who had been alone for several years. His wife had passed away, and (a) he was confined to a wheelchair.

(B)

I smiled at David's sense of duty and kindness. I watched from the window as David gathered the branches into a pile and cleaned up the mess on the lawn. Then, (b) he knocked on Albert's door. I couldn't hear what he said, but I watched in silence as he ran back to our house, grabbed two plates and the stack of pancakes that was waiting for him, and returned to Albert's. The two of them sat on the porch and chatted as they shared breakfast. After that, they became good friends.

* porch: (건물 입구에 지붕이 얹혀 있는) 현관

(C)

Some years later, Albert passed away. David was heartbroken and emotional over the loss of his friend. Then one day, Albert's lawyer visited us and handed us an envelope. In it, there was a large sum of money in the form of a check. Albert had saved money for David's first year of college! And there was a letter — the last letter (c) he wrote to David. "The world is a better place because you're in it. Keep spreading kindness, and continue to do good. We need people like you now more than ever. Thank you for being a friend to a lonely old man."

(D)

One evening, there was an awful thunderstorm. When we woke up the next morning, David noticed that Albert's house had been hit particularly hard. Branches were scattered across the grass, and things that had been on the porch were scattered all over (d) his front lawn. "Mom, I've got to do something to help Mr. Albert," David said as I made pancakes. "I think I'll wait to have breakfast until after I'm finished helping (e) him." And he ran into the garage to get his tools.

26 주어진 글 (A)에 이어질 내용을 순서에 맞게 배열한 것으로 가장 적절한 것은?   25007-0170

① (B) − (D) − (C)   ② (C) − (B) − (D)   ③ (C) − (D) − (B)
④ (D) − (B) − (C)   ⑤ (D) − (C) − (B)

27 밑줄 친 (a)~(e) 중에서 가리키는 대상이 나머지 넷과 다른 것은?   25007-0171

① (a)   ② (b)   ③ (c)   ④ (d)   ⑤ (e)

28 윗글에 관한 내용으로 적절하지 않은 것은?   25007-0172

① Albert는 아내와 사별하고 혼자 살고 있었다.
② I는 David가 나뭇가지를 모아서 쌓는 것을 창문으로 지켜보았다.
③ David는 집에 돌아와 팬케이크를 먹은 후에 다시 Albert의 집으로 갔다.
④ Albert의 변호사가 건넨 봉투 안에는 거액의 수표가 들어 있었다.
⑤ Albert의 집은 특히 심하게 뇌우의 타격을 입었다.

01 다음 글의 목적으로 가장 적절한 것은?　　　　　　　　　　　　　　　25007-0173

Dear Ruth Allen,

　I hope this message finds you well. My name is Orville Rivera and I recently joined the team to work on various corporate strategy initiatives. One of my current responsibilities is to lead the quality systems project. As I embark on this task, I'm in the initial stages of gathering information about the existing systems and understanding the requirements from the users' perspective. Given your role and expertise, I believe your insights would be invaluable to ensure the success of this endeavor. I would greatly appreciate the opportunity to discuss the project with you in more detail, including current workflows, user expectations, and any other insights you may have. Would it be possible for us to schedule a meeting later this week at your convenience? Thank you for considering my request, Ruth.

Best regards,
Orville Rivera

* initiative: (새로운 중요) 기획[계획]　** embark on: ~을 시작하다

① 새롭게 바뀐 기업 전략을 소개하려고
② 시스템 관리 분야 전문가를 추천하려고
③ 전략 회의 날짜가 변경되었음을 알리려고
④ 프로젝트 신청을 위한 개인 정보 제공 동의를 구하려고
⑤ 프로젝트를 함께 논의하기 위해 만나 줄 것을 요청하려고

02 다음 글에 드러난 'I'의 심경 변화로 가장 적절한 것은?

25007-0174

The day came for the operation, and we arrived at the hospital early in the morning. Sal was taken into surgery, and I went down to the chapel to pray. Tears were streaming down my face. I prayed like I had never prayed before. I just wanted to see this man whom I loved so much come out of the surgery and be free of this threatening disease. I went back up to the waiting room and sat there with our daughter for what seemed like an eternity. The doctor finally came into the room. All I could see was his eyes. He said that the surgery went well, and they removed any possible lingering cancer cells. I started to cry again, but with joy this time that this man I loved would be by my side again, healthy and smiling and looking forward to all the beautiful things that life had in store for us.

* chapel: 예배당 ** eternity: 영원 *** lingering: 계속 남아 있는

① bored → surprised
② anxious → relieved
③ doubtful → confident
④ grateful → frightened
⑤ thrilled → disappointed

03 다음 글에서 필자가 주장하는 바로 가장 적절한 것은?

25007-0175

Avoiding scientific jargon is not as hard as it seems, as articles written for the public, for government and even industry usually focus on the application of the science, not on the science itself. It is nearly always possible to describe the application of science in plain language. Nevertheless, scientists sometimes complain that the translation of science into plain language 'devalues' it or 'dumbs it down'. However, if the use of scientific terminology will only cause the audience to misunderstand — or, worse, completely misinterpret what is being said — then it makes no sense to use it, as the result will only be confusion. Scientists should never expect people outside their discipline to understand the exact meaning they ascribe to a specialised term — even an apparently simple one like 'model'. Every effort should be made to re-phrase the language so that it has meaning to the audience. This sometimes takes more time and effort than some researchers can spare, and is the reason for the growing value of the skilled communicator as a messenger and interpreter between science and society.

* jargon: (특정 집단 내의) 전문어 ** dumb ~ down: ~을 지나치게 단순화하다 *** terminology: 용어

① 과학자는 자신의 연구에서 각 개념을 정확하게 정의해야 한다.
② 새로운 과학 이론을 제시할 때는 충분한 근거를 제시해야 한다.
③ 어려운 과학 용어를 알기 쉽게 정비한 용어집을 만들어야 한다.
④ 과학은 일반 독자들이 이해할 수 있도록 쉬운 언어로 설명되어야 한다.
⑤ 논문은 각 전공 분야별로 정해진 엄격한 형식에 따라 작성되어야 한다.

25007-0176

04 밑줄 친 failing to dip into the thought processes that go through the black boxes가 다음 글에서 의미하는 바로 가장 적절한 것은?

Creativity is significantly different from pursuit of excellence. When limits to excellence begin to operate, only creativity and innovation can take companies to different trajectories. In a competitive and uncertain world, companies that are creative would most certainly have advantage over companies that are not. Nevertheless, creativity figures low in corporate agenda as a work or organizational ethic that needs to be institutionalized. For most organizations, only the visible working of an employee mind to follow the prescribed practices is relevant. The employee mind is considered similar to a black box ("one can never fathom what happens inside a human brain, nor is it necessary")! For most organizations, their environmental view rarely goes beyond an explored business canvas into unexplored white spaces ("one has to have deep pockets to lose money on white spaces")! By failing to dip into the thought processes that go through the black boxes, companies miss out on the opportunities of white-space growth that lies ahead of them.

* trajectory: 궤도 ** agenda: 의제 *** fathom: 헤아리다

① carelessly overlooking the prescribed practices of workplace safety
② consistently not hiring employees who meet the essential qualifications
③ discouraging the organizational environment of openness and transparency
④ not properly equipping systems to effectively evaluate employees' creativity
⑤ not adequately exploring creative thinking processes within employees' minds

05 다음 글의 요지로 가장 적절한 것은?

25007-0177

Life is perhaps 5% (or less) what happens to us and 95% (or more) how we interpret and respond to what is happening. It is our interpretation of external events that gives them meaning. Situations and circumstances are neutral until we decide what they mean whether positive, negative, or insignificant. Individuals in the excellence category refuse to accept the limiting biases and judgments of others. They make their own interpretations, and decide accordingly. There is an often-noted parable about a shoe factory that sends two marketing scouts to a distant country to study the possibilities of expanding the shoe business. The first scout sends back a message that the situation is hopeless, no one wears shoes, and there is no market. The other scout responds enthusiastically that this is an outstanding business opportunity since no one has shoes. In any situation, the meaning we give to circumstances determines what we see and how we think and act.

* parable: 비유

① 과거 경험은 상황에 대한 해석 방식에 큰 영향을 준다.
② 상황에 대한 해석이 개인의 생각과 결정에 영향을 미친다.
③ 문제를 정확히 인식하는 것이 문제 해결 과정의 첫 단계이다.
④ 부정적 상황을 긍정적으로 바꾸는 경험을 통해 개인은 발전한다.
⑤ 문화에 대한 편견은 사업상 중요한 결정에 부정적인 영향을 준다.

06 다음 글의 주제로 가장 적절한 것은?   25007-0178

One area of research that's become increasingly popular among food scientists is the arena of phantom aromas. Once in a while you might imagine a smell that isn't actually around you. Aroma is the sense that is most strongly associated with memory, and this relationship can be manipulated during cooking to trick our brain to reconstruct what it perceives the food should taste like. Ham is a salt-cured meat, and we've learned to associate the aroma of ham with saltiness. In a taste experiment, the presence of "ham aroma" in food samples convinced a group of people that their food tasted saltier. You can play with this yourself. If you repeatedly use aromatic spices such as cinnamon, rosewater, and vanilla in desserts, you start to associate those aromas with sweetness. The next time you make a dessert such as Polenta Kheer, cut back on the amount of added sweetener and add a bit more of the "sweeter" aromatic spice. Your dinner companions will probably find that the dessert tastes very sweet.

* **phantom**: 착각에 의한, 환상의   ** **salt-cured**: 소금 절임 된   *** **Polenta Kheer**: 옥수수 키르(인도 지역의 디저트)

① the use of popular tastes and aromas in the field of marketing
② the cause of the sensitivity of some individuals to specific flavors
③ the manipulation of aroma-memory associations in food perception
④ the scientific methods applied to boost the nutritional value in daily food
⑤ the technique used to create more flavorful desserts with artificial flavorings

07 다음 글의 제목으로 가장 적절한 것은?

25007-0179

Overoptimism may be partially explained by "representativeness bias." Representativeness bias is the result of a heuristic whereby individuals generalize from particular "representative" cases, even when better general statistical information exists. It is easy to see why the representativeness heuristic might be useful in everyday reasoning — specific information is often more accurate or more useful than general information, and favoring it may represent a useful cognitive shortcut when assessing probabilities. However, when this assumption fails, the representativeness heuristic can lead to bizarre and erroneous judgments. Overoptimism may result from representativeness when, for example, individuals infer from knowledge about a specific car accident involving a bad driver that bad driving correlates with accidents more highly than it does in fact. In this case, even if they believe they are average drivers, they will systematically underestimate the probability that they will be involved in an accident, because the representativeness heuristic will have generated an underestimation of the probability that average drivers are involved in accidents.

* representativeness: 대표성 ** heuristic: 경험칙, 어림짐작 *** bizarre: 기이한

① How Mental Shortcuts Help Us Make Quick Decisions
② The Power of a Positive Attitude to Drive Forward Progress
③ Obstacles as Stepping Stones: Turning Challenges into Victories
④ Thinking Mistakes: How Mental Shortcuts Cause Overoptimism
⑤ Wearing Many Hats: How Managing Multiple Tasks Can Boost Productivity

08 다음 도표의 내용과 일치하지 <u>않는</u> 것은?   25007-0180

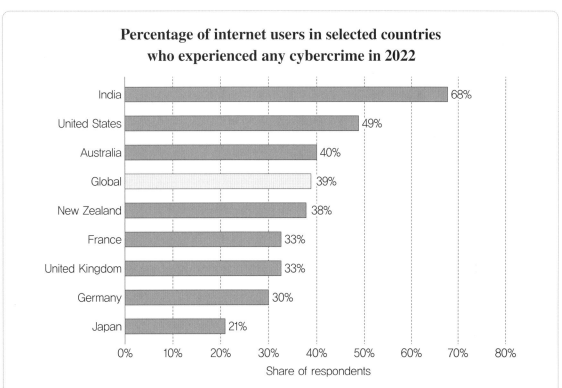

**Percentage of internet users in selected countries
who experienced any cybercrime in 2022**

India — 68%
United States — 49%
Australia — 40%
Global — 39%
New Zealand — 38%
France — 33%
United Kingdom — 33%
Germany — 30%
Japan — 21%

0%   10%   20%   30%   40%   50%   60%   70%   80%

Share of respondents

   The graph above shows the percentage of internet users who experienced any cybercrime in a selection of countries in 2022, including a reference to the global average. ① Among the surveyed countries, India reported the highest percentage of internet users experiencing cybercrime, followed by the United States, where nearly half of the internet users encountered such incidents. ② Conversely, Japan reported the lowest rate of cybercrime encounters among its internet users, and Germany had the second lowest rate, with more than a quarter of respondents acknowledging experiencing cybercriminal activities. ③ Australia recorded the third highest percentage of internet users among the selected countries, who experienced any cybercrime, at 1 percentage point above the global average. ④ In the three countries above the global average, at least four out of ten respondents said that they experienced any cybercrime. ⑤ In all of the countries where the proportion of respondents reporting cybercrime was below the global average, at least more than one-third of their internet users still experienced some form of cybercrime.

09 Francis William Aston에 관한 다음 글의 내용과 일치하지 <u>않는</u> 것은?

25007-0181

Francis William Aston, a British chemist and physicist, was born in Harborne, England, the son of a metal merchant. He was educated at Mason College, the forerunner of Birmingham University, where he studied chemistry. From 1898 until 1900, he did research under P. F. Frankland on optical rotation. He left Birmingham in 1900 to work in a Wolverhampton brewery for three years. During this time he continued with scientific research in a home laboratory, where he worked on the production of vacua for x-ray discharge tubes. This work came to the notice of J. H. Poynting of the University of Birmingham, who invited Aston to work with him. He remained at Birmingham until 1910, when he moved to Cambridge as research assistant to J. J. Thomson. He became a research fellow at Cambridge in 1920 and stayed there for the rest of his life, apart from the war years spent at the Royal Aircraft Establishment, Farnborough. Aston's main work, for which he received the Nobel Prize for chemistry in 1922, was on the design and use of the mass spectrograph, which was used to clear up several outstanding problems and became one of the basic tools of the new atomic physics.

* forerunner: 전신(前身) ** brewery: 맥주 공장 *** mass spectrograph: 질량 분석기

① Mason College에서 화학을 공부했다.
② J. H. Poynting으로부터 함께 연구하자는 요청을 받았다.
③ 1910년에 Cambridge로 가서 J. J. Thomson의 연구 조수로 일했다.
④ 전쟁 기간 내내 Cambridge를 떠나지 않았다.
⑤ 1922년 노벨 화학상을 수상했다.

10 Caremaxy Shoe Dryer에 관한 다음 안내문의 내용과 일치하지 <u>않는</u> 것은?

25007-0182

Caremaxy Shoe Dryer

This device removes moisture by drying your shoes and eliminates bacteria using Active Oxygen (Ozone) technology.

Operation:
1. Place your shoes over the holders.
2. Press the power button to start. If necessary, you can adjust the timer, which is initially set to 120 minutes.
3. Press the Dry button. The drying light will come on, starting the drying process. When the process has finished, the dryer will stop automatically.

Using the Ozone Function:
- Press the Ozone button to switch it on. It automatically turns off after 15 minutes.
- Press the Ozone button again to restart the ozone function if necessary.
- Note that when the ozone function is on, the unit will not generate heat. Heat is only produced after the ozone function is turned off.

Safety Instructions:
- Excessive ozone can lead to poor air quality, causing headaches.
- Keep the dryer away from water to prevent damage.

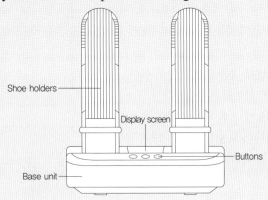

① 타이머를 조정할 수 있으며 최초 120분으로 설정되어 있다.
② 건조 과정이 끝나면 건조기가 자동으로 정지된다.
③ 오존 기능은 15분 후 자동으로 꺼진다.
④ 오존 기능이 작동 중일 때 기기에서 열이 발생한다.
⑤ 과도한 오존은 공기질 저하로 이어져 두통을 유발할 수 있다.

11 Woodlaw School Book Fair에 관한 다음 안내문의 내용과 일치하는 것은?   25007-0183

Woodlaw School Book Fair

The Woodlaw School Book Fair provides an opportunity for students of all ages to enrich their home libraries and cultivate a love for reading.

What to expect:
- The book fair will take place from March 10 to March 14 in the Woodlaw library.
- Students will have the chance to explore and purchase books during designated class times.
- Additionally, there will be a Family Night on Wednesday, March 12, from 4:45 pm to 7:30 pm.

How to pay:
- The book fair offers eWallet, a secure alternative to cash.
- Simply create a free account to add funds and invite family and friends to contribute.
- All sales amounts and any unused funds will be donated to the local community center's children's library building fund.

For more information, email us at bookfair@woodlawschool.ac.us.

① 참가 대상 학생에 대한 연령 제한이 있다.
② 3월 10일부터 일주일간 진행된다.
③ 3월 12일 수요일 오후 가족의 밤이 예정되어 있다.
④ 올해 행사에서는 eWallet이 제공되지 않는다.
⑤ 책 판매 금액을 제외한 미사용 기금은 전액 기부된다.

12 다음 글의 밑줄 친 부분 중, 어법상 틀린 것은?

25007-0184

Requests to take part politically arise from a variety of sources: through social media, or in a sermon at church, or from the newsletter of an organization ① of which one is a member, or directly from a friend, neighbor, or co-worker. Because those who attempt to get others ② involved in politics act as "rational prospectors," the ordinary processes through which people are asked to take part do not ameliorate the class bias in individual political voice. Seeking to get results as efficiently as possible, rational prospectors aim their requests at those who are likely not only to say yes but also, on assenting, ③ to participate effectively. Often they target people who have been active in the past. The result is that participation undertaken in response to requests from others ④ bring in a disproportionate share of previous activists, which in turn exaggerates the socioeconomic status bias of political participation. In this way, ordinary processes of recruitment to political involvement do not simply replicate the socioeconomic structuring of political participation: they actually amplify ⑤ it.

* ameliorate: 개선하다 ** assent: 동의하다 *** replicate: 복제하다

13 다음 글의 밑줄 친 부분 중, 문맥상 낱말의 쓰임이 적절하지 <u>않은</u> 것은?

25007-0185

Storing too many memories could slow down the process of retrieval, so memories that are deemed ① <u>irrelevant</u> are quietly erased. This is why you probably cannot remember what you had for breakfast two Fridays ago as such information is unlikely to be useful. This of course raises the question as to how we can determine in advance whether or not a memory is likely to be useful in the future. One factor is ② <u>emotion</u>. In times of great stress, people will often have very detailed memories of the event; in extreme cases this can manifest itself as post-traumatic stress disorder (PTSD). Sufferers of PTSD can be overwhelmed by memories of traumatic events and can show hypervigilance in ③ <u>threatening</u> situations. Although PTSD is clearly maladaptive, it may be an extreme form of a process that is evolutionarily ④ <u>useful</u>. It would surely be good design to ensure that an animal that has had, for example, a near-death experience would learn from that experience and ⑤ <u>suppress</u> vigilance in similar situations.

* retrieval: (기억) 인출 ** deem: (~이라고) 여기다 *** hypervigilance: 과도한 경계

14 다음 빈칸에 들어갈 말로 가장 적절한 것은?

25007-0186

One example of information that has greatest value when it is in the _____ of a human operator can be found in the context of driving. Vehicles today are designed with increasingly sophisticated sensor packages aimed at detecting a variety of aspects of the driving environment. For example, forward-looking cameras and forward-looking radar systems can judge the distance to vehicles in front of a driver. Computations that measure this information over time reveal changes in distance. This information can be used to alert a driver when the change in distance for a given vehicle speed is rapid enough to suggest a collision might take place. That information is important, but it is only valuable if the driver acts on it. (Unless, of course, the vehicle itself acts on it without driver intervention.) The key here is that the driver must have the ability to pay attention to the information for the information to have value to the driver. If the driver is distracted by a phone, for example, they might fail to process the important information the vehicle is presenting.

* collision: 충돌

① criticism
② imagination
③ conscience
④ awareness
⑤ patience

15

다음 빈칸에 들어갈 말로 가장 적절한 것은?

25007-0187

A biologist, Tyler Volk, points out that cells are self-generated dynamic entities that at any given moment are always on the cusp between persisting and dying. They manage to survive by using their metabolism to stay ahead in this game. When metabolic wastes are expelled, the result is a loss of molecules. To compensate, cells also use metabolism to grow new molecules. If the exchange is at least equal, the cell can persist in its present form. If more molecules are generated than are lost, which adds protection against dying, net growth results, and the cell gets bigger. But a cell can only grow so much, as larger cells require more nutrients, and the cell runs up against a basic principle of physics — as a sphere gets bigger, its interior increases to a greater degree than its surface area. For a cell, this makes it harder for the surface to keep the flow of nutrients high enough to sustain the ever-larger interior. So what's a cell to do? It divides in half and starts the process all over as it approaches its useful size limit. This _____.

* entity: 독립체 ** on the cusp: ~의 경계에 있는 *** metabolism: 신진대사

① achieves a balance between growth and persistence
② makes it difficult for a new cell to regrow its damaged parts
③ causes a cell to find a different source of nutrients to rely on
④ boosts the energy intake for cells to become indefinitely bigger
⑤ creates a channel for cells to exchange molecules between each other

16 다음 빈칸에 들어갈 말로 가장 적절한 것은?

25007-0188

In a study, Iowa State University researchers read people lists of words, and then asked for each list to be recited back either right away, after fifteen seconds of rehearsal, or after fifteen seconds of doing very simple math problems that prevented rehearsal. The subjects who were allowed to reproduce the lists right after hearing them did the best. Those who had fifteen seconds to rehearse before reciting came in second. The group distracted with math problems finished last. Later, when everyone thought they were finished, they were all surprised with a pop quiz: write down every word you can recall from the lists. Suddenly, the worst group became the best. Short-term rehearsal gave purely short-term benefits. Struggling to hold on to information and then recall it had helped the group distracted by math problems transfer the information from short-term to long-term memory. The group with more and immediate rehearsal opportunity recalled nearly nothing on the pop quiz. Repetition, it turned out, was _____.

① essential for memorization
② less important than struggle
③ required before intensive practice
④ helpful in early language development
⑤ ineffective for understanding complex ideas

17

다음 빈칸에 들어갈 말로 가장 적절한 것은?   25007-0189

Psychological safety plays a powerful role in the science of failing well. It allows people to ask for help when they're in over their heads, which helps eliminate preventable failures. It helps them report — and hence catch and correct — errors to avoid worse outcomes, and it makes it possible to experiment in thoughtful ways to generate new discoveries. Think about the teams that you've been a part of at work or at school. These groups probably varied in psychological safety. Maybe in some you felt completely comfortable speaking up with a new idea, or disagreeing with a team leader. In other teams you might have felt it was better to hold back — to wait and see what happened or what other people did and said before sticking your neck out. That difference is now called psychological safety — and I have found in my research that it's an emergent property of a group, not a personality difference. This means your perception of whether it's safe to speak up at work is unrelated to whether you're an extrovert or an introvert. Instead, it's shaped by

_____.

* emergent property: 창발성(구성 요소에는 없으나 상위 구조에서 출현하는 특성)

① whether you have been in a similar situation before
② how much interest you have in the issue being discussed
③ how much more you know than others about the matter at hand
④ whether your thoughts are valued by your family members or not
⑤ how people around you react to things that you and others say and do

18 다음 글에서 전체 흐름과 관계 없는 문장은?

25007-0190

　　We live in a highly mobile world, where people are regularly uprooted from their home locales. Some do so by choice, but for others political or economic conditions force them to relocate to vastly different physical and cultural places. ① In these circumstances, aside from their private realms, public space plays a role in creating a sense of belonging to a new place. ② The sensory qualities of public space — the smell of the sea, a familiar tree, the design of a fence, the quality of light, a soccer field — have an utmost capacity to create a connection to "home" for the new arrivals. ③ These seemingly ephemeral qualities of public space can bring deep psychological comfort by triggering a familiar memory, creating an interconnectedness, or forming a continuum between the past and the present. ④ In a way, cities are massive public spaces divided into private spaces, whereas the suburbs are a bunch of private territories that make up mostly residential areas. ⑤ These experiences in and of public space can also establish a common or shared experience between the locals and the newcomers.

* ephemeral: 일시적인

19 주어진 글 다음에 이어질 글의 순서로 가장 적절한 것은?

25007-0191

The brain does not try to use all of its billions of neurons to represent everything; different brain regions represent different kinds of sensory stimuli.

(A) Thus, different parts of the brain have groups of neurons that fire when different kinds of visual, olfactory (smell), taste, auditory, and touch stimuli are presented. The human brain can do a lot more than just represent stimuli presented to it, because a group of neurons can respond to inputs from many groups of neurons.

(B) This can produce a combined representation of what the input neurons represent. For example, there are regions in the frontal cortex of monkeys where the sensory modalities of taste, vision, and smell come together, enabling the representation of fruits and their key properties. It is clear, therefore, that the brain is a superb representational device.

(C) For example, the visual cortex at the back of the brain has neurons that respond to different visual inputs. There are neuronal groups whose firing patterns correspond spatially to the structure of the input — for example, when a column of neurons fires together to represent the fact that a line is part of the visual stimulus.

* frontal cortex: 전두엽 피질 ** sensory modality: 감각 양상

① (A) – (C) – (B)　　　② (B) – (A) – (C)　　　③ (B) – (C) – (A)
④ (C) – (A) – (B)　　　⑤ (C) – (B) – (A)

20 주어진 글 다음에 이어질 글의 순서로 가장 적절한 것은?

25007-0192

For the last few centuries, experts have worried ceaselessly that we aren't exercising enough.

(A) An especially influential proponent of this movement was Friedrich Jahn, the "Father of Gymnastics." Following Napoleon's humiliating string of victories over German armies in the early nineteenth century, Jahn argued that educators had a responsibility to restore the physical and moral strength of his nation's youth with gymnastics, hiking, running, and more.

(B) Nationalism is one major source of this anxiety. Just as ancient Spartans were required and Romans were urged to be fit enough to fight as soldiers, flag-waving leaders and educators increasingly encouraged ordinary citizens to participate in sports and other forms of exercise as preparation for military service.

(C) Later, similar worries in America were spurred by the embarrassing lack of fitness among many men who enlisted or were drafted for World Wars I and II and by the poor state of fitness among schoolchildren at the start of the Cold War. National movements to drum up fitness for the sake of the state still occur in China and elsewhere.

* proponent: 지지자 ** spur: 불러일으키다 *** enlist: 입대하다

① (A) − (C) − (B) 　　② (B) − (A) − (C) 　　③ (B) − (C) − (A)
④ (C) − (A) − (B) 　　⑤ (C) − (B) − (A)

21 글의 흐름으로 보아, 주어진 문장이 들어가기에 가장 적절한 곳은?   25007-0193

> It may be, however, that warming will be experienced as a reduction in the fall of temperature at night, due to increased cloudiness, with little or no change in daytime temperatures.

Many climatologists accept there is a real possibility of global climatic warming due to an enhanced greenhouse effect. As with many large-scale changes, there would be winners and losers. (①) Were climate belts to shift toward higher latitudes, which seems the most likely overall result, parts of the Sahara and southern Russia would receive increased rainfall. (②) They would benefit and their agricultural output would increase. (③) On the other hand, southern Europe and the United States cereal belt might become drier. (④) If warming produced a rate of evaporation that exceeded the increase in the rate of precipitation, soils would become more arid. (⑤) In that event, nighttime frosts would become less frequent, soils would become somewhat moister, and agriculture would benefit.

* latitudes: 위도 지역   ** precipitation: 강수량   *** arid: 메마른

22 글의 흐름으로 보아, 주어진 문장이 들어가기에 가장 적절한 곳은?

25007-0194

For example, many social media platforms do not fit well with skills such as writing essays and giving formal presentations.

Social media platforms enhance learning experiences, increase student interaction, and foster engagement. However, just like with other pedagogical activities, understand the why behind what you do with social media. (①) Knowing this reason will guide you with choosing the most effective digital tool for reaching your instructional goals. (②) Be sure to match your instructional goals with digital tools that are accessible on websites and apps. (③) When selecting digital tools, refrain from being attracted to apps with fireworks and flying colors because such highly attractive apps may not necessarily serve your academic purposes nor your students' learning needs. (④) Selecting a social media platform is a practical decision based on making a good fit with a targeted skill. (⑤) Yet, these platforms can enhance other pedagogical activities such as classroom assessment techniques by allowing students to provide virtual input about a topic through social interaction.

* pedagogical: 교육적인

23 다음 글의 내용을 한 문장으로 요약하고자 한다. 빈칸 (A), (B)에 들어갈 말로 가장 적절한 것은?   25007-0195

The only way to produce crystal-clear writing is to know how a reader will respond to the choices you make in composing text and graphics. You need to know which sentence structures are most easily understood, which organization of material into sections is most easily followed, and so on. It's certainly possible to offer some general rules along these lines: for example, "use the active voice," "divide the paper into Introduction, Methods, Results, and Discussion sections," and "use a figure instead of a table when quantities are to be compared." In principle, you could tape a long list of such rules above your computer and treat it as the voice of authority on how to reach readers. But long lists of rules are boring. Besides, using them makes writing mechanical, and good writing sometimes entails knowing when to bend the rules instead of following them. Furthermore, using a list of rules is oddly indirect: instead of relying on rules you've been told will produce clear text, surely it would be more effective to understand how readers think, and write to that understanding.

In order to produce clear writing, you need to grasp the viewpoint of your _____(A)_____ and apply general writing rules _____(B)_____ .

	(A)		(B)
①	audience	······	flexibly
②	audience	······	consistently
③	editor	······	systematically
④	editor	······	consistently
⑤	peers	······	flexibly

[24~25] 다음 글을 읽고, 물음에 답하시오.

In the early 1970s, the Peace Corps office in Botswana was concerned by the number of volunteers who seemed to be "burned out," failing in their assignments, leaving the assigned villages, and increasingly hostile to their Tswana hosts. The Peace Corps asked American anthropologist Hoyt Alverson, who was (a) familiar with Tswana culture and society, for advice. Alverson discovered that one major problem the Peace Corps volunteers were having involved exactly the issue of (b) similar actions having very different meanings. The volunteers complained that the Tswana would never leave them alone. Whenever they tried to get away and sit by themselves for a few minutes to have some private time, one or more Tswana would quickly join them. This made the Americans (c) angry. From their perspective, everyone is entitled to a certain amount of privacy and time alone. To the Tswana, however, human life is (d) social life; the only people who want to be alone are witches and the insane. Because these young Americans did not seem to be either, the Tswana who saw them sitting alone naturally assumed that there had been a breakdown in hospitality and that the volunteers would (e) reject some company. Here, one behavior — a person walking out into a field and sitting by himself or herself — had two very different meanings.

* the Peace Corps: (미국의) 평화 봉사단

24 윗글의 제목으로 가장 적절한 것은?　　　25007-0196

① Social Evolution: Adapting to Change in a Connected World
② Creative Solutions for Coping with and Managing Challenges
③ The Heart of Volunteering: Making a Difference Where It Counts
④ Uncovering Cross-cultural Misunderstandings: Action vs. Meaning
⑤ Universal Values and Diverse Expressions of Hospitality Across Cultures

25 밑줄 친 (a)~(e) 중에서 문맥상 낱말의 쓰임이 적절하지 않은 것은?　　　25007-0197

① (a)　　② (b)　　③ (c)　　④ (d)　　⑤ (e)

[26~28] 다음 글을 읽고, 물음에 답하시오.

(A)

In the heart of the forest, there lived a young eagle named Edge and his twin brother, Jo. Jo could soar high in the sky and experience the freedom of flight, while Edge was born with a damaged wing and was unable to fly. Although this could have discouraged (a) him greatly, he never lost hope and always believed that he would one day achieve his dream of flying alongside Jo.

* soar: 날아오르다

(B)

Finding the mystical herb would be difficult and dangerous, so Jo tried to persuade him not to go. However, Edge was determined to find the herb. In the end, Jo understood and wished him a safe and successful journey. Edge set off on his journey, which was especially challenging because of his damaged wing, forcing (b) him to climb the mountain on foot. At one point, a sudden landslide wiped out the path he had been following, making him start the climb all over again. He encountered many other obstacles along the way, but he never gave up. Finally, after several days of climbing, he reached the top of the mountain and found the mystical herb.

* mystical: 신비로운 ** landslide: 산사태

(C)

When Edge returned to the sage, the sage used the mystical herb to fix Edge's wing. With Jo cheering him on, Edge spread his wings and took off into the sky for the first time. He felt the wind rushing past (c) him and the sun shining on his feathers. From that day forward, the two brothers often flew together and explored the vast skies. Edge became a symbol of hope and determination. And even though he never forgot the sage who had helped him achieve his dream, Edge also knew that his true spirit came from within, which gave (d) him the strength to overcome any obstacle.

* sage: 현자

(D)

One day, Jo and Edge met an old sage while they were looking for food. Known for his wisdom, the sage lived alone in a small hut hidden among the trees. Wanting his twin brother to achieve his dream, Jo eagerly asked if the sage could help Edge fly. The sage took pity on the young eagle and told Edge about a mystical herb. The herb grew on the top of the tallest mountain in the forest and glowed blue. He said that if Edge was able to find the mystical herb and bring it back to (e) him, he would be able to fix his wing so he could fly.

26 주어진 글 (A)에 이어질 내용을 순서에 맞게 배열한 것으로 가장 적절한 것은?　　　25007-0198

① (B) − (D) − (C)　　　② (C) − (B) − (D)　　　③ (C) − (D) − (B)

④ (D) − (B) − (C)　　　⑤ (D) − (C) − (B)

27 밑줄 친 (a)~(e) 중에서 가리키는 대상이 나머지 넷과 <u>다른</u> 것은?　　　25007-0199

① (a)　　　② (b)　　　③ (c)　　　④ (d)　　　⑤ (e)

28 윗글에 관한 내용으로 적절하지 <u>않은</u> 것은?　　　25007-0200

① Edge는 손상된 날개를 가지고 태어났다.
② 갑작스러운 산사태는 Edge가 따라가던 길을 쓸어 버렸다.
③ Edge는 희망과 결단력의 상징이 되었다.
④ 현자는 나무들 사이에 숨겨진 작은 오두막에서 혼자 살았다.
⑤ 약초는 산의 가장 깊숙한 곳에서 자라며 푸른 빛을 발했다.

01 다음 글의 목적으로 가장 적절한 것은?

25007-0201

[Monday, January 20, 2025]

Dear Parents/Guardians and Students of TBS High School,

　Thank you for your patience during the heavy snowstorm that dropped more than 40 cm of snow in Calbary. This evening, the City of Calbary has issued a snowstorm alert and is encouraging residents to stay home. At TBS High School, we are also facing several concerns: students may not be able to walk to school safely, and many school buses will likely be delayed or, in some cases, canceled. With all of this in mind, we have made the decision to close our school on Tuesday, January 21. Students may either work on previously assigned tasks or spend time reviewing material previously covered at home. Thank you for your patience and understanding.

Sincerely,
Caroline R. Robbins
TBS High School Principal

① 과제 평가 결과를 통지하려고
② 폭설로 인한 휴교를 공지하려고
③ 폭설 시 대피 요령을 안내하려고
④ 눈 치우기 운동 참여를 요청하려고
⑤ 학교 축제 일정 변경 이유를 설명하려고

02 다음 글에 드러난 'I'의 심경 변화로 가장 적절한 것은?

25007-0202

By the time our family arrived at the Church of Santa Teresa, I tried to soak up the chapel's beauty in silence. However, I found a group of people singing loudly in the corner of the chapel. *Why are these people ruining my trip? Do they even know what this place is?* To make matters worse, they started dancing, too! My moment with the chapel was completely ruined by their inconsiderate behavior. Suddenly, some women in the group began motioning for me to join in. I attempted to decline by taking backward steps. Still, they just kept stretching their hands toward me. I looked down at my daughter, who looked back at me with curiosity. I shrugged and smiled. The next thing I knew, we were dancing together with huge smiles on our faces. As I squeezed my daughter's hand and did my best to keep time with the lively beat, my heart was filled with joy. I knew that this would be my favorite memory of the trip.

① hopeful → stressed
② ashamed → relaxed
③ bored → frightened
④ annoyed → delighted
⑤ excited → disappointed

03 다음 글에서 필자가 주장하는 바로 가장 적절한 것은?

25007-0203

For better or worse, we see the world through a singular lens that often causes us to fall victim to certain biases. I've seen a number of entrepreneurs who built successful businesses and attributed all of their achievements to their own ingenuity. These people tend to lack the self-reflection and objectivity to see that their accomplishments likely would not have been possible without the assistance of various people and institutions external to themselves. It's all too easy to see failures in our lives as purely or predominantly due to outside (external) forces, and successes in our lives as purely or predominantly due to inside (internal) forces. Therefore, we must know that credit must be given and shared. Apple's success wasn't exclusively Steve Jobs's, nor was Microsoft's success exclusively Bill Gates's. Santa has thousands of little helpers who allow him to achieve the impossible. It takes a team of people (or elves) and a huge network for us to accomplish anything.

* entrepreneur: 사업가 ** ingenuity: 재간

① 목표 설정 시 조직의 취약점을 분석하는 것이 우선되어야 한다.
② 자신의 성취 뒤에는 다른 사람들의 도움이 있음을 인정해야 한다.
③ 리더는 조직원들과의 소통을 통해 끊임없이 상황을 개선해야 한다.
④ 조직의 발전을 위해서는 자신의 이익과 타인의 이익을 조율해야 한다.
⑤ 자신의 분야에서 성공하기 위해서는 그 분야에 대한 열정을 가져야 한다.

04 밑줄 친 Everyone loves a good story.가 다음 글에서 의미하는 바로 가장 적절한 것은? 25007-0204

We tend to believe people who share our beliefs without taking the time to fact-check all the information. With the emergence of 'identity politics', this has given rise to the recent phenomenon of fake news. Anything that doesn't conform to your tribe's views is rejected and labelled as fake news. It is not always falsified information, but it is often opinions that differ hugely from our own that are labelled fake news by politicians and the media alike. Jonathan Freedland in the *Guardian* terms it a new kind of cognitive bias called Tribal Epistemology, which is when the truth no longer corresponds to facts or evidence but rather when a specific assertion agrees with the viewpoint of the tribe or social group one belongs to. The boundaries between 'works for us' or 'good for us' and 'true' have blurred. Of course, this isn't a new phenomenon; we have been sorting ourselves into tribes since the beginning of human evolution. It has just been ignited by the recent political climate, media and technology, and by our tendency to believe in stories and narratives rather than facts. Everyone loves a good story.

* epistemology: 인식론 ** blur: 흐릿해지다 *** ignite: 불을 붙이다

① A good story is the only way to unite everyone in a society.
② We live in a world of fake news because of information overload.
③ We blindly follow the story that fits our social group's viewpoints.
④ The art of storytelling involves finding ways to connect with others.
⑤ Combining old and new experiences helps us create an appealing story.

05 다음 글의 요지로 가장 적절한 것은?

25007-0205

Ambiguity is an uncomfortable feeling for most people. As such, it's one that we tend to quickly misattribute, concluding that something must be wrong. But in fact, ambiguity can be a productive, even positive state. When we conduct research, we try to encourage ourselves, our students, and our colleagues to enjoy a state of ambiguity. The logic is that when we don't know the answer, real knowledge can arise, so it is best to go slowly, think carefully, and enjoy the process of trying to find out what is going on. Sadly, this love of ambiguity is an uncommon mindset. Usually we want quick and clear answers, especially when we are stressed. For some people, this antipathy to ambiguity contributes to their descent into misbelief. The ability not to rush to conclusions, to keep multiple hypotheses in mind, and to remain open to new information and possibilities is key to not getting sucked into misbelief. We tend to admire and seek out conviction and confidence. But we would be better served if we learned to admire and enjoy a state of ambiguity.

* **antipathy**: 반감, 혐오 ** **descent**: 전락, 하락

① 사건 간 인과 관계를 명확하게 파악하는 것이 문제 해결의 출발점이다.
② 나와 생각이 다른 사람과 꾸준히 교류해야 진정한 지식에 도달할 수 있다.
③ 명확한 정보가 부족한 상황에서는 다양한 시나리오를 만들어 대비해야 한다.
④ 연구 결과를 보고할 때 모호한 표현을 사용하면 연구의 신뢰성을 떨어뜨릴 수 있다.
⑤ 모호함의 상태를 기꺼이 받아들이는 사고방식을 가져야 잘못된 믿음에 빠지지 않는다.

06 다음 글의 주제로 가장 적절한 것은?

25007-0206

When leaders in companies, nonprofits, or governments invest in artificial intelligence, much of their attention goes to hiring machine learning experts or paying for tools. But this misses a critical opportunity. For organizations to get the most that they can from AI, they should also be investing in helping all of their team members to understand the technology better. Understanding machine learning can make an employee more likely to spot potential applications in their own work. Many of the most promising uses for machine learning will be humdrum, and this is where technology can be at its most useful: saving people time, so that they can concentrate on the many tasks at which they outperform machines. An executive assistant who has a better understanding of machine learning might suggest that calendar software learn more explicitly from patterns that develop over time, reminding them when their boss has not met with a team member for an unusually long time. A calendar that learns patterns could give an executive assistant more time for the human specialties of the job, such as helping their boss to manage a team.

* humdrum: 단조로운, 평범한

① benefits of introducing AI-based management tools
② difficulty of hiring competent machine learning experts
③ reasons for retaining human employees in the age of AI
④ ways of overcoming management reluctance to embrace AI
⑤ necessity of investing in enhancing AI literacy for all employees

07 다음 글의 제목으로 가장 적절한 것은?

25007-0207

It's a useful cliché, repeated by practitioners and radio educators alike, that radio is 'theatre of the mind'. By telling stories using words, music and sound effects, radio can engage the imagination to communicate ideas and images that create a kind of narrative uniquely experienced by each individual listener. Through the omission of visual cues and by embracing the openness of the work, radio storytelling has the capacity to make personal connections, paint pictures with sound, and indeed create scenes that would be impossible in another context. The capacity of audio as a medium for imaginative and compelling storytelling is undiminished in the digital age, although the production, distribution and consumption cultures and technologies through which those stories are mediated have radically changed. In fact the possibilities for radiophonic narrative are in many ways expanded in the digital age, as research engineers, professionals and enthusiasts explore the parameters of new production processes, platforms and interactive opportunities, as well as opportunities for radio storytelling to be taken outside the realm of the radio professional.

* cliché: 상투적인 표현   ** omission: 생략, 빠짐   *** radiophonic: 라디오의 소리 효과에 의한

① How Radio Storytelling Differs from Written Storytelling
② The Enduring Value of Radio Storytelling in the Digital Era
③ The Challenges Faced by Radio in a Changing Media Landscape
④ Beyond Entertainment: Radio's Lasting Impact on Shaping Culture
⑤ From Signals to Streaming: Exploring the Evolution of Radio Technology

08 다음 도표의 내용과 일치하지 <u>않는</u> 것은?

25007-0208

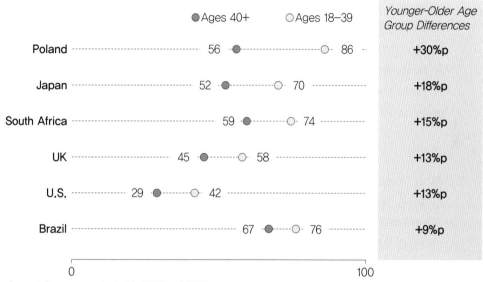

Perceptions on Social Media's Influence on Democracy Among Adults in Six Selected Countries

% of adults who say social media has been a good thing for democracy in their country

● Ages 40+   ○ Ages 18–39

	Ages 40+	Ages 18–39	Younger-Older Age Group Differences
Poland	56	86	+30%p
Japan	52	70	+18%p
South Africa	59	74	+15%p
UK	45	58	+13%p
U.S.	29	42	+13%p
Brazil	67	76	+9%p

0 — 100

Source: Surveys conducted in 2022 and 2023

   The graph above shows adults' views on social media's positive impact on democracy in six selected countries. ① For all the six countries in the graph, the percentages of younger adults (ages 18-39) saying social media has benefited democracy are higher than those of older adults (ages 40 and older). ② Among the six selected countries, the percentage point difference between the two age groups is the largest in Poland and the smallest in Brazil. ③ In both Japan and South Africa, over two-thirds of adults ages 18-39 say that social media has been beneficial for democracy. ④ In the UK, more than half of adults in the older age group say social media has benefited democracy in their country. ⑤ Among the six countries, the U.S. shows the lowest percentage of adults ages 40 and older who say social media has been positive for democracy in their country, at below 30 percent.

09 Barbara Hillary에 관한 다음 글의 내용과 일치하지 <u>않는</u> 것은?

25007-0209

Barbara Hillary was born in 1931 in New York City and raised in Harlem. Her family was poor, but she was encouraged to read. She majored in gerontology at the New School in New York City, and after graduating she became a nurse and worked for fifty-five years before retiring. At age 67, Hillary was diagnosed with lung cancer, which required surgery to remove. The surgery resulted in a 25 percent reduction in her breathing ability, but she would not let this stop her. She became interested in the North Pole after visiting Canada and taking pictures of polar bears. In the spring of 2007, at the age of 75, she became the oldest woman and the first African-American woman to ski to the North Pole. She reached the South Pole in 2011 at the age of 79, becoming the first African-American woman to reach both poles.

* gerontology: 노인학

① New York City에서 태어나 Harlem에서 자랐다.
② 간호사가 되어 은퇴 전까지 55년간 일했다.
③ 67세에 폐암 진단을 받았다.
④ 캐나다를 방문해 북극곰의 사진을 찍은 후 북극에 대한 관심이 생겼다.
⑤ 2007년에 북극과 남극에 모두 도달한 최초의 아프리카계 미국인 여성이 되었다.

10 2025 Greenfields Horse Camp에 관한 다음 안내문의 내용과 일치하지 <u>않는</u> 것은?   25007-0210

2025 Greenfields Horse Camp

Join us for an unforgettable experience! This camp is for children aged 6–10, offering a safe and fun environment to make a life-long connection with horses.

Dates
- Session 1: July 7th – 11th
- Session 2: July 14th – 18th

Activities
During camp, participants will:
- learn how to ride in our indoor and outdoor facilities
- learn how to safely handle, feed, and take care of horses
- enjoy a variety of games and activities focused on learning about horses

Notes
- Camp runs from 9 a.m. to 2 p.m. Optional activities are available until 5 p.m. for an extra fee.
- Drinks are provided, but participants should bring a bag lunch.
- The fee for each session is $250, with a separate $50 sign-up fee.
- No refunds for missed days.

Call 805-222-3990 or email greenhorseclub@abc.com to save your spot!

---

① 6세에서 10세의 어린이를 대상으로 한다.
② 실내 및 실외 시설에서 말을 타는 법을 배울 것이다.
③ 추가 요금 지불 시 오후 5시까지 선택 활동을 이용할 수 있다.
④ 참가자는 점심 도시락을 가져와야 한다.
⑤ 기별 비용과 별도로 250달러의 등록비가 있다.

11 Smart Automatic Pet Feeder에 관한 다음 안내문의 내용과 일치하는 것은?   25007-0211

Smart Automatic Pet Feeder User Manual

This automatic pet feeder is designed to help you feed your pet at specific times. It is for indoor household use only.

Filling the Food Compartment
1. Slide the food compartment upwards to remove it, without pressing or twisting it.
2. Fill the compartment with pet food and slide the food compartment back into place.

Managing Mealtime
1. Press the SET button. Use the arrow keys to select the desired feeding time.
2. Press SET again to save.
3. Repeat the steps to set additional mealtimes. Up to 4 mealtimes can be set for one day.

Recording a Personal Voice Message
1. Press and hold the REC button.
2. Speak into the Recording Port and record your message. Personal voice messages can be up to 10 seconds.
3. Release the REC button.

Precautions
- Do not use wet food.
- The food tray and food compartment can be removed and handwashed but should not be cleaned in dishwashers.

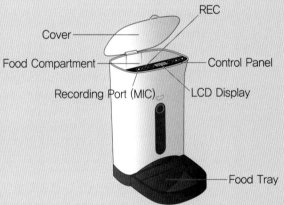

* compartment: 통, 칸

① 실외에서도 사용 가능하다.
② 사료 통을 분리하려면 한번 누른 후 위로 밀어 올린다.
③ 식사 시간을 하루에 최대 세 번까지 설정할 수 있다.
④ 개인 음성 메시지는 최대 10초까지 녹음할 수 있다.
⑤ 사료 트레이는 식기세척기로 세척 가능하다.

12 다음 글의 밑줄 친 부분 중, 어법상 틀린 것은?

25007-0212

For livestock, researchers are typically measuring temperament traits in groups of animals on experimental or commercial farms. Compared to a more controlled experimental laboratory environment, these environments offer both opportunities and constraints to ① measuring and understanding temperament traits. Farms are a source of large numbers of animals whose pedigree links with others on the farm and elsewhere ② are frequently known. Within any given farm, animals may also have experienced a fairly standardized rearing process. However, one farm differs from another in many respects, such as the quality of human interaction, the level of nutrition provided, and the social groups the animals are kept in, among others, all of ③ them could affect temperament. For instance, ④ to assess aggressiveness in individual pigs when groups of pigs are mixed with others, a test must be used in which all animals are mixed with a standardized number of unfamiliar animals, within a controlled weight range, and into pens of the same size. Such standardization can be achieved only on some farms. All of these factors mean ⑤ that farms cannot be treated as statistical replicates, and each farm must be regarded as unique.

* temperament: 기질 ** pedigree: 혈통 *** replicate: 복제본, 반복되는 것

13

다음 글의 밑줄 친 부분 중, 문맥상 낱말의 쓰임이 적절하지 <u>않은</u> 것은?

25007-0213

Storing and retrieving information is so key to survival that nature has evolved thousands of techniques to do so. For instance, Venus flytraps can ① <u>count</u>. To make sure that their traps don't get sprung with a single raindrop, they require three touches to a trap within a certain short period before they expend the energy needed to ② <u>close</u>. Thus, they must be able to count the touches as well as ③ <u>store</u> the time interval between them. We actually know how this is done. When a trap is touched, some calcium is released by the plant, which slowly — over seconds — dissipates. Subsequent touches ④ <u>accumulate</u> more calcium, and if it hits a certain threshold, the trap snaps shut. Since the calcium is constantly dissipating, the ⑤ <u>sequential</u> touches have to be within a certain time window to hit the threshold. This capability is clearly encoded in DNA somewhere, but again, information is being acted upon with hardly any cognition.

* **Venus flytrap**: 파리지옥(덫을 닫아 곤충을 잡아먹는 식충 식물) ** **dissipate**: 소멸되다
*** **threshold**: 임계치

14 다음 빈칸에 들어갈 말로 가장 적절한 것은?

25007-0214

Philosopher of science Hans Reichenbach and others use the colour of crows to illustrate why inference based upon hypothesis tests cannot demonstrate with certainty that a circumstance always occurs. Consider the hypothesis that 'all crows are black'. Even if one has seen a million black crows and never seen a crow of another colour, there is no guarantee that the next crow will be black. Even if one thinks one has seen every crow, one may be mistaken. Furthermore, one cannot see every future or past crow. Thus, no matter how many crows have been studied, even if all of them have been consistent with the hypothesis that all crows are black, there is no way to be certain, to prove, that all crows are black. Given the many crows I have seen and heard about, I am confident the great majority of crows are black, but I do not know for a fact that all crows are black. This is the inherent _____ of inductive reasoning.

① flexibility
② limitation
③ inclusivity
④ practicality
⑤ intentionality

15

다음 빈칸에 들어갈 말로 가장 적절한 것은?

25007-0215

Our pets are very different from wild animals; millennia of domestication have changed them almost beyond recognition from their wild ancestors. But not completely. At least with dogs and cats we can sit down, observe, and interact with them, and they are happy for us to observe and interact. Real communication with our pets is not about teaching them human words ('sit', 'lie down', 'drop it'); real communication is about _____. If you forget about any intention to train your dog to 'sit', what do you see when you spend time with him? Very much what you would see were you to spend time with a pack of wolves. Behaviours like expressing annoyance (perhaps just getting up and moving to another spot in the room) or expressing a desire for social interaction (licking your hand or lying next to you). These concepts are the true vocabulary of animals — not words and sentences — and the more you spend time with your animals, the more of this true vocabulary you will notice.

① observing the animal in them
② recognizing them as human beings
③ restoring and testing their wildness
④ maximizing their intellectual potential
⑤ making them understand nonverbal signals

16 다음 빈칸에 들어갈 말로 가장 적절한 것은?

25007-0216

Dias & Ressler conducted an experiment with mice where they trained male mice to be afraid of the scent of cherry blossoms. Every time they introduced the smell of cherry blossoms, they would electrocute the mice. Three days later, they had the mice mate with female mice. The resultant pups were not electrocuted or introduced to the scent of cherry blossoms until adulthood. When they eventually introduced the offspring to the scent of cherry blossoms, the pups started shaking and displaying other symptoms of anxiety. These pups could also pick up subtle traces of cherry blossoms in the air, which changed the structure of their brains and certain areas lit up when monitored digitally. This same behaviour was observed in the following generation of the mice even though they had never been electrocuted. This led researchers to understand how we, as humans, _____.

* electrocute: 전기 충격을 가하다   ** pup: (여러 동물들의) 새끼

① share relatively limited evolutionary similarities with other animals
② devise specialized treatments to address specific medical conditions
③ evolve and adapt by unlocking the amazing capabilities within our genes
④ carry our ancestors' emotional reactions to traumatic situations and events
⑤ develop emotional traits through environmental influences, not through heredity

17 다음 빈칸에 들어갈 말로 가장 적절한 것은?   25007-0217

In many cases, information about a target location is stored in one's long-term memory. Such information is more likely to involve an association of a target to a particular landmark than an association of a target to a particular viewer position. For example, a frequent location of one's keys may be on the chest of drawers near the door. However, the viewer probably does not have a habitual location relative to the chest, so obtaining the keys involves knowing the position of the keys relative to the chest as well as one's own position relative to that chest. Of course, in some cases, a person does have a habitual position relative to a landmark in a certain spatial context and that position is directly associated with the target object. For example, in a kitchen, the chef may have a habitual location relative to the stove, and the condiments may be in a specific location relative to that stove. In this context, _____, and an inference about the relation of the condiments to the self via the stove will not be necessary.

* condiment: (보통 복수로) 조미료, 양념

① a viewer's position may well change
② ego-centered coding will be sufficient
③ the location of a specific landmark is given
④ the habitual location is stored in short-term memory
⑤ knowing the spatial arrangement might be inadequate

18 다음 글에서 전체 흐름과 관계 없는 문장은?

25007-0218

Tough times and changing market conditions can severely affect the performance and survival of enterprises. Some enterprises view tough times as an opportunity to strengthen their businesses. ① Therefore, they try to establish their advantage over their competitors by investing aggressively in marketing activities, research and development, and developing new products, whereas some enterprises cut their marketing expenditures and wait for the recession to pass. ② In challenging times, the most frequent response of enterprises to a recession is retrenchment of marketing investments in order to survive until the economy recovers. ③ Reducing the marketing expenses such as advertising and communication may give short-term relief to the enterprise. ④ The use of social media for advertising during a recession can enhance customer independence, enabling them to exercise their free will in purchasing products. ⑤ However, in the long run, enterprises may risk losing their customer base and market share.

* expenditure: 지출 ** recession: 불황 *** retrenchment: 삭감

19

주어진 글 다음에 이어질 글의 순서로 가장 적절한 것은?   25007-0219

> In the case of self-driving cars that get into an accident, we are dealing with the following phenomenon: In the situation immediately before the accident, no more decisions can be made.

(A) This would not, however, lead to self-driving vehicles acquiring the status of "moral agents." Their behavior would not be considered an action in the sense of a result of genuine decision-making. An autonomous vehicle merely implements the rules programmed into its software.

(B) The decision about the behavior of an autonomous car was made when a decision was made about its programming. This can be a lengthy process involving both the creation of appropriate legal regulations and their implementation by the manufacturer down to the individual programmer. Now, in addition to attempts to program machines to apply certain moral theories to particular situations, there are also those that aim to mimic human judgment as best as possible.

(C) This is also true when forms of self-learning Artificial Intelligence are used. Here, too, humans will select the training examples and decide what the correct answer is in each case. They decide what the program should "learn" and when it has "learned" enough.

* autonomous: 자율적인   ** mimic: 모방하다

① (A) – (C) – (B)   ② (B) – (A) – (C)   ③ (B) – (C) – (A)
④ (C) – (A) – (B)   ⑤ (C) – (B) – (A)

20 주어진 글 다음에 이어질 글의 순서로 가장 적절한 것은?

25007-0220

Multiculturalism is an ideology advocating that society should consist of, or at least allow and include, distinct cultural and religious groups, with fairness. Some countries have official multiculturalism policies aimed at preserving the cultural identities of immigrant groups.

(A) Which term is used depends on what language is spoken by the people. For example, English-speaking European researchers usually use the term multicultural, while non-English-speaking researchers use the term intercultural. It is also argued that multicultural describes the nature of the society whose members are from different ethnic and religious groups, while intercultural describes their interactions, negotiations, and processes.

(B) In this context, multiculturalism advocates a society that extends fairness to distinct cultural and religious groups, no one culture predominating. However, the term is more commonly used to describe a society consisting of minority immigrant cultures existing alongside a predominant, indigenous culture. Often, multiculturalism is interchangeably used with the term *interculturalism.*

(C) Another view is that intercultural refers to two culturally different groups of people, and multicultural refers to more than two culturally different groups of people. Therefore, the term multicultural is acceptable when referring to multiple cultures.

* predominate: 우위를 차지하다   ** indigenous: 토착의

① (A) – (C) – (B)   ② (B) – (A) – (C)   ③ (B) – (C) – (A)
④ (C) – (A) – (B)   ⑤ (C) – (B) – (A)

21 글의 흐름으로 보아, 주어진 문장이 들어가기에 가장 적절한 곳은?

25007-0221

Clearly, by the light of our current knowledge, most of what people in the past thought they knew was false.

Looking at science's history suggests that the answer to the question, what is this thing called science, is that there is no single thing called science. Science is certainly not a unified and continuous body of beliefs. (①) Neither is it captured by a single scientific method. (②) If we want to understand what science is — the different ways knowledge has been produced both now and in the past, and how different peoples and cultures have come to hold the beliefs they have about the natural world — we simply have no choice but to look at history. (③) To do this successfully, we need to overcome that modern condescension for the past that is often particularly prevalent in the history of science. (④) It is worth remembering, though, that by the same pessimistic induction, most of what we think we know now will turn out to be wrong by the future's standards. (⑤) The history of science should not be a game of rewarding winners and losers in the past.

* condescension: 우월감

22 글의 흐름으로 보아, 주어진 문장이 들어가기에 가장 적절한 곳은?

25007-0222

A work must be "fixed in a tangible medium of expression," to quote U.S. copyright law, to qualify for copyright protection.

Generally a work is eligible for copyright protection if it is "original," that is, it contains markers of creative decision making by the author or artists. Something raw or common like an alphabetical listing of names would not be eligible for copyright protection. (①) But a song melody that used an alphabetical list of names as lyrics would be. (②) This means that if I stand up in public and speak a poem into the air, it does not enjoy copyright protection. (③) But if I record the sound of speaking it into the air or I type the poem out on a keyboard into a computer hard drive, it immediately enjoys protection. (④) My computer hard drive is a "tangible medium of expression." (⑤) So are film, photographic paper, concrete, cloth, and the huge slabs of steel that the sculptor Richard Serra uses to make his sculptures.

* tangible: 유형(有形)의   ** eligible: 자격이 있는   *** slab: 평판

23 다음 글의 내용을 한 문장으로 요약하고자 한다. 빈칸 (A), (B)에 들어갈 말로 가장 적절한 것은?   25007-0223

You might think that one reason that some individuals are successful at recognizing others' emotions from their facial expressions is because they typically show emotions clearly on their own faces. Surprisingly, though, research findings related to this idea are mixed. A review of available evidence indicates that both reading and expressing emotions are related but only when individuals are *trying* to communicate their feelings to others. When these expressions occur spontaneously — for instance, an expression of joy when something wonderful suddenly happens — then being able to recognize others' facial expressions and displaying those cues clearly oneself are *not* related. To put it in other terms, people who express their emotions openly and easily are not necessarily accurate at recognizing the facial expressions of others. Accuracy appears to be tied to people's intentional focus on showing their own emotions in their facial expressions. Perhaps they gain greater insight into the nature of others' expressions by doing so, and this helps them to recognize others' underlying feelings more accurately.

Research findings show that one's tendency to clearly convey their own emotions through facial expressions doesn't necessarily ___(A)___ an accurate reading of others' emotions, as the accuracy of reading seems to be linked to the ___(B)___ display of one's own emotions.

	(A)		(B)
①	guarantee	······	private
②	guarantee	······	deliberate
③	prevent	······	unintentional
④	generate	······	frequent
⑤	prevent	······	exaggerated

[24~25] 다음 글을 읽고, 물음에 답하시오.

If you look at mammals, the (a) larger the animal, generally speaking, the longer its life span. This makes evolutionary sense. A small animal is more (b) vulnerable to predators, and there would be no point in having a long life span if it is going to be eaten long before it dies of old age. But the more fundamental reason for the relationship between size and life span is that size is related to metabolic rate, which is roughly the rate at which an animal burns fuel in the form of food to provide the energy it needs to function. Small mammals have more surface area for their size and so (c) lose heat more easily. To compensate, they need to generate more heat, which means maintaining a (d) higher metabolic rate and eating more for their weight. This means that the total number of calories burned per hour by an animal increases more slowly than the mass of the animal. An animal that is ten times as large burns only four to five times as many calories per hour. So for their weight, smaller animals burn (e) fewer calories than larger animals. The relationship between how fast an animal burns calories and its mass is named Kleiber's law after Max Kleiber, who showed in the 1930s that an animal's metabolic rate scales to the 3/4 power of its mass. The exact power is a matter of dispute and some show that for mammals, a 2/3 power fits the data better.

* power: 제곱

24 윗글의 제목으로 가장 적절한 것은?   25007-0224

① Survival Strategies of Small Mammals
② Size Variation: Evidence of Genetic Diversity
③ Life Span: An Indicator of Successful Adaptation
④ Different Food Types, Different Amount of Calories
⑤ Size Matters When It Comes to the Life Span of Animals

25 밑줄 친 (a)~(e) 중에서 문맥상 낱말의 쓰임이 적절하지 않은 것은?   25007-0225

① (a)   ② (b)   ③ (c)   ④ (d)   ⑤ (e)

[26~28] 다음 글을 읽고, 물음에 답하시오.

(A)

Larry's little brother had a lot to live up to. Larry was the star basketball player in the family. Though he was only five feet eight, Larry had a miraculous ability. College scouts were struck by his jaw-dropping play. Larry's brother was a year younger, even shorter, and not nearly as good. Larry always won their endless games of horse and one-on-one, and that was okay for his hero-worshiping brother. But it also drove the kid to push (a) himself harder to up his game.

* horse: 호스(농구 슈팅 게임)

(B)

He was devastated, but this failure caused him to work harder until the next year, when he got the privilege of riding the bench while Larry starred on the court. Then Larry's kid brother miraculously grew five inches between (b) his sophomore and junior year. He went on to be a McDonald's High School All-American, and he led his college team to the NCAA national championship, his Chicago Bulls to six NBA titles, and the US Olympic team to gold. (c) He set more individual records than any other player. Larry's kid brother was worth a cool $1.31 billion.

* sophomore: 2학년(의)

(C)

We sometimes forget that Larry Jordan's brother has also made his share of errors on and off the court. Michael famously confessed in a CBS interview, "I've missed more than nine thousand shots in my career. I've lost almost three hundred games. Twenty-six times, I've been trusted to take the game-winning shot and missed." He will also tell you that, inch for inch, he is the second-best player in his family. He says, "When you say 'Air Jordan,' I'm number two, (d) he's one."

(D)

Just before the skinny boy's sophomore year, the coach invited him to join his older brother at a basketball camp. He figured that Larry's DNA must be in his little brother. The coach was disappointed. He admired the kid's raw speed and developing skills but figured that (e) he was destined to remain small like the rest of his family. When the coach announced the varsity roster, all of the kid's friends who were six feet or taller were chosen. But Larry's brother was cut.

* varsity: 대표팀 ** roster: 명단

26 주어진 글 (A)에 이어질 내용을 순서에 맞게 배열한 것으로 가장 적절한 것은? 25007-0226

① (B) − (D) − (C)   ② (C) − (B) − (D)   ③ (C) − (D) − (B)
④ (D) − (B) − (C)   ⑤ (D) − (C) − (B)

27 밑줄 친 (a)~(e) 중에서 가리키는 대상이 나머지 넷과 <u>다른</u> 것은? 25007-0227

① (a)   ② (b)   ③ (c)   ④ (d)   ⑤ (e)

28 윗글에 관한 내용으로 적절하지 <u>않은</u> 것은? 25007-0228

① Larry에게는 한 살 어린 남동생이 있었다.
② Larry의 남동생은 미국 올림픽 팀을 금메달로 이끌었다.
③ Larry의 남동생은 자신을 가족 중 두 번째로 가장 훌륭한 선수라고 말했다.
④ Larry의 남동생은 학교 농구 캠프에 참가했다.
⑤ Larry의 남동생은 코치가 발표한 대표팀에 선발되었다.

한눈에 보는 **정답**

Exercise 강	1	2	3	4	5	6	7	8	9	10	11	12
01	②	③	③	①	④	⑤	④	②	①	①	①	④
02	①	③	④	①	①	⑤	⑤	⑤	④	⑤	②	④
03	②	②	④	①	①	②	②	③	①	①	②	⑤
04	①	③	①	①	①	⑤	③	④	⑤	②	④	⑤
05	②	①	⑤	①	④	④	④	①	④	③	④	⑤
06	④	⑤	④	①	④	②	⑤	③	⑤	⑤	②	③
07	⑤	⑤	④	③	④	④	④	⑤	④	③	④	②
08	④	④	⑤	⑤	③	⑤	⑤	③	⑤	③	③	②
09	③	③	④	①	④	③	③	⑤	②	③	④	⑤
10	⑤	⑤	①	③	①	③	②	⑤	②	④	⑤	⑤
11	⑤	④	①	③	③	④	④	②	③	④	②	⑤
12	④	③	④	③	③	⑤	④	②	②	④	①	⑤

Mini Test

Mini Test 1

01	02	03	04	05	06	07	08	09	10
⑤	④	⑤	⑤	⑤	⑤	⑤	③	③	⑤

11	12	13	14	15	16	17	18	19	20
⑤	④	④	②	⑤	①	②	③	②	③

21	22	23	24	25	26	27	28		
②	④	①	⑤	④	④	②	③		

Mini Test 2

01	02	03	04	05	06	07	08	09	10
⑤	②	④	⑤	②	③	④	⑤	④	④

11	12	13	14	15	16	17	18	19	20
③	④	⑤	④	①	②	⑤	④	④	②

21	22	23	24	25	26	27	28		
⑤	⑤	①	④	⑤	④	⑤	⑤		

Mini Test 3

01	02	03	04	05	06	07	08	09	10
②	④	②	③	⑤	⑤	②	④	⑤	⑤

11	12	13	14	15	16	17	18	19	20
④	③	④	②	①	④	②	④	②	②

21	22	23	24	25	26	27	28		
④	②	②	⑤	⑤	④	④	⑤		

고1~2, 내신 중점

고교

구분	고교 입문 >		기초 >	기본 >	특화	+ 단기
국어	고등예비과정	내 등급은?	윤혜정의 개념의 나비효과 입문 편 + 워크북 / 어휘가 독해다! 수능 국어 어휘	**기본서** 올림포스 ―― 올림포스 전국연합학력평가 기출문제집 ―― **유형서** 올림포스 유형편	**국어 특화** 국어 독해의 원리 / 국어 문법의 원리	단기 특강
영어			정승익의 수능 개념 잡는 대박구문 / 주혜연의 해석공식 논리 구조편		**영어 특화** Grammar POWER / Reading POWER / Listening POWER / Voca POWER — **영어 특화** 고급영어독해	
수학			**기초** 50일 수학 + 기출 워크북 / 매쓰 디렉터의 고1 수학 개념 끝장내기		**고급** 올림포스 고난도 — **수학 특화** 수학의 왕도	
한국사 사회				**기본서** 개념완성 ―― 개념완성 문항편	고등학생을 위한 多담은 한국사 연표	
과학			50일 과학		**인공지능** 수학과 함께하는 고교 AI 입문 / 수학과 함께하는 AI 기초	

과목	시리즈명	특징	난이도	권장 학년
전 과목	고등예비과정	예비 고등학생을 위한 과목별 단기 완성		예비 고1
	내 등급은?	고1 첫 학력평가 + 반 배치고사 대비 모의고사		예비 고1
국/영/수	올림포스	내신과 수능 대비 EBS 대표 국어·수학·영어 기본서		고1~2
	올림포스 전국연합학력평가 기출문제집	전국연합학력평가 문제 + 개념 기본서		고1~2
	단기 특강	단기간에 끝내는 유형별 문항 연습		고1~2
한/사/과	개념완성&개념완성 문항편	개념 한 권 + 문항 한 권으로 끝내는 한국사·탐구 기본서		고1~2
국어	윤혜정의 개념의 나비효과 입문 편 + 워크북	윤혜정 선생님과 함께 시작하는 국어 공부의 첫걸음		예비 고1~고2
	어휘가 독해다! 수능 국어 어휘	학평·모평·수능 출제 필수 어휘 학습		예비 고1~고2
	국어 독해의 원리	내신과 수능 대비 문학·독서(비문학) 특화서		고1~2
	국어 문법의 원리	필수 개념과 필수 문항의 언어(문법) 특화서		고1~2
영어	정승익의 수능 개념 잡는 대박구문	정승익 선생님과 CODE로 이해하는 영어 구문		예비 고1~고2
	주혜연의 해석공식 논리 구조편	주혜연 선생님과 함께하는 유형별 지문 독해		예비 고1~고2
	Grammar POWER	구문 분석 트리로 이해하는 영어 문법 특화서		고1~2
	Reading POWER	수준과 학습 목적에 따라 선택하는 영어 독해 특화서		고1~2
	Listening POWER	유형 연습과 모의고사·수행평가 대비 올인원 듣기 특화서		고1~2
	Voca POWER	영어 교육과정 필수 어휘와 어원별 어휘 학습		고1~2
	고급영어독해	영어 독해력을 높이는 영미 문학/비문학 읽기		고2~3
수학	50일 수학 + 기출 워크북	50일 만에 완성하는 초·중·고 수학의 맥		예비 고1~고2
	매쓰 디렉터의 고1 수학 개념 끝장내기	스타강사 강의, 손글씨 풀이와 함께 고1 수학 개념 정복		예비 고1~고1
	올림포스 유형편	유형별 반복 학습을 통해 실력 잡는 수학 유형서		고1~2
	올림포스 고난도	1등급을 위한 고난도 유형 집중 연습		고1~2
	수학의 왕도	직관적 개념 설명과 세분화된 문항 수록 수학 특화서		고1~2
한국사	고등학생을 위한 多담은 한국사 연표	연표로 흐름을 잡는 한국사 학습		예비 고1~고2
과학	50일 과학	50일 만에 통합과학의 핵심 개념 완벽 이해		예비 고1~고1
기타	수학과 함께하는 고교 AI 입문/AI 기초	파이선 프로그래밍, AI 알고리즘에 필요한 수학 개념 학습		예비 고1~고2

고2~N수, 수능 집중

구분	수능 입문 >	기출/연습 >	연계 + 연계 보완 >	고난도 >	모의고사
국어	윤혜정의 개념/패턴의 나비효과 / [기본서] 수능 빌드업	윤혜정의 기출의 나비효과	수능특강 문학 연계 기출	수능특강 사용설명서 / 수능완성 사용설명서	하루 3개 1등급 국어독서	FINAL 실전모의고사
영어	수능특강 Light / [강의노트] 수능 개념	수능 기출의 미래	수능연계교재의 VOCA 1800 / 수능연계 기출 Vaccine VOCA 2200 / 수능 영어 간접연계 서치라이트	하루 6개 1등급 영어독해 / 수능연계완성 3주 특강	만점마무리 봉투모의고사 시즌1 / 만점마무리 봉투모의고사 시즌2
수학	수능 감(感)잡기	수능 기출의 미래 미니모의고사	수능 연계교재 [감수] 수능특강	[감수] 수능완성		만점마무리 봉투모의고사 고난도 Hyper
한국사 사회	수능 스타트	수능특강Q 미니모의고사	eBook 전용 수능완성R 모의고사 / 수능 등급을 올리는 변별 문항 공략	박봄의 사회·문화 표 분석의 패턴	수능 직전보강 클리어 봉투모의고사
과학					

구분	시리즈명	특징	난이도	영역
수능 입문	윤혜정의 개념/패턴의 나비효과	윤혜정 선생님과 함께하는 수능 국어 개념/패턴 학습		국어
	수능 빌드업	개념부터 문항까지 한 권으로 시작하는 수능 특화 기본서		국/수/영
	수능 스타트	2028학년도 수능 예시 문항 분석과 문항 연습		사/과
	수능 감(感) 잡기	동일 소재·유형의 내신과 수능 문항 비교로 수능 입문		국/수/영
	수능특강 Light	수능 연계교재 학습 전 가볍게 시작하는 수능 도전		영어
	수능개념	EBSi 대표 강사들과 함께하는 수능 개념 다지기		전 영역
기출/연습	윤혜정의 기출의 나비효과	윤혜정 선생님과 함께하는 까다로운 국어 기출 완전 정복		국어
	수능 기출의 미래	올해 수능에 딱 필요한 문제만 선별한 기출문제집		전 영역
	수능 기출의 미래 미니모의고사	부담 없는 실전 훈련을 위한 기출 미니모의고사		국/수/영
	수능특강Q 미니모의고사	매일 15분 연계교재 우수문항 풀이 미니모의고사		국/수/영/사/과
	수능완성R 모의고사	과년도 수능 연계교재 수능완성 실전편 수록		수학
연계 + 연계 보완	수능특강	최신 수능 경향과 기출 유형을 반영한 종합 개념 학습		전 영역
	수능특강 사용설명서	수능 연계교재 수능특강의 국어·영어 지문 분석		국/영
	수능특강 문학 연계 기출	수능특강 수록 작품과 연관된 기출문제 학습		국어
	수능완성	유형·테마 학습 후 실전 모의고사로 문항 연습		전 영역
	수능완성 사용설명서	수능 연계교재 수능완성의 국어·영어 지문 분석		국/영
	수능 영어 간접연계 서치라이트	출제 가능성이 높은 핵심 간접연계 대비		영어
	수능연계교재의 VOCA 1800	수능특강과 수능완성의 필수 중요 어휘 1800개 수록		영어
	수능연계 기출 Vaccine VOCA 2200	수능 - EBS 연계와 평가원 최다 빈출 어휘 선별 수록		영어
고난도	하루 N개 1등급 국어독서/영어독해	매일 꾸준한 기출문제 학습으로 완성하는 1등급 실력		국/영
	수능연계완성 3주 특강	단기간에 끝내는 수능 1등급 변별 문항 대비		국/수/영
	박봄의 사회·문화 표 분석의 패턴	박봄 선생님과 사회·문화 표 분석 문항의 패턴 연습		사회탐구
	수능 등급을 올리는 변별 문항 공략	EBSi 선생님이 직접 선별한 고변별 문항 연습		수/영
모의고사	FINAL 실전모의고사	EBS 모의고사 중 최다 분량 최다 과목 모의고사		전 영역
	만점마무리 봉투모의고사 시즌1/시즌2	실제 시험지 형태와 OMR 카드로 실전 연습 모의고사		전 영역
	만점마무리 봉투모의고사 고난도 Hyper	고난도 문항까지 국·수·영 논스톱 훈련 모의고사		국·수·영
	수능 직전보강 클리어 봉투모의고사	수능 직전 성적을 끌어올리는 마지막 모의고사		국/수/영/사/과

memo

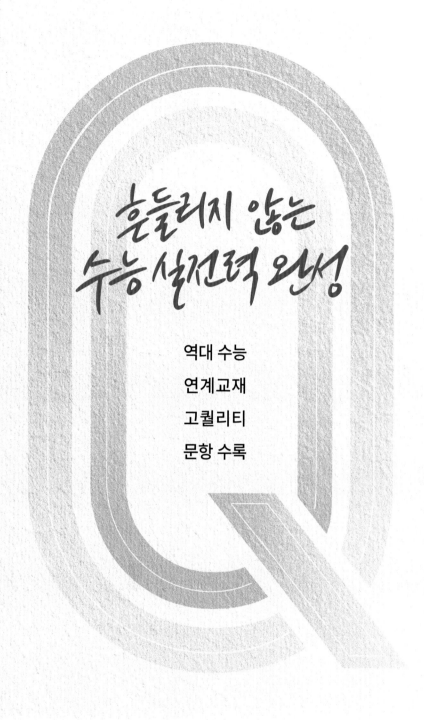

훈들리지 않는
수능 실전력 완성

역대 수능

연계교재

고퀄리티

문항 수록

14회분
수록

미니모의고사로 만나는 수능연계 우수 문항집

수능특강Q
미니모의고사

국 어	Start / Jump / Hyper
수 학	수학Ⅰ / 수학Ⅱ / 확률과 통계 / 미적분
영 어	Start / Jump / Hyper
사회탐구	사회 · 문화
과학탐구	생명과학Ⅰ / 지구과학Ⅰ

내가 뭘 하고 싶은지
뭘 잘 하는지
나도 모를 때가 있어
그럴때 나는

우리에게 물어봐
서울여대

SWU

*전공자율선택제 「자유전공」 운영

**2026학년도
신·편입학 모집**
● 입학처: https://admission.swu.ac.kr
● 입학상담 및 문의: 02-970-5051~4

 서울여자대학교

수능특강

정답과 해설

2026학년도
수능 연계교재

Lucky Box!

한국교육과정평가원
감 수
본 교재는 2026학년도 수능
연계교재로서 한국교육과정
평가원이 감수하였습니다.

Good Luck!

영어영역
영어독해연습

본 교재는 대학수학능력시험을 준비하는 데 도움을 드리고자 영어과 교육과정을 토대로 제작된 교재입니다.
학교에서 선생님과 함께 교과서의 기본 개념을 충분히 익힌 후 활용하시면 더 큰 학습 효과를 얻을 수 있습니다.

세계와 같이
성신같이

혁신도, 미래도 세계와 같이!
리더십도, 글로벌도 성신같이!
같이를 가치로 만드는 인재를 키웁니다

중국어문·문화학과
조윤솔

성신!
BEYOND THE BEST

2026학년도 성신여자대학교 신입학 모집

입학관리실 | ipsi.sungshin.ac.kr   입학상담 | 02-920-2000

성신여자대학교
SUNGSHIN WOMEN'S UNIVERSITY

수능특강

영어영역 | **영어독해연습**

정답과 해설

틀리기 쉬운 유형편

| 1 ② | 2 ③ | 3 ③ | 4 ① | 5 ④ | 6 ⑤ |
| 7 ④ | 8 ② | 9 ① | 10 ① | 11 ① | 12 ④ |

Exercise 1 정답 ②

소재 전형성에 의존하지 않는 범주 판단

해석 전형성에 의존하지 '않을' 때 범주 판단은 어떻게 이루어지는가? 이 질문에 대한 접근 방법으로 한 가지 예에 대해 충분히 생각해 보자. 레몬을 생각해 보자. 레몬에 빨간색과 흰색 줄무늬를 칠해 보라. 그것은 여전히 레몬인가? 대부분의 사람은 그렇다고 말한다. 이제 레몬에 달콤한 맛이 나도록 설탕물을 주입하라. 그런 후 팬케이크처럼 편평해지도록 트럭으로 레몬 위를 달려 보라. 이 시점에서 우리는 무엇을 얻었는가? 줄무늬가 있으며, 인위적으로 달콤하고, 편평해진 레몬인가? 아니면 우리는 레몬이 아닌 것을 얻었는가? 대부분의 사람은 여전히 이 가엾고 학대받은 과일을 레몬으로 인정하지만, 이러한 판단이 무엇을 포함하는지 생각해 보라. 우리는 이 물체를 (레몬의) 원형으로부터 점점 더 멀어지게 하고, 또한 여러분이 이제껏 접해 본 어떤 특정한 레몬과도 매우 다르게 만드는 조치를 취했다. 하지만 이것이 그 물체가 여전히 레몬이라는 여러분의 믿음을 흔들지는 않는 것 같다. 분명히, 우리는 쉽게 알아볼 수 없는 레몬, 즉 예외적인 레몬을 가지고 있지만, 그것은 여전히 레몬이다. 보아하니 어떤 것이 다른 레몬과 거의 닮지 않아도 레몬일 수 있는 것 같다.

문제 해설
어떠한 물체가 원형에서 점점 더 멀어지게 되어 전형적인 형태를 취하지 않아도 여전히 그 물체로 판단될 수 있다는 내용이므로, 빈칸에 들어갈 말로 가장 적절한 것은 ② '전형성'이다.
① 전통
③ 분류
④ 합리성
⑤ 기능성

구조 해설
■ But this seems not to shake your faith [that the object remains a lemon].
[ ]는 your faith의 구체적 내용을 설명하는 동격절이다.

어휘 및 어구
category 범주, 구분
think through ~에 대해 충분히 생각하다
inject 주입하다
artificially 인위적으로, 인공적으로
abuse 학대하다, 혹사하다
object 물체
distant 멀리 있는
encounter 접하다, (우연히) 만나다
exceptional 예외적인
apparently 보아하니, 외관상으로는
virtually 거의, 사실상
resemblance 닮음, 비슷함

Word Search
| 정답 | 1. category 2. encountered 3. exceptional
1. 이 새로운 제품은 전자 제품 범주에 속한다.
2. 나는 지난주에 시내 커피숍에서 옛 친구를 우연히 만났다.
3. 제출 마감일은 예외적인 상황에서만 연장될 것이다.

Exercise 2 정답 ③

소재 유인책이 동기에 미치는 영향

해석 유인책이 동기를 저해한다는 것을 발견한 연구들의 정말 많은 것들이 아이들을 대상으로 수행된 데에는 이유가 있다. 아이들은 자신이 좋아하는 것과 싫어하는 것을 알아내느라 바쁘다. 내가 8살짜리 아들에게 특정한 학교 과목을 좋아하는지 물어보면, 그는 그것에 대해 생각해야 하고, 여러분이 알 수도 있는 것처럼 직관적으로 답을 알지 못한다. 아이들은 주로 어른들에 의해 통제되는 세상에 비교적 익숙하지 않기 때문에, 그들의 하루를 차지하는 활동 중 많은 것이 설명을 필요로 한다. 그들은 "내가 그림 그리는 것을 좋아해서 그림을 그리고 있는 건가, 아니면 선생님이 그림을 그리게 시켜서 그림을 그리고 있는 건가?" 또는 "이 음식이 내 입맛에 맞는 건가, 아니면 이 음식을 먹지 않으면 디저트를 못 먹기 때문에 먹고 있는 건가?"라고 자신에게 물어볼 수도 있다. 유인책은 그들에게 자신이 좋아하는 것과 싫어하는 것을 종합하여 이해하기 시작할 단서를 제공한다. 그리고 만약 여러분이 아이인데, 한 어른이 여러분에게 기꺼이 돈을 주며 무언가를 하게 한다면, 그것은 그렇지 않으면 여러분이 그것을 하는 것을 즐기지 않을 것이라는 단서이다.

문제 해설
아이들은 어렸을 때 자신이 좋아하는 것과 싫어하는 것을 알아내는 과정을 거치는데, 어른이 아이에게 무언가를 하도록 할 때 돈과 같은 유인책을 제시하면 아이는 유인책이 제시되지 않는다면 그 일을 하는 것을 즐기지 않으리라고 인식하여 결국 하려는 동기가 저해된다는 내용의 글이다. 따라서 빈칸에 들어갈 말로 가장 적절한 것은 ③ '유인책이 동기를 저해한다'이다.
① 우리의 직관을 믿는 것이 득이 된다
② 우리는 우리 자신의 사리사욕을 위해 행동한다
④ 오래된 선호는 평생 지속될 수 있다
⑤ 보상은 처벌보다 더 효과가 있다

구조 해설

■ And [if you're a child, and an adult is willing to pay you to do something], that's a clue [that you wouldn't otherwise enjoy doing it].
첫 번째 [ ]는 조건의 부사절이다. 두 번째 [ ]는 a clue와 동격 관계를 이룬다.

어휘 및 어구

conduct 수행하다, 실시하다
figure out ~을 알아내다
dislike 싫어하는 것, 혐오
intuitively 직관적으로
relatively 비교적
occupy 차지하다
incentive 유인책
clue 단서
piece together ~을 종합하여 이해하다

Word Search

| 정답 | **1.** dislike **2.** occupy **3.** incentive
1. 그는 무례하고 버릇없는 사람들에 대한 혐오가 있다.
2. 직장에서의 문제가 그의 마음을 한동안 계속 차지했다.
3. 현금 보너스는 직원들에게 있어 강력한 유인책임이 드러났다.

Exercise 3
정답 ③

소재 경쟁에서의 과장

해석 경쟁이 폭력으로 격화될 때 보통 두 당사자 모두 비용을 치르기 때문에, 같은 종의 구성원 간의 경쟁은 보통 상대방에게 물러나도록 설득하려고 의도된 각 당사자에 의한 진실과 과장의 조합이다. 만약 자신의 능력을 상대방의 능력에 견주어 시험하는 데 비용이 발생하지 않는다면 과장은 사라질 것이다. 만약 우리가 마지막 남은 케이크 한 조각을 두고 경쟁하는데 내가 여러분보다 더 강할 수도 있다고 생각하면, 나는 그냥 여러분을 주먹으로 가격하여 알아낼 것이다. 하지만 이 시험에는 현저한 비용이 존재하는데, 여러분이 나를 되받아 가격할 가능성이 있기 때문이다. 이것은 최선의 상황에서도 불쾌한 경험이겠지만, 여러분이 나보다

강하다면 특히 그렇다. 속이는 개체가 반드시 들키지 않고도 그들의 강점을 과장하고 약점을 축소할 수 있도록 하는 것은 바로 이러한 경쟁에 대한 확실한 비용이다. 이러한 유형의 과장은 동물계 전체에 걸쳐 목격되는데, 예컨대 무스나 하이에나가 등 뒤에 목덜미 털을 세워 (몸집을) 더 커 보이게 하는 경우나, 게가 근육으로 채우지 않은 불필요하게 큰 집게발 껍질을 키우는 경우가 그러하다.

문제 해설
같은 종의 구성원 간의 경쟁이 폭력으로 격화될 때 비용을 치러야 하므로 개체가 강점을 과장하고 약점을 축소하여 상대방이 물러나도록 한다는 내용의 글이다. 따라서 빈칸에 들어갈 말로 가장 적절한 것은 ③ '경쟁에 대한 확실한 비용'이다.
① 행동하지 않음의 막대한 대가
② 조화에 대한 끊임없는 추구
④ 물리적 갈등에 대한 타고난 선호
⑤ 상대에 대한 과도한 과소평가

구조 해설

■ **It is** [this guaranteed cost of competition] **that** allows deceptive individuals [to {exaggerate their strengths} and {play down their weaknesses} without necessarily getting caught].
「It is ~ that ...」 강조 구문을 사용하여 첫 번째 [ ]를 강조하고 있다. 두 번째 [ ]는 allows의 목적격 보어로 그 안에 동사구인 두 개의 { }가 and로 연결되어 to에 이어진다.

어휘 및 어구

party 당사자
escalate 격화되다, 확대되다
contest 경쟁
blend 조합, 혼합
exaggeration 과장
opponent 상대방
compete 경쟁하다
deceptive 속이는, 기만적인
play down ~을 축소하다, ~을 경시하다
moose 무스(사슴과에서 가장 큰 동물)
crab 게
claw 집게발

Word Search

| 정답 | **1.** escalate **2.** exaggeration **3.** deceptive
1. 두 경쟁 국가 간의 긴장은 계속 격화되었다.
2. 그녀가 잡은 물고기가 고래만큼 크다는 그녀의 주장은 명백한 과장이었다.
3. 그 광고의 기만적인 주장은 많은 소비자를 오도했다.

Exercise 4

정답 ①

소재 타인의 반응이 도움 행동에 미치는 영향

해석 타인의 반응은 어떤 한 개인이 도움을 주기로 결정할지 여부에 영향을 미칠 수 있다. 아무도 결국 응급 상황이 아닐 수도 있는 경우에 어리석게 서둘러 돕고 싶어 하지 않는다. 사실, 사람들은 때때로 타인 앞에서 어리석어 보이는 것이 두려워 행동에 나서지 못한다. 그래서 우리는 보통 침착함을 유지하며 주변에 있는 타인이 무엇을 하고 있는지 확인해 본다. 당연히, 다른 모든 사람들도 타인의 반응을 확인하면서 침착함을 유지하고 있다면, 모두가 도움이 필요하지 않거나 규범상 도움을 주는 것이 부적절하다고 결론을 내릴 것이다. 한 일련의 연구에서 실험자들은 학생들이 앉아서 설문지를 작성하고 있는 실험실에 연기가 쏟아져 들어오도록 설계했다. 학생들만 있을 때, 그 특이한 상황에서의 그들의 우려는 곧 그들이 도움을 구하게 만들었다. 하지만 방에 있는 두 명의 (실험) 공모자가 연기에 대해 아무런 반응을 보이지 않을 때는, 참가자들도 아무런 행동을 취하지 않았다. 사람들이 구경하는 사람들이나 지나가는 사람들이 반응하지 않는 것을 인지하게 될 때, 그러한 관찰은 그들이 도움을 줄 가능성을 줄인다. 따라서 방관하는 사람들의 존재가 도움을 주는 것에 영향을 미칠 수 있는 한 가지 방식은 도움을 주는 것이 규범에 어긋난다고 암시하는 것이다.

문제 해설

사람들은 도움을 주는 행동을 하고자 할 때 타인의 반응을 확인하는데, 구경하는 사람들이 아무 반응을 보이지 않을 경우 그것은 도움을 주는 것이 규범에 어긋난다고 사람들에게 암시하여 도움 행동이 일어나지 않도록 유도할 수 있다는 내용의 글이다. 따라서 빈칸에 들어갈 말로 가장 적절한 것은 ① '도움을 주는 것이 규범에 어긋난다고 암시하는 것'이다.
② 타인이 불필요한 위험을 감수하는 것을 막는 것
③ 안전을 확보할 수 있는 규범을 따르는 것
④ 사람들에게 보상에 주목하도록 장려하는 것
⑤ 규범은 모든 사람에게 적용되어야 한다고 강조하는 것

구조 해설

- In one series of studies, experimenters arranged for smoke to pour into a laboratory room [in which students were sitting {completing questionnaires}].
  [ ]는 a laboratory room을 수식하는 관계절이고, 그 안의 { }는 students의 부수적인 동작을 나타내는 분사구문이다.

어휘 및 어구

rush 서두르다
keep calm 침착함을 유지하다
present 주변에 있는
norms 규범
pour into ~에 쏟아져 들어오다
laboratory room 실험실
questionnaire 설문지
bystander 구경하는 사람, 방관자
passerby 지나가는 사람, 행인
unresponsive 반응하지 않는
likelihood 가능성

Word Search

| 정답 | **1.** norms **2.** passerby **3.** unresponsive
1. 각 문화는 고유한 사회 규범을 발전시킨다.
2. 그 행인은 붐비는 거리에서 자동차 사고를 목격한 후 재빨리 도움을 요청했다.
3. 고객 서비스 담당자가 내 문의에 계속 응답하지 않아 나를 불만스럽게 했다.

Exercise 5

정답 ④

소재 자연 철학이 더 오래된 문화권에서 시작되지 못한 이유

해석 많은 업적에도 불구하고 더 오래된 문화권(이집트, 바빌로니아, 인도, 중국)이 아니라 고대 그리스인들에서부터 자연 철학을 시작하는 데에는 심오한 이유가 있다. 이 더 오래된 문화권은 기술적 지식, 예리한 관찰력, 그리고 자료와 정보의 방대한 자산을 가지고 있었음에도 불구하고, 자연 세계와 초자연 세계를 분리하지 않았기 때문에 자연 철학을 만들어 내지 못했다. 고대 제국의 종교는 초자연적인 존재와 힘이 물질세계를 지배하고 거기에 깃들어 있으며, 이러한 초자연적인 힘의 작용에 대한 이유는 대체로 알 수 없다는 믿음에 근거했다. 4대 강 문화권 사회에는 많은 기술 발전이 있었지만, 지적 유산은 사제들이 지배했으며, 물질세계에 대한 그들의 관심은 그들의 신학 개념을 확장한 것이었다. 이집트, 바빌로니아, 아즈텍 제국과 같은 많은 고대 문명은 종교 활동에 (사회의 시간, 부, 기술 및 공공 공간과 같은 것을 포함하는) 사회 자본의 상당 부분을 사용했다.

문제 해설

이집트, 바빌로니아, 인도, 중국과 같은 더 오래된 문화권이 아니라 고대 그리스인들에서부터 자연 철학을 시작하는 이유는, 더 오래된 문화권은 초자연적인 존재와 힘이 물질세계를 지배하고 거기에 깃들어 있다는 믿음에 근거한 종교를 가지고 있었고, 물질세계에 대한 사제들의 관심은 신학 개념을 확장한 것이었으며, 종교 활동에 사회 자본의 상당 부분을 사용하여 자연 철학을 만들어 내

는 데 실패했기 때문이라는 내용의 글이다. 따라서 빈칸에 들어갈 말로 가장 적절한 것은 ④ '자연 세계와 초자연 세계를 분리하지 않았기'이다.

① 기술적 지식을 자신들의 종교적 틀에 통합시킬 수 없었기
② 철학적 노력을 실용적인 추구에 비해 부차적인 것으로 간주했기
③ 자신의 지적 유산을 전달할 수 있는 체계가 없었기
⑤ 추상화와 영성의 중요성을 경시했기

구조 해설

- [Many ancient civilizations, such as the Egyptian, Babylonian, and Aztec empires], spent [a large proportion of social capital {(covering such things as the time, wealth, skill, and public space of the society)}] on religious activity.

  첫 번째 [ ]는 문장의 주어부이다. 두 번째 [ ]는 spent의 목적어인데, 그 안의 { }는 social capital을 수식하는 분사구이다.

어휘 및 어구

profound 심오한, 의미심장한
natural philosophy 자연 철학
keen 예리한
vast 방대한
inhabit 깃들다, 거주하다
supernatural 초자연적인
heritage 유산
dominate 지배하다
priest 사제
extension 확장
proportion 부분, 비율
religious 종교의

Word Search

| 정답 | **1.** philosophy **2.** heritage **3.** religious
1. René Descartes는 근대 철학의 창시자로 여겨진다.
2. 부족의 공유된 문화 유산은 깊은 공동체 의식과 소속감을 조성했다.
3. 그 고대 사원은 지역 사회에 깊은 종교적 의미를 지니고 있었다.

Exercise 6 ————————————— 정답 ⑤

소재 문서를 매개로 한 통치 형태의 등장

해석 근대 초기 유럽에서 국가가 형성되고 통치 체제가 중앙 집권화되는 과정에는 점점 더 많은 양의 정보 사용이 수반되었다. 역사학자들은 편지 쓰기, 보고서 작성하기 및 주석 달기, 서식 및 설문지 발부하기 등과 같은, 캐나다의 사회학자 Dorothy Smith가 '문서를 매개로 한 통치 형태'라고 지칭했으며, 정보 국가, 기록 보관 국가 또는 문서 국가라고 다양하게 알려져 있고 현재는 디지털 국가로 변모하는 과정에 있는 것과 관련이 있는 통치 형태의 등장을 주목해 왔다. 이 과정은 용어의 원래 의미로서의 '관료제', 즉 관청의 국(局), 다시 말해 사무실과 그곳의 관리들에 의한 통치의 등장으로 설명될 수도 있다. 이 관리들은 서면으로 된 명령을 발부하고 따랐으며, 이러한 명령을 의사 결정에 도움을 주는 국내외 정치 상황 보고와 함께 서류철에 기록했다. 16세기 스페인의 펠리페 2세와 17세기 프랑스의 루이 14세의 유명한 사례에서처럼, 말을 탄 통치자는 점차 책상에 앉아 있는 통치자로 변모했다.

문제 해설

근대 초기 유럽에서 국가가 형성되고 정부가 중앙 집권화되는 과정에서 점점 더 많은 양의 정보 사용이 수반됨에 따라 문서를 매개로 한 관료제적 통치 형태가 등장했다는 내용이므로, 밑줄 친 부분이 의미하는 바로 가장 적절한 것은 ⑤ '국가를 통치하기 위해 문서를 통한 의사소통에 의존하는 지도자'이다.

① 결정에 도달하기 전에 폭넓은 토론에 참여하는 통치자
② 반대 세력을 억압하기 위해 정보의 흐름을 통제하는 독재자
③ 중앙 집권화된 권력과 공식적인 관료 체계가 없는 왕
④ 무력 충돌보다 평화적인 협상을 선호하는 국가 원수

구조 해설

- The ruler on horseback was gradually transformed into the ruler [sitting at his desk], [as in the famous cases of {⟨Philip II of Spain in the sixteenth century⟩ and ⟨Louis XIV of France in the seventeenth⟩}].

  첫 번째 [ ]는 the ruler를 수식하는 분사구이다. 두 번째 [ ]는 '~처럼'이라는 의미를 나타내는 부사구이고, 그 안의 { }는 of의 목적어인데, 명사구인 두 개의 ⟨ ⟩가 and로 연결되어 있다.

어휘 및 어구

formation 형성
centralization 중앙 집권화, 집중화
note 주목하다, 언급하다
sociologist 사회학자
issue 발부하다, 발급하다
questionnaire 설문지
archive 기록 보관
official 관리, 공무원
assist 도움을 주다
Philip II 펠리페 2세
Louis XIV 루이 14세

Word Search

| 정답 | **1.** centralization **2.** issued **3.** assist
1. 지나친 집중화는 개인의 창의성과 혁신을 가로막는 경우가 많다.
2. 신입 회원에게는 임시 신분증이 발급될 예정이다.
3. 새로운 소프트웨어는 사용자가 재정을 관리하는 것을 돕도록 설계되었다.

Exercise 7
정답 ④

소재 반복 실험의 의의

해석 반복 실험의 주요 목표는 관찰된 관계가 연구 가설에 대한 여러 다양한 실험에 걸쳐 일반화하는 정도를 알아내는 것이다. 그러나 어떤 연구 결과가 일반화하지 않는다고 해서 그것이 흥미롭지 않거나 중요하지 않다는 의미는 아니다. 실제로, 과학은 <u>이전에 입증된 관계에 대한 제한 조건을 발견함</u>으로써 나아간다. 모든 환경에서 그리고 모든 사람에게 적용되는 관계는 거의 없다. 시간이 지나면서 그 한계에 대한 더 많은 정보가 발견되면서 과학 이론은 수정된다. 예를 들어, 폭력적인 자료를 접하는 것이 공격성에 미치는 영향을 조사하는 연구에서 흥미로운 문제 중 하나는 폭력을 보는 것이 평균적으로 공격성을 높이는 경향이 있다는 것은 잘 알려져 있지만, 모든 사람에게 이런 현상이 일어나는 것은 아니라는 사실에 관한 것이다. 따라서 폭력적인 자료를 접하면 어떤 사람이 영향을 받고 어떤 사람이 영향을 받지 않을지 알아내기 위해 참가자 반복 실험을 실시하는 것이 매우 중요하다.

문제 해설
관찰된 관계가 모든 환경과 모든 사람에게 적용되는 경우는 거의 없고, 과학 이론은 그 한계에 대한 정보가 발견되면서 수정된다고 하면서 폭력적인 자료를 접하면 평균적으로 공격성이 높아진다고 하더라도 참가자 반복 실험을 통해 그것에 영향을 받는 사람과 더불어 영향을 받지 않는 사람을 알아내는 것이 중요하다고 하였다. 따라서 빈칸에 들어갈 말로 가장 적절한 것은 ④ '이전에 입증된 관계에 대한 제한 조건을 발견함'이다.
① 특정 그룹을 위해 설계된 전문 지식을 개발함
② 일반화될 수 있는 가설의 발견을 강조함
③ 서로 다른 연구 분야 전체에서 수많은 가설을 생성함
⑤ 인류가 환경에 의해 부과된 한계에 적응할 수 있도록 도움

구조 해설
- The primary goal of replication is [to determine the extent {to which an observed relationship generalizes across different tests of the research hypothesis}].

[　]는 주격 보어 역할을 하는 to부정사구이다. 그 안의 {　}는 the extent를 수식하는 관계절이다.

어휘 및 어구
primary 주요한
determine 알아내다, 결정하다
extent 정도, 범위
generalize 일반화하다, 보편화하다
hypothesis 가설
proceed 나아가다
modify 수정하다
investigate 조사하다
exposure to ~을 접(하게)함, 직접 체험(하게)함
aggression 공격성
conduct 실시하다

Word Search

| 정답 | **1.** primary **2.** exposure **3.** aggression
1. 지구 온난화의 주요 원인은 무엇인가?
2. 부모는 자녀에게 다른 문화를 접하게 하기 위해 아이와 함께 해외 여행을 한다.
3. 야생에서 동물들은 자신의 영역을 지키기 위해 공격성을 자주 드러낸다.

Exercise 8
정답 ②

소재 운전 중 휴대 전화 통화 시 방해 요소

해석 운전 중 휴대 전화로 대화하는 것이 예를 들어 차 안 동승자와 대화하는 것보다 훨씬 더 방해가 되는 이유가 무엇인지 물을지도 모른다. 그럴듯한 이유 하나는 휴대 전화로 통화할 때 <u>상황에 대한 통제력 상실</u>이다. 차 안의 동승자는 후자가 까다로워질 때 운전자가 운전이라는 주요 과업에 집중할 필요가 있다는 것을 비언어적 신호로부터 으레 알아챈다. 원격 대화자는 이러한 신호를 알아차릴 가능성이 훨씬 더 적으며, 따라서 주요 운전 과업에 자원을 쏟도록 부수적인 과업이 종료될 필요가 있을 때 인지적으로 부담이 되는 대화를 으레 계속한다. 인지적으로 부담이 되는 대화, 특히 운전자가 자신의 인지 자원의 배분을 역동적으로 조정하는 면에서 통제력이 거의 또는 전혀 없는 대화는 아마도 감각 입력에 대한 주의 감소의 결과로 운전의 과업에 필요한 속도, 거리 및 폭 계산에 방해가 되는 것으로 보인다. 또한 휴대 전화를 사용하면 키패드에 전화번호를 입력하는 것과 같은 다른 부수적인 과업이 필요한데, 이는 또한 주요 운전 과업에 방해가 되는 경향이 있을 것이다.

문제 해설

운전을 하면서 차 안의 동승자와 대화를 하는 경우 운전이라는 주요 과업에 집중할 필요가 있다는 것을 비언어적 신호로부터 으레 알아채지만, 운전을 하면서 휴대 전화로 대화를 하게 되면 원격 대화자가 운전자의 비언어적 신호를 알아채기 어려워 인지적으로 부담이 되는 대화가 계속 이어지고, 운전자가 통제하지 못하는 이러한 대화는 운전에 필요한 사항을 처리하는 데 방해가 된다고 하였다. 따라서 빈칸에 들어갈 말로 가장 적절한 것은 ② '상황에 대한 통제력 상실'이다.
① 배경 소음의 영향
③ 네트워크 안정성에 대한 의존성
④ 즉각적인 상호 작용의 억제
⑤ 언어적 단서 해석의 어려움

구조 해설

■ A remote interlocutor [is much less likely to pick up these cues] and [therefore will continue to make cognitively demanding conversation at a time {when the secondary task needs to be shut down to devote resources to the main driving task}].

두 개의 [ ]는 and로 대등하게 연결되어 문장의 술부 역할을 한다. 두 번째 [ ] 안의 { }는 a time을 수식하는 관계절이다.

어휘 및 어구

disruptive 방해가 되는
passenger 동승자
non-verbal 비언어적인
cue 신호, 단서
concentrate on ~에 집중하다
tricky 까다로운, 다루기 힘든
cognitively 인지적으로
demanding 부담이 되는, 힘든
secondary 부수적인, 부차적인
devote ~ to ... ~을 …에 쏟다[기울이다]
in terms of ~ 면에서
allocation 배분
interfere with ~을 방해하다
computation 계산
diminished 감소된

Word Search

| 정답 | 1. disruptive 2. cue 3. concentrate
1. 시끄러운 음악이 시험을 치르는 학생들에게 방해가 되는 영향을 주었다.
2. 그의 올라간 눈썹은 그가 계획에 대해 확신이 없다는 신호였다.
3. 선생님이 말씀하시는 것에 완전히 집중하도록 노력해 주세요.

Exercise 9
정답 ①

소재 묶음 서비스

해석 기업은 편의성이나 마케팅 목적으로 상품이나 서비스를 자주 묶음으로 제공한다. 신발 판매 회사는 왼발과 오른발을 따로 판매할 수도 있겠지만, 거의 모든 소비자들은 묶음을 구매하고 싶어 할 것이다. 묶음 제공은 소비자들이 관련 상품들에 대해 불완전하게 상관된 선호도를 가질 때 판매자가 더 높은 수익을 얻는 데에도 도움이 될 수 있다. 예를 들어, 유선방송 사업체는 일반적으로 스포츠, 음식, 드라마, 그리고 뉴스를 전문으로 하는 채널을 포함하여 다양한 프로그램을 제공한다. 유선방송 사업체는 고객들에게 채널을 '따로 골라' 구매하는 것을 허용할 수도 있을 것인데, 스포츠 팬은 스포츠 채널만 구매할 수 있고, 다른 팬은 다른 채널만 구매할 수 있는 것 등이다. 하지만 유선방송 사업체는 대신에 개별로 따로 골라 사는 가격보다 지나치게 더 높지는 않은 단일 묶음 제공 가격을 책정한다. (예를 들어, '스포츠+음식+드라마+뉴스' 묶음의 가격은 그 사업체가 스포츠 패키지에 대해서만 청구할 가격보다 훨씬 더 높지는 않다.) 유선방송 사업체는 묶음 판매에 따른 한계 비용이 본질적으로 제로이므로 이러한 관행이 수익 증대에 도움이 된다.

문제 해설

유선방송 사업체의 경우 고객이 채널을 따로 골라 구매하도록 하는 것보다 여러 채널의 묶음을 제공하고, 개별로 따로 골라 사는 가격보다 지나치게 더 높지는 않은 묶음 제공 가격을 책정하는 것이 수익 증대에 도움이 된다는 내용의 글이므로, 빈칸에 들어갈 말로 가장 적절한 것은 ① '판매자가 더 높은 수익을 얻는 데에도 도움이 될'이다.
② 더 효율적인 자원 배분으로 이어질
③ 그럼에도 불구하고 고객의 혼란을 가중시킬
④ 때로는 묶음 상품에 대한 불만을 야기할
⑤ 소비자의 요구에 제품을 더 잘 맞추는 것을 용이하게 할

구조 해설

■ For example, the price for the "sports+food+drama+news" bundle is not **much** more than the price [the service would charge for the sports package alone].

much는 비교급 more를 수식하는 부사이고, [ ]는 the price를 수식하는 관계절이다.

어휘 및 어구

bundle 묶음으로 제공하다; 묶음
convenience 편의성
specialize in ~을 전문으로 하다
essentially 본질적으로, 기본적으로

practice 관행
profit 수익

| 정답 | **1.** convenience **2.** profit(s) **3.** specialize

1. 우리는 고객의 편의성을 위해 온라인 주문 선택을 확대했다.

2. 투자자들은 항상 주식시장에서 자신의 수익을 극대화할 기회를 찾고 있다.

3. 그 변호사는 자유와 정의를 위해 싸우기 위해 인권법을 전문으로 하기로 선택했다.

Exercise 10
정답 ①

소재 상상을 통한 현실 연습

해석 많은 연구에 따르면 뇌는 잘 상상된 경험과 실제 경험의 차이를 인식하지 못한다. 이 실험을 해 보라. 즙이 많은 보기 좋은 노란색 레몬을 여러분의 손에 갖고 있다고 상상해 보라. 자신이 레몬을 반으로 자르고 레몬의 즙이 가득한 둥근 부분을 바라보고 있다고 상상해 보라. 이제 자신이 레몬을 깨물고 있다고 상상해 보라. 많은 사람들과 마찬가지라면, 여러분은 침이 나오기 시작한다. 신맛 때문에 목에 약간의 긴장감을 느낄지도 모른다. 하지만 여러분은 실제 레몬이 없기 때문에 상상 속의 경험에 대한 생리적 반응이 일어나고 있다는 것을 알 수 있다. 준비하는 것도 마찬가지인데, 차분하고 편안하게 정시에 도착하는 것을 생생하게 상상할 수 있을수록 여러분의 몸은 그것이 실제 경험이라는 신호를 뇌로부터 더 많이 받게 된다. 마음속에 그려봄으로써 여러분은 현실을 위해 연습하고 있는 것이다.

문제 해설

레몬을 깨무는 상상만으로도 실제 침이 나오기 시작하게 되는 것처럼 상상이 실제 반응을 일으키고, 준비하는 것도 마찬가지로 차분하고 편안하게 정시에 도착하는 것을 생생하게 상상할수록 실제 경험이라는 신호를 뇌로부터 더 많이 받게 된다는 내용이므로, 빈칸에 들어갈 말로 가장 적절한 것은 ① '여러분은 현실을 위해 연습하고 있는 것이다'이다.

② 뇌는 새로운 정보를 더 잘 흡수한다

③ 여러분은 전통적인 사고에 도전하고 있는 것이다

④ 거짓을 분별하는 능력이 발달한다

⑤ 여러분은 성급하게 결론을 내리고자 하는 본능을 제어한다

구조 해설

▪ Many studies have shown [that the brain cannot recognize the difference between {a well-imagined experience} and {the real thing}].

[ ]는 shown의 목적어 역할을 하는 명사절이다. 그 안에서 명사구인 두 개의 { }는 and로 대등하게 연결되어 between의 목적어 역할을 하고 있다.

어휘 및 어구

recognize 인식하다
experiment 실험
juicy 즙이 많은
slice 자르다
throat 목, 목구멍
sourness 신맛
reaction 반응
organize (일정 등을) 준비하다
vividly 생생하게
visualize 마음속에 그리다

| 정답 | **1.** slicing **2.** reaction **3.** vividly

1. 나의 어머니는 수프에 넣기 위해 채소를 잘게 자르고 있다.

2. 자신의 승진 소식에 대한 그의 반응에는 놀라움과 기쁨이 섞여 있었다.

3. 나는 그 햇볕의 따스함을 마치 어제 일처럼 생생하게 기억한다.

Exercise 11
정답 ①

소재 연구 가설의 수정

해석 한 변수의 어느 정도의 양이 무슨 영향을 미쳤는지를 더 구체화함으로써 가설을 변경하는 것과 아울러, 한 변수의 어느 측면이 무슨 영향을 미쳤는지를 더 구체화함으로써 가설을 변경할 수 있다. 그러므로 만약 여러분의 가설이 일반적인 개념을 포함한다면, 그 다차원적인 개념을 개별적인 차원으로 나눈 다음 그 개별 구성 요소와 관련된 가설들을 세움으로써 여러분의 가설을 개선할 수 있을지도 모른다. 예를 들어, 사랑은 시간이 지남에 따라 더 커질 것이라는 가설을 세우기보다는, 사랑의 특정 측면들(헌신, 친밀감)은 시간이 지남에 따라 더 커지지만 반면에 다른 부분들(열정적인 사랑)은 그렇지 않을 것이라는 가설을 세울 수도 있다. 마찬가지로, 스트레스가 기억력을 방해한다고 말하기보다는, 기억의 어떤 부분이 스트레스의 영향을 가장 많이 받는지 찾아보려고 할 수도 있을 것이다. 그것(가장 큰 영향을 받는 부분)이 부호화, 되뇌기, 조직화, 또는 인출인가? 구성 요소 전략은 편견을 의식과 무의식 차원으로 세분화한 사회 심리학자들과 전반적인 (전체적인) 자존감을 여러 유형(신체적 자존감, 학업적 자존감, 사

회적 자존감 등)으로 세분화한 성격 심리학자들에게 성공적이었다.

(문제 해설)

사랑이라는 구성 개념을 헌신, 친밀감, 열정적인 사랑이라는 부분으로 나누거나, 기억을 부호화, 되뇌기, 조직화, 또는 인출이라는 부분으로 나눠서 생각하는 것처럼, 일반적 구성 개념을 개별 측면으로 세분화한 다음 그 개별 구성 요소와 관련된 가설을 세워 가설을 개선할 수 있다는 내용의 글이다. 가설의 변경 방법으로 앞에 제시된 '한 변수의 어느 정도의 양이 무슨 영향을 미쳤는지를 더 구체화함'에 이어 빈칸에는 추가적인 변수의 구체화 방법의 내용이 들어가야 할 것이므로, 빈칸에 들어갈 말로 가장 적절한 것은 ① '한 변수의 어느 측면이 무슨 영향을 미쳤는지'이다.
② 어떤 다른 변수들이 같은 영향을 미치는지
③ 주요 변수를 식별하는 데 있어 표본 크기의 역할
④ 어떤 방법론이 변수를 측정하는 데 사용되었는지
⑤ 변수의 문화적, 역사적 배경

(구조 해설)

- Thus, [if your hypothesis involves a general construct], you may be able to improve your hypothesis by [breaking that multidimensional construct down into its individual dimensions] and then [making hypotheses {involving those individual components}].

  첫 번째 [ ]는 조건을 나타내는 부사절이고, 두 번째와 세 번째 [ ]는 and로 대등하게 연결되어 전치사 by의 목적어를 이룬다. 세 번째 [ ] 안의 { }는 hypotheses를 수식하는 분사구이다.

(어휘 및 어구)

hypothesis 가설(pl. hypotheses)
variable 변수
involve 포함하다
construct (구성) 개념
break ~ into ... ~을 …으로 나누다
multidimensional 다차원적인
dimension 차원
component 구성 요소
aspect 측면
commitment 헌신
intimacy 친밀감
interfere with ~을 방해하다
encoding 부호화(기억의 첫 단계)
rehearsal 되뇌기, 시연, 암송
prejudice 편견
conscious 의식의

(Word Search)

| 정답 | 1. variable　2. dimensions　3. prejudice
1. 그는 식단이 기분에 영향을 미치는 변수가 될 수 있는지 알아보기 위해 연구를 수행했다.
2. 환경 문제에는 정치적 차원이 있는데, 이는 많은 경우 그것이 정부 정책을 수반한다는 것을 의미한다.
3. 그녀는 장애인에 대한 자신의 편견이 부당하다는 것을 깨달았다.

Exercise 12　　　　　　　정답 ④

(소재) 리더십과 직원의 태도

(해석) 대부분의 조직과 리더는 리더십 과정의 실행 단계에서 문제를 겪는다. 자기 잇속만 차리는 리더가 주도권을 잡으면 전통적인 계층적 피라미드식 체제가 건재하게 된다. 그런 일이 일어나면 사람들은 자신이 누구를 위해 일한다고 생각하는가? 자신의 위에 있는 사람들이다. 여러분이 실행에서 윗사람을 위해 일한다고 생각하는 순간, 여러분은 그 사람, 즉 여러분의 상사에게 '책임이 있고' 여러분의 업무는 그 상사와 그 상사의 변덕이나 바람에 '즉각 반응하는' 것이라고 가정하고 있는 것이다. 이제 '상사 눈치 보기'는 인기 있는 스포츠가 되고, 사람들은 위쪽으로 영향력을 행사하는 기술을 바탕으로 승진하게 된다. 그 결과, 조직의 모든 에너지가 고객 그리고 현장 활동에 가장 가까이 있는 최전선 직원으로부터 멀어져 계층 구조를 따라 위쪽으로 올라간다. 얻는 것은 오리 연못이다. 고객이 원하는 것과 상사가 원하는 것 사이에 충돌이 발생할 때 상사가 승리한다. 사람들이 "그건 저희 방침입니다." "저는 그냥 여기서 일할 뿐이에요." "제 상사를 불러드릴까요?"라며 오리처럼 꽥꽥거리는 일을 겪게 된다.

(문제 해설)

직원이 상사를 위해 일한다고 생각하는 순간, 자신의 업무는 상사와 그 상사의 바람에 즉각 반응하는 것이 되고, 조직의 모든 에너지가 고객보다는 상사에 집중된다고 하였다. 그러한 직원들이 '그건 저희 방침이다', '전 그냥 여기서 일할 뿐이다' 등의 말을 하는 것을 오리가 꽥꽥거리는 것에 비유하였으므로, 밑줄 친 부분이 글에서 의미하는 바로 가장 적절한 것은 ④ '고객에게 서비스를 제공하는 것보다 상사를 기쁘게 하는 것을 우선시하는 직원들로 가득 찬 조직'이다.
① 평화로워 보이지만 고객이 제기한 숨겨진 어려움이 있는 회사
② 상사가 소속감을 강조하고 어떤 갈등이라도 없애려고 노력하는 회사
③ 상사가 더 높은 생산성을 위해 직원들 간의 경쟁을 강화하는 조직

⑤ 직원들이 반복적인 업무를 의문 없이 수행함으로써 상사를 모방하는 조직

구조해설

- [When there is a conflict between {what the customers want} and {what the boss wants}], the boss wins.

  [ ]는 시간을 나타내는 부사절이고, 그 안에서 명사절인 두 개의 { }는 and로 대등하게 연결되어 between의 목적어 역할을 하고 있다.

어휘 및 어구

implementation 실행
phase 단계
self-serving 자기 잇속만 차리는
hierarchical 계층적인
alive and well 건재한
assume 가정하다, 추정하다
responsive 즉각 반응하는
promote 승진시키다
frontline 최전선의
conflict 충돌
supervisor 상사, 관리자

Word Search

| 정답 | **1.** assume **2.** responsive **3.** conflict(s)
1. 명확한 의사소통 없이 다른 사람의 의도를 추정하지 말라.
2. 좋은 지도자는 직원의 요구와 제안에 항상 즉각 반응한다.
3. 양국 간의 충돌은 전면전으로 발전했다.

Week 1 **02**강   본문 18~29쪽

| **1** ① | **2** ③ | **3** ④ | **4** ① | **5** ① | **6** ⑤ |
| **7** ⑤ | **8** ⑤ | **9** ④ | **10** ⑤ | **11** ② | **12** ④ |

Exercise 1   정답 ①

소재 언어의 준규칙성

해석 우리는 언어를, 그 준규칙성에도 불구하고—실은 그 때문에—매우 능숙하게 말하고 이해할 수 있다. 의사소통에는 공유된 지식이 필요하므로 언어는 자의적이기보다는 체계적이어야 한다. 그러나 언어를 이해하고 만들어 내는 요구 사항에는 추가적인 유연성이 필요한데, 화자가 표준 양식에서 벗어나는 형태를 만들어 내고 청자가 그것을 이해할 수 있어야 하기 때문이다. 말을 유창하게 하도록 촉진하는 많은 줄임말, 예를 들어 원 단어와 일부 겹치는 'gonna', 'hafta', 'tryna'와 같은 말이 결국 언어에 입력된다. 이러한 서로 상충하는 압력의 산물이 준규칙성이다. 이러한 패턴은 폭넓은 연습으로 숙달할 수 있는데, 이는 여러분이 언어를 말하면서 자라 유창하게 읽을 수 있게 되었다면 쉽게 얻을 수 있다. 강세 패턴에 숙달하는 것은 영어를 제2의 언어로 배우는 사람들에게 훨씬 더 어려운데, 이들은 흔히 '강세 난청'을 보인다.

문제해설

의사소통에는 공유된 지식이 필요하지만, 말을 유창하게 할 수 있게 하는 많은 줄임말 같이 화자가 표준 양식에서 벗어나는 형태를 만들어 내고 청자는 그것을 이해할 수 있어야 하는 상황이 생기는데 이를 위해 필요한 것은 유연성이라고 볼 수 있다. 따라서 빈칸에 들어갈 말로 가장 적절한 것은 ① '유연성'이다.
② 보안
③ 통합
④ 위임
⑤ 서류화

구조해설

- However, the demands of comprehending and producing language require additional flexibility [because speakers produce forms {that deviate from standard patterns} and listeners must be able to comprehend them].

  [ ]는 이유의 부사절이다. 그 안의 { }는 forms를 수식하는 관계절이다.

어휘 및 어구

comprehend 이해하다
systematic 체계적인

additional 추가적인

shortcut 줄임말, 쉬운 방법

fluent 유창한

overlap 겹치다

conflicting 상충하는

extensive 폭넓은, 광범위한

obtain 얻다

stress 강세

deafness 난청

Word Search

| 정답 | **1.** comprehend **2.** fluent **3.** overlaps

1. 그녀는 이해할 수 없어서 그 사고를 바라보고 서 있었다.

2. 그는 5개 외국어를 공부했지만 그중 3개만 유창하다.

3. 9월에는 축구 시즌이 야구 시즌과 겹친다.

Exercise 2 ──────── 정답 ③

소재 훌륭한 광고의 본질

해석 사람들이 항상 여러분에게 '상자 밖에서(새로운) 사고를 하라'고 말하는 것을 알고 있는가? 글쎄, 나는 그 표현을 싫어한다. 나는 그 문구의 더 넓은 의미, 즉 관습에 도전하는 예상치 못한 해결책을 찾으라는 뜻은 이해한다. 그것에는 아무런 문제가 없다. 하지만 나에게 광고는 전부 상자 '안에서' 생각하는 것과 관련이 있다. 그리고 광고는 상자, 즉 제한, 틀, 구체적 현실로 가득 차 있다. 예산이 상자이다. 페이지의 크기가 상자이다. 제품 속 성분이 상자이다. 모든 것들 중 가장 중요한 상자는 전략이다. 만약 전략의 제한 내에 꼭 맞는 훌륭한 창의적 아이디어를 생각해 낼 수 있다면, 여러분은 천재이다. 아주 빗나가고 전략적이지 '않은' 훌륭한 아이디어를 생각해 내면, 여러분은 예술가이지 광고인이 아니다. 이것은 여러분이 상자에 대고 투덜댈 수 없다는 말이 아니다. 또는 상자의 크기를 바꾸려 노력할 수 없다는 말이 아니다. 하지만 본질적으로, 광고는 그것이 <u>상자의 분명한 결과물</u>일 때만 진정으로 광고가 될 수 있다. 우리 중 가장 똑똑한 사람은 광고의 가장 큰 재미가 아이디어, 실행, 새로운 미디어 배치를 가지고 우리가 어느 정도까지 성공할 수 있으면서도 '여전히 상자 안에 있을' 수 있는지를 보는 것이라는 점을 깨닫는다.

문제 해설

전략의 제한 내에 꼭 맞는 훌륭한 창의적 아이디어를 생각해 낼 수 있다면 천재라고 하였고, 광고의 가장 큰 재미는 아이디어, 실행, 새로운 미디어 배치를 가지고 성공하면서도 여전히 상자 안에 있는 것이라고 하였으므로, 빈칸에 들어갈 말로 가장 적절한 것은

③ '상자의 분명한 결과물'이다.

① 신기원을 이룩하는

② 관객의 경험

④ 다른 유형의 상자에 영감을 받는

⑤ 기존의 관습에 대한 도전

구조 해설

■ [If you can come up with a great creative idea {that fits within the confines of the strategy}], then you're a genius.

[ ]는 조건의 부사절이고, { }는 a great creative idea를 수식하는 관계절이다.

어휘 및 어구

phrase 문구, 어구

convention 관습

limitation 제한, 한계

framework 틀

concrete 구체적인

budget 예산

dimension 크기, 차원

ingredient 성분, 재료

strategy 전략

come up with ~을 생각해 내다

genius 천재

off the mark 빗나간

essence 본질

execution 실행

placement 배치

Word Search

| 정답 | **1.** concrete **2.** budget(s) **3.** genius

1. "그것은 단지 의심일 뿐이고요, 구체적인 것은 없습니다."라고 경찰관이 말했다.

2. 그 가족은 그들의 가계 예산의 균형을 맞추기 위해 고군분투한다.

3. 아인슈타인은 수학 천재였다.

Exercise 3 ──────── 정답 ④

소재 지구의 에너지 사용에 대한 인간의 진화

해석 우리 행성의 생명체는, 대략 독립 영양 생물과 종속 영양 생물, 즉 태양이나 화학 반응으로부터 얻는 에너지를 이용하는 유기체와 이미 에너지를 획득한 것들로부터 그것을 얻는 유기체로 정리될 수 있다. 우리 종의 특이한 점은 우리가 다른 종으로 진화

할 필요 없이 점점 더 많은 에너지를 사용할 수 있었다는 것이다. 우리는 사회적 학습, 복잡한 문화, 기술의 조합을 통해 이것을 성취해 왔다. 우리는 알로사우루스의 갈고리 발톱을 얻기 위해 새로운 종으로 분화할 필요가 없고, 정보를 공유해 탄두나 발전소를 설계할 수 있다. 다시 말해서, 우리는 우리의 몸보다는 도구를 바꾼다. 수십만 년 동안 불과 창이 문제를 해결해 주었는데, 마침내 우리는 식량 공급원의 가축화를 고안했다. 다음 큰 변화는 우리에게 산업혁명을 준 공정의 기계화로 왔다. 이를 통해 우리는 지구에서 고대의 유기 에너지 매장물을 끌어내어 태울 수 있게 되었다.

(문제 해설)
인간은 다른 종으로 진화할 필요 없이 점점 더 많은 에너지를 사용할 수 있었는데, 이것을 사회적 학습, 복잡한 문화, 기술의 조합을 통해 성취해 왔다고 했다. 그리고 그 예로 불과 창의 이용, 가축화, 공정의 기계화를 들고 있는데, 이는 인간이 이용한 도구의 예를 보여 주는 것이다. 따라서 빈칸에 들어갈 말로 가장 적절한 것은 ④ '우리의 몸보다는 도구를 바꾼다'이다.
① 우리가 마주하는 모든 미개척 영역을 훼손한다
② 장기적인 위협에 대처하기 위해 진화한다
③ 우리의 도구를 최적으로 사용할 것 같지 않다
⑤ 우리가 가는 곳마다 유전 정보를 남긴다

(구조 해설)
- [What is unusual about our species] is [that we've been able to use more and more energy without {having to evolve into a different species}].
  첫 번째 [ ]는 주어 역할을 하는 명사절이고, 두 번째 [ ]는 주격 보어 역할을 하는 명사절이다. { }는 without의 목적어 역할을 하는 동명사구이다.

(어휘 및 어구)
arrange 정리하다, 배열하다
autotroph 독립 영양 생물
heterotroph 종속 영양 생물
exploit 이용하다, 개발하다
capture 획득하다
evolve 진화하다
combination 조합
claw (새·짐승의) 갈고리 발톱
power station 발전소
spear 창
do the trick 문제를 해결하다, 바라는 결과를 만들어 내다
devise 고안하다
domestication 가축화
mechanisation 기계화
deposit 매장물, 매장층

(Word Search)
| 정답 | 1. exploit 2. devised 3. deposits
1. 많은 기업들이 천연자원을 개발하기 위해 오고 있다.
2. 도시의 대기 오염을 통제하기 위해 새로운 시스템이 고안되었다.
3. 그 지역에는 귀중한 석유의 매장층이 많다.

Exercise 4 ——————————————— 정답 ①

(소재) 전 세계 미디어 기관이 직면한 어려움

(해석) 전 세계의 미디어 기관은 자금 조달, 신뢰성, 대표성, 책임성, 그리고 합법성의, 복합적인 위기에 직면해 있다. 자본주의의 핵심을 구성하는 국가 중 많은 국가에서, 신문과 잡지 산업이 심각하게 쇠퇴하고 있는데, 이는 대형 디지털 중개업체가 광고 수익의 대부분을 차지하기 때문이다. 뉴스 산업의 지속 가능성에 대한 논쟁의 상당 부분은 이 '깨진 비즈니스 모델'에 관한 논쟁을 중심으로 전개된다. 특히 지역 뉴스는 점점 더 위협받고 있다. 영국에서는 인구의 대다수(57.9%)에게 지역 일간지가 더 이상 제공되지 않는다. 높은 수준의 수익성을 유지하기 위해, 미디어 회사들은 출판물을 폐간하거나 합치고, 일자리를 줄였으며, 기자들을 그들이 근무하는 지역 사회에서 멀리 떨어진 곳으로 자주 이동시켜 더 이상 그 지역 사회에 적절한 콘텐츠를 제공할 수 없게 되었다. 간단히 말해서, 수익을 추구하는 대응은 미디어가 점점 더 지속 불가능해진다는 것을 의미한다.

(문제 해설)
전 세계의 미디어 기관들이 복합적인 위기에 직면하고 있는 상황에서 특히 지역 뉴스는 점점 더 위협을 받고 있는데, 높은 수준의 수익성을 유지하기 위한 조치들로 인해 적절한 콘텐츠를 제공할 수 없는 상황에 이른다고 했다. 따라서 빈칸에 들어갈 말로 가장 적절한 것은 ① '미디어가 점점 더 지속 불가능해진다는 것'이다.
② 미디어에는 지적할 실수가 많다는 것
③ 기자들의 사회적 책무를 강화하는 것
④ 향상된 판매 수익이 편집의 독립성을 가능하게 한다는 것
⑤ 뉴스와 정보의 질과 이용 가능성을 향상시키는 것

(구조 해설)
- [To retain high levels of profitability], media corporations have closed or merged titles and cut jobs, [{often moving journalists long distances away from the communities ⟨they serve⟩} and {no longer being able to provide content of relevance to them}].
  첫 번째 [ ]는 목적을 나타내는 to부정사구이다. 두 번째 [ ]는 주어 media corporations의 부수적인 상황을 설명하는 분사

구문이고, 두 개의 { }가 and로 연결되어 분사구문을 구성하고 있다. 〈 〉는 the communities를 수식하는 관계절이다.

어휘 및 어구

institution 기관
multiple 복합적인, 다수의
crisis 위기(*pl.* crises)
representation 대표(성), 표현
accountability 책임성, 책무
capitalism 자본주의
decline 쇠퇴
take over ~을 차지하다[장악하다]
sustainability 지속 가능성
circulate 전개되다, 유포되다, 순환하다
retain 유지하다, 보유하다
profitability 수익성
corporation 회사
merge 합치다, 합병하다
relevance 적절성, 관련성

Word Search

| 정답 | **1.** multiple/many **2.** relevance **3.** profitability
1. 그녀는 그 충돌 사고에서 복합적인/많은 부상을 입었다.
2. 이 문제는 현재 상황과 관련이 없다.
3. 수익성을 높이기 위해 효율적인 프로세스를 도입해야 한다.

Exercise 5

정답 ①

소재 인간이 개입한 자연환경

해석 자연에 대한 인간 개입의 복잡성은 생태계가 적응하거나 많은 경우에 멸종해야 했다는 것을 의미한다. 오늘날 영국에는 고대의 삼림 지대가 작은 구획으로만 존재한다. 이러한 잔존물의 많은 부분은 자연 보호 구역과 국립 공원에 둘러싸여 있다. 그것은 특별한 보호가 필요하다. 그 자체의 생태계를 가진 새로운 서식지가 생겨났다. 지형의 도시화와 도로 및 철도 회랑의 건설은 우리에게 많은 종이 번성하는 정원 서식지를 주었다. 더 많은 염분의 퇴적물이 있는 고속도로변은 다른 상황에서는 해안가에서나 발견되는, 염분을 좋아하는 식물들이 살 수 있게 한다. 도로는 죽은 동물을 먹는 동물들에게 자동차에 치여 죽은 동물을 풍부하게 제공한다. 이러한 것은 그것이 대체하는 것에 비해 미약한 대체물일 수도 있지만, 적절하게 관리된다면 더 많은 것을 더해 줄 수 있는 서식지이다. 우리가 보존하고자 할 수도 있는 자연환경은 <u>부분적으로는 우리가 만들어 낸</u> 자연환경이다.

문제 해설

자연에 대한 인간의 개입으로 새로운 생태계가 형성되고 그 생태계에 맞는 새로운 서식지가 생겨난다는 내용이다. 이러한 새로운 서식지와 새로운 생태계는 결국 인간이 보존하고자 하는 자연환경 속에 포함되어 있기 때문에 인간이 보존하고자 할 수도 있는 자연환경은 결국 부분적으로 인간이 만들어 낸 자연환경이라고 할 수 있다. 따라서 빈칸에 들어갈 말로 가장 적절한 것은 ① '부분적으로는 우리가 만들어 낸'이다.
② 생존이 시험이 될
③ 아직 부정적으로 영향을 받지 않은
④ 외래종을 심는 것으로 인해 우리가 파괴되는
⑤ 노력을 요구하지 않고 관심을 끄는

구조 해설

■ These may be poor substitutes for [what they replace], but they are habitats [that can add more {if properly managed}].
첫 번째 [ ]는 for의 목적어 역할을 하는 명사절이다. 두 번째 [ ]는 habitats를 수식하는 관계절이고, { }는 조건의 의미를 나타내는 부사절로 if 다음에 they are가 생략된 것으로 이해할 수 있다.

어휘 및 어구

complexity 복잡성
intervention 개입
ecosystem 생태계
adapt 적응하다
die out 멸종하다
enclose 둘러싸다
reserve 보호 구역
habitat 서식지
urbanization 도시화
landscape 지형, 경관
thrive 번성하다
motorway 고속도로
deposit 퇴적물
abundant 풍부한
substitute 대체물, 대용물
properly 적절하게
conserve 보존하다

Word Search

| 정답 | **1.** urbanization **2.** enclosed **3.** substitute
1. 도시화 속도는 미래 도시가 직면한 난제이다.
2. 그의 마당은 철제 울타리로 둘러싸여 있다.
3. 많은 국가들이 지구를 보호하기 위해 석탄의 대체재를 찾아야 한다.

Exercise 6

정답 ⑤

소재 마음을 프로그램하는 주체들

해석 우리들 중 대부분에게, 우리의 마음은 우리 친구, 우리 부모님, 대중매체, 그리고 광고주 등의 요소들의 조합에 의해 프로그램되어 왔다. 이러한 프로그래밍 주체 중 일부는 여러분을 진정으로 알고 있으며, 여러분의 특별한 강점을 강화하고 고질적인 약점을 극복하도록 도우면서 여러분의 최선의 이익을 염두에 두는데, 그들은 여러분을 더 행복하게 하고 여러분의 삶을 더 좋게 하려고 노력하고 있기 때문이다. 프로그래밍의 다른 주체들은 그들의 목표를 달성하기 위한 도구로 여러분을 사용하려고 하고 있는데, 그것은 흔히 여러분 자신의 목표와 매우 다르다. 이런 일이 일어나는 경우 그 프로그래밍은 여러분을 점점 덜 행복하게 하는데, 그것이 여러분에게 없는 문제를 해결하도록 여러분을 '도와주고', 여러분이 실제로 가진 문제는 더 악화시키기 때문이다. 다른 사람들이 여러분 마음의 프로그래밍을 지배하도록 허용할 때, 그런 다음 여러분의 마음이 자동 조종 장치로 작동할 때, 여러분은 결국 자신을 더 행복하게 만들 방식으로 행동하기보다는 그 프로그래머의 목표를 달성하는 방식으로 행동하게 된다. 그러므로, 여러분의 마음에 프로그램된 코드를 주기적으로 점검하는 것이 중요하다.

문제해설
우리의 마음은 친구, 부모님, 대중매체, 광고주와 같은 요소들의 조합에 의해 영향을 받는데, 그러한 요소들이 좋은 영향을 미칠 수도 있고 나쁜 영향을 미칠 수도 있다는 내용이므로, 밑줄 친 부분을 포함하는 문장은 자신의 마음에 영향을 미치는 요소들을 주기적으로 점검하는 것이 중요하다는 의미이다. 따라서 밑줄 친 부분이 의미하는 바로 가장 적절한 것은 ⑤ '다른 사람들이 여러분의 사고와 행동에 미치는 영향'이다.
① 여러분이 다른 사람들에 대해 가지고 있는 편견
② 여러분의 목표를 달성하기 위한 핵심 수단
③ 여러분의 뇌 활동을 향상시키는 방법
④ 진정한 친구와 그런 척하는 사람을 구별해 주는 자질

구조해설
- When you allow others to dominate the programming of your mind, then when your mind runs on automatic pilot, you end up [behaving in ways {that achieve the goals of those programmers}] **rather than** [behaving in ways {that would make you happier}].
  '…이라기보다는 ~'의 의미를 나타내는 「~ rather than …」 구문이 사용되어 두 개의 [ ]로 표시된 부분을 연결하고 있고, 그 안에서 두 개의 { }는 각각 선행하는 ways를 수식하는 관계절이다.

어휘 및 어구
combination 조합
agent 주체, 동인
reinforce 강화하다
overcome 극복하다
troublesome 고질적인, 성가신
weakness 약점
dominate 지배하다
automatic pilot 자동 조종 장치
periodically 주기적으로

Word Search

| 정답 | 1. reinforce 2. troublesome 3. weaknesses
1. 그런 농담은 해로운 고정관념을 강화하는 경향이 있다.
2. 그녀는 고질적인 발 부상을 치료할 수술이 필요했다.
3. 자신의 강점과 약점을 아는 것이 중요하다.

Exercise 7

정답 ⑤

소재 구전 문화의 정보 전달의 정확성

해석 우리는 구전 문화의 유효성을 과소평가하는 경향이 있다. 우리는 모두 한 사람이 다른 사람에게 메시지를 속삭여 그것이 방 안을 한 바퀴 돌 때까지 전달하는 어린이 놀이를 잘 알고 있다. 원래의 메시지와 마지막 메시지가 비교될 때, 메시지는 항상 왜곡되고, 때로는 아주 재미있는 결과를 낳는다. 하지만 이는 오해의 소지가 있다. 중요한 경우, 구전 문화는 먼 거리를 가로지르고 여러 세대에 걸쳐 정보를 정확하게 전달할 수 있다. 예를 들어, 미국 작가 Alex Haley는 아프리카에서 자신의 조상에 대한 구전 기록을 발견할 수 있었고, 그의 탐색은 1976년 출간된 책 *Roots: The Saga of an American Family*에 묘사되어 있다. 마찬가지로, *Odyssey*와 *Iliad*는 원래 그리스 문화권의 영웅에 관한 구전 역사로서, 지어진 지 수 세기가 지난 후에야 비로소 기록되었다.

문제해설
우리는 메시지를 귓속말로 전달하는 어린이 놀이를 통해 구전되는 정보가 부정확할 수 있다는 것을 알지만, 중요한 경우 구전 문화는 먼 거리와 세대를 넘어 정보를 정확하게 전달한다는 내용이므로, 빈칸에 들어갈 말로 가장 적절한 것은 ⑤ '유효성'이다.
① 창의성
② 미묘함
③ 즉시성
④ 복잡성

14 EBS 수능특강 영어독해연습

구조 해설

- We're all familiar with the children's game [in which a message is whispered from one person to another {until it goes around a room}].

[ ]는 the children's game을 수식하는 관계절이고, 그 안에서 { }는 시간의 부사절이다.

어휘 및 어구

familiar 잘 아는, 익숙한, 친숙한

whisper 속삭이다

invariably 항상, 반드시

distorted 왜곡된

misleading 오해의 소지가 있는

accurately 정확하게

transmit 전달하다

ancestor 조상

heroic 영웅에 관한

compose (시·글을) 짓다. (음악 작품을) 작곡하다

Word Search

| 정답 | 1. ancestor(s) 2. composed 3. misleading

1. 여러 세대 전, 나의 조상(들)은 이 땅에 도착하여 용기와 모험이라는 유산을 남겼다.

2. 그 피아니스트는 오늘날 여전히 찬미되는 유명한 음악 작품을 작곡했다.

3. 그 헤드라인은 오해의 소지가 있어, 기사의 내용에 대한 잘못된 인상을 주었다.

Exercise 8   정답 ⑤

소재 이로운 적응인 감정

해석 감정은 이로운 적응이 되는 기준을 충족시킨다. 먹이를 두고 다툼에서 싸울 자세를 취하는 두 동물을 예로 들어 보자. 말 그대로든, 비유적으로든, 그것들이 서로 뿔을 맞댈[다툴] 준비를 할 때, 그것들의 강렬한 감정은 할 수 있는 모든 신체 반응을 촉발한다. 동물의 등이 활처럼 구부러지고 털이 쭈뼛 서 있을 때, 그것은 더 크고 더 강해 보인다. 이빨을 드러내거나, 눈썹을 찌푸리거나, 격렬한 소리를 내거나, 뿔을 과시할 때, 그 동물은 그렇게 강한 적과 싸우는 것은 그만한 가치가 없을지도 모른다는 신호를 상대방에게 보내는 것이다. 이러한 신호는, 공격성의 표시인데, 상대방이 물러날 가능성을 직접적으로 높여서, 폭력을 예방하고 잠재적인 부상이나 죽음을 피할 수 있게 할 것이다. 이러한 신호를 보내는 것은 이러한 메시지를 해석하는 능력과 마찬가지로 종에게 이익이 된다. 그것은 서로에게 유리하다.

문제 해설

동물이 먹이를 두고 싸울 때, 강력한 감정이 신체 반응을 촉진하여 더 크고 더 강해 보이게 하는데, 이는 상대방에게 강한 상대와는 싸울 가치가 없다는 신호를 보내어 상대를 물러나게 해서 폭력을 예방하고 부상이나 죽음을 피하게 하기에, 이러한 신호 생성과 해석 능력은 종에게 이익이 된다는 내용의 글이다. 따라서 빈칸에 들어갈 말로 가장 적절한 것은 ⑤ '이로운 적응'이다.

① 이기적인 성향

② 해로운 반응

③ 공격적인 본능

④ 정보원

구조 해설

- [When it {bares its teeth}, {frowns its brows}, {makes fierce noises}, or {displays its horns}], it signals to the other animal [that fighting such a strong adversary may not be worth it].

첫 번째 [ ]는 시간을 나타내는 부사절인데, 그 안에서 네 개의 { }가 콤마와 or로 연결되어 술부를 이루고 있다. 두 번째 [ ]는 signals의 목적어 역할을 하는 명사절이다.

어휘 및 어구

criterion 기준(pl. criteria)

prepare 준비하다, 대비하다

literally 말 그대로

figuratively 비유적으로

intense 강렬한

prompt (어떤 일이 일어나도록) 촉발하다, 자극하다

repertoire (한 사람이 할 수 있는) 모든 것

arch (몸이) 활처럼 구부러지다

bare 드러내다, 노출시키다

frown (눈살을) 찌푸리다

brows 눈썹

fierce 격렬한

horn 뿔

adversary 적, 상대방

aggression 공격성

withdraw 물러나다

interpret 해석하다

Word Search

| 정답 | 1. criteria 2. prepare 3. prompted/prompts

1. 그 취업 지원자들은 엄격한 기준에 근거하여 평가되었다.

2. 그들은 집을 단단히 고정하여 폭풍에 대비하기 시작했다.

3. 그 회사에 대한 뉴스는 그가 즉시 사무실로 돌아가도록 했다[한다].

Exercise 9

정답 ④

소재 이탈리아의 문화적 정체성과 생활 방식의 변화

해석 15세기부터 17세기까지 이탈리아인들은 실내 공간의 건축과 물품 비치를 통해 다른 생활 방식을 발전시킬 수 있고 새로운 문화가 정의되는 세계를 창조했다. 이것이 물건에 그렇게 많은 비용이 소비되었던 이유이고, 상당히 많은 새로운 종류의 물건들이 생겨나게 된 이유이며, 예술이 과거의 종교계에서 그랬듯이 이제 가정 세계에서도 번성했던 이유이다. 소비는 문화적 정체성을 구축할 수 있는 창조적인 힘이었다. 도자기에서 그림에 이르기까지 모든 종류의 새로운 비품을 발명하고, 그것들의 형태를 정교하게 만들어 내고, 생산을 개선하고, 집 안의 새로운 공간 배치로 조직하는 데 있어 이탈리아인들은 자신들을 위한 새로운 가치와 즐거움을 발견했고, 품위에 대한 새로운 기준으로 삶을 재정비했고, 자신에 대한 무언가를 타인에게 전달했다. 즉, 요컨대 문화를 창출했고 그 과정에서 스스로 정체성을 만들어 냈다. 이러한 문화적 발전에는 사람과 물리적 물체 사이의 상호 작용에서 생긴 변화의 원동력이 존재했다.

문제 해설
15세기부터 17세기까지 이탈리아인들은 모든 종류의 새로운 비품을 발명하고, 그것들의 형태를 정교하게 다듬고, 생산을 개선하고, 집 안의 새로운 공간 배치로 조직하면서 새로운 문화와 자신들의 정체성을 창출했다는 내용의 글이므로, 빈칸에 들어갈 말로 가장 적절한 것은 ④ '사람과 물리적 물체 사이의 상호 작용'이다.
① 혁신과 현대성에 대한 추구
② 부에 의한 사회 계층의 형성
③ 의도적이고 체계적인 문화 교육
⑤ 예술의 중요성에 대한 커지는 인식

구조 해설

- This is [why so much was spent on objects], [why so many new kinds of objects came into existence], [why the arts flourished now in the domestic world {as they **had** earlier in the ecclesiastical world}].

세 개의 [ ]는 주격 보어 역할을 하는 명사절이다. { }는 '~듯이'라는 의미의 접속사 as가 이끄는 부사절로 had는 had flourished에서 반복되는 flourished가 생략된 형태이다.

어휘 및 어구
construction 건축
furnish (가구나 물품을) 비치하다
interior 실내의, 내부의
flourish 번성하다
domestic 가정의
consumption 소비

construct 구축[구성]하다, 건설하다
identity 정체성
furnishings 비품, 가구
pottery 도자기
elaborate 정교하게 만들어 내다
refine 개선하다, 다듬다
spatial 공간의
arrangement 배치, 배열
dynamic 원동력, 힘

Word Search

| 정답 | **1.** flourished **2.** construct **3.** consumption

1. 르네상스 시대에는 예술이 번성하여 몇몇 유명한 걸작들이 탄생했다.
2. 엔지니어들은 새로운 다리를 건설하기 위해 끊임없이 노력했다.
3. 에너지 소비 감축을 목표로 한 정책들은 환경과 관련된 이익으로 이어졌다.

Exercise 10

정답 ⑤

소재 개척 정착민에게 지급된 자영 농지의 분배 방식

해석 개척 정착민에게 지급된 미국 농가 자영 농지의 분배는 자연 경계와 일치하지 않는 인위적 경계의 전형적인 사례이다. 토지는 지형과 상관없이 변의 길이가 같은 남북 및 동서의 경계를 가진 측량치를 근거로 할당되었다. 이는 자영 농지의 경계가 유역의 경계와 관련이 없었다는 것을 의미했다. 일부 자영 농지는 높고 건조했다. 다른 자영 농지는 더 낮고 더 습했지만 홍수의 피해를 입기 쉬웠다. 하류의 토지 소유자는 상류 토지에서 발생하는 침식과 유수(流水)를 제어할 수 없었다. 따라서, 고지대의 토지 소유자가 환경에 미치는 영향은 저지대의 다른 토지 소유자에게 피해를 줄 수 있었지만, 저지대의 토지 소유자는 선택의 여지가 없었다. 이러한 19세기의 결정은 오늘날까지도 영향을 미치고 있다. 예를 들어, Austin College의 Sneed 초원 복원 프로젝트에서 가장 어려운 문제 중 하나는 부실하게 관리된 상류의 토지로부터 오는 침식성 돌발 홍수에 의한 유수(流水)이다.

문제 해설
미국 개척 정착민에게 할당된 자영 농지는 지형에 상관없이 변의 길이가 같은 인위적인 경계에 따라 분배되었고, 이는 농장 토지의 경계가 유역의 경계와 일치하지 않게 만들었으며, 하류의 토지 소유자는 상류 토지에서 발생하는 침식과 유수로 인해 피해를 입었지만, 대처할 방법이 없었다는 내용의 글이므로, 빈칸에 들어갈 말로 가장 적절한 것은 ⑤ '자연 경계와 일치하지 않는 인위적 경계'이다.

① 개척자들의 이익에 따라 경계를 바꾸는 것
② 수로에 따른 경계를 반영하는 토지 경계
③ 경계선을 둘러싼 원주민과 개척자 사이의 갈등
④ 자연적 경계에 따른 비효율적인 경계 설정

[구조 해설]

■ The distribution of US farm homesteads [granted to pioneer settlers] is a classic case of artificial boundaries [not matching natural boundaries].
두 개의 분사구 [ ]는 각각 US farm homesteads와 artificial boundaries를 수식한다.

[어휘 및 어구]

distribution 분배, 배급
grant 지급하다, 수여하다
pioneer 개척자의; 개척자
property 토지, 소유지, 재산
allocate 할당하다
survey 측량치, 측량
regular 등변의(변의 길이가 같은)
boundary 경계
regardless of ~에 상관없이
lay of the land 지형, 형세
subject to (피해 등을) 입기 쉬운
erosion 침식, 부식
consequence (보통 복수로) 영향(력)
restoration 복원, 부활
flash-flood (호우 등으로 인한 협곡, 경사로의) 돌발 홍수

Word Search

| 정답 | **1.** grant **2.** pioneer(s) **3.** allocate
1. 그 자선 단체는 재난 구호 활동을 위한 자금을 지급하기로 결정했다.
2. 그 개척자는 새로운 영토에 정착하는 동안 많은 어려움에 직면했다.
3. 관리자는 기한을 맞추기 위해 업무를 효율적으로 할당해야 한다.

Exercise 11 ──────────── 정답 ②

[소재] 출판 편향의 오류

[해석] 출판 편향은 동료 검토 문헌에 보고된 많은 행동 현상에 대해 효과의 크기가 과장될 수 있음을 의미한다. 예를 들어, 우울증에 대한 새로운 행동 요법이 환자의 우울 증상을 상당히 감소시킨다는 것을 보여 주는 몇 가지 연구를 읽는다고 가정해 보자. 만약 어떤 연구자가 이 동일한 행동 요법의 유효성을 시험해 보고 어떤 효과도 발견하지 못한다면, 어떤 동료 검토 저널도 그 원고를 받아들이지 않을 가능성이 있기 때문에, 여러분은 결코 그것을 찾거나 읽을 수 없게 된다. 그러므로 이 요법의 유효성이 과장되어 있을 가능성이 있는데, 이는 효과를 보여 주지 못하는 연구는 출판되는 동료 검토 문헌에 포함되지 않기 때문이다. 연구자들은 '과학적 진보는 현재 지식의 대부분을 신뢰함으로써 이루어진다'고 말하였는데, 출판 편향은 이러한 신뢰를 훼손한다. 동료 검토 문헌에 보고된 긍정적인 결과는 분명 신뢰할 수 있지만, 또한 많은 부정적인 결과가 검색에 포함되지 않을 수도 있다는 점을 인지함에 주의해야 한다는 것을 염두에 두라.

[문제 해설]

동료 검토 문헌에서 긍정적이지 않은 연구 결과는 받아들여지지 않고 출판되지 않을 가능성이 있다고 했으므로, 빈칸에 들어갈 말로 가장 적절한 것은 ② '많은 부정적인 결과가 검색에 포함되지 않을 수도 있다'이다.
① 그 문헌은 단지 한 번의 실험에 토대를 두고 있을 수 있다
③ 편견은 어떤 결과도 반대로 해석되게 할 수 있다
④ 실험의 조건이 의도적으로 바뀌었을 수도 있다
⑤ 그 문헌의 다음 판은 전혀 다른 결과를 드러낼 수도 있다

[구조 해설]

■ For example, suppose [you read a few studies showing {that a new behavioral therapy for depression significantly reduces symptoms of depression in patients}].
[ ]는 suppose의 목적어 역할을 하는 명사절로 접속사 that이 생략된 것으로 이해할 수 있고, 그 안에서 { }로 표시된 that 절은 showing의 목적어 역할을 하는 명사절이다.

[어휘 및 어구]

effect 효과, 영향
overstate 과장하다
literature 문헌, 문학
therapy 요법, 치료
depression 우울증
significantly 상당히, 의미심장하게
symptom 증상
effectiveness 유효성
accept 받아들이다, 수락하다
manuscript 원고
bulk 대부분, 규모
compromise 훼손하다

Word Search

| 정답 | **1.** therapy **2.** accept **3.** effect
1. 부상 후, 그는 일주일에 두 번 물리 치료 활동에 참석했다.

2. 그녀는 새로운 기회에 흥분하여 그 일자리 제의를 수락하기로 결정했다.

3. 새로운 광고 캠페인은 판매에 즉각적인 영향을 미쳤다.

Exercise 12

정답 ④

[소재] 경제에서의 긍정적인 피드백 함정

[해석] 경제 문제를 극복하기 위해 경제 성장에 의존하는 것은 긍정적인 피드백 함정에 빠진다. 정부는 가난한 사람들을 빈곤에서 벗어나게 하고 부자들의 요구를 충족시키는 수단으로 경제 생산의 증가를 장려하지만, 그 결과로 생겨나는 새로운 제품의 매력은 훨씬 더 많은 생산에서 발생하는 새로운 수입으로 충족되는 새로운 욕구를 만들어 낸다. 다시 말해, 사람들은 욕구(그리고 물론 필요)를 충족시킬 돈을 벌기 위해 일하지만, 그들의 노력은, 새로운 혁신을 포함하여, 판매되는 경우 다른 사람들의 욕구를 증가시키는 상품의 생산으로 이어진다. 그러면 그 다른 사람들이 자신의 새로운 욕구를 충족시키기 위해 일하여 다른 사람들에게 판매되는 훨씬 더 많은 상품을 생산하며, 이런 것이 반복되어 지구 자원의 재고를 고갈시키고 더 많은 폐기물과 우리가 그 결과를 부분적으로만 이해하는 새로운 유형의 폐기물을 발생시키는 긍정적인 피드백이 된다. 이런 상황을 되돌아보면서, Daniel Quinn의 동명 소설에 나오는 지혜로운 고릴라 Ishmael은 현대인을 어머니 문화의 재소자로, 감옥 산업에 종사하며 세상을 소비한다고 묘사한다.

[문제 해설]

경제적인 욕구를 충족시키기 위한 생산 장려는 더 많은 생산과 새로운 욕구를 창출하며, 이 과정이 반복되면서 자원 고갈과 같은 부정적인 결과를 초래하는데, 이러한 상황을 빗대어 소설 Ishmael에서는 현대인이 감옥 산업에 종사하며 세상을 소비하는 것으로 묘사한다고 했으므로, 밑줄 친 부분이 의미하는 바로 가장 적절한 것은 ④ '소비와 자원 고갈의 끝없는 순환에 갇혀 있는'이다.

① 소비자의 수요를 초과하는 과잉 생산에 갇혀 있는

② 정글의 법칙이 적용되는 시나리오에 휘말린

③ 시장 지배를 위한 기업 간 경쟁에 참여하는

⑤ 자유가 박탈된 정부 통제 경제 속에서 억압받는

[구조 해설]

- In other words, people work to earn money [to satisfy wants (and of course needs)], but their effort results in production of goods, including new innovations, [that, when marketed, increase others' desires].

  첫 번째 [ ]는 목적을 나타내는 to부정사구이고, 두 번째 [ ]는 goods를 수식하는 관계절이다.

[어휘 및 어구]

overcome 극복하다

encourage 장려하다

poverty 빈곤

satisfy 충족시키다, 만족시키다

appeal 매력

earn 벌다, 얻다

innovation 혁신

grind away at ~을 고갈시키다, ~을 갈아 없애다

stock 재고

consequence (보통 복수형으로) (좋지 못한) 결과

reflect on ~을 되돌아보다, ~을 반성하다

prisoner 재소자, 죄수

Word Search

| 정답 | **1.** earn **2.** satisfy **3.** Innovation

1. 그는 우수한 작업 결과를 내놓음으로써 간신히 승진할 수 있었다.

2. 안전 규정을 충족시키기 위해 그 제조업체는 장비를 업그레이드했다.

3. 혁신은 산업이 발전하여 새로운 도전에 적응하도록 이끈다.

Tips

mother culture(어머니 문화): Daniel Quinn의 소설 Ishmael에 등장하는 개념으로 사람들에게 당연시되며 의문 없이 받아들여지는 사회적이고 관습적인 패러다임을 가리킨다. 이 글에서 mother culture는 현대 인간 사회가 자연스럽게 받아들이는 경제적 성장과 소비 중심적인 생활 방식을 형성하고 강화하는 기본적인 문화적 배경을 의미한다.

본문 30~41쪽

1 ②	2 ②	3 ④	4 ①	5 ①	6 ②
7 ②	8 ③	9 ①	10 ①	11 ②	12 ⑤

Exercise 1
정답 ②

소재 개인화되고 고유한 말

해석 말하여진 것은 결코 단순한 반복이 아니다. 설령 한 사람이 이전에 말했던 것과 정확히 말한[글자] 그대로를 일치시킨다 하더라도, 세상, 화자, 상황, 청자, 그리고 말했던 것의 의미 등 모든 것이 변했다. 그러므로 사람이 말하는 것은 각각의 경우에 고유하다. 비록 우리의 발언 대부분이 딴 데서 빌려온 것이라 할지라도, 그것(발언)은 (단조롭거나, 추하거나, 진부할 수 있는) 그것의 특별하고 심지어 고유한 스타일을 통해 누군가가 그것을 자기 것으로 만들고, 그와 함께 (그것을) 개인화했다는 것을 보여 준다. 모든 문장은 필자가 받은 선물을 자신의 생각으로 바꾸었다는 것을 증명한다. 자신의 것[말이나 글]인 것에 대한 분석은, 그렇다면, 화자가 부모, 스승, 친구, 책, 유행 등으로부터 빌린 것의 많은 부분을 드러내 보일 수도 있으며, 동시에 이러한 모든 영향이 어떻게 필자가 말했던 것의 고유한 결과물로 한데 모아졌는지 밝힐 수도 있다.

문제 해설 똑같은 말이라도 세상, 화자, 상황, 청자, 그리고 말했던 것의 의미 등이 변하고, 우리가 말하는 것은 각각의 경우에 고유하고 개인화했다는 것을 보여 주며, 필자가 받은 영향들이 고유한 결과물로 모아졌다는 내용의 글이다. 따라서 설령 한 사람이 이전에 말했던 것과 정확히 말한[글자] 그대로를 일치시킨다 해도 모든 것이 변했으므로, 빈칸에 들어갈 말로 가장 적절한 것은 ② '단순한 반복'이다.
① 숨겨진 진실
③ 편향된 발언
④ 진정한 발견
⑤ 정확한 표현

구조 해설
■ An analysis of what is one's own might then reveal [much of what the speaker has borrowed from parents, guides, friends, books, fashions, etc.], [while at the same time disclosing {how all these influences have converged into the unique results of what the author said}].
첫 번째 [ ]는 동사 reveal의 목적어 역할을 한다. 두 번째 [ ]는 접속사 while과 함께 쓰인 분사구문으로, 그 안의 { }는 disclosing의 목적어 역할을 하는 명사절이다.

어휘 및 어구
agree ~ with ... ~을 …과 일치시키다
circumstance (보통 복수로) 상황
addressee 청자, 수신인
unique 고유한
statement 발언, 진술
unoriginal 딴 데서 빌려온, 본래의 것이 아닌
dull 단조로운
appropriate 자기 것으로 만들다
personalize 개인화하다
analysis 분석
reveal (보이지 않던 것을) 드러내 보이다

Word Search

| 정답 | 1. circumstance(s) 2. unique 3. revealed
1. 현재 상황에서 회사는 모든 직원의 안전을 보장할 필요가 있다.
2. 이 종은 다른 모든 종과 구별되는 고유한 노래를 가지고 있다.
3. 그것(감정)을 숨기려는 그의 시도에도 불구하고 그의 표정은 그의 진정한 감정을 드러내 보였다.

Exercise 2
정답 ②

소재 불확실성에 대한 조바심

해석 걱정은 흔히 불완전함에 대한 인식이다. 불확실성으로 손상된 세상에서, 의심이 끊임없이 여러분의 주위에서 소용돌이치는 가운데, 하나의 물음이 여러분의 의식 표면으로 떠오른다. 그리고 그 인식의 순간에, 여러분은 그 수수께끼를 풀기로 결정할 수도 있다. 여러분은 여러분의 삶을 괴롭히는 풀 수 없는 문제를 마침내 해결하길 바라는 희망을 버리지 않으면서 걱정을 한다. 불확실성, 더 정확하게는 불확실성에 대한 조바심은 걱정의 많은 측면에 걸쳐 있는 공통된 맥락이다. 깔끔하고 만족스러운 해결책을 찾으려는 추구가 자주 이 행동을 하도록 만든다. 아마도 더 잘하고자 하는 지칠 줄 모르는 전념에서 비롯, 여러분의 삶에서 중요한 것들을 추구하려는 의지에 의해 주도되었을 것 같기에, 그것은 이해할 만하지만 걱정은 그 길에서 벗어나기 시작한다. 여러분의 좋은 의도에도 불구하고, 여러분은 행동에 나서기보다는 생각에 갇히게 된다.

문제 해설 걱정은 불확실성에 대한 조바심에서 비롯되며, 해결책을 찾으려는 추구가 자주 걱정을 하게 하고, 행동에 나서기보다 생각에 갇

히게 만든다는 내용이므로, 빈칸에 들어갈 말로 가장 적절한 것은 ② '불완전함에 대한 인식'이다.

① 자신의 의도에 대한 깨달음
③ 타인의 완전함을 인정하는 데에 대한 망설임
④ 불확실성 속에서 행동을 취하는 것에 대한 주저
⑤ 불확실성을 통제하는 것보다 중요한 문제를 우선시하는 것

구조 해설

- You engage in worrying, [hoping against hope that you'll finally nail down the unsolvable questions {troubling your life}].

  [ ]는 주절의 부수적인 상황을 표현하는 분사구문이며, 그 안의 { }는 the unsolvable questions를 수식하는 분사구이다.

어휘 및 어구

uncertainty 불확실성
doubt 의심, 의구심
awareness 의식
mystery 수수께끼
impatience 조바심, (고통 · 압박 등을) 참을 수 없음
thread 맥락, 가닥
aspect 측면
quest 추구, 탐구
resolution 해결책
drive (사람을 특정한 방식의 행동을 하도록) 만들다[몰아붙이다]
tireless 지칠 줄 모르는
commitment 전념, 헌신
willingness 의지

Word Search

| 정답 | **1.** doubt   **2.** impatience   **3.** commitment
1. 그는 그들의 주장에 대해 상당한 의심[의구심]을 드러냈다.
2. 그 아이의 조바심은 그가 그들이 언제 도착할지를 계속 물어보면서 명백히 드러났다.
3. 자원봉사 활동에 대한 그녀의 헌신적인 전념은 지역 사회에 눈에 띄는 변화를 가져왔다.

Exercise 3   정답 ④

소재 코드를 작성하는 엔지니어와 그 영향력

해석 상업적 인터넷의 초기 시절, 학자들은 사이버 공간에서 컴퓨터 코드가 일종의 '법'으로서 작동한다는 것을 발견했다. 우리가 알고 있는 법, 즉 입법자와 판사들이 결정하는 공공 규칙이 아니라, 기술 자체에 내장된 다른 종류의 법이었다. 우리가 앱, 플랫폼, 스마트폰 또는 컴퓨터를 사용할 때마다, 우리는 이러한 기술에 코드화된 엄격한 규칙을 따를 수밖에 없다. '당신은 맞는 비밀번호 없이는 이 시스템에 접근할 수 없습니다'라는 규칙과 같이 어떤 규칙들은 아주 흔하다. 이런 이유로 자신의 가상 화폐 지갑의 비밀번호를 기억하지 못해 2억 달러 넘게 잃어버린 한 젊은 남자가 있다. 다른 규칙들은 더 논란의 여지가 있다. 2020년 말, 한 소셜 미디어 플랫폼은 어느 유명 인사의 아들에 대한 부패 혐의를 담은 논란이 되는 기사를 사용자들이 공유할 수 없게 했는데, 해킹된 자료를 공유하는 것을 금지하는 플랫폼 규칙을 위반했다는 것을 근거로 하였다. 우리의 행동, 상호 작용과 거래 중 점점 더 많은 것들이 디지털 기술을 통해 영향을 받게 됨에 따라, 코드를 작성하는 사람들이 갈수록 더 나머지 우리가 살면서 따라야 하는 규칙을 작성하게 된다. 소프트웨어 엔지니어가 사회적 엔지니어로 되어가고 있다.

문제 해설

컴퓨터 코드는 일종의 '법'으로서 작동하며 우리는 앱이나 컴퓨터 등의 기술에 내장된 엄격한 규칙을 따를 수밖에 없는데, 우리 삶의 많은 부분이 점점 더 디지털 기술의 영향을 받게 됨에 따라 코드를 작성하는 소프트웨어 엔지니어의 사회적 영향력이 커지고 있다는 내용의 글이다. 따라서 빈칸에 들어갈 말로 가장 적절한 것은 ④ '나머지 우리가 살면서 따라야 하는 규칙을 작성하게 된다'이다.

① 자신의 기술을 발전시키기 위해 의식적인 노력을 할 필요가 있다
② 오로지 웹사이트 플랫폼 개선에만 집중하게 된다
③ 코딩 규칙을 통제하기 위해 법적 원칙에 의존하게 된다
⑤ 사회적 규범에 대해 힘을 덜 행사하게 된다

구조 해설

- [As more and more of our actions, interactions and transactions are mediated through digital technology], those [who write code] increasingly write the rules [by which the rest of us live].

  첫 번째 [ ]는 '~함에 따라'라는 의미를 나타내는 부사절이며, 두 번째와 세 번째 [ ]는 각각 선행사 those와 the rules를 수식하는 관계절이다.

어휘 및 어구

commercial 상업적인
scholar 학자
legislator 입법자
strict 엄격한
commonplace 아주 흔한, 평범한
access 접근하다
virtual currency 가상 화폐
controversial 논란의 여지가 있는
corruption 부패
violate 위반하다

transaction 거래

mediate 영향을 주다, 가능하게 하다

Word Search

| 정답 | **1.** legislator(s) **2.** controversial **3.** violate

1. 그 입법자는 의료 개혁을 목표로 한 새로운 계획을 제안했다.

2. 새로운 정책은 논란의 여지가 있는 것이 입증되어서, 소셜 미디어 플랫폼 전반에 걸쳐 열띤 논쟁을 불러일으켰다.

3. 여러분이 그 법을 위반하면 여러분은 벌금을 포함한 심각한 처벌에 직면할 것이다.

Exercise 4 ━━━━━━━━━━ 정답 ①

소재 의미를 창출하는 브랜드

해석 사실 70년대 후반과 80년대 초반이 되어서야 비로소 브랜드 개념이 모든 사업 분야로 확장되기 시작했다. 기업이 민영화되고, 시장이 개방되며, 경쟁이 더 치열해짐에 따라 여러분의 사업을 차별화할 필요성이 더 커졌다. 공익 사업체, 통신사, 은행, 보험사, 그리고 항공사가 모두 이제 브랜드의 힘을 열렬히 받아들이고 있었다. 브랜드 소유자들이 여러분의 마음속에 자리잡기 위해 경쟁하면서, 광고는 엄청나게 영향력이 커졌다. 초점은 고객으로서의 여러분이 자신을 가장 잘 대표하는 브랜드에 둘러싸일 수 있도록 제품과 서비스에 의미를 불어넣는 데 맞추어져 있었는데, 이는 쇼핑을 개인적인 표현의 한 형태로 여기는 바로 그 개념이었다. 심지어 다른 기업에 서비스를 제공하는 기업도 브랜드를 갖는 것이 중요한 사업 지원이라는 것을 깨닫기 시작했다. 브랜드는 이제 단순한 로고나 태그라인 그 이상으로 여겨졌으며, 그것은 의미를 창출할 수 있는 기회로 여겨졌다.

문제 해설

70년대 후반과 80년대 초반, 브랜드 개념이 모든 사업 분야로 확장되었으며, 기업은 경쟁 속에서 브랜드의 힘을 받아들이고, 광고를 통해 제품과 서비스에 의미를 부여했다는 내용의 글이다. 빈칸바로 앞에 브랜드가 이제 단순한 로고나 태그라인 그 이상이었다는 내용이 나오므로, 빈칸에 들어갈 말로 가장 적절한 것은 ① '의미를 창출할 수 있는 기회로 여겨졌다)'이다.

② 아주 높은 가격을 정당화하는 핵심 요소가 되어 가고 (있었다)

③ 지속 가능한 제품을 만드는 데 집중하기 시작하고 (있었다)

④ 직원의 사기를 높이는 효과적인 방법으로 인식되(었다)

⑤ 기능적 이점에 연관된 고객 만족에 반영되(었다)

구조 해설

▪ Even businesses [serving other businesses] began to realize [that {having a brand} was an important

business support].

첫 번째 [ ]는 businesses를 수식하는 분사구이다. 두 번째 [ ]는 동사 realize의 목적어로 쓰인 명사절이고, 그 안의 { } 는 명사절의 주어 역할을 하는 동명사구이다.

어휘 및 어구

concept 개념

privatize (기업·산업 분야를) 민영화하다

competition 경쟁

fierce 치열한

differentiate 차별화하다

enthusiastically 열렬히, 열광적으로

embrace 받아들이다

influential 영향력이 큰

surround 둘러싸다, 에워싸다

notion 개념

Word Search

| 정답 | **1.** competition **2.** influential **3.** notion

1. 두 회사 간의 치열한 경쟁은 빠르게 혁신하도록 그들을 밀어붙였다.

2. 그 철학자의 영향력이 큰 저술은 사람들이 현대 철학에 대해 생각하는 방식을 형성해 왔다.

3. 평등의 개념은 민주주의 원칙의 기본이다.

Exercise 5 ━━━━━━━━━━ 정답 ①

소재 인간을 오도할 수 있는 왜곡된 감각 현상

해석 다음 예시와 같은 평범한 일상 상황이 많이 있다. 물에 담근 곧은 막대기는 구부러진 것처럼 보이지만, 우리는 단지 그 막대기가 쉽게 뚫고 들어갈 수 있는 액체인 물에 담겼다고 해서 그것이 구부러졌다고 믿지는 않는다. 철로는 먼 곳에서 수렴하는 것처럼 보이지만, 그것이 겉보기에 합쳐진 것처럼 보이는 지점까지 걸어가 보면, 평행하다는 것을 알게 된다. 텔레비전에서 보이는 자동차의 바퀴는, 자동차가 앞으로 움직이는 것으로 보일 때, 뒤로 가는 것 같다. 하지만 이것은 불가능하다. 이러한 왜곡된 인식 사례는 끝없이 제시될 수도 있다. 따라서 각각의 이러한 감각 현상은 어떤 면에서 오해의 소지가 있다. 인간이 세상을 보이는 그대로 받아들인다면, 그들은 사물이 실제로 어떤지에 대해 속게 될 것이다. 그들은 물에 있는 막대기가 정말로 구부러져 있고, 페이지에 있는 글씨가 정말로 거꾸로 되어 있으며, 바퀴가 정말로 뒤로 가고 있다고 생각할 것이다.

문제 해설

물에 담근 막대기가 구부러져 보이거나 철로가 멀리서 수렴하는 것과 같은 감각 현상은 왜곡된 인식의 사례들로, 실제로는 그렇지 않다는 내용의 글이다. 따라서 빈칸에 들어갈 말로 가장 적절한 것은 ① '그들은 사물이 실제로 어떤지에 대해 속게 될 것이다'이다. ② 자신의 감각이 공상 세계에 이르는 길로 여겨질 것이다 ③ 물리적 세계를 이해하는 것이 수월하고 간단해질 것이다 ④ 그들은 세상이 환상이 없고 객관적인 사실로 가득 차 있다는 것을 깨달을 것이다 ⑤ 그들의 왜곡된 인식은 빠르게 사라질 것이다

구조 해설

- Railroad tracks seem to converge in the distance, and yet [when we walk to the spot {where they apparently merged}] we find them to be parallel.

  [ ]는 시간의 부사절이고, 그 안의 { }는 the spot을 수식하는 관계절이다.

어휘 및 어구

ordinary 평범한
apparently 겉보기에
merge 합쳐지다, 융합하다
parallel 평행한
distorted 왜곡된
perception 인식
multiply (예 따위를 얼마든지) 제시하다, 증가시키다
endlessly 끝없이
phenomenon 현상(pl. phenomena)
misleading 오해의 소지가 있는
reverse 거꾸로 하다, 뒤집다

Word Search

| 정답 | 1. parallel 2. distorted 3. perception
1. 그 도로들은 몇 마일 동안 서로 평행하게 달리며, 항상 같은 거리를 유지하며 떨어져 있다.
2. (놀이공원의) 도깨비 집 거울에 비친 왜곡된 모습은 아이들을 즐겁게 했다.
3. 그 상황에 대한 그녀의 인식은 잘못된 정보에 의해 영향을 받았다.

Exercise 6　　　　　정답 ②

소재 선배 학자들의 지원과 안내

해석 고대 그리스 속담에 따르면 "노인들이 그들은 그것(나무)의 그늘에 결코 앉지 못할 나무를 심을 때 그 사회는 크게 성장한다"는 말이 있다. 마찬가지로, 선배 학자들이 결코 자신의 이름조차 모를 수 있는 후배 학자들을 위해 관대한 행위를 할 때 학문 문화는 크게 성장한다. 문학 학자이자 시인인 Lesley Wheeler는 자신의 첫 책 원고에 대한 그들(두 익명의 독자)의 사려 깊은 반응이 그녀가 성공적인 학자가 되는 길을 시작하게 해 준 것에 대해 두 익명의 독자에게 '끝없이 감사하고' 있는데, "그들은 나에게 책의 문제점을 직설적으로 지적해 주면서도 또한 시간을 내어 칭찬해 주었고, 그것은 충분한 격려가 되었습니다"(라고 말한다). 다른 사람들이 심어준 그늘 나무의 따뜻한 쉼터로부터 혜택을 받았기 때문에 Wheeler는 잔인한 리뷰로 공기를 오염시키는 '까다로운' 심사위원에 대해서는 참을성이 거의 없다. 그녀는 동료와 학생들에 대한 그녀 자신의 피드백이 항상 정중하고 건설적인 것이 되도록 확실히 하는 데 주의를 기울이며, "제 작업을 향한 제가 본 성실성과 관대함은 제가 돌려주고 싶은 것입니다"(라고 말한다).

문제 해설

후배 학자들이 자신의 이름조차 모를지라도 그들에게 관대한 행위를 하는 선배 학자들처럼, Lesley Wheeler는 자신의 첫 책 원고에 대한 두 익명의 독자 덕분에 성공적인 학자가 되는 길을 시작할 수 있었다고 말하고 있다. 또한 선배 학자로부터 혜택을 받은 것처럼 자신도 정중하고 건설적인 피드백을 주는 데 주의를 기울이겠다고 말하고 있으므로, 밑줄 친 부분이 글에서 의미하는 바로 가장 적절한 것은 ② '선배 학자로부터 후배 학자에게 주어지는 지원과 안내'이다.
① 학문적 환경의 암묵적인 단점과 영향
③ 학문적 성취에서 얻는 만족감과 성취감
④ 자신의 학문적 기여에 대한 인정과 확인
⑤ 출판을 통해 학계에서 다른 사람들을 능가함으로써 얻는 이점

구조 해설

- [**Having benefited** from the welcoming shelter of shade trees planted by others], Wheeler has little patience for "cranky" referees [who poison the air with mean-spirited reviews].

  첫 번째 [ ]는 완료분사구문으로 주절보다 이전 시제를 나타내기 위해 「having+p.p.」 형태를 사용하였다. 두 번째 [ ]는 "cranky" referees를 수식하는 관계절이다.

어휘 및 어구

proverb 속담
shade 그늘
scholar 학자
generosity 관대함
manuscript 원고
shelter 쉼터
referee 심사위원, 심판
mean-spirited 잔인한, 비열한
gracious 정중한
constructive 건설적인

conscientiousness 성실성

Word Search

| 정답 | **1.** generosity **2.** anonymous **3.** constructive

1. 병원 재건을 돕기 위해 거액을 기부했을 때 그의 관대함이 분명하게 드러났다.

2. 그 편지는 익명이었기 때문에 우리는 그것을 누가 보냈는지 알 수 없었다.

3. 그녀의 건설적인 피드백은 프로젝트를 개선하고 그것을 더 효과적으로 만드는 데 도움이 되었다.

Exercise 7 ───────── 정답 ②

소재 수학 교육에서 창의성을 기르는 방법

해석 소그룹에서 공부하는 것이 학생들의 창의성을 키울 수도 있지만, 단순히 학생들을 소그룹으로 묶어 공부하게 하면 창의성이 저절로 꽃피리라는 것을 의미하는 것은 아니다. 생산적인 팀워크의 다른 측면의 경우와 마찬가지로 이 과정에서도 학습이 필요하다. 교육 심리학 분야의 저명한 연구자인 Meissner에 따르면, 수학 교육에서 창의적 사고를 발전시키기 위해서는 개인적 능력과 사회적 능력 모두를 발전시킬 필요가 있다. 학생들은 창의성에 영향을 미치는 부정적인 요인을 피하는 방법을 배워야 하는데, 이러한 요인에는 산출 방해, 과제와 무관한 행동, 인지 과부하가 포함된 '인지적 간섭' 그리고 사회적 불안, 무임승차, 생산성의 착각이 포함된 '사회적 억제'가 있다. 그들은 또한 집단의 아이디어 창출 잠재력을 강화할 수 있는 요인을 인식하는 법도 배워야 하는데, 예컨대 그 요인에는 개인의 책임감 증가와 팀 성과에 대한 공유된 기준 개발을 포함하는 '사회적 자극'과 연상에 대한 자극, 타인의 기여에 대한 관심, 아이디어를 생각해 내는 기회 등을 포함하는 '인지적 자극'이 있다.

문제 해설 수학 교육에서 학생들은 창의성을 기르기 위해 창의성에 영향을 미치는 부정적인 요인인 인지적 간섭과 사회적 억제를 피하는 방법을 배워야 하며, 사회적 자극과 인지적 자극이라는 집단의 아이디어 창출 잠재력을 강화하는 요인을 인식하는 법도 배워야 한다는 내용의 글이다. 따라서 빈칸에 들어갈 말로 가장 적절한 것은 ② '개인적 능력과 사회적 능력 모두'이다.
① 다양한 교육 기술들
③ 지지하고 포용하는 교실 문화
④ 특히 방과 후 활동에서 맞춤 교육
⑤ 협업 프로젝트와 구조화된 팀워크 연습

구조 해설
■ Students need to learn [how to avoid the negative factors {that affect creativity}]: ~.
[ ]는 learn의 목적어 역할을 하는 명사구이고, 그 안의 { }는 the negative factors를 수식하는 관계절이다.

어휘 및 어구
foster 키우다, 기르다
creativity 창의성
flourish 꽃피다, 발현되다
aspect 측면
further 발전시키다
negative 부정적인
cognitive 인지적인
interference 간섭
task-irrelevant 과제와 무관한
inhibition 억제
anxiety 불안
free riding 무임승차
illusion 착각
productivity 생산성
potential 잠재력
accountability 책임
opportunity 기회

Word Search

| 정답 | **1.** negative **2.** productivity **3.** potential

1. 뉴스 보도가 환경과 관련된 그 회사의 관행을 부정적으로 조명하여 환경 운동가들의 항의가 증가했다.

2. 개선된 자동화 때문에 공장의 생산성이 증가했다.

3. 그녀는 젊은 선수로서 큰 잠재력을 보여 주었다.

Tips

cognitive interference(인지적 간섭): 한 작업을 수행하는 것에 대한 다른 작업의 방해를 말하며, 학습 과정에서 새로운 정보가 기존에 있는 정보와 충돌하거나 상충하는 경우가 인지적 간섭의 대표적인 예이다.

Exercise 8 ───────── 정답 ③

소재 분산 학습의 현실적인 문제

해석 비록 분산 학습 시간이 집중 학습 시간보다 보통 더 효과적인 것으로 입증되었지만, 실제 환경에서는 때때로 이 장점이 현실적인 고려 사항으로 인해 손상될 수 있다. 예를 들어, 공부할 수 있는 시간이 제한된 경우(예를 들어, 시험 전에 복습할 시간이 한 시간밖에 남지 않은 경우)가 있는데, 그런 경우에는 사용 가능한 시간의 일부를 낭비할 휴식을 취하기보다 전체 시간을 사용하는 것이 더 나을 수도 있다. 또한, 매우 바쁜 사람은 하루 일과에 다수의 개별 학습 시간을 끼워 넣기가 어려울 수 있다. 또 다른 문제로서, 분산 학습은 분명히 집중 학습보다 전체적으로 더 많은 시간(즉, 휴식 시간을 포함한 총 시간)을 요구하며, 따라서 휴식 시간이 어떤 가치 있는 일에 사용될 수 없다면 그 시간을 가장 효율적으로 사용하는 것에 해당하지 않을 수도 있다. 분산 학습은 이러한 종류의 현실적인 문제를 일으킬 수 있으므로 학교 교실과 같은 실제 학습 환경에서는 그것의 가치에 대한 명확한 합의가 없다.

〔문제 해설〕
공부 시간이 제한된 경우, 바쁜 일정으로 다수의 개별 학습 시간을 확보하기 어려운 경우, 휴식 시간이 가치 있는 일에 사용되지 않는 경우와 같이 분산 학습이 현실적으로 문제가 있을 수 있다는 내용의 글이다. 따라서 빈칸에 들어갈 말로 가장 적절한 것은 ③ '현실적인 고려 사항으로 인해 손상될'이다.
① 시간 제약으로 인해 어려움을 겪을
② 급하게 준비할 경우 제한될
④ 일관성 없는 학습 방법으로 인해 약화될
⑤ 휴식의 가치가 낮게 인식되기 때문에 무시될

〔구조 해설〕
▪ [Although **it** has been demonstrated {that spaced learning sessions are usually more effective than massed learning sessions}], in real-life settings this advantage may sometimes be compromised by practical considerations.
[ ]는 양보의 의미를 나타내는 부사절이며, 그 안의 it은 형식상의 주어이고, { }는 내용상의 주어로 쓰인 명사절이다.

〔어휘 및 어구〕
session 시간, 학기
compromise 손상하다
available 할 수 있는, 사용 가능한
revise 복습하다, (시험에 대비해) 공부하다
break 휴식
obviously 분명히, 명백하게
overall 전체적으로
represent (~에) 해당하다, 상당하다
efficient 효율적인
worthwhile 가치 있는
practical 현실적인, 실제적인
agreement 합의

〔Word Search〕

| 정답 | **1.** available **2.** worthwhile **3.** practical

1. 그 도서관은 매일 오전 8시부터 오후 8시까지 학생들이 이용할 수 있다.
2. 그 회의 참석은 귀중한 통찰력을 얻었으므로 가치 있는 경험이었다.
3. 그 과정에는 이론적인 지식과 실제적인 훈련이 모두 포함되어 있다.

〔Tips〕

spaced learning(분산 학습): 일정한 시간을 나누어 중간에 휴식을 취하면서 학습하는 방식으로, 특정 과목의 시험을 준비하는 데 6시간 정도 학습 시간이 필요하다면 2시간씩 3일 동안 학습하는 것을 말한다.

massed learning(집중 학습): 주어진 분량을 일정한 시간에 연속적으로 공부하여 마무리하는 학습 방법으로, 중간에 휴식을 취하지 않고 계속해서 학습하는 것을 말한다.

Exercise 9
정답 ①

〔소재〕 물의 낭비와 가격 책정 문제

〔해석〕 최근 몇 년간, 더 많은 H_2O가 저가 작물(면화와 알팔파)에서 고가 작물(견과류와 딸기류)로 흘러가고 있다. 침체된 농장은 물 사용권을 생산적인 산업과 급성장하는 도시에 팔고 있다. 습한 녹색 기후 지역(미국 북동부, 브라질)에서 재배된 식품이 건조한 갈색 기후 지역(애리조나, 인도)으로 점점 더 수출되고 있어, 그 지역의 물이 식수 공급을 위해, 대수층 수준 유지를 위해 또는 기타 우선순위가 높은 용도를 위해 보존될 수 있게 한다. 그러나 사람들은 희소성을 반영하는 방식으로 물 가격을 책정하는 문제에 대해 말하기를 회피해 온 경향이 있다. 물은 필수 자원이기 때문에, 그것은, 말하자면 기름이 가지고 있는 '시장 가치'가 없다. 그러나 물을 효율적으로 사용하기 위한 가격 유인이 없기 때문에, 사람들은 에너지와 광물 프로젝트에 막대한 양의 물을 사용하고, 물을 오염시킴으로써 물을 낭비하는 경우가 많다. 많은 곳에서 물은 공짜이거나, 가격이 너무 낮게 책정되어, 그것이 창출하는 수익이 저수지나 배수관 및 처리장을 유지하거나 개선하기에 충분하지 않다. 시민들이 물 민영화를 주장하는 사람들에 대해 경계해야 할 충분한 이유가 있지만, 값싼 물은 낭비를 초래한다.

〔문제 해설〕
시장 가치가 없고 가격 유인이 없어 물이 낭비되고 있으며, 많은 곳에서 물은 공짜이거나 가격이 너무 낮게 책정되어 물의 유지 관리 비용을 충당하기 어렵다는 내용의 글이다. 따라서 빈칸에 들어갈 말로 가장 적절한 것은 ① '값싼 물은 낭비를 초래한다'이다.
② 물 가격 책정을 서두르는 움직임은 둔화될 수 있다
③ 공공 통제는 모두를 위한 물의 흐름을 유지한다
④ 기후 변화는 수자원을 회복시킬 것이다
⑤ 물 절약은 도시 거주자에게 덜 필수적이다

구조해설

- In many places water is free, or priced **so** low **that** the revenue [it generates] is not enough to maintain, or upgrade, reservoirs, distribution pipes, and treatment plants.

'너무 ~해서 (그 결과) …하다'라는 의미의 「so ~ that …」 구문이 사용되었으며, [ ]는 the revenue를 수식하는 관계절이다.

어휘 및 어구

flow 흐르다
low-value 저가의
right 권리
productive 생산적인
climate 기후 지역, 기후
export 수출하다
conserve 보존하다
dance around ~에 대해 말하기를 회피하다
reflect 반영하다
scarcity 희소성
essential 필수적인
price incentive 가격 유인
revenue 수익, 소득

Word Search

| 정답 | **1.** reflect **2.** essential **3.** revenue
1. 경제 정책은 모든 시민의 요구와 우선순위를 반영해야 한다.
2. 효과적인 팀워크를 위해서는 원활한 의사소통 능력이 필수적이다.
3. 새로운 제품 라인을 출시한 후 그 회사의 총 수익이 크게 증가했다.

Exercise 10 정답 ①

소재 괴물에 대한 두려움의 중요성

해석 가장 기본적인 수준에서, 괴물은 사회가 가진 두려움, 즉 주변 세계에서 인식되는 위험과 관련이 있는 두려움을 상징한다. 이러한 두려움은 자멸을 초래하는 싸움을 하는 대신에 도망치도록 사람들을 장려함으로써 강력한 진화적 역사를 가지고 있다. 고대의 사냥꾼들은 우연히 검치호랑이를 만났을 때 도망쳤다. 인간 조상인 '호모 에렉투스'는 화난 동굴 곰을 갑자기 마주쳤을 때 도망쳤다. 현대 인간의 가장 가까운 유전적 친척인 침팬지와 보노보는 야생에서 큰 포식자를 마주칠 때 도망친다. 할리우드 영웅들로 인해 도망치는 것은 (영화의) 은막에서 확실히 인기가 없게 되었지만, 어떤 끔찍한 공포에 맞서 자기 자리를 지킨 영웅을 연기해 온 모든 배우 한 사람 한 사람은 위험에 직면했을 때 도망친 겁쟁

이들의 긴 유전적 혈통의 출신이다. 그것이 그들이 오늘날 여기서 연기할 수 있는 이유이다. 만약 그들의 조상이 할리우드 영웅들이 항상 하는 것처럼 자신들보다 훨씬 더 강력한 괴물들과 싸웠더라면, 그들의 혈통은 오래전에 포식자들에 의해 파괴되었을 것이다. 간단히 말해, 두려움은 사람들을 살아 있게 한다.

문제해설

괴물은 사람들이 주변 환경에서 인식하는 위험과 관련된 두려움을 상징하는 존재로, 이러한 두려움이 인간이 생존을 위협하는 위험에서 도망치도록 했고, 그렇게 살아 남았기에 그 후손들 중 영웅을 연기하는 배우도 있을 수 있다는 내용의 글이다. 따라서 빈칸에 들어갈 말로 가장 적절한 것은 ① '사람들을 살아 있게 한다'이다.
② 영웅적 행동에 영감을 준다
③ 사람들을 싸우게 만든다
④ 힘이 있다는 느낌을 만들어 낸다
⑤ 개인을 더 강하게 만든다

구조해설

- [When chimpanzees and bonobos, {the nearest genetic relatives to modern humans}, encounter large predators in the wild], they run.

[ ]는 시간의 부사절이고, 그 안의 { }는 chimpanzees and bonobos와 동격 관계를 이룬다.

어휘 및 어구

basic 기본적인
represent 상징하다, 나타내다
evolutionary 진화의, 진화론적인
encourage 장려하다, 고무하다
flee 도망치다
ancient 고대의
encounter 만나다, 마주치다
ancestor 조상
cave bear 동굴 곰
genetic 유전적인
predator 포식자
distinctly 확실히, 분명히
coward 겁쟁이

Word Search

| 정답 | **1.** Ancient **2.** encounter **3.** predator
1. 고대의 신화와 전설은 흔히 상상력이 풍부한 방식으로 자연 현상을 설명한다.
2. 숲에서 하이킹을 하다 보면 예기치 않게 다양한 야생 동물과 마주치는 경우가 드물지 않다.
3. 치타는 먹이를 사냥할 때의 놀라운 속도로 알려진 숙련된 포식자이다.

Exercise 11

정답 ②

소재 효모 연구의 중요성과 성과

해석 1950년대에는 거의 아무도 효모에 관심을 가지지 않았다. 대부분의 사람들에게 작은 균류 연구를 통해 복잡한 우리 자신에 대해 많은 것을 배울 수 있을 것 같지 않았다. 효모가 빵을 굽고 맥주를 양조하며 와인을 빚는 것보다 더 많은 어떤 유용성을 지닐 수 있다는 것을 과학계에 설득하는 것은 어려운 일이었다. Mortimer와 Johnston이 인식하고, 이후 많은 사람들이 깨닫기 시작했던 것은 그 작은 효모 세포들이 우리 자신과 크게 다르지 않다는 것이었다. 그 크기에 비해, 효모의 유전적, 생화학적 구성은 매우 복잡하여, 우리 자신과 같은 크고 복잡한 생물에서 생명을 유지하고 수명을 조절하는 생물학적 과정을 이해하는 데 특별히 좋은 모델이 된다. 만일 여러분이 효모 세포가 암, 알츠하이머병, 희귀 질환 또는 노화에 대해 우리에게 무엇인가를 알려 줄 수 있다는 것이 미심쩍다면, 세포가 노화의 특징 중 하나인 텔로미어 단축에 어떻게 대응하는지 발견하여 주어진 2009년 수상을 포함하여, 효모의 유전 연구로 5개의 노벨 생리의학상이 수여되었다는 것을 고려해 보라.

문제 해설
1950년대에 효모 연구는 주목받지 못했으나, 그 후 효모가 인간의 복잡한 생물학적 과정 이해에 중요한 모델임이 밝혀졌고, 효모에 관한 유전 연구가 노화를 포함한 인간의 여러 질환 연구에 실마리가 될 수 있다는 내용이다. 따라서 빈칸에 들어갈 말로 가장 적절한 것은 ② '그 작은 효모 세포들이 우리 자신과 크게 다르지 않다'이다.
① 새로운 발견은 자주 오랫동안 존재해 온 것을 재발견한다
③ 과학은 복잡한 현상을 파악하기 위해 여러 분야를 결합해야 한다
④ 우리는 효모를 포함함으로써 우리의 식품 품질을 향상할 수 있다
⑤ 새로운 치료법을 발견하기 위해서는 세포를 이해하는 것이 대단히 중요하다

구조 해설

▪ [If you are doubtful that a yeast cell can tell us anything about cancer, Alzheimer's disease, rare diseases, or aging], consider [that there have been five Nobel Prizes in Physiology or Medicine awarded for genetic studies in yeast, including {the 2009 prize for discovering how cells counteract telomere shortening, one of the characteristics of aging}].
첫 번째 [ ]는 조건의 부사절이고, 두 번째 [ ]는 consider의 목적어 역할을 하는 명사절이다. { }는 전치사 including의 목적어 역할을 하는 명사구이다.

어휘 및 어구
hardly 거의 ~ 않다
struggle 어려운 일
scientific community 과학계
recognize 인식하다
genetic 유전적인
biochemical 생화학적인
makeup 구성
complex 복잡한
biological 생물학적인
process 과정
sustain 유지하다
lifespan 수명
doubtful 미심쩍은
rare 희귀한
characteristic 특징

Word Search

| 정답 | 1. complex 2. genetic 3. rare
1. 열대우림의 생태계는 매우 복잡하며, 수많은 종들이 섬세한 균형을 이루어 상호 작용하고 있다.
2. 과학자들은 유전 공학을 사용하여 질병과 해충에 대한 작물의 회복력을 높이기를 희망한다.
3. 그 박물관에는 고대 문명의 희귀한 유물 수집품이 있다.

Exercise 12

정답 ⑤

소재 효과적인 의사소통 규칙의 중요성

해석 규칙은 우리의 일상생활을 지배한다. 이러한 규칙 중 일부는 명시적이며 정부에 의해 시행되는데, '속도 제한 준수', '주차 금지', '4월 15일은 납세의 날' 같은 것들이다. 그러나 대부분은 비공식적이고 말로 표현되지 않은 경우가 많은 문화적 규범인데, 예의범절, 비즈니스 세계의 행동강령, 사람들 간의 상호 작용 규칙 같은 것들이다. 대부분은 시간이 지나면서 형성된 일반적으로 이해되는 전통, 즉 우리가 보통 생각조차도 하지 않을 만큼 일상적인 습관이다. 안타깝게도, 이러한 자기도 모르게 나오는 습관과 잠재의식적인 관습들이 모두 긍정적이거나 생산적인 것은 아니다. 미국의 비즈니스 및 정치적 의사소통에는 그들이 홍보하고자 하는 회사와 대의에 심각한 피해를 입힐 수 있는 나쁜 습관과 도움이 되지 않는 경향이 만연해 있다. 다른 모든 분야와 마찬가지로, 좋고 효과적인 의사소통에는 규칙이 있다. 그것은 속도위반이나 세금 회피에 대한 규칙만큼 엄격하고 절대적이지는 않을 수도

있지만, 여러분의 주머니에 돈이 있는 채로 목적지에 안전하게 도착하고 싶다면 그만큼 중요하다.

문제 해설

규칙은 일상생활을 지배하며, 명시적 규칙과 비공식적 문화 규범이 있고, 효과적인 의사소통에도 규칙이 필요하며, 이는 나쁜 습관을 피하고 목표를 달성하는 데 중요하다는 내용의 글이다. 따라서 밑줄 친 부분이 글에서 의미하는 바로 가장 적절한 것은 ⑤ '부정적인 결과를 겪는 것을 피하기 위해 효과적으로 의사소통하고'이다.

① 빠르게 변화하는 업계 유행에 발맞추고
② 공동체 구성원과의 장기적인 관계를 조성하고
③ 위험에 구애받지 않고 국제적인 기회를 활용하고
④ 규제 범위 내에서 최소한의 세금을 유지하고

구조 해설

- American business and political communication is rife with bad habits and unhelpful tendencies [that can do serious damage to the companies and causes {they seek to promote}].

첫 번째 [ ]는 bad habits and unhelpful tendencies를 수식하는 관계절이고, 그 안의 { }는 the companies and causes를 수식하는 관계절이다.

어휘 및 어구

govern 지배하다
explicit 명시적인
impose 시행하다, 부과하다
cultural norms 문화적 규범
conduct 행동
ordinary 일상적인, 평범한
involuntary 자기도 모르게 나오는, 무의식적인
subconscious 잠재의식적인
convention 관습
tendency 경향
promote 홍보하다, 촉진하다
inflexible 엄격한
absolute 절대적인
destination 목적지

Word Search

| 정답 | 1. Ordinary 2. involuntary 3. tendency
1. 평범한 사람들이 작은 친절한 행위로 큰 변화를 만들 수 있다.
2. 눈을 깜빡이는 것은 우리의 시력을 보호하는 무의식적인 반응이다.
3. 사람들은 변화에 저항하는 경향이 있다.

Week 2 04강 본문 42~53쪽

| 1 ① | 2 ③ | 3 ① | 4 ① | 5 ① | 6 ⑤ |
| 7 ③ | 8 ④ | 9 ⑤ | 10 ② | 11 ④ | 12 ⑤ |

Exercise 1 정답 ①

소재 기후 변화에서 인간의 화석 연료 사용의 책임 규명 필요성

해석 "인간이 화석 연료를 태움으로써 기후 변화를 일으키고 있다."라는 진술로 시작해 보자. 그것은 나를 포함하여 전 세계 사람들이 화석 연료 사용의 신속한 중단과 탄소 배출이 없는 에너지원으로의 전환을 요구하고 있는 근거이다. 그것은 꽤 대담한 진술이며, 그것은 기후가 변화하고 있다는 '것'을 말하는 것, 곧 과학자들이 '탐지'라고 부르는 것과는 사뭇 다르다. 인간 사회에서 대대적인 변화, 이는 우리의 화석 연료 사용을 끝내기 위해 요구될 것인데, 이를 주장하고자 한다면, 우리가 탐지를 넘어서야 한다고 요구하는 것이 합리적으로 보인다. 어쨌든 화석 연료는 그것의 문제점에도 불구하고 20세기에 걸쳐 사회에 엄청난 혜택을 제공해 왔다. 우리(기후를 걱정하는 대중)가 화석 연료 사용을 중단해야 한다고 주장하고자 한다면, 화석 연료가 제공해 온 매우 실제적인 장점보다 단점이 더 크다는 것을 입증하는 것이 우리의 의무이다. 우리는 기후가 변화하고 있으며, '그리고' 다른 것이 아니라 인간의 화석 연료 사용이 우리가 목격하고 있는 기후 변화의 원인이라는 것을 합리적 의심의 여지 없이 증명해야 한다. 우리는 탐지 외에도 원인 규명이 필요하다.

문제 해설

기후가 변화하고 있다는 것의 탐지를 넘어서 인간의 화석 연료 사용이 기후 변화의 원인이라는 것을 입증해야 한다는 내용의 글이다. 따라서 빈칸에 들어갈 말로 가장 적절한 것은 ① '우리는 탐지 외에도 원인 규명이 필요하다'이다.

② 모든 것은 기후가 변화하고 있다고 말하는 것으로 시작하고 끝난다
③ 우리는 기후 변화에 맞서기 위해 주관성에서 벗어나야 한다
④ 기존의 문제를 탐지하는 것은 협력적인 해결책으로 이어질 수 있다
⑤ 모든 변화는 과학적 정밀함으로 탐지되어야 한다

구조 해설

- [If we're going to argue for a massive change in human society, {which is what will be required to end our use of fossil fuels}], **it** seems reasonable [to ask that we move beyond detection].

첫 번째 [ ]는 조건의 부사절이고, 그 안의 { }는 a massive

change in human society에 대해 부가적으로 설명하는 관계절이다. it은 형식상의 주어이고, 두 번째 [ ]가 내용상의 주어이다.

어휘 및 어구

statement 진술
fossil fuel 화석 연료
call for ~을 요구하다
rapid 신속한, 빠른
transition 전환
carbon 탄소
emission 배출
bold 대담한
detection 탐지
argue 주장하다
massive 대대적인
reasonable 합리적인
tremendous 엄청난
insist 주장하다
downside 단점
doubt 의심
observe 목격하다, 관찰하다

Word Search

| 정답 | 1. rapid 2. transition 3. bold
1. 그 환자는 빠르게 회복했다.
2. 그 나라는 독재에서 민주주의로의 평화로운 전환을 겪었다.
3. 그 팀은 스타 선수를 트레이드하는 대담한 조치를 취했다.

Exercise 2 —————————— 정답 ③

소재 인간의 정신에 존재하는 문법 규칙

해석 언어의 문법 구조는 개인 화자에게 외재적이고 제약을 가한다는 Durkheim의 의미에서의 '사회적 사실'이다. 그것은 그들의 주관적인 선호와 무관하며, 이해받고자 한다면 그들은 규칙을 따라야 한다. 그러나 성, 일치, 수, 주어와 목적어, 소유격 등의 문법 규칙은 일반적으로 서로 대화하는 개인들에 의해 의식적으로 지켜지고 적용되지는 않는다. 화자들은 일반적으로 자신들만의 문법 규칙에 대해 매우 제한적이고 부분적인 인식만을 가지고 있으며, 문법적으로 말하는 것은 의식적인 규칙 준수보다는 깊이 생각하지 않는 습관의 문제이다. 그렇다면 언어의 문법 규칙은 개인 화자들의 정신으로부터 떨어져서 존재하지 않는다. 그것들은 문법책에 공식화될 수도 있지만, 그러한 책은 문법을 거의 불완전

하게 '기록할' 뿐 문법을 '구성하지' 않는다. '올바른' 발화와 잘 구성된 담화를 형성하는 데 따르는 규칙은 개인 화자들의 뇌의 신경생리학적 기억 흔적에 보유된 학습된 성향으로서 그들의 정신 속에만 존재한다.

문제 해설

우리가 사회적 사실인 문법 규칙을 따르는 것 같지만, 일반적으로 문법 규칙은 개인들이 의식적으로 따르고 적용하는 것이 아니며, 그 규칙은 개인 화자들의 뇌의 신경생리학적 기억 흔적에 보유된 학습된 성향으로서 그들의 정신 속에만 존재한다는 내용이므로, 빈칸에 들어갈 말로 가장 적절한 것은 ③ '개인 화자들의 정신으로부터 떨어져서 존재하지 않는다'이다.
① 상대방에 대한 각 화자의 이해를 형성한다
② 효과적인 의사소통을 위한 핵심적 틀로서 역할을 한다
④ 주관적인 감정이 표현되는 방식에 영향을 미친다
⑤ 과거 대화를 기억하는 데 엄청난 제약을 가한다

구조 해설

■ The rules [that are followed in forming a 'correct' utterance and a well-organised discourse] exist only in the minds of the individual speakers as learned dispositions [held in the neurophysiological memory traces of their brains].
첫 번째 [ ]는 The rules를 수식하는 관계절이고, 두 번째 [ ]는 learned dispositions를 수식하는 분사구이다.

어휘 및 어구

structure 구조
external 외재적인, 외부의
constrain 제약을 가하다, 억제하다
be independent of ~과 무관하다, 독립적이다
subjective 주관적인
preference 선호
subject 주어
object 목적어
possessive 소유격
consciously 의식적으로
apply 적용하다
partial 부분적인
awareness 인식
unreflective 깊이 생각하지 않는, 무분별한
formulate 공식화하다, 기술하다
comprise 구성하다
utterance 발화
discourse 담화
disposition 성향, 성질

trace 흔적

Word Search

| 정답 | **1.** external **2.** constrain **3.** awareness
1. 이 로션은 외용으로만 쓸 수 있다.
2. 그 도로들의 수용 능력은 차량 통행량에 제약을 가할 것이다.
3. 관련 문제들에 대한 인식이 완전히 결여되어 있었다.

Tips

Émile Durkheim(에밀 뒤르켐): 프랑스의 사회학자. 카를 마르크스, 막스 베버와 더불어 사회학이라는 학문이 등장하는 데 큰 영향력을 미친 학자이다.

Exercise 3 ──── 정답 ①

소재 인간 참여자를 포함하는 장으로서의 환경

해석 환경이 인간의 이해관계를 수반할 때, 그것은 반드시 독립적인 사물의 집합으로서가 아니라 인간과 관련지어 이해되어야만 한다. 우리는 Kurt Lewin과 J. J. Gibson 같은 사회 심리학자들의 연구에서 이에 대한 근거를 찾을 수 있다. Lewin은 참여자들과 이들이 상호 작용하는 사물 및 환경 사이의 힘의 벡터로 구성된 사회적 세계를 마음속에 그렸다. 이러한 벡터는 특정 행동을 초래하며, 이 때문에 Lewin은 그것들을 독일어 용어 *Affördungsqualitäten* 이라고 부르게 되었는데, 영어로 번역하면 'invitational qualities(초래하는 특성)'이다. 더 최근에는 지각 심리학자인 J. J. Gibson이 환경 구성 및 사물의 디자인과 외관이 인간 행동에서 특정 반응을 부추기는 방식을 연구했다. 그는 이러한 연결 관계를 행동에 대한 '유도성'이라고 불렀는데, 분명히 Lewin의 용어에 의해 영향을 받았으며 그의 의견과 유사했다. Lewin과 Gibson의 연구는 중요하고 교훈적인데 왜냐하면 그것은 환경이 단순히 독립적인 사물들의 배열로 채워진 열린 공간이 아니라 끌어당김, 반발 및 중립 또는 무관심의 강력한 관계 속의 힘의 장(場)이라는 것을 시사하기 때문이다. 그렇다면 환경은 <u>인간 참여자를 포함하는 장</u>이다.

문제 해설
환경은 독립적인 사물의 집합으로서가 아니라 인간과 관련지어 이해되어야만 하고, 단순히 독립적인 사물들의 배열로 채워진 열린 공간이 아니라 다양한 관계 속의 힘의 장이라고 하고 있으므로, 빈칸에 들어갈 말로 가장 적절한 것은 ① '인간 참여자를 포함하는 장'이다.
② 사회 내 갈등 이면에 있는 이유
③ 사물을 통해 우리의 감정을 억누르는 영역
④ 개인의 부정적인 감정 반응의 도화선
⑤ 인간을 사물로부터 단절시키는 보이지 않는 힘

구조 해설
■ More recently, the perceptual psychologist J. J. Gibson studied the ways [in which {the design and appearance of ⟨environmental configurations and objects⟩} encourage particular responses in human behavior].
[ ]는 the ways를 수식하는 관계절이고, { }가 관계절의 주어 역할을 한다. ⟨ ⟩는 전치사 of의 목적어 역할을 하는 명사구이다.

어휘 및 어구
involve 수반하다
in relation to ~과 관련지어
assemblage 집합
social psychologist 사회 심리학자
envision 마음속에 그리다
comprised of ~로 구성된
vector 벡터(방향적 행동을 일으키는 추진력)
interact 상호 작용하다
translate 번역하다
quality 특성
terminology 용어
resemble 유사하다
observation (관찰에 근거한) 의견, 진술
instructive 교훈적인, 유익한
filled with ~로 채워진
arrangement 배열
compelling 강력한
neutrality 중립
indifference 무관심

Word Search

| 정답 | **1.** assemblage **2.** instructive **3.** compelling
1. 인간은 단순히 화학물질의 집합이나 생각 없는 기계가 아니다.
2. 그 의사들이 일하는 모습을 보는 것은 매우 유익했다.
3. 그는 자신의 불행에 대해 이야기하고 싶은 필요성이 강한 슬픈 사람이었다.

Exercise 4 ──── 정답 ①

소재 저널리스트들의 전문성 확립의 어려움

해석 많은 저널리즘 학자들이 주장해 왔듯이, 인쇄 시대의 저널리스트들은 독자 조사를 거부했는데 왜냐하면 그렇게 하는 것(독자 조사를 거부하는 것)이 자신들의 항상 불안정한 전문직의 지위를 보호하는 유일한 수단 중 하나였기 때문이었다. 사회학자 Andrew Abbott는 전문직의 특징을 '다소 추상적인 지식을 특정 사례에 적용하는 다소 독점적인 개인 집단'으로 묘사해 왔다. 일반적으로 전문직으로 분류되지만, 저널리즘은 이 정의에 수월하게 자리 잡기 위해 오랫동안 고군분투해 왔다. 심지어는 인터넷의 부상이 제도적인 게이트키핑의 힘을 언론사에서 기술 플랫폼으로 이동시키는 데 도움을 주기 전에도 저널리스트들은 '다소 독점적인 개인 집단'으로 자리매김하는 데 어려움을 겪었다. 실제로 의학과 법률과 같은 전통적인 전문직은 직업으로의 진입을 제한하기 위해 엄격한 면허 요건에 의존하지만, 미국 헌법 수정 제1조는 미국 저널리즘이 그러한 어떤 것도 설정하는 것을 금지한다. 또한 저널리스트들은 추상적인 지식 형태에 대한 관할권을 강력하게 주장할 수도 없다. 저널리즘 학자 Matt Carlson이 주장했듯이, "추상화는 나쁜 저널리즘을 야기한다. 명확성은 특히 복잡한 주제에 대한 설명에서 좋은 저널리즘을 야기한다." 저널리스트의 언어의 접근성은 대중에게 정보를 제공하는 데 도움이 되지만, 그것은 또한 <u>전문적 지식에 대한 저널리스트들의 주장</u>을 잠재적으로 의심스럽게 만든다.

문제 해설

저널리즘이 추상적인 지식을 적용하는 독점적인 집단이라는 전문직의 정의에 수월하게 자리 잡기 위해 오랫동안 고군분투해 왔지만, 저널리스트들은 추상적인 지식 형태에 대한 관할권을 강력하게 주장할 수는 없다는 내용이므로, 빈칸에 들어갈 말로 가장 적절한 것은 ① '전문적 지식에 대한 저널리스트들의 주장'이다.
② 사회적 쟁점에 대한 저널리스트들의 다양한 의견
③ 진실을 전달하는 매개체로서의 저널리즘
④ 저널리스트들이 자신들이 수행한다고 주장하는 도덕적 의무
⑤ 독자 조사를 수용하는 저널리스트들의 의도

구조 해설

- Even before the rise of the internet helped shift [institutional gatekeeping power] away from news organizations and toward technology platforms, journalists had difficulty establishing **themselves** as a "somewhat exclusive group of individuals."
  [ ]는 shift의 목적어 역할을 하고, themselves는 establishing의 목적어로서, 주절의 주어인 journalists를 가리킨다.

어휘 및 어구

argue 주장하다
unstable 불안정한
professional 전문직의, 전문가의
status 지위

sociologist 사회학자
profession 전문직, 직업
exclusive 독점적인, 전용의
apply 적용하다
abstract 추상적인
struggle 고군분투하다
inhabit ~에 자리 잡다, 존재하다
shift 이동시키다
institutional 제도적인
strict 엄격한
licensing 면허
entry 진입
prohibit 금지하다
clarity 명확성
accessibility 접근성
potentially 잠재적으로
suspect 의심스러운

Word Search

| 정답 | 1. status 2. professions 3. exclusive
1. 의사는 전통적으로 높은 사회적 지위를 누려 왔다.
2. 의료 분야의 대부분의 직업은 수년간의 훈련이 필요하다.
3. 그 방은 투숙객 전용이다.

Tips

gatekeeping(게이트키핑): 뉴스 미디어 조직 내에서 기자나 편집자와 같은 뉴스 결정권자가 뉴스를 취사선택하는 과정을 말한다.

Exercise 5

정답 ①

소재 교통 시스템 관리의 어려움

해석 특히 교통 시스템과 같이 복잡한 것은 무엇이든지 그 관리가 매우 어렵다. 여러 다른 조직이 관련되어 있는 경우가 흔하며, 각 조직에는 여러 수준의 권한을 가진 여러 부서와 흔히 많은 장황하고 문서화된 절차 매뉴얼이 있다. 설상가상으로 그 매뉴얼은 좀처럼 최신 상태로 유지되지 않고, 어떤 경우에도 발생할 수도 있는 모든 요인들의 조합을 도저히 고려할 수 없다. 사건이 벌어지는 동안 통제를 유지할 책임이 있는 사람들(대부분의 상업 항공 사고의 경우 조종사들)은 관련된 사례를 찾으려고 애쓰면서 서로 다른 매뉴얼을 검토하느라 귀중한 시간을 낭비한다. 현대의 컴퓨터 시스템은 자동으로 상황을 진단하고 독자적으로 대응하거나 운영자에게 따를 지침을 제공함으로써 도움을 주려고 시도하지만, (각각의 복잡한 시스템에서는 대부분의 사고가 다른 고유한

요인을 포함하기 때문에) 그 진단이나 권장되는 행동 방침이 항상 적절한 것은 아니다. 서로 다른 조직들이 관련될 수도 있을 터인데, 이를테면 경찰과 소방관, 회사의 안전 담당자, 회사의 다른 부서에서 온 여러 팀, 그리고 자신들의 결정과 조치를 조율해야 하는 다른 회사나 정부 및 규제 기관 등이다. 그 결과가 원활하고 완벽한 관리인 경우는 드물다.

문제 해설

여러 조직이 관련되어 있고, 각 조직에는 여러 수준의 권한을 가진 여러 부서와 장황하고 문서화된 절차 매뉴얼이 있는 경우가 흔하기 때문에 교통 시스템과 같이 복잡한 것의 관리는 매우 어렵다는 내용이므로, 빈칸에 들어갈 말로 가장 적절한 것은 ① '그 결과가 원활하고 완벽한 관리인'이다.
② 시스템 내에서 비전이 보편적으로 공유되는
③ 관리자가 미래의 위험을 예측하기 위해 과거를 들여다보는
④ 기술이 상충하는 관리 문제를 유발하는
⑤ 사람들이 자신들의 의사 결정에서 편견 없는 추론을 사용하는

구조 해설

- Modern computer systems attempt to help **by** [automatically **diagnosing** the situation] and [either {**responding** autonomously} or {**offering** operators the instructions ⟨to be followed⟩}], but the diagnosis or recommended course of actions is not always appropriate (because in each complex system, most accidents involve different unique factors).

두 개의 [ ]는 동명사구로서 and로 대등하게 연결되어 by에 이어져 '~함으로써'라는 의미의 「by + -ing」 구문을 이룬다. 두 번째 [ ] 안의 두 개의 { }는 「either ~ or ....」로 연결되어 있고, ⟨ ⟩는 the instructions를 수식하는 to부정사구이다.

어휘 및 어구

management 관리
transportation 교통
involve 관련시키다, 포함하다
division 부서
authority 권한
procedure 절차
up to date 최신의
factor 요인
maintain 유지하다
relevant 관련된
automatically 자동으로
diagnose 진단하다
operator 운영자
instruction (보통 복수로) 지침, 지시

regulatory agency 규제[관리] 기관
coordinate 조율하다

Word Search

| 정답 | **1.** management **2.** transportation **3.** division
1. 그 보고서는 잘못된 관리를 탓한다.
2. 캠퍼스는 일반적으로 대중교통으로 접근할 수 있다.
3. 나는 행정 부서에서 사무 관리자로 일하고 있다.

Exercise 6 ——————— 정답 ⑤

소재 성인이 되면서 점차 사라지게 되는 길 찾기 능력

해석 나이는 공간적 능력의 유일한 결정 요인이 아니다. 열세 살짜리 아이들은 길 찾기에 능숙해지기 위해 필요한 모든 인지적 자질을 갖추고 있기는 하지만, 어떤 아이들은 다른 아이들보다 길 찾기에 더 능숙하다. 이 시점에 이르면 부모의 태도, 이동의 자유, 인지적 차이 및 삶의 경험이 이미 그것들의 각인을 남기기 시작했으며, 그것들은 결코 약해지지 않는다. 우리 모두 태어날 때는 탐험가일지 모르지만, 우리 중 그 상태로 남는 사람은 거의 없다. 우리는 결국 우리의 어린아이 본성을 억누르게 되고, 틀에 빠져들어 항상 다니던 경로를 따라가게 된다. 최근 캐나다 심리학자들의 연구에 따르면 8세 어린이의 84퍼센트가 자신의 주변을 세심히 살피고 정신적 지도를 만들어서 길을 찾는데, 이는 능숙하게 길을 찾는 거의 모든 성인이 또한 사용하는 이른바 '공간적' 전략이다. 이를 대체하는 것은 더 폐쇄적이고 '자기중심적인' 전략인데, 이는 방향 전환의 순서를 배우고 따르는 것을 수반한다. 20대에는 우리 중 46퍼센트만이 여전히 공간적 접근 방식을 사용하며, 60대에는 39퍼센트만이 사용한다. 우리 모두 자유롭게 떠돌며 시작하지만, 우리 대부분이 결국 곧고 좁은 길을 택하게 되는 것 같다. 인생은 흔히 우리 날개를 꺾게 된다.

문제 해설

우리는 모두 태어날 때는 탐험가이지만, 부모의 태도, 이동의 자유, 인지적 차이 및 삶의 경험이 각인을 남기기 시작하면서 어린아이의 본성을 억누르고 틀에 빠져들어 항상 다니던 경로를 따라가게 된다고 말하며, 우리 모두 대부분은 결국 곧고 좁은 길(즉, 제한적인 길)을 택한다는 내용이다. 따라서 밑줄 친 부분이 글에서 의미하는 바로 가장 적절한 것은 ⑤ '시간이 지나며, 우리는 우리의 모험심을 잃고, 대담하게 길을 찾는 것을 자제하게 된다.'이다.
① 삶은 우리가 외부 세계가 아닌, 우리의 내면의 것을 탐구하도록 강요한다.
② 우리가 나이 들면서, 우리의 공간적 인식은 인지적 편향에 의해 왜곡된다.

③ 아이들이 길을 찾는 데 가장 큰 장애물은 실패에 대한 자신의 두려움이다.

④ 정해진 길을 따르는 우리의 경향은 사회적 압박에 의해 억제된다.

구조 해설

- We **end up suppressing** our childish natures, [{slipping into routines} and {following the routes ⟨we always take⟩}].

「end up+-ing」는 '결국 ~하게 되다'라는 뜻이다. [ ]는 주절의 부수적인 상황을 나타내는 분사구문이다. 그 안의 두 개의 { }는 and로 대등하게 연결되어 있다. ⟨ ⟩는 the routes를 수식하는 관계절이다.

어휘 및 어구

determinant 결정 요인

spatial 공간적인

cognitive 인지적인

attribute 자질, 속성

proficient 능숙한

ease off 약해지다, 완화되다

suppress 억누르다

slip into ~에 빠져들다

navigate 길을 찾다

surroundings 주변

competent 능숙한

alternative 대체하는 것, (2개의 것 중) 다른 하나

egocentric 자기중심적인

wander 떠돌다, 돌아다니다

have a way of -ing 흔히 ~하게 되다

clip 꺾다, 자르다

Word Search

| 정답 | **1.** determinant(s) **2.** spatial **3.** attribute(s)

1. 토양과 기후는 토지 이용 방식의 주요 결정 요인이다.

2. 이 과제는 아이들의 공간 인식을 시험하기 위해 고안되었다.

3. 그는 저널리스트의 필수적 자질을 가지고 있다.

Exercise 7 정답 ③

소재 인간처럼 구문을 사용하는 일본 박새의 노래

해석 우리가 매일 듣는 새소리는 아름답기만 한 것이 아니다. 그것은 실용적인 목적을 수행한다. 새들은 자기 짝을 부르고, 자기 무리를 찾고, 영역을 주장하고, 침입자를 겁주어 쫓아내고, 포식자에 대해 다른 새들에게 경고하기 위해, 그리고 수많은 다른 기능을 위해 자신의 목소리를 사용한다. 예를 들어, 일본과 스위스의 연구자들은 최근에 뚜렷한 흰색 뺨과 함께 새까만 머리와 목을 가진 작은 새인 일본 박새가 사람이 말을 할 때 그러는 것처럼 노래에 구문을 사용한다는 것을 발견했다. 구문은 언어에서 매우 중요하다. 예를 들어, "나는 저 식당을 좋아해요."라고 말하면 메시지가 명확하다. 그러나 "식당 좋아해요 저 나는."이라고 하면 스타워즈의 요다 마스터도 이해할 수 없을 것이다. 최근까지 과학자들은 인간만이 그러한 발성을 연결할 수 있다고 믿었다. 밝혀진 바에 따르면, 음운론적 구문, 즉 개별적으로 의미가 없는 소리를 결합하여 집합적인 소리로 만드는 능력을 사용할 수 있는 인간을 제외하고는, 일본 박새가 그렇게 하는 최초의 동물이다. 자기 무리의 다른 구성원들에게 포식자를 찾도록 지시하거나 짝을 유혹하기 위해, 일본 박새는 반드시 올바른 순서로 여러 개의 구별되는 음을 불러야 하며, 그 연구에 의하면, 음을 다르게 부르면, 다른 새들이 반응하지 않는다.

문제 해설

언어에서는 구문이 중요하고 구문에 어긋나게 말을 하면 이해할 수 없는데, 일본 박새도 인간과 같이 노래에 음운론적 구문을 사용한다는 내용이므로, 빈칸에 들어갈 말로 가장 적절한 것은 ③ '올바른 순서로 여러 개의 구별되는 음을 불러야'이다.

① 예상치 못한 패턴으로 음을 발성해야

② 역동적으로 노래의 음량을 달리 해야

④ 각 노래를 자기 자신만의 독특한 리듬으로 전달해야

⑤ 다른 새 종의 소리를 정확하게 흉내 내야

구조 해설

- The Japanese great tit, [it turns out], is the first animal apart from humans [who can use phonological syntax — the ability to combine sounds {that individually have no meaning} into a collective sound] — [that does].

첫 번째 [ ]는 삽입절이다. 두 번째 [ ]는 humans를 수식하는 관계절이고, { }는 sounds를 수식하는 관계절이다. 세 번째 [ ]는 the first animal을 수식하는 관계절이며, does는 uses phonological syntax를 대신한다.

어휘 및 어구

employ 사용하다

mate 짝

flock 무리

claim 주장하다

territory 영역, 영토

intruder 침입자

predator 포식자

prominent 뚜렷한

vocalization 발성
combine 결합하다
instruct 지시하다
attract 유혹하다, 유인하다

Word Search

| 정답 | **1.** attract **2.** combine **3.** prominent
1. 그 빵집에서 나는 갓 구운 빵의 냄새는 지나가는 사람들을 유혹하기에 충분했다.
2. 예술가들은 자주 다양한 재료를 결합하여 독특하고 혁신적인 작품을 만든다.
3. 그 산맥의 뚜렷한 특징들은 몇 마일 떨어진 곳에서 명확하게 보였다.

Exercise 8 —— 정답 ④

소재 불안에 대한 유전자의 역할

해석 어떤 종류의 위험에 직면할 때, 생물은 일반적으로 경계하고, 위험의 본질에 대한 정보를 얻기 위해 탐색하며, 효과적인 대처 반응을 찾으려고 애쓴다. 그리고 일단 신호가 안전을 나타내면, 예컨대 그 사자를 피했거나 그 교통경찰관이 설명을 믿어 딱지를 떼지 않으면, 그 생물은 긴장을 풀 수 있다. 그러나 이것이 불안한 사람에게 일어나는 일은 아니다. 대신에, 대처 반응 사이에 긴장된 쟁탈전이 있어서, 어떤 것이 효과가 있었는지를 확인하지 않은 채 하나의 반응에서 다른 반응으로 갑자기 전환하는데, 이는 모든 것을 감안해 다양한 반응을 동시에 해 보고자 하는 동요된 시도이다. 또는 안전 신호가 언제 발생하는지 감지하지 못하여 불안한 경계가 계속된다. 정의상, 불안은 환경이 개인에게 영향을 가하는 상황을 벗어나면 거의 의미가 없다. 그러한 틀에서 보면, 뇌 화학물질 그리고 결국 불안과 관련된 유전자가 여러분을 불안하게 하는 것이 아니다. 그것들은 불안을 유발하는 상황에 여러분이 더 민감하게 반응하게 한다. 즉, 환경에서 안전 신호를 감지하기 더 어렵게 된다.

문제 해설

위협이 가해질 때의 일반적인 대응 방식과는 달리, 불안한 개인은 적절히 대처하지 못하거나 안전 신호를 감지하지 못하는데, 뇌의 화학물질과 유전자는 불안을 야기하는 상황에 더 민감하게 반응하게 만들어 제대로 대응하게 하지 못한다는 내용이므로, 빈칸에 들어갈 말로 가장 적절한 것은 ④ '환경에서 안전 신호를 감지하기'이다.
① 다른 사람들의 개인적 경계를 존중하기
② 확정된 목표에 집중을 유지하기
③ 가능한 탈출 계획을 떠올리기

⑤ 상황을 긍정적으로 바라보기

구조 해설

- Instead, there is a nervous scrambling among coping responses — [abruptly shifting from one to another without checking {whether anything has worked}], [an agitated attempt to {cover all the bases} and {attempt a variety of responses simultaneously}].
첫 번째 [ ]는 a nervous scrambling among coping responses를 추가적으로 설명하고, 그 안의 { }는 checking의 목적어 역할을 하는 명사절이다. 두 번째 [ ]는 앞 문장 전체와 동격을 이루는 어구로서 그 내용을 요약하여 서술한다. 그 안에 있는 두 개의 { }는 and로 대등하게 연결되어 to에 이어진다.

어휘 및 어구

organism 생물, 유기체
confront 직면하게 하다
struggle 애쓰다
evade 피하다
issue a ticket 딱지를 떼다
scrambling 쟁탈전, 서로 밀치기
shift 전환하다
attempt 시도; 해 보다
cover all the bases (야구 경기에서 내야의 모든 베이스를 지키듯이) 모든 것을 감안하다
simultaneously 동시에
detect 감지하다
restless 불안한, 안절부절못하는
by definition 정의상
ultimately 결국, 궁극적으로
gene 유전자
relevant 관련된

Word Search

| 정답 | **1.** evade **2.** shift **3.** detect
1. 그 도둑은 버려진 건물에 숨어서 가까스로 경찰을 피했다.
2. 그는 부상 후에 자신의 경력을 스포츠에서 학문으로 전환해야 했다.
3. 그 화재 경보기는 열기와 연기의 급격한 증가를 감지할 수 있다.

Exercise 9 —— 정답 ⑤

소재 확률을 예측하는 아기

해석 아기들은 통계적인 학습을 이용해 세상에 대해 예측하여,

자신의 행동을 유도한다. 꼬마 통계학자처럼 그들은 가설을 세우고, 자신의 지식에 기반하여 확률을 평가하며, 환경의 새로운 증거를 통합하고, 테스트를 수행한다. 발달 심리학자 Fei Xu가 수행한 한 창의적인 연구에서, 10개월에서 14개월 된 아기들이 먼저 분홍색 또는 검은색 막대 사탕에 대한 선호를 드러냈으며, 그런 다음 이들에게 두 개의 사탕 병을 보여 주었는데, 분홍색보다 검은색 막대 사탕이 더 많이 들어 있는 것과 검은색보다 분홍색을 더 많이 가진 것이었다. 그다음에 실험자는 눈을 감고 아기들이 막대기만 보고 색깔은 볼 수 없도록 각각의 병에서 막대 사탕을 하나씩 뽑았다. 각 막대 사탕은 서로 다른 불투명한 컵에 넣고 막대기만 보이게 했다. 아기들은 통계적으로 자신이 선호하는 색깔이 들어 있을 가능성이 더 높은 컵으로 기어갔는데, 왜냐하면 그것이 그 색깔이 다수인 병에서 나온 것이기 때문이다. 이와 같은 실험은 아기들이 단지 세상에 반응하기만 하는 존재가 아니라는 것을 보여 준다. 심지어 아주 어린 나이부터, 아기들은 원하는 결과를 최대화하기 위하여, 적극적으로 <u>자신이 관찰하고 학습한 패턴을 바탕으로 하여 확률을 추정한다</u>.

(문제 해설)
자신이 좋아하는 색의 막대 사탕이 많이 들어 있는 병에서 꺼내면 그 색의 막대 사탕이 나올 확률이 높기에, 아기가 그 병에서 꺼낸 막대 사탕이 있는 컵 쪽으로 기어간다고 했으므로, 빈칸에 들어갈 말로 가장 적절한 것은 ⑤ '자신이 관찰하고 학습한 패턴을 바탕으로 하여 확률을 추정한다'이다.
① 자기 주변에서 선택적으로 정보를 받아들인다
② 관찰된 예외를 사용해 창의적인 과정을 이해한다
③ 성인의 조언을 활용하여 의사 결정 과정을 개선한다
④ 현재의 감정 상태에 의지하여 잠재적인 결과를 예측한다

(구조 해설)
- Like little statisticians, they [form hypotheses], [assess probabilities based on their knowledge], [integrate new evidence from the environment], and [perform tests].
  네 개의 [ ]가 콤마와 and로 연결되어 주어 they에 대등하게 이어져 술부의 역할을 한다.

(어휘 및 어구)
statistical 통계적인
prediction 예측
hypothesis 가설(*pl*. hypotheses)
probability 확률
integrate 통합하다
evidence 증거
developmental psychologist 발달 심리학자
contain ~이 들어 있다. ~을 담고 있다

infant 아기
crawl 기어가다
majority 다수
demonstrate 보여 주다
reactive 반응하는
maximize 최대화하다
desire 원하다

(Word Search)
| 정답 | 1. prediction(s) 2. evidence 3. contain
1. 선거 결과에 대한 그의 예측은 정확한 것으로 판명되었다.
2. 그 변호사는 자기 의뢰인의 무죄를 입증하기 위한 증거를 제출했다.
3. 구급상자에는 붕대, 진통제, 거즈가 들어 있어야 한다.

Exercise 10 —————————— 정답 ②

(소재) 냄새의 기능

(해석) 인류학 교수 David Howes는 서로 다른 문화권에서 냄새와 장례식이나 통과 의례와 같은 전환기 의식이 자주 연관된다는 점에 주목한다. 그가 언급하기를, 냄새는 사회적 전환의 순간에 상징적으로 적절하다고 느껴지는데, 왜냐하면 요리하는 냄새가 원재료가 음식으로 전환되는 것을 알릴 때처럼 냄새는 다른 유형의 물리적 전환을 수반하고 표시하는 경우가 아주 많기 때문이라고 한다. 냄새는 공간을 벗어나 인간의 통제를 벗어나 퍼지는 경향이 있지만, 냄새에 대한 우리의 경험은 많은 경우 전환적인데, 왜냐하면 우리가 냄새의 범위에 처음으로 들어가는 경우 그것을 훨씬 더 강하게 인식하기 때문이다. 집에 들어서면 빵 굽는 냄새를 강하게 느끼지만, 집 안에서 몇 분이 지나면 심지어 의도적으로 노력해도 여러분은 더는 냄새를 맡지 못할 수도 있는데, 곧 후각 적응 또는 피로라고 알려진 신체적 과정이다. 한동안 계속하여 노출된 후에도 우리의 인식을 유지하기 위해서는 압도적인 냄새가 필요하다. 냄새는 <u>상태의 변화뿐만 아니라 공간의 전환</u>을 나타내며, 이 때문에 사회적으로 중요한 변화의 순간을 표시하는 데 사용된다.

(문제 해설)
냄새는 원재료가 음식으로 전환될 때처럼 상태의 변화를 표시한다고 했고, 집에 들어서면서 갓 구운 빵 냄새를 강하게 느끼는 것이 집 밖에서 안으로의 전환에 대한 신호라고 했으므로, 빈칸에 들어갈 말로 가장 적절한 것은 ② '상태의 변화뿐만 아니라 공간의 전환'이다.
① 시간에 따른 감각적 인지의 적응

③ 이벤트 분위기를 조성하기 위한 환경 조정
④ 사회가 개인들에게 부여한 사회적 지위
⑤ 사회적 공간과 집단 기억의 보이지 않는 구조

구조 해설

- He suggests [that scent is felt to be symbolically appropriate for moments of social transition {because it so frequently accompanies and marks other types of physical transition, ⟨as when cooking smells signal the transformation of raw ingredients into food⟩}].

[ ]는 suggests의 목적어 역할을 하는 명사절이고, { }는 이유를 나타내는 부사절이며, ⟨ ⟩ 안에 쓰인 as when은 '~할 때처럼'이라는 뜻이다.

어휘 및 어구

association 연관
scent 냄새, 향기
transition 전환(기), 변화
funeral 장례식
symbolically 상징적으로
accompany 수반하다
mark 표시하다
raw ingredient 원재료
deliberate 의도적인
adaptation 적응
overwhelming 압도적인
retain 유지하다
exposure 노출

Word Search

| 정답 | **1.** scent(s) **2.** deliberate **3.** retain
1. 신선한 꽃의 향기가 방 안을 가득 채워 기분 좋은 분위기를 만들었다.
2. 그녀는 스트레스가 많은 상황에서도 침착함을 유지하려고 의도적인 노력을 기울였다.
3. 벽들은 겨울에 열을 유지할 수 있을 만큼 충분히 두껍다.

Exercise 11 ─────── 정답 ④

소재 다양한 접근 방식의 필요성

해석 최고의 판단이 보상받을 경쟁 상황에서는 판단의 다양성이 기대되고 환영받는다. 여러 기업(또는 같은 조직 내 여러 팀)이 동일한 고객 문제에 대한 혁신적인 해결책을 만들어 내기 위해 경쟁할 때, 우리는 그들이 동일한 접근 방식에 집중하기를 원하지 않는다. 백신 개발과 같이 여러 연구자 팀이 하나의 과학적 문제에 대처할 때도 마찬가지이다. 우리는 그들이 다양한 각도에서 그 문제를 바라보길 매우 바란다. 심지어 예측가들도 때때로 경쟁하는 선수처럼 행동한다. 어느 누구도 예상하지 못한 불경기를 정확히 예측하는 분석가는 분명 명성을 얻게 될 것이지만, 일치된 의견에서 결코 벗어나는 적이 없는 분석가는 주목받지 못한 채로 남아 있다. 이러한 환경에서 아이디어와 판단의 다양성이 다시 환영받는데, 왜냐하면 차이는 첫 단계에 불과하기 때문이다. 두 번째 단계에서 이러한 판단의 결과가 서로 맞붙어 최고가 승리하게 된다. 자연과 마찬가지로 시장에서도 선택은 차이 없이는 작동할 수 없다.

문제 해설

문제에 대한 혁신적인 해결안을 마련하기 위해서는 다양한 방식으로 접근해야 한다는 내용이므로, 빈칸에 들어갈 말로 가장 적절한 것은 ④ '차이 없이는 작동할 수 없다'이다.

① 일반적인 경향을 반영한다
② 협력으로 이어질 수 있다
③ 유연성을 허용해서는 안 된다
⑤ 단계별 과정이 필요하다

구조 해설

- [Variability in judgments] is expected and welcome in a competitive situation [in which the best judgments will be rewarded].

첫 번째 [ ]는 문장의 주어이다. 두 번째 [ ]는 a competitive situation을 수식하는 관계절이다.

어휘 및 어구

variability 다양성
judgment 판단
generate 만들어 내다, 생성하다
innovative 혁신적인
solution 해결책
angle 각도
forecaster 예측가
call 예측[예상]하다
fame 명성
triumph 승리하다

Word Search

| 정답 | **1.** judgment(s) **2.** solution **3.** generate
1. 그의 판단은 감정에 의해 흐려져 불공정한 결정으로 이어졌다.
2. 그 엔지니어는 기계의 효율성을 개선하기 위해 혁신적인 해결책을 제안했다.
3. 그 앱은 게임을 위한 임의의 숫자를 생성하도록 설계되었다.

Exercise 12

정답 ⑤

소재 사전 조사의 중요성

해석 똑똑한 실패는 준비에서 시작된다. 이전에 실행했다가 실패한 실험에 시간이나 재료를 낭비하고 싶어 하는 과학자는 없다. 숙제를 하라. 전형적인 똑똑한 실패는 가설에 의해 주도된다. 여러분은 어떤 일이 일어날지, 즉 왜 일어날 일에 대해 자신이 옳을 수 있다고 믿을 만한 이유가 있는지 시간을 들여 충분히 생각했다. 기업가 정신 전문가인 나의 하버드 동료 Thomas Eisenmann은 많은 신생 업체의 실패가 기본적인 숙제를 건너뜀으로써 야기된다고 생각한다. 예를 들어, 온라인 데이팅 신생 업체인 Triangulate는 시장의 어떤 요구에도 맞지 않는 완전한 기능을 갖춘 서비스를 서둘러 출시했다. 빨리 출시하기를 열망한 나머지, 창업자들은 연구, 즉 충족되지 않은 요구를 조사하기 위한 고객 인터뷰를 생략했다. 그 중요한 준비 단계에 주의를 기울이지 않아 그 회사는 대가를 치렀다. Thomas는 이러한 통상적인 실패의 원인이 부분적으로 "'빨리 실패하라'는 주문" 때문이라고 여기는데, 그것은 행동을 지나치게 강조하여 준비를 소홀히 하게 만든다. 나아가 자명해 보일 수도 있지만, 일단 숙제를 마쳤으면 그것이 알려 주는 내용에 주의를 기울여야 한다.

문제해설
충분한 준비 없이 행동해 실패하지 말고 철저하게 사전 조사를 하라는 내용이므로, 밑줄 친 부분이 의미하는 바로 가장 적절한 것은 ⑤ '실제 작업 전에 시간을 들여 철저한 연구를 하라.'이다.
① 경쟁 우위를 확보하기 위해 신속하게 움직이라.
② 장애물을 만났을 때 조언을 구하라.
③ 소중한 교훈을 얻기 위해 실수를 반성하라.
④ 진행 상황을 정기적으로 모니터링하고 조정하라.

구조해설
- Thomas **attributes** this common failure, in part, **to** "the 'fail fast' mantra," [which overemphasizes action, {shortchanging preparation}].
「attribute ~ to ...」는 '~의 원인이 …에 있다고 여기다'라는 뜻이다. [ ]는 "the 'fail fast' mantra"를 부가적으로 설명하는 관계절이고, 그 안의 { }는 which ~ action의 내용에 부가적으로 이어지는 결과를 나타내는 분사구문이다.

어휘 및 어구
hypothesis 가설(*pl.* hypotheses)
start-up 신생 업체
skip 건너뛰다
launch 출시하다
founder 창업자
probe 조사하다
crucial 중요한
overemphasize 지나치게 강조하다
shortchange (기대에 못 미치는 수준으로) 소홀히 하다
self-evident 자명한

Word Search

| 정답 | 1. hypothesis(hypotheses) 2. skip
3. overemphasize
1. 그 가설을 테스트하기 위해 그들은 여러 통제된 실험을 수행했다.
2. 그녀는 아침 식사를 건너뛰고 대신에 이른 점심을 먹기로 결정했다.
3. 배움의 가치를 소홀히 하면서 성적의 중요성을 지나치게 강조하기 쉽다.

Tips

intelligent failure(똑똑한 실패): 발전에 도움이 되지 않는 일반적인 실패와 구분되는 개념으로, 창조와 혁신을 달성하는 데 도움이 되는 실패를 의미한다.

punish those ⟨who do not do the same⟩ even when doing so is costly to the punisher}].

[ ]는 *strong reciprocity*와 동격을 이루는 명사구이고, 그 안에서 두 개의 to부정사구 { }는 and로 대등하게 연결되어 a tendency의 내용을 구체적으로 설명한다. 두 번째 { } 안의 ⟨ ⟩는 those를 수식하는 관계절이다.

어휘 및 어구

cooperation 협동, 협력
hallmark 특징
considerately 사려 깊게
substantial 상당한
evolution 진화
arise 발생하다, 생기다
interact 상호 작용을 하다
sociality 사회성
foundation 기초
morality 도덕성
discredited 신빙성 없는, 불신받는
promote 촉진하다
prosocial 친사회적인
reciprocity 상호주의, 호혜성(두 사람이나 두 집단 간에 서로 도움이나 혜택을 주고받는 것)
costly 큰 손실[타격]이 되는

Word Search

| 정답 | 1. Cooperation 2. morality 3. promote
1. 국가 간 협동은 세계 기아와 싸우기 위해 음식과 물자를 나누는 데 도움이 된다.
2. 잃어버린 지갑을 돌려주기로 한 그의 결정은 그의 강한 도덕성을 보여 주었다.
3. 그 단체는 환경 문제에 대한 인식을 촉진하기 위한 캠페인을 주도했다.

Week 3 05강

본문 54~65쪽

| 1 ② | 2 ① | 3 ⑤ | 4 ① | 5 ④ | 6 ④ |
| 7 ④ | 8 ① | 9 ④ | 10 ③ | 11 ④ | 12 ⑤ |

Exercise 1

정답 ②

소재 협동심의 진화

해석 협동은 인간 사회의 특징이다. 사람들은 완전히 낯선 이에게 사려 깊게 행동할 뿐만 아니라 때로는 그들을 위해 상당한 금전적, 심지어 신체적 희생을 한다. 이러한 행동은 서로 무관한 타인 간의 협동이, 그 동일한 개인들이 반복적으로 상호 작용을 할 때만 발생할 수 있다는 이타주의의 진화에 대한 고전적인 모델로는 설명이 불가능해 보인다. 그러나 인간 사회성의 진화에 대한 새로운 모델과 그것을 시험하기 위한 실험 결과는 공정성에 대한 감각과 도덕성의 다른 기초가 깊은 진화적 뿌리를 가지고 있음을 시사한다. 그러한 모델은 개인으로 이루어진 집단 수준에서 과정을 고려하지만, 개인이 '집단의 이익을 위해' 행동한다는 집단 선택이라는 신빙성 없는 개념에는 의존하지 않고 그렇게 한다. 개인에게 친사회적 성향을 촉진하는 유전자는 인류 진화 초기 단계의 특징이라고 주장되는 방식으로 집단이 경쟁할 때 발생할 수 있다. 특히 그러한 상황은 '강한 상호주의'의 진화를 촉진했을 수도 있는데, 그것은 곧 익명의 관계없는 타인과 협동하며, 똑같이 협동하지 않는 사람을 처벌하되, 그렇게 하는 것이 처벌하는 사람에게 큰 손실이 되는 경우조차도 그리하는 경향이다.

문제 해설
② which로 이어지는 절이 필수 요소를 갖추고 있는 완전한 절이므로 관계부사 where 혹은 「전치사+관계사」 형태의 in which로 고쳐야 한다.
① 부정어구 Not merely가 맨 앞에 쓰였으므로 주어(people) 앞에 조동사(do)가 온 것은 어법상 적절하다.
③ 맥락상 new models of the evolution of human sociality를 가리키는 말이므로 복수형 대명사 them은 어법상 적절하다.
④ 분사구의 수식을 받는 ways는 주장하는(argue) 행위의 대상이므로 수동의 의미를 나타내는 과거분사 argued는 어법상 적절하다.
⑤ 맥락상 to부정사구인 to cooperate ~ others와 and로 대등하게 연결되어 a tendency 뒤에 이어지는 구조이므로 to부정사구를 이끄는 to punish는 어법상 적절하다.

구조 해설
- In particular, such conditions may have promoted the evolution of *strong reciprocity*, [a tendency {to cooperate with anonymous unrelated others} and {to

Exercise 2

정답 ①

소재 삼림 벌채의 원인 규명

해석 삼림 벌채의 동인(動因)은 매우 다양하며 국가와 주마다 다르다. 많은 중요한 연구가 수많은 지역 연구를 일반화하고 통합하려고 시도해 왔다. 그러나 이러한 작업은 동인에 대한 대부분의 정보가 정량적이지 않아서 특정 양의 삼림 벌채와 특정 활동을 연결할 수 있는 직접적인 정량적 상관관계가 거의 없다는 사실로 인해 어려움을 겪고 있다. Angelsen과 Kaimowitz는 열대 삼림 벌채의 원인을 분석한 140개의 경제 모델을 검토했다. 그들은 직접적인 원인을 살펴볼 때 삼림 벌채는 많은 경우 더 많은 도로의 존재, 농산물 가격 상승, 임금 하락 및 고용 부족과 관련되어 있다는 것을 발견했다. 또한 그들은 경제 자유화 및 관련 조정과 연관된 정책 개혁이 삼림에 대한 압박을 증가시킬 것 같다고 생각했다. 그러나 그들은 많은 연구가 부실한 방법론과 낮은 품질의 데이터를 채택해 왔으며, 그 때문에 거시경제적 요인의 역할에 대한 명확한 결론을 도출하는 것이 어렵다는 점을 지적했다.

[문제 해설]

삼림 벌채를 명확하게 특정 원인의 결과로 보는 것이 어렵다는 내용의 글이다. 삼림 벌채의 원인에 대한 정보가 정량적이지 않아 특정 양의 삼림 벌채와 특정 활동을 연결하는 정량적 상관 관계를 찾기 어렵고, 열대 삼림 벌채의 원인에 대한 연구 역시 명확한 결론을 도출하지 못했다고 하였으므로, 글의 주제로 가장 적절한 것은 ① '삼림 벌채를 명확하게 특정 원인의 결과로 보는 것의 어려움'이다.
② 삼림 벌채가 지역 경제에 미치는 영향을 평가하는 방법
③ 삼림 벌채 연구에서 학제 간 접근의 중요성
④ 지속 가능한 생태계를 위한 산림 생물 다양성 보존의 필요성
⑤ 경제 발전과 환경 보호의 균형 유지에 있어서의 어려움

[구조 해설]

- They pointed out, however, [that many research studies have adopted poor methodology and low quality data, {which makes ⟨the drawing of clear conclusions about the role of macroeconomic factors⟩ **difficult**}].
  [ ]는 pointed out의 목적어 역할을 하는 명사절이고, 그 안의 { }는 앞 절의 내용을 부가적으로 설명하는 관계절이다. 그 안의 ⟨ ⟩는 makes의 목적어이고, difficult는 목적격 보어이다.

[어휘 및 어구]

deforestation 삼림 벌채
driver 동인, 추진 요인
generalise 일반화하다
quantitative 정량적인, 양적인
correlation 상관관계
analyse 분석하다
be associated with ~과 관련되어 있다
shortage 부족
reform 개혁
liberalisation 자유화
adjustment 조정
adopt 채택하다
methodology 방법론

macroeconomic 거시경제적인

Word Search

| 정답 | **1.** generalise **2.** quantitative **3.** adjustment
1. 한 사람의 경험을 모두에게 일반화하지 마라.
2. 공학에서는 정확한 측정과 계산을 위해 정량적 데이터가 이용된다.
3. 그 문제를 더 잘 다루기 위해서 정책에 대한 조정이 필요하다.

Exercise 3 ───────── 정답 ⑤

[소재] 과학의 우세한 사회적 위치

[해석] 한편으로는 세상을 예측하고 설명하는 현대 과학의 놀라운 성공이 그것에 의존하고 과학 기관에 막강한 권한을 (특히 공공 자금 지원의 형태로) 부여하는 것을 정당화하는 것처럼 보인다. 다른 한편으로 과학의 우세한 사회적 위치는 다른 형태의 지식 습득을 소외시킬 위험이 있다. 과학과 그 대안들 사이에 여전히 공정한 경쟁이 이루어지고 있는지 확실하지 않으며, 21세기에 과학이 과학 혁명 시기(16~18세기)와 비슷한 방식으로 자체의 인식론적 우월성을 증명해야 하는 것도 간단하지 않다. 이것이 Paul Feyerabend가 *Science in a Free Society*에서 한 주요 주장인데, 그는 그것을 "오늘날 과학이 우세한 것은 그것의 비교 우수성 때문이 아니라 과학에 유리하도록 쇼가 조작되었기 때문이다."라고 제시한다. 그러므로 Feyerabend의 주장에 따르면, 과학이 과거에 문제를 다루는 데 가장 적절한 수단이었다는 사실이 우리의 현재와 미래의 문제에 대해서도 그것이 가장 신뢰할 수 있는 지식 습득 방식이라고 보장하지는 않는다.

[문제 해설]

과학은 세상을 예측하고 설명하는 데 성공했지만 현대 사회에서 과학에 우세한 위치를 부여하는 것이 부작용을 가져올 수 있고 과학이 과거의 문제를 다루는 데 가장 적절한 수단이었다고 해서 현재와 미래의 문제에 대해서도 그 신뢰성이 보장되지는 않는다는 내용의 글이므로, 글의 제목으로 가장 적절한 것은 ⑤ '과학은 여전히 지식을 얻는 최고의 길인가?'이다.
① 과학과 다른 분야의 통합: 효율적인 교육을 위한 필수 요소
② 변화하는 시대에서 번창하기 위해 축적된 과학 지식을 활용하라
③ 실험실에서 사회로: 과학 커뮤니케이션의 격차 해소
④ 공공 자금 지원은 과학의 신뢰성을 어떻게 훼손하는가?

[구조 해설]

- It's not clear [that there still is a fair competition

between science and its alternatives]; and **it**'s not straightforward [that science has to prove its epistemic superiority in the 21st century in a way {comparable to the period of the scientific revolution (16th-18th centuries)}].

It과 it은 형식상의 주어이고, 두 개의 [ ]는 각각 내용상의 주어이다. 두 번째 [ ] 안의 { }는 a way를 수식하는 형용사구이다.

(어휘 및 어구)

astonishing 놀라운
predict 예측하다
justify 정당화하다
dominant 우세한, 지배적인
societal 사회적인
marginalize 소외시키다
alternative 다른; 대안
acquisition 습득
straightforward 간단한, 수월한
superiority 우월성
comparable to ~과 비슷한, ~에 필적하는
argument 주장, 논거
cast ~ as ... ~을 …이라고 제시하다[묘사하다]
prevail 우세하다
merit 우수성, 장점
warrant 보장하다
reliable 신뢰할 수 있는

Word Search

| 정답 | **1.** justify **2.** dominant **3.** prevail
1. 그녀는 자신의 결정에 이르게 된 상황을 설명함으로써 자신의 행동을 정당화하려고 노력했다.
2. 사자는 아프리카 대초원에서 우세한 동물이다.
3. 어려운 시기에는 선이 우세할 것이라고 믿는 것이 중요하다.

Exercise 4

정답 ①

소재 이민자 부모와 청소년의 특징

해석 사회경제적 집단과 출신 국가를 통틀어 발견되는 이민자 부모의 특징은, 이민 1세대 부모가 청소년기 자녀의 교육적 성취에 높은 가치를 둔다는 것이다. 이민자 부모는 좋은 성적 받기, 고

등학교 졸업하기, 대학 다니기를 자녀가 목적지 국가에서 자리 잡을 수 있는 주요 수단으로 여기는 경향이 있다. 비슷한 교육 수준을 가진 현지 출생 부모에 비해, 이민자 부모가 자녀에 대한 교육적 열망이 더 높은 것으로 나타났다. 이민 청소년은 교육의 중요성에 대한 부모의 가치관을 내면화하여 비이민 청소년이 그러는 것보다 더 많이 공부하고 학업에 더 많은 노력을 기울이는 경향이 있다. 청소년기에 이민자 또래 집단은 자주 비이민자 또래 집단과 더 분리되며, 이민자 청소년들은 특히 사회경제적 지위(SES)가 더 높은 청소년들의 경우 또래의 학업 성취를 응원한다.

(문제 해설)

(A) 주어의 핵심어구가 A hallmark이므로 단수형 동사 is가 어법상 적절하다.
(B) 관계사 뒤가 필수 요소를 갖추고 있는 완전한 절이므로 「전치사+관계사」 형태의 through which가 어법상 적절하다.
(C) 맥락상 and로 대등하게 연결된 일반동사 study와 devote가 이끄는 어구를 대신하므로 대동사 do가 어법상 적절하다.

(구조 해설)

■ During adolescence, immigrants' peer groups often become more segregated from nonimmigrants' peer groups, [with the immigrant adolescents supporting their peers' academic achievement, particularly for higher SES (socioeconomic status) adolescents].

[ ]에는 앞 절의 부수적인 상황을 표현하는 「with+명사구+분사구」의 구조가 사용되었다.

(어휘 및 어구)

immigrant 이민자의, 이민의; 이민자
socioeconomic 사회경제적인
place a high value on ~에 높은 가치를 두다, ~을 중시하다
achievement 성취
adolescent 청소년기의; 청소년
primary 주요한
aspiration 열망
internalize 내면화하다
devote ~ to ... …에 ~을 기울이다
segregated 분리된

Word Search

| 정답 | **1.** immigrant **2.** aspiration(s) **3.** internalize
1. 많은 이민자 가족은 새로운 나라에서의 생활에 적응하면서 고유한 어려움에 직면한다.
2. 그 젊은 선수는 어렸을 때부터 올림픽에 출전하고 싶다는 열망을 가지고 있었다.
3. 아이들은 흔히 부모로부터 관찰하는 가치와 행동을 내면화한다.

────── 정답 ④

소재 즉흥 말하기

해석 대중 연설에서 준비가 거의 또는 전혀 없는 연설을 '즉흥 말하기'라고 한다. 여러분은 준비나 연습이 전혀 없이 그 순간에 떠오르는 생각과 아이디어를 전달할 때 매일 즉흥 말하기를 한다. 예를 들어, 수업 시간에 질문에 답하거나 캠퍼스 내 단체의 회의에서 발언할 때 즉흥 말하기를 사용하는 것이다. 이런 점에서, 즉흥 말하기는 여러분이 이미 가지고 있고 흔히 사용하는 기본적인 의사소통 기술을 사용하는 또 다른 방법일 뿐이다. 조사나 광범위한 준비 또는 메모에 의존하지 않고 즉석에서 자신을 표현하는 방법을 배우면 대중 연설 수업과 교실 외에도 덜 구조화된 말하기 상황에서 여러분이 잘하는 데 도움이 될 것이다. (이러한 전문성은 발표자가 특별히 그 연설을 위해 수행된 세심한 조사를 통해 준비한 관련 정보를 제공한다는 사실에서 비롯된다.) 아울러 즉흥 말하기 기술을 개발하면 어떤 발표 상황에서도 자신감이 높아지고 연설 불안이 줄어든다.

문제 해설

준비가 거의 또는 전혀 없이 하는 즉흥 말하기에 대해 소개하면서 조사나 광범위한 준비 또는 메모에 의존하지 않고 즉석에서 자신을 표현하는 방법을 배우면 도움이 된다고 설명하는 내용의 글인데, ④는 발표자가 그 연설을 위해 수행된 세심한 조사를 통해 준비한 관련 정보를 제공한다는 내용이므로, 글의 전체 흐름과 관계가 없다.

구조 해설

■ [Learning how to express yourself on the spot without relying on research, extensive preparation, or notes] will help you [do well {in your public speaking class} and {in less-structured speaking situations beyond the classroom}].

첫 번째 [ ]는 주어 역할을 하는 동명사구이다. 두 번째 [ ]는 「help+목적어+목적격 보어」 구문에서 목적격 보어 역할을 하는 원형부정사구이고, 그 안에서 두 개의 전치사구 { }가 and로 대등하게 연결되어 있다.

어휘 및 어구

preparation 준비
impromptu 즉흥의
engage in ~을 하다
in this respect 이런 점에서
on the spot 즉석에서
extensive 광범위한
expertise 전문성
relevant 관련된

specifically 특별히, 특히
confidence 자신감
decrease 줄이다
anxiety 불안
context 상황, 맥락

Word Search

| 정답 | 1. extensive 2. confidence 3. decrease
1. 그녀의 연구는 매우 광범위해서 여러 다른 분야를 다뤘다.
2. 철저히 준비했기 때문에 그녀는 자신감을 가지고 입사 면접에 임했다.
3. 규칙적인 운동은 심장 질환의 위험을 줄이는 데 도움이 될 수 있다.

Exercise 6 ────── 정답 ④

소재 수익과 팀의 경기력

해석 원칙적으로 수익과 팀 전력 사이에는 직관적으로 그럴듯한 관계가 있다. 팬들은 결과가 불확실한 흥미진진한 경기를 보고 싶어 한다. 이는 대항팀 간의 경쟁적 균형의 결과이다. (C) 결과가 결코 불확실하지 않은 한쪽으로 기운 경기보다 막상막하이고 격전을 벌이는 경기가 보기에 더 재미있다. 그러나 팬들은 또한 자신의 팀이 승리하는 것을 보기를 선호한다. 그 결과, 팀의 성적이 향상되면 관중 수는 증가할 것이다. 아마도 더 나은 팀이 더 약한 팀보다 더 많이 이길 것이다. (A) 따라서 향상된 팀 전력은 더 많은 승리와, 그로 인해 관중의 증가라는 결과로 이어질 것이다. 일련의 정해진 가격에서 관중 증가는 입장료, 주차, 구내 매점에서 수입 증가를 가져온다. (B) 이러한 수입 증가는 수익 증가로 이어져야 하지만, 전력이 낮은 팀보다 전력이 높은 팀을 출전시키는 데는 더 많은 비용이 든다. 결과적으로 트레이드오프가 있으며, 따라서 팬 지지 기반의 차이에 따라 팀마다 다를 최적의 팀 전력이 있다.

문제 해설

팬들이 보고 싶어 하는 결과가 불확실한, 흥미진진한 경기는 대항팀 간의 경쟁적 균형의 결과라는 주어진 글 다음에는, 이에 대해 부연 설명을 하면서 한쪽으로 기운 경기보다는 막상막하인 경기가 더 재미있다고 말하는 (C)가 이어져야 한다. (C)의 뒷부분에서는 팬들이 동시에 자신의 팀이 승리하는 것을 보고 싶어하므로 팀의 성적이 향상되면 관중 수가 늘 것이고, 전력이 더 높은 팀이 더 약한 팀보다 더 많이 이길 것이라고 하였으므로, 그 뒤에는 그에 대한 결과로 팀 전력이 향상되면 더 많은 승리와 관중 증가, 그리고 수입 증가로 이어질 것이라는 내용의 (A)가 이어져야 한다. 그 뒤에는 (A)의 수입 증가에 대한 내용을 이어받아, 이러한 수입 증가가 수익 증가로 이어져야 하지만, 전력 높은 팀의 출전에는 비용이 들게 되므로 팬 지지 기반의 차이에 따라 팀마다 다를 최적의 팀 전력이 있다는 내용의 (B)가 이어지는 것이 자연스럽다. 따라서 주어진 글 다음에 이어질 글의 순서로 가장 적절한 것은 ④ '(C) − (A) − (B)'이다.

구조 해설

- Consequently, there is a trade-off and therefore an optimal level of team quality [that will vary from team to team depending on variation in the fan base].
  [ ]는 an optimal level of team quality를 수식하는 관계절이다.

어휘 및 어구

in principle 원칙적으로
intuitively 직관적으로
profit 수익
outcome 결과
uncertain 불확실한
competitive 경쟁적인
opponent 대항팀, 상대, 적수
concession 구내 매점
field 출전시키다
trade-off 트레이드오프(어느 것을 얻으려면 반드시 다른 것을 희생하여야 하는 경제 관계)
optimal 최적의
vary 다르다, 달라지다
variation 차이, 변동
in doubt 불확실한, 의심스러운
improvement 향상
performance 성적, 성과
presumably 아마도

Word Search

| 정답 | **1.** outcome **2.** vary **3.** improvement
1. 오늘 선거의 결과는 내일 오전에 발표될 것이다.
2. 사막에서는 낮과 밤의 기온이 크게 다를 수 있다.
3. 그의 시험 성적은 지난번보다 확실한 향상을 보였다.

Exercise 7 ————————————— 정답 ④

소재 화산이 날씨에 미치는 영향

해석 화산이 날씨에 미치는 영향은 분출 지역을 훨씬 넘어 확대될 수 있으며, 지질학적 시간 규모에 걸쳐 발생하는 기후 변화의 가장 큰 동인 중 하나이다. 지구상에 있는 65조 5천억 톤의 탄소

거의 대부분은 암석 안에 붙잡혀 있다. 나머지는 바다, 대기, 식물, 토양 및 화석 연료에 존재한다. 지구 내부의 이산화 탄소는 화산 분화 중에 지속적으로 방출되며, 산업 혁명 이전에는 화산이 지구 대기로 유입되는 이산화 탄소의 가장 큰 공급원이었다. 그러나 탄소 방출은 자연적인 탄소 순환에 의해 주로 조절되는데, 이는 화산이 대기 중으로 방출하는 만큼의 탄소를 대기에서 감소시켜 지구적 규모의 자동 온도 조절 장치 역할을 한다. 만약 격렬한 화산 활동 시기로 인해 기온이 상승하면, 더 많은 탄소가 대기에서 감소될 것이고 그것은 기온을 이전 수준으로 되돌릴 것이다. 하지만 이러한 화학 반응 중 일부의 느린 속도를 고려해 볼 때, 그 시스템이 안정되는 데 수십만 년이 걸릴 수 있다.

문제 해설

④ 문맥상 앞 절과 연결되어 앞 절의 내용을 부가적으로 설명하는 관계절이 이어져야 하므로, 대명사 it을 관계사 which로 고쳐야 한다.

① 주어인 The vast majority of the 65,500 billion tons of carbon on Earth가 hold의 대상이므로 과거분사 held가 쓰여 수동태의 술어를 이루는 것은 어법상 적절하다.

② 동사구 is released를 수식하는 부사 continually는 어법상 적절하다.

③ the natural carbon cycle을 의미상 주어로 하는 분사구문을 유도하는 현재분사 acting은 어법상 적절하다.

⑤ '~하는 데 시간이 걸리다'라는 의미를 나타내는 「it + takes + 시간 + to부정사(구)」 구문에서 의미상의 주어를 나타내는 「for + 목적격」 뒤에 사용된 to stabilize는 어법상 적절하다.

구조 해설

- [If temperatures increase {due to a period of intense volcanism}], more carbon will be drawn down from the atmosphere, [which will take temperatures back to their previous levels].
  첫 번째 [ ]는 조건의 부사절이고 { }는 이유의 부사구이다. 두 번째 [ ]는 앞 절의 내용을 부가적으로 설명하는 관계절이다.

어휘 및 어구

extend 확대되다
driver 동인, 추진 요인
geological timescale 지질학적 시간 규모
remainder 나머지
reside in ~에 존재하다, ~에 거주하다
draw down ~을 감소시키다
planetary 지구(상)의, 행성의
intense 격렬한, 강렬한
volcanism 화산 활동

stabilize 안정되다, 안정시키다

| 정답 | **1.** geological **2.** resides **3.** intense
1. 이 지역의 지층은 지구의 역사에 대한 귀중한 통찰을 제공한다.
2. 그 나라 인구의 대부분은 도시 지역에 거주한다.
3. 강렬한 폭염은 전국적으로 광범위한 정전 사태를 일으켰다.

Exercise 8
정답 ①

소재 음반을 통한 음악 전수

해석 부에노스아이레스의 탱고 오케스트라 El Arranque의 Ignacio Varchausky는 다큐멘터리 *Si Sos Brujo*에서 자신과 다른 사람들이 옛 오케스트라들이 그들이 한 일을 어떻게 해냈는지를 음반으로 배우려고 시도했지만, 그것은 어렵고 거의 불가능했다고 말한다. 결국 El Arranque는 그 합주단의 아직 생존한 연주자들을 찾아 그들에게 어떻게 그렇게 했는지를 물어봐야 했다. 나이 든 연주자들은 젊은 연주자들에게 그들이 낸 효과를 어떻게 재현하는지, 어떤 음표와 박자를 강조해야 하는지를 실제로 보여 주어야 했다. 그러므로 음악은 어느 정도는 여전히 한 사람에게서 다른 사람에게 전수되는 구전의(그리고 유형의) 전통이다. 음반은 음악을 보존하고 전파하는 데 많은 역할을 할 수 있지만, 직접적인 전수가 하는 것을 할 수는 없다. 같은 다큐멘터리에서 Wynton Marsalis는 학습, 즉 배턴을 넘겨주는 것은 연주대에서 일어난다고, 즉 보고 모방함으로써 배우기 위해 다른 사람들과 함께 연주해야 한다고 말한다. Varchausky에게 있어 이 나이 든 연주자들이 사라질 때, 만약 그들의 지식이 직접 전달되지 않는다면 전통(그리고 기법)은 사라질 것이다. 역사와 문화는 과학 기술만으로는 실제로 보존될 수 없다.

문제 해설
오케스트라 El Arranque가 음반을 통해 옛날 오케스트라들이 어떻게 연주했는지를 배우려고 시도했지만 매우 어려웠으며, 음반은 음악을 보존하고 전파하는 데 많은 역할을 할 수 있지만, 직접적인 전수가 하는 것을 할 수는 없다는 내용의 글이므로, 글의 주제로 가장 적절한 것은 ① '음반을 통해 간접적으로 음악을 배우는 것의 한계'이다.
② 예술적 가치가 있는 저평가된 음악을 보존하려는 노력
③ 음악계에서 혁신과 전통 사이의 갈등
④ 직접적인 전수에 지나치게 중점을 둔 음악 교육의 영향
⑤ 음악 학습에 있어 모방의 유효성에 대한 상충하는 견해

구조 해설
■ The older players had to physically show [the younger players] [{how to replicate the effects they got}, and {which notes and beats should be emphasized}].
첫 번째 [ ]는 show의 간접목적어이다. 두 번째 [ ]는 show의 직접목적어로 두 개의 { }가 and로 연결되어 있다.

어휘 및 어구
record 음반, 레코드
surviving 생존한
physically 실제로, 눈에 보이는 모양으로
replicate 재현하다, 복제하다
note 음표
to some extent 어느 정도
oral 구전의, 구두의
hand down ~을 전수하다
transmission 전수, 전달
baton 배턴
bandstand 연주대

| 정답 | **1.** replicate **2.** oral **3.** transmission
1. 과학자들은 다른 실험실에서 실험 결과를 재현하는 데 어려움을 겪었다.
2. 그 회의 중에 그녀는 기후 변화가 해양 생태계에 미치는 영향에 대해 인상적인 구두 발표를 했다.
3. 한 세대에서 다음 세대로의 문화적 전통의 전수는 그것의 보존을 보장한다.

Exercise 9
정답 ④

소재 의사소통 목적에 따른 적절한 기술 사용

해석 팀은 이메일 및 채팅 플랫폼부터 웹 회의 및 화상 회의에 이르기까지 많은 의사소통 기술을 원하는 대로 사용할 수 있다. 사람들은 흔히 자신에게 가장 편리하거나 친숙한 도구를 사용하는 것을 기본적으로 선택하지만, 일부 기술은 다른 기술보다 특정 과업에 더 적합하며, 잘못된 기술을 선택하면 문제가 발생할 수 있다. 의사소통 도구는 정보의 풍부함(즉, 사람들이 의미를 해석하는 데 도움이 되는 비언어적 신호 및 기타 신호를 전달하는 능력)과 가능한 실시간 상호 작용의 수준을 포함하여 여러 차원에서 다르다. 팀의 의사소통 과업 또한 마찬가지로 서로 다른 관점을 조화시키거나, 피드백을 주고받거나, 혹은 오해가 발생할 가능성을 피해야 하는 필요성에 따라 복잡성이 각기 다르다. 의사소통의 목적이 전달 방법을 결정해야 한다. 따라서 목표를 신중하게 고려하라. 정보를 한 방향으로 전달하는 경우, 예컨대 정례적인 정보 및 계획을 배포하고, 아이디어를 공유하며, 간단한 데이터를 수집

하는 경우에는, 이메일, 채팅, 게시판과 같은 더 간결한 텍스트 기반 매체를 사용하라. 웹 회의 및 화상 회의는 더 풍부하고 더 상호 작용적인 도구로서, 문제 해결 및 협상과 같이 서로 다른 아이디어와 관점을 조율할 것을 요구하는 복잡한 과업에 더 적합하다.

(문제 해설)

이메일, 채팅, 웹 회의 및 화상 회의와 같이 팀 내 의사소통에 활용되는 기술의 경우에는 특정 과업에 더 적합한 것이 있을 수 있으므로 팀의 의사소통 목적에 맞는 기술을 사용하라는 내용의 글이다. 따라서 글의 제목으로 가장 적절한 것은 ④ '팀의 의사소통 목표에 적합한 기술을 맞추라'이다.

① 목적 중심 팀워크의 힘을 발견하라!
② 가상 회의는 정말 팀의 상호 작용을 약화시키는가?
③ 업무의 미래: 디지털 시대에 재정의되는 협업
⑤ 혁신에 활력을 불어넣기: 다양한 개성을 가진 사람으로 이루어진 팀의 마법

(구조 해설)

▪ Web conferencing and videoconferencing are richer, more interactive tools [better suited to complex tasks such as problem solving and negotiation], [which require squaring different ideas and perspectives].
   첫 번째 [ ]는 richer, more interactive tools를 수식하는 형용사구이다. 두 번째 [ ]는 complex tasks such as problem solving and negotiation을 부가적으로 설명하는 관계절이다.

(어휘 및 어구)

have ~ at one's disposal ~을 원하는 대로 사용할 수 있다
range from ~ to ... ~부터 …에 이르다
platform 플랫폼(사용 기반이 되는 컴퓨터 시스템)
videoconferencing 화상 회의
suited 적합한, 적당한
dimension 차원, 관점
nonverbal 비언어적인
cue 신호, 단서
determine 결정하다, 알아내다
lean 간결한, (비용을) 절감한
bulletin board (전자) 게시판
circulate 배포하다, 유통하다
routine 정례적인, 일상의

(Word Search)

| 정답 | 1. platform 2. determine 3. circulate

1. 새로운 온라인 플랫폼은 사용자들 간의 다양한 주제에 대한 토론을 촉진했다.
2. 그 환자의 질병에 대한 정확한 원인을 알아내기 위해 추가 검사가 필요

하다.
3. 중앙은행은 경제를 활성화하기 위해 더 많은 돈을 유통할 예정이다.

Exercise 10 ──────────── 정답 ③

(소재) 동물의 미래 지향적인 행동

(해석) 동물들은 보금자리 짓기부터 동면까지 규칙적으로 미래 지향적인 행동을 한다. 분명히 이러한 행동은 기능적으로 미래를 내다보는 것이지만, 그것들이 인지 과정에 의해 어느 정도 통제되는지는 여전히 해결되지 않은 문제로 남아 있다. 예를 들어, 철새들은 겨울, 더 따뜻한 기후, 또는 이동의 위험을 한 번도 경험한 적이 없음에도 불구하고 추운 겨울을 피하고자 장거리를 이동한다. 이 새들은 자신들이 도대체 무엇을 피하고 있는지, 그리고 왜 이러한 행동 방침이 자신들에게 이로운지 '알' 수 없다. 실제로, 서로 다른 이동 경로를 가진 두 조류 개체군이 이종 교배되면 그 결과로 생기는 새끼는 부모의 이동 방향 간의 중간 방향으로 이동하는데, 이는 이동 방향이 유전적으로 결정된다는 점을 시사한다. 이 예시를 통해 우리는 모든 미래 지향적인 행동이 미래에 대한 인식을 포함하는 것으로 간주될 수는 없으며, 대신 자연계의(예를 들어, 낮 길이의 계절적 변화 또는 호르몬) 또는 학습된 신호에 대한 반사적인 반응일 수도 있다는 것을 알 수 있다.

(문제 해설)

③ 문맥상 know의 목적어 역할을 하는 명사절을 유도해야 하는데 avoiding의 목적어가 없는 불완전한 구조가 이어지고 있으므로, that을 what으로 고쳐야 한다. 강조 구문 형태의 What is it (that) they are avoiding?이 know의 목적어가 되기 위해 「의문사+주어+동사」의 순서가 되어 what it is (that) they are avoiding으로 전환된 것으로 볼 수 있다.
① 주어의 핵심어가 the extent이므로 단수형 동사 remains를 쓰는 것은 어법상 적절하다.
② 문맥상 주절의 동사인 travel보다 앞선 시제를 나타내야 하므로, 동명사의 완료형인 having experienced를 쓰는 것은 어법상 적절하다.
④ 앞 절의 내용을 부가적으로 설명하는 분사구문을 유도해야 하므로, 현재분사 suggesting을 쓰는 것은 어법상 적절하다.
⑤ all prospective behavior가 consider의 주체가 아닌 대상이므로 수동태인 be considered를 쓰는 것은 어법상 적절하다.

(구조 해설)

▪ Clearly these behaviors are functionally prospective, but the extent [to which they are controlled by cognitive processes] remains an open question.

[ ]는 the extent를 수식하는 관계절이다.

어휘 및 어구

future-oriented 미래 지향적인
functionally 기능적으로
prospective 미래 지향적인, 예측하는
cognitive 인지의
migrate 이동하다
resulting 결과로 생기는
determine 결정하다
awareness 인식
cue 신호, 단서

Word Search

| 정답 | **1.** prospective **2.** cognitive **3.** migrate
1. 그 젊은 운동선수의 훈련 프로그램은 그녀의 미래의 대회 성공을 예측하는 지표이다.
2. 규칙적인 운동은 인지 기능을 향상시키고 기억력을 개선하는 것으로 밝혀졌다.
3. 매년 겨울, 제왕나비는 더 따뜻한 기후의 지역으로 수천 마일을 이동한다.

Exercise 11

정답 ④

소재 토양과 식물 간의 상호 작용

해석 토양과 거기서 자라는 식물은 여러 방식으로 지속해서 상호 작용하는데, 직접적이기도 하고, 또 항상 존재하는 여러 종류의 박테리아, 곰팡이류, 그리고 다른 작은 유기체의 영향을 통하기도 한다. 무엇보다도 특히, 식물은 끊임없이 잎과 다른 부분들을 떨어뜨리고 있으며, 이것들이 땅에 떨어져 썩어 그 자리에서 영양분을 방출하지만, 일부 더 건조한 기후 지역에서는 식물 잔해에 존재하는 무기 영양분이 빠져나오게 하기 위해 번개로 인한 화재가 필요하다. 어느 경우든, 죽은 식물 물질에 존재하는 대부분의 영양분은 결국 근처 토양에 남게 되고, 한편 식물 뿌리와 균사(菌絲)의 광대한 네트워크는 박테리아, 지의류 및 끈적한 부식토와 결합하여 심지어 폭우와 강풍에도 토양을 효과적으로 고정한다. 그 결과, 토양과 식물 사이를 거의 무한히 왔다 갔다 이동할 수 있는 식물 영양분의 효과적이고 지속적인 재활용이 이루어진다. (영양분이 부족한 토양의 질을 개선하는 방법에는 돌려짓기와 유기물 추가가 포함된다.) 토양은 또한 근처의 식물이 생산하는 종자와 포자의 거대한 저장고인데, 이는 교란 후 식물이 재생할 수 있도록 하는 핵심 요소이다.

문제 해설

토양과 그 안에서 자라는 식물은 직접적으로 혹은 주변에 존재하

는 박테리아, 곰팡이류, 그리고 작은 유기체의 영향을 통해 지속적으로 상호 작용한다는 내용이므로, 영양분이 부족한 토양의 질을 개선하는 방법에 관한 ④는 글의 전체 흐름과 관계가 없다.

구조 해설

▪ The result is effective and continuous recycling of plant nutrients [which are capable of {moving back and forth between the soil and the plants almost indefinitely}].

[ ]는 plant nutrients를 수식하는 관계절이고, 그 안의 { }는 전치사 of의 목적어 역할을 하는 동명사구이다.

어휘 및 어구

fungus 곰팡이류(*pl.* fungi)
notably 특히
rot 썩다
release 방출하다
nutrient 영양분
liberate 빠져나오게 하다, 유리(遊離)하다
litter 잔해, 쓰레기
vast 광대한, 방대한
fungal thread 균사(菌絲)
sticky 끈적한
indefinitely 무한히
vegetation 식물
regenerate 재생하다
disturbance 교란

Word Search

| 정답 | **1.** nutrients **2.** vast **3.** disturbance
1. 토양에 필수 영양소가 부족하여 식물의 성장이 저해되었다.
2. 대양은 방대했고, 끝이 보이지 않게 넓게 펼쳐져 있었다.
3. 생태 교란은 섬세한 생태계를 파괴하여 상당한 생태적 불균형으로 이어질 수 있다.

Exercise 12

정답 ⑤

소재 진화의 속도

해석 다윈이 자신의 자연 선택 이론을 전개할 때, 그는 진화가 지질학적 시간 규모에 따라 일어나는 것을 상상했다. 즉 빙하가 계곡의 형태를 만들어 내거나 비바람이 바위에 거세게 내리치는 방식처럼 그렇게 시간이 한 종을 다른 종으로 만든다고 상상했던 것이다. (C) 그러한 것이 1세기 넘도록 지속된 일반적 통념이었다. 하지만 20세기 후반에 진화에 대한 우리의 이해가 바뀌기 시

작했는데, 이는 상당 부분 *Beagle*호 항해에서 다윈이 직접 관찰했던 갈라파고스핀치에 관한 지속적인 연구 덕분이었다. (B) 진화가 다윈이 상상했던 것보다 훨씬 더 빠르게 이루어진다는 것이 밝혀졌다. 갈라파고스의 핀치는 우리가 관찰할 수 있는 시간 규모에 따라 진화한다. 예를 들어, 엄청나게 많은 비가 내린 해인 1983년에 작은 씨앗을 가진 덩굴 식물 종이 Daphne 섬의 식물군에 급속히 퍼졌다. (A) 가장 작고 가장 끝이 뾰족한 부리를 가진 새들이 갑자기, 그리고 명백히 유리해졌다. 그 새들의 유전자는 빠르게 퍼졌다. 형질에 대한 선택은 꼭 수십억 년이 아니라 몇 년이나 수십 년에 걸쳐서도 일어날 수 있다. 그리고 다윈의 핀치는 예외적인 것이 아니라 전형적인 사례이다. 그것들은 생물학자 John Thompson이 '끊임없는 진화'라고 부른 패러다임의 특권적 사례이다.

문제 해설

다윈이 자연 선택 이론을 전개할 때 진화가 지질학적 시간 규모에 걸쳐 일어난다고 믿었다는 내용의 주어진 글은 그러한 다윈의 생각을 Such로 지칭하여 이것이 1세기 넘도록 지속된 일반적 통념이었지만, 20세기 후반에 진화에 대한 이해가 변화하기 시작했다는 내용의 (C)로 이어져야 한다. 그다음은 (C)에서 이해의 변화가 발생한 이유로 언급된 갈라파고스핀치에 관한 연구에서 핀치가 우리가 관찰할 수 있는 시간 규모에 따라 진화한다는 (B)로 이어지고, 그것에 대한 자세한 부연 설명으로 1983년에 Daphne 섬에서 폭우로 인해 특정 덩굴 식물 종이 서식지에 퍼지면서 생존에 유리한 형질을 가진 새들의 유전자가 빠르게 퍼졌다는 내용의 (A)로 이어지는 것이 자연스럽다. 따라서 주어진 글 다음에 이어질 글의 순서로 가장 적절한 것은 ⑤ '(C) − (B) − (A)'이다.

구조 해설

- They are a privileged case of the paradigm [that the biologist John Thompson has called "relentless evolution."]

  [ ]는 the paradigm을 수식하는 관계절이다.

어휘 및 어구

natural selection 자연 선택
unroll 일어나다, 펼쳐지다
sculpt 형태를 만들다, 조각하다
beat 거세게 내리치다
mold 만들다, 빚어내다
pointy 끝이 뾰족한
distinctly 명백하게
trait 형질, 특징
exemplary 전형적인, 모범적인
relentless 끊임없는, 집요한
superabundant 엄청나게 많은

vine 덩굴 식물
overrun (~에) 급속히 퍼지다
conventional 일반적인, 전통적인
observe 관찰하다
voyage 항해

Word Search

| 정답 | 1. mold 2. conventional 3. observing
1. 그 조각가의 능숙한 손은 점토를 깜짝 놀랄 만큼 생생한 형상으로 빚어낼 수 있다.
2. 그 회사는 새로운 접근 방식을 실험하기보다는 전통적인 방법을 고수하기로 결정했다.
3. 그 어린 소녀는 나비 하나가 꽃에서 꽃으로 날아다니는 것을 관찰하고 있었다.

1 ④	2 ⑤	3 ④	4 ①	5 ④	6 ②
7 ⑤	8 ③	9 ⑤	10 ⑤	11 ②	12 ③

Exercise 1 ——————————— 정답 ④

소재 마케팅 성과 측정의 중요성

해석 이익은 수익을 극대화하고 비용을 통제하기 위한 가격 전략을 수립함으로써 관리될 뿐만 아니라 마케팅 성과를 측정하려는 책무에 의해서도 영향을 받는다. 조직이 자신의 목표를 달성하는 데 있어서 (자신이) 어떤 상황에 있는지 알지 못하면 조직은 그 목표를 이룰 수 없다. 마케팅 활동에 대한 측정은 향후 마케팅 전략을 수립하기 위해서뿐만 아니라 과거의 성과를 평가하기 위해서도 필요하다. 예를 들어, 마이너리그 야구팀은 토요일 밤 불꽃놀이 홍보 활동의 유효성[효과성]을 추가로 판매된 티켓의 수와 해당 고객들의 관련 지출 측면에서 이해하는 것이 중요하다. 이 경우 성과를 추적하면 발생된 수익과 그 홍보를 실행하기 위해 누적된 비용을 비교할 수 있을 것이다. 마케팅 투자와 수익의 관계를 이해하고 향후 마케팅 활동을 위한 의사 결정을 안내하기 위해 이러한 투자 수익률 분석이 가능할 때마다 수행되어야 한다.

문제 해설
④ revenue(수익)는 generate(발생시키다, 만들어 내다)와 의미상 수동의 관계에 있으므로 현재분사 generating을 과거분사 generated로 고쳐야 어법상 적절하다.
① know의 목적어 역할을 하는 의문절을 이끄는 의문사가 필요한데, 이어지는 절이 주어(it)와 자동사(stands)로 이루어진 완전한 형태이므로 의문부사 where는 어법상 적절하다.
② assess와 as well as로 연결되어 to에 이어져 to부정사구를 이루고 있으므로 set이 오는 것은 어법상 적절하다.
③ additional tickets(추가 티켓)와 sell(판매하다)은 의미상 수동의 관계에 있으므로 어법상 적절하다.
⑤ 앞에 동사구 should be performed가 존재하고 접속사가 없으므로 동사가 아닌 부사적 용법의 to부정사가 쓰인 것은 어법상 적절하다.

구조 해설
■ An organization cannot achieve its objectives [if it does not know where it stands in meeting those objectives].
[ ]는 조건의 의미를 나타내는 부사절로 절의 주어인 it은 an organization을 대신[의미]한다.

어휘 및 어구
profit 이익, 수익
manage 관리하다, 운영하다
strategy 전략
maximize 극대화하다
revenue 수익
commitment 책무, 의지, 전념
measure 측정하다
organization 조직
achieve 이루다, 달성하다
assess 평가하다
fireworks 불꽃놀이
track 추적하다
accumulate 누적하다, 축적하다
execute 실행하다
return on investment 투자 수익률

Word Search

| 정답 | **1.** organization   **2.** achieved   **3.** manage
1. 그 비영리 조직은 가난한 지역 사회를 지원하기 위해 일한다.
2. 그는 마침내 출판 작가가 되겠다는 평생의 꿈을 이루었다.
3. 자원을 관리하는 그의 능력 덕분에 프로젝트가 예산 내에서 완료될 수 있었다.

Exercise 2 ——————————— 정답 ⑤

소재 완결감에 대한 유아의 욕구

해석 몇몇 어린아이들을 포함한 일부 사람들이 어떻게 몇 시간 동안 직소 퍼즐을 하는 것을 즐기는지 생각해 보라. 큰 놀라움은 없다. (마지막 조각이 장난꾸러기 형제나 자매에 의해 숨겨져 있지 않은 한 그렇다.) 놀이 참가자들은 최종 결과가 정확히 어떻게 나올지 알고 있으며 중간에 깜짝 놀랄 일이 거의 없다. 그러나 완성하자마자 그럼에도 불구하고 큰 만족감이 있을 수 있다. 재미있는 이야기를 들을 때 결말에 이르지 않으면 크게 실망할 수 있다. 집이나 공원 같은 새로운 장소를 탐험할 때, 우리는 자주 전체적인 배치에 대해 명확히 파악하고 싶어 하고 서로 다른 장소들이 어떻게 연결되어 있는지 알지 못하면 낙담할 수 있다. 우리 모두와 마찬가지로 유아는 완결감을 얻으려고 분투한다. 그들은 상자 안에 들어 있는 '모든' 장난감을 식별하거나 새로운 공간의 '모든' 방을 탐험하려는 욕구를 가지고 있다. 그들은 일이 어떻게 일어나는지에 대한 인과 관계 모델에서도 비슷한 완결감의 욕구를 가지고 있다. 완결감에 대한 욕구들의 이러한 집합체는 특히 설명상의 공백을 메울 필요가 있을 때, 놀라운 호기심을 유발하는 데 자주 도움이 된다.

문제 해설

우리가 퍼즐을 완성하거나 이야기의 결말을 듣기 원하는 것과 마찬가지로 유아들은 장난감 식별, 새로운 장소 탐험 및 인과 관계 파악에서 완결감의 욕구를 가지고 있으며 그들의 이러한 욕구가 호기심을 유발한다는 것이 핵심 내용이다. 따라서 글의 주제로 가장 적절한 것은 ⑤ '어린 시절부터 두드러지는 완결에 대한 인간의 타고난 욕구'이다.

① 익숙한 작업을 완결하는 데 있어서 예측 불가능성의 영향
② 어린아이들이 새로운 도전에 직면하는 것을 피하는 이유
③ 교육의 이면에 있는 원동력으로서의 마무리의 결여
④ 직소 퍼즐을 즐기는 데 있어서 완결의 사소한 역할

구조 해설

■ They have a need [to {identify *all* the toys ⟨that are in a box⟩}, or {explore *every* room of a new space}].
[ ]는 선행하는 a need의 구체적 내용을 설명하는 to부정사구로, 그 안에서 두 개의 { }가 or로 연결되어 to에 이어진다. 첫 번째 { } 안의 ⟨ ⟩는 *all* the toys를 수식하는 관계절이다.

어휘 및 어구

sibling 형제자매
startled 깜짝 놀란
satisfaction 만족감
completion 완성, 완결
disappointed 실망한
explore 탐험하다
overall 전체적인
layout 배치
frustrated 낙담한, 좌절한
location 장소, 위치
infant 유아, 아기
strive 분투하다
identify 식별[확인]하다
wonder 놀라운 호기심, 경탄

Word Search

| 정답 | 1. frustrated 2. location 3. infant
1. 그는 복잡한 퍼즐을 풀지 못해 좌절감을 느꼈다.
2. 리조트의 위치는 그것을 관광객들에게 인기 있는 여행지가 되게 하였다.
3. 그 부모는 그들의 아기가 첫걸음을 내딛는 모습을 보며 기뻐했다.

Exercise 3 정답 ④

소재 미국 일간지의 특성

해석 유럽의 전국적이고 당파적인 언론과 달리 미국의 일간지들은 대개 도시 수준에서 경쟁했다. 일간지의 배포 범위는 주로 다른 도시와의 거리와 발행 시간에 의해 제한되었다. 19세기 말에는 노면 전차가, 20세기에는 트럭이 신문을 더 넓은 지역으로 적시에 배달할 수 있게 했다. 같은 소도시에서 경쟁하는 서로 다른 신문들은 신문의 물리적 배포와 기사 편집 범위를 중심 도시로부터의 서로 다른 반경 거리로 제한하거나 확장하는 것을 선택했다. 물론 밤 동안 인쇄되는 조간 신문은 석간 신문보다 더 넓은 범위에서 배포할 수 있었다. 그러나 대체로 사건이 지역 뉴스로서 정기적으로 보도되는 지리적 영역은 신문이 대부분의 독자를 끌어들이는 영역과 일치했다.

문제 해설

미국의 일간지는 주로 도시 단위에서 경쟁하였는데, 신문들은 배포 범위를 조정하면서 주로 지역 뉴스가 주된 독자층을 형성하는 지역에 맞추어 보도했다는 내용이므로, 글의 제목으로 가장 적절한 것은 ④ '지역 집중: 미국 일간지의 특성'이다.

① 유럽 신문에는 공통점이 없다
② 왜 모든 미국 신문은 같은 내용을 보도할까?
③ 글로벌 경쟁: 미국 신문사의 편집 전략
⑤ 교통 혁명, 국제 신문을 탄생시킨 것

구조 해설

■ Different papers [competing in the same towns] chose [to limit or spread out their physical distribution and editorial coverage to different radial distances from the central city].
첫 번째 [ ]는 Different papers를 수식하는 분사구이고, 두 번째 [ ]는 chose의 목적어 역할을 하는 to부정사구이다.

어휘 및 어구

compete 경쟁하다
predominantly 대개
circulation 배포, 발행 부수, 순환
primarily 주로
bound 제한[억제]하다
publication 발행, 출판
streetcar 노면[시내] 전차
region 지역
distribution 배포
editorial coverage 기사 편집의 범위
geographical 지리적인
coincide 일치하다

Word Search

| 정답 | **1.** circulation **2.** compete(d) **3.** coincide(d)

1. 규칙적인 운동은 혈액 순환을 개선할 수 있다.
2. 재능 있는 선수들이 국제 대회에서 경쟁한다[경쟁했다].
3. 축제 날짜는 국경일과 일치한다[일치했다].

Exercise 4
정답 ①

소재 바람과 지구의 물순환

해석 바람은 지구의 물순환에서 중요한 역할을 한다. 최근에, 한때 받아들여졌던 물순환에 대한 관점이 변화하였고, 대기 순환에서 바람의 역할이 더 많은 인정을 받게 되었다. 전통적으로는 지구의 물순환이 대개 대형 수역에서 증발한 수분이 바람에 의해 지구 표면을 가로질러 분포되는 것을 수반한다고 생각되었다. 지난 50년 동안 그 관점은 변했고, 이제는 숲이 강우를 유발하는 증산 작용을 생성하는 데 중요한 역할을 한다는 것이 알려져 있다. 숲속에 사는 다양한 식물들은 뿌리에서 물을 흡수한 후 증산 작용을 통해 그것을 수증기로 방출한다. 탁월풍이 이 물을 공기를 통해 운반하여 다른 지역에 강우를 제공한다. 이 물이 바람을 통해 대량으로 이동할 때, 기상학자 José Marengo가 '하늘을 나는 강'이라고 부른 것이 된다.

문제 해설
(A) involved의 목적어 자리에 동명사의 의미상 주어인 evaporated moisture from large bodies of water와 동명사의 수동형 being distributed가 연결된 형태이므로 being이 어법상 적절하다.
(B) 글의 흐름상 water를 가리켜야 하므로 it이 어법상 적절하다.
(C) called의 목적어 역할을 하는 관계사가 와야 하는데 선행사가 존재하지 않으므로 선행사를 포함하는 관계사 what이 어법상 적절하다.

구조 해설
- Traditionally, **it** was thought [that the planet's water cycle mostly involved {⟨evaporated moisture from large bodies of water⟩ being distributed by wind across the Earth's surface}].
  it은 형식상의 주어이고, that이 이끄는 명사절 [ ]가 내용상의 주어이다. 그 안에서 { }는 동사 involved의 목적어 역할을 하는 동명사구이며 ⟨ ⟩는 동명사구의 의미상의 주어이다.

어휘 및 어구
cycle 순환
atmospheric circulation 대기 순환

appreciation 인정
evaporate 증발시키다[하다]
moisture 수분, 습기
distribute 분포시키다, 분배하다
surface 표면
significant 중요한
vapour 수증기

Word Search

| 정답 | **1.** appreciation **2.** moisture **3.** surface

1. 그 팀은 그들의 노력에 의해 널리 인정을 받았다.
2. 비가 온 후 공기가 습기로 가득 찼다.
3. 호수의 표면이 너무 맑아서 주변을 완벽하게 비추었다.

Exercise 5
정답 ④

소재 사회적 변화와 언어

해석 양육자와 아이의 주변 환경에 있는 다른 사람들은 그 아이가 습득하고 있는 언어에 영향을 미친다. 나중에는 사회 집단이 아이의 언어에 영향을 미친다. 이는 계층, 민족성, 연령은 물론 성별과 성별 표현에 따라 차이가 있는 언어의 사회적 변종을 생기게 한다. 그러나 사회 집단이 다른 유형의 의사소통의 필요성을 일으키는 여러 세대에 걸친 큰 변화를 겪을 때는 이전의 변종은 사용을 멈추게 될 수도 있다. 이것이 Cosme와 Nueva Australia에서 처음 사용된 영어에서 발생했는데, 그곳들은 1890년대 호주 이민자들에 의해 세워진 파라과이의 시골 공동체였다. (새로운 문화 규범에 적응하고 자신의 의사소통과 이해를 향상시키기 위한 수단으로 영어는 차용을 광범위하게 활용한다.) 영어를 사용한 그 이민자들로부터 남아 있는 유일한 언어 흔적은 Kennedy, Smith, Stanley, 그리고 Wood와 같은 흔한 파라과이의 성씨뿐이다.

문제 해설
사회 집단의 영향으로 다양한 언어 변종이 생기는데, 세대를 거쳐 사회 집단에 큰 변화가 발생하면 이전의 변종은 사용되지 않는다는 내용으로, 이에 대한 예시로 파라과이에 정착한 호주 이민자들의 영어를 들고 있다. 따라서 문화 규범에 적응하고 이해를 향상시키기 위해 영어가 차용을 활용한다는 내용의 ④는 글의 전체 흐름과 관계가 없다.

구조 해설
- This occurred with the English initially spoken in Cosme and Nueva Australia, [which were rural Paraguayan communities {founded by Australian immigrants in the 1890s}].

[ ]는 Cosme and Nueva Australia를 부가적으로 설명하는 관계절이며, 그 안의 { }는 rural Paraguayan communities 를 수식하는 분사구이다.

어휘 및 어구

caregiver 양육자, 돌보는 사람
surroundings 환경
social 사회의
variety 변종
class 계층
ethnicity 민족성
gender 성별
expression 표현
generation 세대
cease 멈추다
community 공동체
immigrant 이민자

Word Search

| 정답 | 1. social 2. ceased 3. communities
1. 그 디자인은 모든 연령대와 사회 집단에게 매력이 있어야 한다.
2. 사용 부족으로 인해 일부 언어는 결국에는 점차적으로 사용을 멈추게 되었다.
3. 그 지역의 다양한 공동체는 전통을 보존하기 위해 함께 노력했다.

Exercise 6 ── 정답 ②

소재 통증의 신경학적 반응과 뇌로의 전달 과정

해석 통증은 외부 자극에 대한 신경학적 반응이다. (B) 모기에 물리면 매우 적은 수의 통증 수용기가 촉발되어 뇌에 가벼운 인식 감각을 생성할 것이다. 그 신호는 뉴런의 사슬에 의해 개인의 척추를 따라 뇌의 특정 장소로 전달된다. (A) 뇌의 통증 뉴런이 자극되면 결과적으로 우리의 의식 속에 어떤 존재를 강력히 주장하게 된다. 우리는 어떤 것이 팔에 있다는 것을 인식하게 되는 것이다. 그것이 벌에 쏘인 상처라면, 벌이 쏜 부위에 독성의 액체를 남기므로, 모기의 경우보다 벌이 쏜 부위에 있는 더 많은 뉴런이 촉발된다. (C) 따라서 더 강한 신호가 뇌로 보내져 개인의 의식에 더 강하게 삽입된다. 나무가 넘어져 다리를 으스러뜨리면 아주 많은 통증 수용기가 활성화되어 의식이 통증 신호로 압도된다. 찢어진 살은 의료적 처치로 그 문제가 처리될 때까지 계속해서 신호를 보낸다.

문제 해설

통증이 외부 자극에 대한 신경학적 반응이라는 주어진 글에 이어

(B)에서 모기 물림의 자극이 뇌에 전달되는 예를 설명한 후, 이어서 뇌에서 이를 어떻게 인식하는지 설명하고 이보다 더 강한 벌에 쏘인 자극은 더 많은 뉴런을 촉발한다는 내용의 (A)가 이어져야 한다. 그런 다음 다리가 으스러지는 예를 들면서 더 강한 신호일수록 더 강하게 의식한다는 내용의 (C)가 이어지는 것이 자연스럽다. 따라서 주어진 글 다음에 이어질 글의 순서로 가장 적절한 것은 ② '(B)−(A)−(C)'이다.

구조 해설

■ A mosquito bite, [triggering a very small number of pain receptors], will create a mild sensation of awareness in the brain.
[ ]는 분사구문으로 A mosquito bite의 부수적인 상황을 나타낸다.

어휘 및 어구

neurological 신경학적인
response 반응, 응답
external 외부의
stimulus 자극
excitement 자극, 흥분
neuron 뉴런[신경 세포]
assert 강력히 주장하다, 확고히 하다
consciousness 의식
bee sting 벌에 쏘인 상처
liquid 액체
trigger 촉발하다
signal 신호
relay 전달하다
insertion 삽입, 주입
crush 으스러뜨리다
overwhelm 압도하다

Word Search

| 정답 | 1. response 2. signal 3. insertion
1. 학생의 질문에 대한 선생님의 대답은 수업을 명확하게 하는 데 도움이 되었다.
2. 신호등이 녹색으로 변하면서 차들에게 움직이기 시작하라는 신호를 보냈다.
3. 정상 유전자를 병든 세포에 삽입하는 것이 그 병을 치료할 수 있을 것이다.

Exercise 7 ── 정답 ⑤

소재 사람을 다루는 업무의 핵심에 있는 갈등

해석 갈등은 사람을 다루는 일의 바로 핵심에 있는데, 특히 사회복지사, 경찰, 보호 관찰관과 같은 전문가들에게 그러하며, 이들은 개인의 권리와 자유를 타인에게 가해질 위험과 피해에 견주어 보도록 요구받는 경우가 많다. 개인이나 가정에 대한 위험 평가를 맡는 바로 그 행위가 관련된 사람들 중 일부가 격렬하게 다툴 수도 있는 결정이 내려질 가능성을 수반한다. '위험에 처한' 아이를 가족으로부터 분리하여 안전한 곳으로 보내거나 정신적 고통을 겪고 있는 누군가가 치료를 위해 입원하도록 주장하는 것은 갈등에 연루되는 것이다. 여러 전문 분야로 구성된 병원 팀 내에서, 특정 개인이나 가정과 가장 효과적으로 일하거나 (그들을) 치료하는 방법에 대한 서로 충돌하는 접근법이 있을 수도 있다. 대부분의 사람을 다루는 업무 종사자와 의료 종사자들은 또한, 화가나 있거나 고통스러워하거나 실망한, 그리하여 그들의 반응이 많은 경우 높은 수준의 갈등을 유발한 사람들을 다루어야 했던 때의 사례를 제시할 수 있다. 그러므로 갈등을 다루기 위한 적절한 의사소통 기술을 개발하는 것이 필수적이다.

문제 해설
⑤ 주어인 To develop ~ conflict에 이어지는 술어동사가 없기 때문에 being을 is와 같은 표현으로 고쳐야 한다.
① 수식받는 the people과 involve(관련시키다) 사이에는 의미상 수동의 관계가 성립하므로 과거분사 involved의 쓰임은 어법상 적절하다.
② 앞의 To remove ~와 함께 or로 대등하게 연결되어 주어 역할을 하고 있으므로 to insist의 쓰임은 어법상 적절하다.
③ 동사 work를 수식해야 하므로 부사 effectively의 쓰임은 어법상 적절하다.
④ 수식을 받는 angry, distressed or disappointed people이 관계절 내에서 response를 수식하는 소유격의 역할을 해야 하므로 소유격 관계대명사 whose의 쓰임은 어법상 적절하다.

구조 해설
- [The very act of undertaking a risk assessment on an individual or family] **carries** with it the likelihood of [decisions being taken {with which some of the people involved may violently disagree}].
첫 번째 [ ]는 문장의 주어이고 carries가 문장의 술어동사이다. 두 번째 [ ]는 of의 목적어 역할을 하는 동명사구이고, decisions가 being taken의 의미상의 주어이다. { }는 decisions를 수식하는 관계절이다.

어휘 및 어구
conflict 갈등
professional 전문가
undertake 맡다, 착수하다
assessment 평가

violently 격렬하게
distress 고통
treatment 치료
multi-disciplinary 여러 전문 분야로 구성된, 다학제의
disappointed 실망한
provoke 유발하다
appropriate 적절한
essential 필수적인

Word Search
| 정답 | 1. undertake 2. assessment(s) 3. distress
1. 교수들은 가르치면서 연구도 맡는다.
2. 여러분은 업무 범위를 정의하는 초기 평가를 수행해야 한다.
3. 우리는 그의 깊은 정서적 고통을 감지했다.

Exercise 8 ———————————— 정답 ③

소재 당분 음료 섭취가 우리 몸에 미치는 영향
해석 마시는 것은 우리의 많은 타고난 음식 통제 시스템을 무시한다. 큰 잔으로 사과 주스 한 잔을 마시는 것이 얼마나 쉬운지를 한번 생각해 보라. 하지만 자리에 앉아 한 자리에서 사과 여덟 개를 먹지는 않을 것인데, 두 번째 사과를 먹고 난 후 통제 시스템이 개입하여 여러분이 충분히 먹었다고 말할 것이다. 통째로 된 사과는 섬유질을 함유하고 있는데, 이것은 사과의 당분이 방출되는 것을 늦추고, 그렇게 하면서 그것을 먹는 것이 우리의 에너지에 미치는 영향에 균형을 잡아준다. 주스에는 섬유질이 없기 때문에, 당분이 우리를 에너지 사이클의 롤러코스터에 태워 보내는데, 급상승으로 시작하여 결국 급강하로 끝난다. 이 급강하는 여러분이 일어나서 아이디어를 발표하도록 요청받거나 할 일 목록에 있는 그 까다로운 작업을 시작해야 하는 바로 그 시점에 발생할 수도 있다. 그것은 정말 그럴 만한 가치가 없다. 당분이 든 카페인 음료도 더 심하지는 않더라도 비슷한 영향을 미치지만, 대형 커피숍 체인들은 추가적인 한 끼에 상당하는데도 전혀 그런 인상을 주지 않아 혈당 급강하 사이클을 처음부터 다시 시작하는 비싸고 얼음을 넣거나 크림을 얹은 음료를 우리에게 파는 데에서 커다란 즐거움을 얻는다.

문제 해설
마시는 것은 우리의 많은 타고난 음식 통제 시스템을 무시한다는 것을 사과 주스와 당분이 든 카페인 음료를 먹었을 때 겪게 되는 혈당 급강하 사이클을 통해 설명하고 있으므로 글의 주제로 가장 적절한 것은 ③ '당분이 든 음료를 마시는 것이 우리 몸에 미치는 영향'이다.
① 인간 생리 기능에서 당분의 필수적인 역할

② 과일이 어떻게 우리의 건강에 도움이 되는지에 대한 일반적인 근거 없는 믿음들
④ 영양소 흡수를 돕는 요소에 대한 이해
⑤ 우리가 흡수하는 에너지와 연소하는 에너지의 균형을 맞추는 방법

구조 해설

- The crash may be just at the point [when {you are asked to stand up and present your ideas} or {you need to kick off that tricky task on your to-do list}].
  [ ]는 the point를 수식하는 관계절이다. 그 안에서 두 개의 { }가 or로 연결되어 있다.

어휘 및 어구

contain 함유하다
fibre 섬유질
release 방출
boost 급상승
crash 급강하, 급락
tricky 까다로운
profound 심한
take delight in ~을 즐기다
simply 전혀
register as ~이라는 인상을 주다, ~로 인식되다

Word Search

| 정답 | 1. fibre 2. boost 3. tricky
1. 충분한 섬유질을 섭취하면 노폐물이 몸을 원활하게 통과하는 데 도움이 될 수 있다.
2. 시험에 합격하고 나니 내 자신감이 정말 급상승했다.
3. 이 장비는 설치하기가 너무 까다로웠다.

Exercise 9 — 정답 ⑤

소재 모호한 전략을 가진 광고와 집중된 전략을 가진 광고의 차이점

해석 광고가 잘못될 수 있는 방법은 많다. 그러나 광고의 성공 가능성은 똑똑하고 집중된 전략으로 시작함으로써 높아진다. 이러한 집중과 함께 광고는 광고가 소통하는 많은 수준에서 두드러진 메시지를 전달할 수 있다. 광고는 스스로를 크게 엉망으로 만들기보다는 파고 들어가 그것의 목적을 달성한다. 독자가 '그것을 이해'할 기회가 엄청나게 증가한다. 이면에는 집중되지 않은 전략은 많은 서로 다른 요점을 모두 동시에 전달해야 하는 광고로 이어진다는 것이 있다. 이것은 효과적이지 않다. 비유하자면 이렇다. 열두 개의 공을 한꺼번에 전부 어떤 사람에게 던져 보라. 단 한 개의 공을 잡는 것도 불가능하다. 사실 포기하고, 얼굴을 가리

며, 다치지 않으려 애쓰는 충동이 일어난다. 완전한 혼란 상태이다. 하지만 단 한 개의 공을 목표하는 대상을 향해 똑바로 던지면, 아마도 그 사람은 잡을 것이다. 그것이 모호한 전략을 가진 효과적이지 못한 광고와 집중된 전략을 가진 효과적인 광고의 차이점이다.

문제 해설 이 글은 분명하지 않은 전략을 가진 효과적이지 못한 광고와 집중된 전략을 가진 효과적인 광고의 차이점을 공을 던지는 상황으로 비유적으로 설명하였다. 따라서 글의 제목으로 가장 적절한 것은 ⑤ '집중된 광고 전략을 공들여 만드는 것의 중요성'이다.
① 광고의 책무 탐색하기
② 광고 융합: 창의성을 데이터와 혼합하기
③ 똑똑한 광고를 통해 브랜드 충성도 구축하기
④ 광고를 통해 소비자 행동 이해하기

구조 해설

- The flip side is [that an unfocused strategy leads to ads {that must deliver many different points all at the same time}].
  [ ]는 is의 보어 역할을 하는 명사절이고, { }는 ads를 수식하는 관계절이다.

어휘 및 어구

odds 가능성, 확률
strategy 전략
deliver 전달하다
singular 두드러진, 뛰어난
bore 파고 들다
make one's point 목적을 달성하다, 주장을 관철하다
make a mess of ~을 엉망으로 만들다
tremendously 엄청나게
impulse 충동
shield 가리다
absolute 완전한
chaos 혼란 (상태), 혼돈

Word Search

| 정답 | 1. impulse 2. delivers/delivered 3. odds
1. 그는 일어나서 춤을 추고 싶은 충동을 갑자기 느꼈다.
2. 이 영상은 수상 안전에 대한 명확한 메시지를 전달한다[전달했다].
3. 주차 공간을 찾을 확률이 매우 낮다.

Exercise 10 — 정답 ⑤

소재 영어를 부가적인 언어로 배우는 사람들을 교육할 때 유의점

해석 우리는 모두 chip, grass, joint 및 hardware와 같은 단어 사용에서의 변화와 같이, 우리 자신의 일생 동안 의미를 바꾸거나 추가적인 의미를 얻은 단어를 떠올릴 수 있다. 이것은 어떻게 단어가 의미 구성체를 위한 단순한 표시에 불과한지를 보여 준다. 만약 이러한 구성체가 사람마다 서로 다른 구조를 가진다면, 그들 사이 의사소통의 이행은 완전하거나 정확하지 않을 것이다. 영어를 부가적인 언어로 배우는 사람들에게, 영어의 형태와 구조, 그리고 그 의미에 대한 이해가 어려움을 야기할 수도 있고, 혹은 심지어 모국어가 가진 특성에서 오는 의미의 양상들로 가미될 수도 있다. 점진적으로, 완전한 유창성과 그 언어 사용의 교류를 통해 이해의 공통성이 대체로 달성될 것이다. 그러나 우리 학교에서 영어 습득의 초기 단계에 있는 아이들을 가르치는 사람들이 가르치고 있는 단어와 어구의 표시 이면에 있는 의미에 대한 이해를 구축하는 것을 처리하는 것이 중요하다.

문제 해설
⑤ it은 형식상의 주어이고, 내용상의 주어가 누락되어 있다. 따라서 attend를 to attend로 고쳐야 한다.
① changed와 or로 연결되어 have에 이어지므로 acquired의 쓰임은 어법상 적절하다.
② 주절에서 주어의 핵심어가 transitivity로 단수이므로 단수형 동사 is의 쓰임은 어법상 적절하다.
③ the characteristics는 소유하는 행위의 주체가 아니라 대상이므로 수동의 의미가 있는 과거분사 owned의 쓰임은 어법상 적절하다.
④ will be achieved를 수식하는 부사 generally의 쓰임은 어법상 적절하다.

구조 해설
■ For those [who are learning English as an additional language], apprehension of English forms and structures, and their meanings [may pose difficulties], or [even may be tinged with aspects of meaning from the characteristics {owned by the native tongue}].
첫 번째 [ ]는 those를 수식하는 관계절이다. 두 번째와 세 번째 [ ]는 or로 연결되어 주절의 술부를 구성한다. 세 번째 [ ] 안의 { }는 the characteristics를 수식하는 분사구이다.

어휘 및 어구
additional 추가적인, 부가적인
demonstrate 보여 주다, 증명하다
construct 구성체
make-up 구조, 만듦새
complete 완전한
accurate 정확한
apprehension 이해

native tongue 모국어
fluency 유창성
commerce 교류
commonality 공통성
acquisition 습득

Word Search

| 정답 | 1. Accurate 2. apprehension 3. acquisition
1. 정확한 측정값은 안전을 유지하는 데 필수이다.
2. 관련된 모든 것을 완전히 이해하는 것은 그 과학자들이 절대적으로 할 수 없는 것이다.
3. 언어 습득은 어린이나 어른이 언어를 학습하고 이해하는 과정을 의미한다.

Exercise 11 ——————— 정답 ②

소재 지연된 만족

해석 우리 중 많은 사람은 즉각적인 만족을 원하는 경향이 있다. 즉 '유행을 따르기' 위해 끊임없이 새 옷을 사거나, '유행에 뒤처지지 않기' 위해서 최신 제품을 원하거나 즉각적인 결과나 즐거움을 추구한다. 비록 지금 어떤 것을 선택하는 것이 당장에는 기분이 좋을지 모르지만, 더 장기적인 관점을 가지고 자제심을 발달시키는 것이 미래에 더 크고 더 나은 보상을 가져다줄 수 있다. 지연된 만족이란 이러한 즉각적인 즐거움의 유혹을 견디는 우리의 능력으로서, 우리가 장기적인 목표에 계속 집중할 수 있도록 한다. (큰 그림의 목표를 가지는 것이 중요하지만, 여러분이 최종 결과에 도달하게 하는 것은 바로 중요한 단기적인 목표이다.) 그것은 충동 소비와 의사 결정을 피하기 위한 절제력과 인내심을 갖기 시작하는 것뿐만 아니라 신중하게 예산을 세워서 소비하는 것을 포함한다. 여러분이 자기 자신의 재정적 목표를 수립하고 있을 때 즉각적인 만족을 지연된 만족과 비교하여 생각해 보라. 여러분의 목표가 단기적인 이익에 초점을 맞추고 있는지 아니면 더 큰 그림을 기대하도록 설계되어 있는지를 생각하라.

문제 해설
장기적인 목표에 계속 집중할 수 있게 해 주는 지연된 만족을 위해 자제심과 인내심을 발달시키고 더 큰 그림을 기대하도록 목표가 설계되어야 한다는 내용의 글이다. 따라서 단기적인 목표의 중요성을 강조하는 내용의 ②는 글의 전체 흐름과 관계가 없다.

구조 해설
■ Consider [whether your goals are {focused on short-term gains} or {designed to look forward to a bigger picture}].

[ ]는 Consider의 목적어 역할을 하는 명사절이다. 두 개의
{ }는 or로 연결되어 are에 이어지면서 수동태를 구성하고 있다.

(어휘 및 어구)

be prone to ~의 경향이 있다
instant 즉각적인
constantly 끊임없이
trendy 유행을 따르는, 최신 유행의
immediate 즉각적인
self-discipline 자제심
reward 보상
delayed 지연된
temptation 유혹, 유혹물
budget 예산을 세우다
mindfully 신중하게
discipline 절제력, 자제심
patience 인내심
impulse 충동
financial 재정적인

Word Search

| 정답 | **1.** instant **2.** temptation **3.** discipline
1. 그의 새 소설은 즉각적인 큰 인기를 끌었다.
2. 값비싼 보석은 도둑에게 유혹물이다.
3. 간식을 멀리하려면 많은 절제력이 필요하다.

Exercise 12
정답 ③

(소재) 새로운 기술과 절차의 불확실성에 대한 과학자들의 태도

(해석) 어떤 도구나 절차를 사용할지에 대한 불확실성과 결과가
보이는 대로가 아닐 위험은 모든 과학 분야에 공통적으로 존재하
는 문제이다. (B) 새로운 도구의 개발은 과학자들이 과거에는 답
할 수 없었던 질문에 답할 수 있게 해 주고, 그 질문에 대한 답은
새로운 질문으로 이어질 것이며, 이런 식으로 계속될 것이다. 따
라서 새로운 기술과 절차는 과학의 발전에 매우 중요하다. (C) 동
시에 새로운 도구에 익숙하지 않은 다른 과학자들은 회의적인 태
도를 보이며 다른 사람들에게 실험을 정확하게 똑같이 반복해 볼
것을 요구할 수도 있다. 이러한 회의적인 태도는 자주 짧고 기억
에 남기 쉬운 한두 문장으로 요약된 말의 형태로 우리에게 오고,
실험 도구에 대한 불확실성이 대부분의 사람들에게, 심지어는 과
학 학사 학위를 가진 사람에게조차도 익숙하지 않은 과학의 한 측
면이기 때문에, 회의적인 태도가 미적거리는 것처럼 보일 수도 있
다. (A) 여러분에게 오는 과학적 정보를 바탕으로 하여 결정을 내

리려고 할 때 미적거리는 것은 짜증스럽다. 그러나 새로운 기법이
그 불확실성의 근원이라면, 시간과 향후 실험을 통해 그것의 유용
성을 확인하거나 그것의 부당성을 증명하여 불확실성을 제거할 것
이다.

(문제 해설)
주어진 글은 도구와 절차의 사용에 대한 불확실성과 결과가 겉보
기와 다를 수 있는 위험성이 과학 분야에 공통으로 존재하는 문제
라는 내용이다. 주어진 글에 이어 새로운 도구의 개발이 과학자들
에게 중요한 역할을 하므로 새로운 기술과 절차가 과학의 발전에
매우 중요하다는 내용의 (B), 그와 동시에 새로운 도구에 익숙하
지 않은 다른 과학자들의 회의적인 태도가 미적거리는 것으로 여
겨질 수 있다는 (C), 이러한 미적거림은 짜증스러우나 그 불확실
성은 시간과 향후 실험을 통해 제거될 것이라는 내용의 (A)로 이
어지는 것이 자연스럽다. 따라서 주어진 글 다음에 이어질 글의
순서로 가장 적절한 것은 ③ '(B)−(C)−(A)'이다.

(구조 해설)

▪ [Because this skepticism often comes to us in the
form of sound bites], and [because uncertainty about
experimental tools is an aspect of science {that is not
familiar to most people, even people with a bachelor's
degree in science}], the skepticism may seem like
waffling.
두 개의 [ ]가 and로 연결되어 이유의 부사절을 이룬다. 두 번
째 [ ] 안의 { }는 an aspect of science를 수식하는 관계절
이다.

(어휘 및 어구)
uncertainty 불확실성, 반신반의
procedure 절차
discipline 학문 분야
annoying 짜증스러운, 짜증 나게 하는, 성가신
come one's way (일이 ~에게) 오다, 닥치다
confirm 확인하다
disconfirm ~의 부당성을 증명하다
crucial 매우 중요한
replicate 정확하게 똑같이 반복하다, 정확하게 똑같이 하다
bachelor's degree 학사 학위

Word Search

| 정답 | **1.** uncertainty **2.** annoying **3.** confirmed
1. 태풍의 정확한 경로에 대해서는 상당한 불확실성이 있다.
2. 그의 가장 짜증 나게 하는 습관은 입을 벌리고 먹는 것이었다.
3. 엑스레이 검사에서 그녀의 뼈가 하나도 부러지지 않은 것으로 확인되
었다.

| 1 ⑤ | 2 ⑤ | 3 ④ | 4 ③ | 5 ④ | 6 ④ |
| 7 ④ | 8 ⑤ | 9 ④ | 10 ③ | 11 ④ | 12 ② |

Exercise 1
정답 ⑤

소재 자기 효능감과 실제 수행 간의 불일치

해석 아이들이 어려운 과제를 수행하는 것에 대한 과신을 말하는 것은 드문 일이 아니다. 그들이 형편없이 수행했다는 것을 지적하는 피드백을 받을 때조차도, 그들의 자기 효능감은 감소하지 않을 수도 있다. 아이들의 자기 효능감과 실제 수행 간의 불일치는 아이들이 과제에 대해 익숙함이 없고 과제를 성공적으로 수행하는 데 무엇이 필요한지 완전히 이해하지 못할 때 발생할 수 있다. 경험이 쌓이면서, 그들의 정확성은 향상된다. 아이들은 또한 특정 과제의 특징들에 과도하게 영향을 받아 자신이 그 과제를 수행할 수 있는지 없는지를 이것들에 근거하여 결정할 수도 있다. 예를 들어, 뺄셈에서 아이들은 문제에 포함된 숫자나 열이 얼마나 많은지에 집중하여, 열이 더 적은 문제를 열이 더 많은 문제보다 덜 어렵다고 판단할 수도 있는데, 전자가 개념적으로 더 어려울 때조차도 그렇다. 이것은 학생들이 비현실적으로 과신하고 도움을 구하거나 자신들의 기량을 향상시키려는 동기를 느끼지 않기 때문에 더 높은 자기 효능감이 문제가 되는 예이다. 여러 특징에 집중하는 아이들의 능력이 향상됨에 따라, 그들의 정확성도 향상된다.

문제 해설

⑤ 맥락상 앞 절에 쓰인 improves를 대신해야 하므로, is를 does로 고쳐야 한다.

① 동사구 have performed를 수식하는 부사 poorly는 어법상 적절하다.

② understand의 목적어 역할을 하는 명사절을 이끄는 what은 어법상 적절하다.

③ 접속사 and로 앞선 동사구와 대등하게 연결되어 Children may에 이어지는 decide는 어법상 적절하다.

④ 문맥상 problems를 대신하는 those는 어법상 적절하다.

구조 해설

■ Children may also [be excessively swayed by certain task features] and [decide {based on these} {that they can or cannot perform the task}].

두 개의 [ ]는 and로 대등하게 연결되어 may also에 이어지며, 첫 번째 { }는 decide를 수식하는 분사구이고, 두 번째 { }는 decide의 목적어 역할을 하는 명사절이다.

어휘 및 어구

overconfidence 과신
accomplish 수행하다
self-efficacy 자기 효능감
decline 감소하다, 줄어들다
lack (~이) 없다, 부족하다
familiarity 익숙함
execute 수행하다
accuracy 정확성
excessively 과도하게
feature 특징
conceptually 개념적으로
problematic 문제가 되는
unrealistically 비현실적으로
capability 능력

Word Search

| 정답 | **1.** decline **2.** lack **3.** capability

1. 반복되는 실패에도 불구하고 그의 열정은 줄어들지 않았다.

2. 어린 아이들은 자주 자신의 능력을 평가하는 데 필요한 경험이 부족하다.

3. 복잡한 문제를 해결하는 능력은 그녀를 팀의 자산으로 만든다.

Exercise 2
정답 ⑤

소재 채용 과정에서 생성형 AI의 활용과 그 한계

해석 채용 과정에서 생성형 AI를 사용하는 것에는 여전히 난제와 한계점이 있다. 예를 들어, 생성형 AI는 채용자나 지원자의 명성이나 신뢰성을 해칠 수도 있는 부정확하거나 오해의 소지가 있는 내용을 생성할 수 있다. 또한 데이터 프라이버시, 데이터 동의 혹은 데이터 소유권과 관련된 윤리적 혹은 법적 문제를 일으킬 수 있다. 따라서 완전한 자동화는 현재 영역에서는 있음직해 보이지 않는데, 이는 채용 관리자, 법무 부서 등이 생성된 구인 광고와 지원자 소통을 여전히 검토하고 승인해야 하기 때문이다. 그러므로 생성형 AI는 채용에서 인간의 판단이나 상호 작용을 대체해서는 안 되며, 오히려 그것을 보완해야 한다. 생성형 AI는 인간의 능력과 창의성을 증강하는 도구로 사용되어야 하며, 이를 완전히 자동화해서는 안 된다. 채용자는 항상 정보를 확인하고 생성형 AI가 생성한 내용을 사용하기 전에 편집해야 한다. 그들은 채용 결과와 지원자 경험에 대한 생성형 AI의 성과와 영향을 추적 관찰해야 한다.

문제 해설

생성형 AI는 부정확하거나 오해의 소지가 있는 내용을 생성할 수

있으며, 윤리적 또는 법적 문제를 일으킬 수 있어서 채용에서 완전한 자동화는 어렵고, 생성형 AI는 인간의 판단과 상호 작용을 보완하는 도구로 사용되어야 한다는 내용의 글이다. 따라서 글의 주제로 가장 적절한 것은 ⑤ '채용 과정에 생성형 AI를 통합할 때 인간 참여의 필요성'이다.

① 미래 일자리 시장을 형성하는 데 있어 AI가 하는 대단히 중요한 역할
② 지원자 선정에서 AI 활용과 밀접한 관련이 있는 진화하는 법적 체계
③ 생성형 AI를 사용한 완전한 채용 자동화로의 불가피한 변화
④ 인간의 역할을 AI로 대체하는 것으로 이어지는 기술의 발전들

(구조 해설)
- Generative AI may, for instance, produce inaccurate or misleading content [that could harm {the reputation or credibility of recruiters or candidates}].
  [ ]는 inaccurate or misleading content를 수식하는 관계절이며, 그 안의 { }는 harm의 목적어 역할을 하는 명사구이다.

(어휘 및 어구)
generative AI 생성형 AI
inaccurate 부정확한
misleading 오해의 소지가 있는
reputation 명성
credibility 신뢰성
recruiter 채용자
candidate 지원자
ethical 윤리적인
legal 법적인
consent 동의
ownership 소유권
review 검토하다
sign off on ~에 대해 승인하다
communication 소통
complement 보완하다
creativity 창의성
verify 확인하다
edit 편집하다
monitor 추적 관찰하다
performance 성과
impact 영향
outcome 결과

Word Search

| 정답 | 1. ethical 2. communication 3. impact

1. 그 회사는 신제품에 동물 실험을 사용할지 여부를 결정할 때 윤리적 딜레마에 직면했다.
2. 위기 상황에서는 잘못된 정보를 방지하기 위해 명확하고 간결한 소통이 필수적이다.
3. 새로운 규정이 회사 운영에 큰 영향을 미쳤다.

Exercise 3 — 정답 ④

(소재) 기다림의 교육적 역할

(해석) 학생들에게 질문을 던진 후, 보통의 미국 교사는 학생을 지명하여 답하라고 하기 전에 보통 1초도 기다리지 않는데, 이는 속도가 복잡한 사고보다 중시된다는 강한 메시지를 보낸다. 하지만 플로리다 대학교의 연구는 교사가 아이를 지명하기 위해 기다리고, 그런 다음 그 아이가 응답에 대해 생각하기 위해 조금만 더 시간을 낼 때(단 3초만이라도) 마법 같은 일이 벌어진다는 것을 발견했다. 가장 즉각적인 이점은 아이들의 답변 길이에서 나타났다. 그 적은 양의 사고 시간은 자신의 견해에 대한 더 많은 증거와 대안 이론에 대한 더 많은 고려를 포함하여 아이들이 자신의 생각을 정교하게 하는 데 3배에서 7배 사이의 더 많은 시간을 보냈음을 의미했다. 기다리는 시간이 늘어난 것은 또한 아이들이 서로의 의견을 듣고 자신의 아이디어를 발전시키도록 장려했다. 고무적인 것은, 그들의 더 정교한 사고가 글쓰기로도 옮겨가, 글에 더 미묘한 차이가 생겼고 복잡해졌다는 것이다. 그것은 교사의 인내심을 발휘하는 단순한 행위에서 비롯된 놀라운 개선이다. 연구자 Mary Budd Rowe가 그녀의 독창적인 논문에서 언급했듯이, '느리게 하는 것이 속도를 높이는 방법일 수 있다.'

(문제 해설)
보통의 미국 교사는 질문 후 1초도 기다리지 않지만, 3초만 기다려도 학생들의 답변이 길어지고 사고력이 향상되며, 이 단순한 교사의 인내가 학생들의 정교한 사고와 글쓰기 능력을 크게 발전시킨다는 내용의 글이다. 따라서 글의 제목으로 가장 적절한 것은 ④ '멈춤의 힘: 기다림이 학생의 사고를 깊게 만들 수 있는 방법'이다.
① 교실 환경이 학습 능력에 어떤 영향을 미치는가?
② 능동적인 참여가 학생의 이해력을 높이는 이유
③ 질문: 학생의 더 깊은 사고를 여는 열쇠
⑤ 빠른 속도의 학습: 학생들의 즉각적인 이해를 장려하기

(구조 해설)
- [Having asked the class a question], the average American teacher typically waits less than a second before picking a child to provide an answer ~.

[ ]는 the average American teacher를 의미상의 주어로 하는 분사구문이며, 주절의 동작(waits)이 일어나는 시점보다 앞선 시점을 나타내고 있다.

어휘 및 어구

average 보통의, 평균적인
typically 보통
strong message 강한 메시지
value 중시하다
complex 복잡한
immediate 즉각적인
elaborate 정교하게 하다, 공들이다
viewpoint 견해
consideration 고려
alternative theory 대안 이론
encourage 장려하다, 격려하다
sophisticated 정교한
exercise 발휘하다
patience 인내심
original 독창적인, 원래의

Word Search

| 정답 | 1. average 2. immediate 3. encourage
1. 이번 달의 평균 기온은 계절에 비해 비정상적으로 높았다.
2. 이 소식에 대한 그녀의 즉각적인 반응은 놀라움과 흥분의 반응이었다.
3. 코치의 동기 부여 연설은 큰 경기 전에 선수단을 격려하기 위한 것이었다.

Exercise 4
정답 ③

소재 광물 자원 원천 탐사의 한계

해석 우리가 이미 알고 있는 것 외에 고품질의 광물의 원천을 찾을 희망은 거의 없다. 지구의 지각은 철저히 탐사되었으며 더 깊이 파는 것은 도움이 되지 않을 것인데, 이는 광물이 지표면 근처에서 작동하는 열수 작용 과정에 의해 주로 형성되기 때문이다. 해저는 광물을 포함하기에는 지질학적으로 너무 최근에 형성되었으며, 대륙 근처의 해저만이 광물 자원의 유용한 원천이 될 수 있다. 해양 자체에는 금속 이온이 포함되어 있지만, 바닷물에서 희귀 금속을 추출하는 것은 에너지 측면에서 그 과정을 매우 비싸게 만드는 그것의 미미한 농도 때문에 불가능하다. 게다가, 용해된 양도 그리 많지 않다. 예를 들어, 해양 속에 있는 구리 농도를 고려할 때, 용해된 총량은 현재 광산 생산의 10년 치에 해당한다고 우리는 계산할 수 있다. 어떤 이들은 광물 자원의 원천으로서 우주를 제안하지만, 지구를 떠나는 에너지 비용이 주요 장벽이다.

게다가, 예를 들어 달과 소행성과 같은 태양계 대부분의 천체는 지구화학적으로 '죽은' 상태이며 광물을 포함하고 있지 않다.

문제 해설

(A) The planet's crust(지구의 지각)는 explore(탐사하다)와 의미상 수동의 관계에 있으므로 has been과 함께 현재완료 수동태 문장을 이루는 explored가 어법상 적절하다.
(B) 명사구 their minute concentrations의 앞이므로 전치사구 because of가 어법상 적절하다. that make the process highly expensive in energy terms는 their minute concentrations를 수식하는 관계절이다.
(C) calculate의 목적어 역할을 하는 명사절에서 뒤에 문장의 필수 요소를 다 갖춘 완전한 절(the total amount dissolved corresponds to 10 years of the present mine production)이 이어지고 있으므로 접속사 that이 어법상 적절하다.

구조 해설

■ The oceanic floor is geologically too recent for containing ores; [only the sea floor {near the continents}] could be a useful source of minerals.
[ ]는 주어 역할을 하는 명사구이며, { }는 the sea floor를 수식하는 전치사구이다.

어휘 및 어구

high grade 고품질의
mineral 광물, 광석
crust 지각
explore 탐사하다, 탐험하다
dig 파다
oceanic floor 해저, 해양의 바닥
geologically 지질학적으로
contain 포함하다
metal ion 금속 이온
extract 추출하다
minute 미미한, 극미한
concentration 농도
dissolve 용해시키다, 용해되다
copper 구리
barrier 장벽
solar system 태양계
geochemically 지구화학적으로

Word Search

| 정답 | 1. floor 2. contain 3. dissolve
1. 해저는 아직 완전히 탐험되지 않은 영역이다.
2. 제출하기 전에 봉투에 필요한 모든 서류가 포함되어야 한다.
3. 설탕은 뜨거운 물에 빠르게 용해되어 달콤한 용액을 만든다.

Exercise 5

정답 ④

소재 생물학적 문제에 대한 기술적 해결책 적용

해석 복잡한 생물학적 시스템 내에 실제로 자리 잡고 있는 문제에 대해 단순한 기술적 해결책을 적용하면, 우리는 그 시스템이 그 기술에 어떻게 적응할지 쉽게 예측할 수 없다. 사실, 그 '문제'가 어떻게 생태계와 관련이 있거나 생태계에 의해 지원되는지를 이해하지 못한다면, 우리는 그것의 생물학적 또는 생태학적 기반을 이해하지 못한 채 인간의 관점에서 그 문제를 규정하고 있는 것이다. 예를 들어, 우리는 1940년대 후반부터 농업에서 원치 않는 해충을 통제하기 위해 합성 살충제를 사용해 왔다. 일반적이고 광범위하게 사용된 모든 살충제는 내성 해충 종의 진화를 초래했으며, 이는 그 대책으로서 대체 살충제를 끊임없이 찾는 결과를 낳았다. (새로운 살충제의 개발이 진행됨에 따라 작물 수확량이 크게 증가하였고, 이는 세계적인 식량 공급 안정을 개선하였다.) 생물학적 문제를 기술적 해결책으로 통제하려는 이 반복적이고 지속적인 싸움은 끝이 없으며 근본적인 문제를 인식하지 않으려는 인간 측의 태도에서 비롯된 것이다.

문제 해설

농업에서 합성 살충제를 사용한 것이 내성 해충의 진화를 초래해 대체 살충제를 끊임없이 찾는 결과를 가져온 것처럼 생물학적 문제에 대한 근본적인 인식 없이 단순한 기술적 해결책을 적용하려는 시도는 예측하기 어려운 결과를 야기할 수 있다는 내용의 글인데, ④는 살충제 저항성에 대한 대체물을 찾는 과정이 성공적으로 이어지면서 도시에서 소규모 유기농 농업이 사라졌다는 내용이다. 따라서 글의 전체 흐름과 관계 없는 문장은 ④이다.

구조 해설

- In fact, [if we don't understand {how the "problem" is linked to or supported by the ecosystem}], then we are defining the problem in human terms without understanding its biological or ecological foundation.
  [ ]는 조건을 나타내는 부사절이고, 그 안에서 how가 이끄는 { }는 understand의 목적어 역할을 한다.

어휘 및 어구

nest (~ 안에) 두다, 끼워 넣다
biological 생물학적인
link 관련짓다
ecological 생태학적인
foundation 기반, 토대
pest 해충
agriculture 농업
evolution 진화
resistant 내성이 있는, 저항하는

constant 끊임없는
countermeasure 대책
underlying 근본적인

Word Search

| 정답 | **1.** link **2.** foundation **3.** evolution
1. 새로운 화석의 발견은 현대의 새들을 그들의 조상들과 관련지을 수 있을 것이다.
2. 신뢰의 기반은 강한 관계를 구축하는 데 필수적이다.
3. 패션의 진화는 변화하는 문화적 가치관을 반영한다.

Exercise 6

정답 ④

소재 정신의학의 복잡성

해석 정신의학은 자주 의학의 '불쌍한 사촌'으로 여겨지는데, 이는 그럴 만한 이유가 있다. (C) 지난 세기 동안, 신체 메커니즘에 대한 우리의 이해는 서로 다른 많은 영역 전체의 과학적 이해를 통해 진보하였고, 심장학, 암 치료, 세균 및 감염성 질환 치료와 같은 다양한 분야에서 중요한 새로운 약물 치료로 이어졌다. (A) 그러나 인간의 뇌가 전체적으로 어떻게 작동하는지에 대한 우리의 이해는 여전히 전혀 명확하지 않으며, 결과적으로 정신 장애에 대한 약물 치료는 흔히 전혀 고무적이지 않았는데, 심지어 정신 건강에 정말로 긍정적인 영향을 미치는 것으로 보이는 그런 약물들의 작용 메커니즘조차도 여전히 모호하다. (B) 이러한 명확성의 결여는 뇌가 신체의 다른 어떤 기관보다 훨씬 더 복잡하기 때문이기도 하지만, 이것은 또한 인간의 의식적 인식이 생물학적 존재인 동시에 사회적 존재이기도 하며, 따라서 인간의 정신 장애가 주요한 사회적 입력 자극을 가지고 있다는 사실도 반영한다.

문제 해설

정신의학이 의학의 '불쌍한 사촌'으로 여겨지며, 이에는 그럴 만한 이유가 있다는 주어진 글 다음에, 신체 메커니즘에 대한 우리의 이해는 과학적 이해를 통해 진보했으며, 이는 다양한 질환에 새로운 약물 치료로 이어졌다는 내용의 (C)가 오고, 반면에 인간의 뇌에 대한 이해는 여전히 명확하지 않으며, 정신 장애에 대한 약물 치료는 고무적이지 않았고, 약물의 작용 메커니즘도 모호하다는 내용의 (A)가 온 다음에, 이러한 명확성 부족의 원인을 언급한 (B)가 마지막에 오는 것이 자연스럽다. 따라서 주어진 글 다음에 이어질 글의 순서로 가장 적절한 것은 ④ '(C)−(A)−(B)'이다.

구조 해설

- This lack of clarity is due to [{the brain} being so much more complex than any other organ in the body], but it also reflects the fact [that human conscious awareness is as much a social as a biological entity, and therefore human mental disorders have a major social input].

첫 번째 [ ]는 due to의 목적어 역할을 하는 동명사구이고, 그 안에서 { }는 동명사구의 의미상의 주어이다. 두 번째 [ ]는 the fact와 동격 관계이다.

Word Search

| 정답 | 1. remain 2. organ 3. treatment
1. 광범위한 연구에도 불구하고 그 질병의 많은 측면이 여전히 제대로 이해되지 않고 있다.
2. 뇌는 인체에서 가장 복잡한 기관이다.
3. 질병을 조기에 치료하면 환자의 (치료) 결과를 크게 개선할 수 있다.

Exercise 7
정답 ④

소재 투쟁 혹은 도피 반응과 의사 결정

해석 우리의 뇌가 낯설거나 잠재적으로 위험한 것을 접하게 되면, 우리는 신속히 '투쟁 혹은 도피 반응'으로 전환하는데, 그 경우 우리의 모든 정신적 (때로는 신체적) 에너지를 인식된 위험을 다루는 데 쏟는다. 하버드 의학전문대학원의 연구자들은 이 반응에 대해 광범위한 연구를 수행하여 낯설거나 위협적으로 느껴지는 정보에 직면하게 되는 경우 우리의 뇌가 자율 신경계를 통해 신체의 나머지 부분에 위험 신호를 보내는 것을 발견했는데, '그것(자율 신경계)은 호흡, 혈압, 심장 박동 그리고 주요 혈관과 작은 기도의 확장 또는 수축과 같은 그런 불수의(不隨意)적인 신체 기능을 제어한다.' 이러한 신체 반응들은 불편하거나 심지어 고통스러울 수 있지만, 더 중요한 것은 그것들이 그러고 나서 결국 우리의 의사 결정을 이끌어 내는 감정을 유발한다는 것이다. 우리는 모두 공포 영화의 주인공이 극도의 위험에 직면하여 당황스럽게 끔찍한 결정을 내리는 것을 보는 동안, '나라면 저것보다는 훨씬 더 똑똑할

텐데'라고 머리를 흔들며 생각하는 것이 어떤 것인지 알고 있다. 그러나 아마도 우리가 더 똑똑하지는 않을 것이다. 투쟁 혹은 도피 모드에서, 우세해지는 우리의 뇌 부분은 공포 영화 수준으로 똑똑할 뿐인데, 이는 전혀 똑똑한 것이 아니다.

문제 해설 ④ 지각동사(watching)의 목적어(the main character in a horror movie) 다음에 목적격 보어 자리에는 원형부정사나 현재분사가 와야 한다. 따라서 makes를 make나 making으로 고쳐야 한다.
① 전치사 to 다음에 '다루다'를 뜻하는 동사 address의 동명사 형태가 쓰여 어법상 적절하다.
② 의미상의 주어인 our brains와 동사 face가 의미상 수동의 관계이므로 과거분사 faced는 어법상 적절하다.
③ uncomfortable과 or로 대등하게 연결되어 These physical responses can be에 이어져 주격 보어 역할을 하고 있으므로, 형용사 painful은 어법상 적절하다.
⑤ 앞 절의 only horror-movie-level smart를 부가적으로 설명하는 관계절을 이끌면서 관계절에서 주어 역할을 하므로, 관계사 which를 쓰는 것은 어법상 적절하다.

구조 해설
■ In fight-or-flight mode, the part of our brain [that takes over] is only horror-movie-level smart, [which isn't very smart at all].

첫 번째 [ ]는 the part of our brain을 수식하는 관계절이며, 두 번째 [ ]는 앞의 내용을 부가적으로 설명하는 관계절이다.

Word Search

| 정답 | 1. threat 2. trigger 3. extreme
1. 증가하는 사이버 보안 공격의 위협은 회사들이 더 나은 보안 조치에 훨

씬 더 많은 돈을 쓰게 만들었다.

2. 밝게 깜빡이는 빛에 노출되면 일부 사람들에게 심한 두통과 불편함을 유발할 수 있다.

3. 그는 극도의 피로 때문에 더 이상 걸을 수 없었다.

Exercise 8

정답 ⑤

소재 인공 지능을 통한 소비자 행동 이해와 그 효과

해석 오늘날 디지털 방식의 역동적인 환경에서 소비자 행동과 경험은 산업 전반에 걸쳐 기업의 성공을 결정하는 데 매우 중요한 역할을 한다. 인공 지능(AI)의 출현은 기업이 소비자의 요구를 이해하고 분석하며 대응하는 방식에 혁신을 일으켰다. 고급 알고리즘, 기계 학습, 그리고 데이터 분석의 융합을 통해 AI는 소비자 행동을 더 깊이 철저하게 조사할 수 있는 엄청난 기회를 제공하여, 개인화된 경험을 촉진하고 장기적인 관계를 조성하는 매우 유용한 통찰력을 제공해 준다. 소비자 행동을 이해하는 것은 모든 성공적인 기업 전략의 핵심을 이루고 있다. AI는 방대한 양의 데이터를 실시간으로 처리할 수 있는 능력을 갖추고 있어 소비자의 선호도, 습관, 그리고 경향에 대한 포괄적인 이해를 가능하게 한다. 과거의 행동과 상호 작용을 분석함으로써 AI는 전통적인 분석을 벗어날 수도 있는 패턴을 알아내어 기업이 미래의 요구를 예측하고 상품을 그에 따라서 맞추어 만들 수 있게 한다. 구매 패턴 예측이든, (온라인) 가게 안의 물건들을 둘러보는 행동의 이해든, 혹은 소셜 미디어 상호 작용을 통한 감정의 이해든 간에, AI 기반의 통찰력은 기업이 최대 효과를 내기 위해 자신의 전략을 세밀하게 조정할 수 있게 한다.

문제 해설
오늘날 AI는 소비자 행동을 분석하고 대응하는 방식을 혁신하여, 개인화된 경험과 장기적 관계를 조성하는 데 유용한 통찰력을 제공하며, 실시간 데이터 처리로 소비자 선호도와 경향을 이해하고, 구매 패턴 예측과 소셜 미디어 감정 분석을 통해 사업 전략을 최적화한다는 내용의 글이다. 따라서 글의 주제로 가장 적절한 것은 ⑤ '소비자 이해도를 높이고 수익성을 추진하는 데 있어 AI의 매우 중요한 역할'이다.
① AI 주도의 분석을 적용할 때의 난제와 제약
② 소셜 미디어 사용자의 진화하는 수요를 충족시키는 방법
③ 기업에서 AI의 사용을 둘러싼 윤리적 규제의 필요성
④ 효과적인 마케팅 전략을 개발하는 데 있어 과대평가된 데이터의 중요성

구조 해설
■ **By analyzing** past behaviors and interactions, AI identifies patterns [that might escape traditional analysis], [enabling businesses to anticipate future needs and tailor offerings accordingly].
「by＋-ing」는 '~함으로써'의 의미이다. 첫 번째 [ ]는 patterns를 수식하는 관계절이며, 두 번째 [ ]는 분사구문으로 앞선 절에 대한 부가적인 내용을 나타낸다.

어휘 및 어구
advent 출현
artificial intelligence 인공 지능(약어 AI, A.I.)
revolutionize 혁신을 일으키다
profound 엄청난, 심오한
invaluable 매우 유용한
insight 통찰력
facilitate 촉진하다
foster 조성하다, 촉진하다
vast 방대한
comprehensive 포괄적인
anticipate 예측하다
tailor (용도·목적에) 맞추어 만들다
browse 가게 안의 물건들을 둘러보다
empower (~을) 할 수 있게 하다, 허용하다
fine-tune 세밀하게 조정하다, 미세 조정하다

Word Search

| 정답 | **1.** artificial **2.** revolutionized **3.** profound

1. 그 과학자는 심부전 환자들을 돕기 위해 인공 심장을 개발했다.

2. 인터넷은 세계적인 의사소통의 혁신을 일으켜 거리를 단축시키고 문화를 연결했다.

3. 그의 연설은 청중에게 엄청난 영향을 미쳤다.

Exercise 9

정답 ④

소재 철도의 발달과 스포츠 관광의 발전

해석 19세기 이전의 스포츠 관광 발전에 대한 주요한 제약 중 하나는 적절한 교통수단의 부족이었다. 15세기부터 계속하여, 더 편안한 마차와, 그리고 18세기에는, 유럽 전역은 아니더라도 적어도 영국에서만큼은 대폭적으로 개선된 도로를 포함한 점진적인 개선이 이루어졌다. 그러나 교통은 주로 느리고 많은 비용이 들었다. 예를 들어, 1680년에 107마일의 거리인 런던에서 바스까지의 여행 시간은 약 60시간이었다. 크게 개선된 도로로 1800년까지 이 시간은 10시간으로 줄었지만, 그 시간과 비용은 여전히 사회의 부유한 사람들만이 상당한 거리를 여행할 수 있음을 의미했

틀리기 쉬운 유형편

다. 19세기에 철도의 발달이 있고 나서야 비로소 비교적 저렴하고 효율적인 형태의 교통수단이 일반 대중에게 제공되어 그때까지 존재했던 소규모의 상류층 전용의 활동을 넘어 스포츠 관광이 성장할 수 있었다. Wray Vamplew가 지적한 것처럼 철도는 '관중의 이동 가능 거리를 넓히고 참가자들이 전국적으로 경쟁할 수 있게 함으로써 영국의 스포츠에 대변혁을 일으켰다.'

(문제 해설)

19세기 이전에는 적절한 교통수단 부족이 스포츠 관광 발전의 주요 제약이었으나, 철도의 발달로 저렴하고 효율적인 교통수단이 제공되면서 스포츠 관광이 상류층을 넘어 발전할 수 있었고, 철도는 관중의 이동 가능 거리를 넓히고 참가자들의 전국적인 경쟁을 가능하게 하여 영국의 스포츠에 대변혁을 일으켰다는 내용의 글이다. 따라서 글의 제목으로 가장 적절한 것은 ④ '교통의 발전이 어떻게 영국에서 스포츠 관광에 변화를 가져왔는가'이다.
① 초기 스포츠 관광 관행의 지역적 차이
② 교통수단의 획기적인 발전: 여행 비용 절감하기
③ 지역 라이벌이 스포츠 팬들의 행동에 미치는 영향
⑤ 비싼 상류층 여행 옵션에서 더 감당할 수 있는 옵션으로의 전환

(구조 해설)

▪ [One of the key constraints on the development of sports tourism prior to the nineteenth century] **was** the lack of suitable transport.
[ ]는 주어부이며, 그 핵심어가 One이므로 술어동사로 단수형 was가 사용되었다.

(어휘 및 어구)

constraint 제약
suitable 적절한
onwards (특정 시간부터) 계속하여, 줄곧
costly 많은 비용이 드는
vastly 크게
substantial 상당한
efficient 효율적인
afford 제공하다
exclusive 전용의, 독점적인
catchment area 이동[통학] 가능 거리

Word Search

| 정답 | 1. constraint 2. substantial 3. efficient
1. 재정적 제약으로 인해 그 비영리 단체는 여러 계획된 활동을 축소해야 했다.
2. 그녀는 신경 과학에서의 선구적인 연구로 상당한 인정을 받았다.
3. 그 새로운 일정 관리 시스템은 더 효율적이어서 회의를 조정하는 데 걸리는 시간을 줄여준다.

Exercise 10 ────────── 정답 ③

(소재) 불임 일개미가 생겨난 이유

(해석) 케임브리지 대학교의 생물학자 William Hamilton은 개미의 사회적 행동을 연구하면서, 개미가 왜 자주 서식처 동료들을 위해 자신을 희생하는지 뿐만 아니라, 암컷 일개미가 번식하지 않고 오히려 이 일을 여왕개미에게 맡기는 것을 고려할 때, 불임 일개미가 대체 어떻게 생겨날 수 있었는지도 궁금해하였다. Hamilton은 특정 번식 방법으로 인해 이러한 불임 일개미가 자신의 유전자의 75퍼센트를 자매들과 공유한다는 사실을 깨달았을 때 이 문제를 해결했다. 그런 다음 Hamilton은 자연 선택의 '유전자적 관점'을 고려하기 시작했고, 그러한 여왕개미가 아닌 일개미는 직접 번식하는 대신 자신의 더 어린 자매들을 키우는 것을 도움으로써 실제로 더 많은 비율의 자신의 유전자를 간접적으로 전달할 수 있다는 것을 즉시 깨달았다. 자매 한 마리를 키울 때마다, 그것은 정상적인 유성생식(有性生殖)으로 전달될 50퍼센트 대신 자신의 유전자의 75퍼센트를 전달하고 있는 것이다. Hamilton이 보기에, 사실상 개미 군체는 일부의 이타적으로 보이는 행동을 통해 전체의 유전적 이익이 충족되는 확대 가족의 한 형태로 인식되어야 한다.

(문제 해설)

③ 전치사 by와 이어져 동명사(구)를 이끌어야 하므로 helped를 동명사 helping으로 고쳐야 한다.
① 동사 sacrificed의 주어인 they(= ants)와 같으므로, 목적어로 재귀대명사 themselves가 쓰인 것은 어법상 적절하다.
② realised의 목적어 역할을 하면서 뒤에 필수 구성 요소를 모두 갖춘 명사절을 이끄는 접속사 that은 어법상 적절하다.
④ the 50 per cent가 pass on의 주체가 아닌 대상이므로 과거분사 passed가 쓰여 앞의 be와 함께 수동태를 이루는 것은 어법상 적절하다.
⑤ a form of extended family를 수식하는 관계절을 이끌어야 하는데 where 다음에 필수 구성 요소를 모두 갖춘 완전한 절이 이어지고 있으므로 관계부사 where는 어법상 적절하다.

(구조 해설)

▪ Cambridge biologist William Hamilton studied the social behaviour of ants and wondered to himself, [**not only** {why they often sacrificed themselves for their nestmates}, **but also** {how it was that sterile worker ants might have arisen, ⟨**given** these females don't breed but rather leave **this** to the queen⟩}].
[ ]는 wondered의 목적어 역할을 하며, 그 안에 두 개의 { }가 「not only ~ but also ...」 구문을 이룬다. ⟨ ⟩는 given(~을 고려할 때)이 이끄는 절로서, 이 경우 보통 given의 뒤에 that이 함께 쓰이지만 비격식 표현에서는 that이 생략되기도 한다.

this는 breeding을 대신한다.

(어휘 및 어구)

sacrifice 희생하다

nestmate 서식처 동료

breed 번식하다

natural selection 자연 선택

pass on ~을 전달하다

proportion 비율

colony 군체

perceive 인식하다

extended family 확대 가족

serve 충족시키다

Word Search

| 정답 | 1. sacrifice 2. breed 3. proportion

1. 그녀는 자신의 가족을 위해 많은 것을 희생해야 했다.

2. 많은 동물들이 일 년 중 특정 시기에만 번식한다.

3. 소득의 많은 비율이 도시 지역에서는 주택에 지출된다.

Exercise 11 ──────── 정답 ④

소재 자율 주행차로의 전환

해석 자율 주행차로의 전환에 대한 가장 흔한 명분은 어쩌면 안전에서 나온 주장이라고 할 수 있는데, 즉 운전 자동화 시스템의 시행은 자동차 사고로 인한 사망 또는 부상의 위험을 줄일 것이다. 그 기본 전제는 충분히 합리적인 것 같은데, 운전석에 있는 인간은 신뢰할 수 없으며, 너무 자주 산만해지거나, 피곤하거나, 혹은 (그 외의) 다른 식으로 영향을 받을 수 있다. 인간 운전자의 이러한 불완전성이 다른 상황에서는 공중 보건 위기로 간주될 수 있는 부상 및 사망률의 원인이 된다는 것은 의심의 여지가 없다. 이에 반해, 운전 자동화 시스템은 (일반적으로) 더 신뢰할 수 있다고 여겨지므로 자율 주행차와 이를 지원하는 기본적 시설의 전반적인 보급에 따라 다른 차량과 협력하여 사고 및 그 밖의 문제를 피할 수 있을 것이다. (안전에 대한 전망에도 불구하고, 자율 주행차 사용에 대한 윤리적 우려도 발생하며, 결과적으로 그것의 사용에 대한 반대가 증가하고 있다.) 이러한 전제 하에서, 그리고 그 새로운 기술이 광고된 대로 작동한다고 가정하면, 불안정하고 신뢰할 수 없는 인간 통제 시스템을 명백히 더 나은 것으로 대체하는 것이 오로지 타당하다.

문제 해설

자율 주행차는 사고로 인한 사망 및 부상을 줄일 수 있으며, 신뢰할 수 없는 인간 운전자와 다르게, 새로운 기술이 광고된 대로 작

동한다면, 인간 통제 시스템을 더 나은 운전 자동화 시스템으로 대체하는 것이 당연하다는 내용의 글이다. 자율 주행차 사용에 대한 반대가 증가하고 있다는 내용은 글의 흐름과 반대되는 내용이므로, 글의 전체 흐름과 관계가 없는 문장은 ④이다.

(구조 해설)

■ This fallibility of human drivers no doubt contributes to a rate of injury and fatality [that, {in other contexts}, would be considered a public health crisis]. [ ]는 a rate of injury and fatality를 수식하는 관계절로, 그 안의 { }는 삽입구이다.

(어휘 및 어구)

justification 명분, 정당화

implementation 시행, 이행

reasonable 합리적인, 합당한

fatality 사망

prevalence 보급

ethical 윤리적인

protest 반대, 항의

assume 가정하다

demonstrably 명백히

Word Search

| 정답 | 1. justification 2. implementation 3. fatality

1. 새로운 정책에 대한 그의 명분은 상세한 데이터 분석과 전문가의 권고를 기반으로 했다.

2. 새로운 교육과정의 시행은 여러 당사자가 참여한 복잡한 과정이었다.

3. 사망률의 증가는 여러 요인에 기인한다.

Exercise 12 ──────── 정답 ②

소재 학습된 사회적 경험을 통해 형성되는 상식

해석 대부분의 학생들은 수많은 상식의 예시들을 제공할 수 있는데, 예를 들어 전기 콘센트에 포크를 꽂지 않기, 얼음으로 덮인 기둥에 혀를 대지 않거나 끓는 물에 손을 넣지 않기, 태양을 직접 바라보는 것을 피하기, 그리고 초대 없이 백악관에서 열리는 국빈 초대 만찬회에 참석을 시도하지 않기와 같은 것이다. (B) 그러나 상식에 대한 이 예시들 중 어느 것도 우리가 태어날 때부터 가지고 있던 지식에 해당하지 않는데, 대신, 우리는 이 모든 것을 배워왔다. 사회학적 관점은 지식이 시행착오, 경험, 그리고 사회적 환경에서 다른 사람들의 영향을 통해 얻어진다는 것을 강조한다. (A) 경험은 우리가 대부분의 사회적 상황에서 관례적으로 행동하도록 가르친다. 얼마 지나지 않아 그러한 기대는 상식으로 여겨지

게 된다. (C) 그러나 우리 각자가 고유한 사회적 경험을 가지고 있으므로 우리는 상식을 다르게 바라보게 되는데, 즉 우리는 자신의 고유한 관점에서 그것을 본다. 그러므로 누군가가 어떤 사람들에 의해 상식적 지식이라고 불리는 특정 행동을 경험하게 되지 않으면, 그 사람은 명백한, 즉 일상적인 방식으로 행동할 수 없다.

[문제 해설]

학생들이 일반적으로 '상식'이라고 생각하는 여러 가지 예시를 다룬 주어진 글 다음에는, 이러한 상식이 타고난 것이 아니라 경험과 사회적 환경의 영향을 통해 학습된 것이라고 설명하는 내용의 (B)가 이어지고, 다음으로 경험은 우리가 대부분의 사회적 상황에서 관례적으로 행동하도록 가르치며, 그러한 기대가 상식으로 여겨지게 된다는 내용의 (A)가 이어진 후, 마지막으로 각 개인은 고유한 사회적 경험을 가지고 있어 상식을 다르게 인식하게 된다는 점을 설명하는 (C)가 이어지는 것이 자연스럽다. 따라서 주어진 글 다음에 이어질 글의 순서로 가장 적절한 것은 ② '(B)−(A)−(C)'이다.

[구조 해설]

- And yet, [**none** of these examples of common sense] represents knowledge [that we were born with]; instead, we have learned all these things.

  첫 번째 [ ]는 문장의 주어에 해당하고, 주어의 핵심어인 none은 부정대명사로 '어느 것도 ~ 않다'라는 의미이다. 두 번째 [ ]는 knowledge를 수식하는 관계절이다.

[어휘 및 어구]

numerous 수많은
common sense 상식
pole 기둥, 막대기
state dinner 국빈 초대 만찬회
routinely 관례적으로
expectation 기대
sociological 사회학적인
gain 얻다
expose 경험하게 하다, 접하게 하다
obvious 명백한

Word Search

| 정답 | 1. expectation 2. expose 3. obvious
1. 팀이 시즌 내내 좋은 성과를 거두고 있어서 팬들 사이에서 기대가 높았다.
2. 그 프로그램은 학생들이 다양한 과학 분야를 경험하게 하도록 설계되었다.
3. 연사의 잦은 멈춤과 긴장한 태도로 미루어 보아, 방 안의 모든 사람에게 그가 준비되지 않았다는 것이 명백했다.

| 1 ④ | 2 ④ | 3 ⑤ | 4 ⑤ | 5 ③ | 6 ⑤ |
| 7 ⑤ | 8 ③ | 9 ⑤ | 10 ③ | 11 ③ | 12 ② |

Exercise 1　　　　　　　정답 ④

[소재] 협력적 행위로서의 감정의 소통

[해석] 내가 보기에 감정의 범주는 집단적 지향성을 통해 현실화된다. 다른 사람에게 화가 났다는 사실을 전달하려면 두 사람 모두 '분노'에 대한 공통된 이해가 필요하다. 특정 상황에서 얼굴 표정과 심혈관 변화의 특정한 조합이 분노라는 데 사람들이 동의한다면, 그것은 그러하다. 이러한 합의에 대해 명시적으로 알 필요는 없다. 특정 사례가 분노인지 아닌지에 대해 동의할 필요도 없다. 분노가 특정 기능을 가지고 존재한다는 것에 원칙적으로 동의하기만 하면 된다. 그 시점에서 사람들은 그 개념에 대한 정보를 그들 사이에서 매우 효율적으로 전달할 수 있어서 분노는 타고난 것처럼 보인다. 만약 여러분과 내가 특정한 상황에서 찌푸린 이마가 분노를 나타낸다는 데 동의하고 내가 이마를 찌푸린다면, 나는 여러분과 효율적으로 정보를 공유하고 있는 것이다. 공기 중의 진동이 소리를 전달하지 않는 것처럼 내 움직임 자체도 여러분에게 분노를 전달하지 않는다. 우리가 개념을 공유한다는 사실 때문에 내 움직임은 여러분의 뇌에 예측을 일으키는데… 이것은 인간만이 가진 독특한 유형의 마법이다. 그것이 협력적 행위로서의 범주화에 해당한다.

[문제 해설]

④ 동사인 transmit을 수식해 주는 역할을 하므로 efficient를 부사인 efficiently로 고쳐야 한다.

① Emotion categories는 make의 대상이므로 수동태인 are made는 어법상 적절하다.

② that절 내에서 a particular combination of facial actions and cardiovascular changes가 주어이고, 주어의 핵심어구인 a particular combination이 단수형이므로, 단수형 동사 is는 어법상 적절하다.

③ agree 다음에 목적어 역할을 하는 명사절이 필요하므로, '~인지 아닌지'라는 뜻을 가진 「whether ~ or not」이 이끄는 명사절이 오는 것은 어법상 적절하다.

⑤ the fact와 동격 관계를 이루는 명사절을 유도하고 있으므로, 접속사 that은 어법상 적절하다.

[구조 해설]

- [If people agree {that ⟨a particular combination of facial actions and cardiovascular changes⟩ is anger in a given context}], then it is so.

[ ]는 조건의 부사절이고, { }는 agree의 목적어 역할을 하는 명사절이다. ⟨ ⟩는 that절 내에서 주어의 역할을 한다.

어휘 및 어구

category 범주
collective intentionality 집단적 지향성
communicate 전달하다
context 상황
explicitly 명시적으로
be aware of ~을 알다, 깨닫다
agreement 합의, 동의
in principle 원칙적으로
function 기능
transmit 전달하다
concept 개념
inborn 타고난
brow 이마, 눈썹
indicate 나타내다
efficiently 효율적으로
by virtue of ~ 때문에, ~의 덕택으로
initiate 일으키다, 시작하게 하다
prediction 예측
brand 유형, 종류, 상표
cooperative 협력적인
act 행위

Word Search

| 정답 | **1.** explicitly **2.** transmit **3.** inborn
1. 지시 사항에는 제출 기한이 금요일이라고 명시적으로 진술되어 있었다.
2. 사람들이 소셜 미디어를 통해 정보를 전달하기 시작하면서 그 뉴스가 빠르게 퍼졌다.
3. 어떤 사람들은 음악적 재능이 타고난 것이라고 믿는 반면, 다른 사람들은 연습을 통해 개발될 수 있다고 생각한다.

Tips

collective intentionality(집단적 지향성): 철학적 개념으로, 사회 구성원이 사실, 사건, 목표, 가치, 물건 등에 대해 공유하고 있는 생각을 의미하며, 공유된 믿음이나 신념, 수용, 감정의 방식으로 나타난다.

Exercise 2 ── 정답 ④

소재 온라인에서의 의견 형성

해석 온라인 상호 작용의 역학 관계와 집단 의사 결정 과정 이면의 심리를 더 잘 이해하기 위해서는, 일반적인 집단 의사 결정에 직접적으로 관련된 한 가지 질문, 즉 '사람들이 웹사이트의 댓글을 어떻게 판단하는가'라는 질문을 생각해 보라. Hebrew University of Jerusalem의 교수 Lev Muchnik과 그의 동료들은 다양한 이야기를 보여 주고 사람들에게 댓글을 게시할 수 있도록 허용하는 웹사이트에서 실험을 수행했는데, 그 댓글은 그다음에 좋아요나 싫어요 투표를 얻을 수 있다. 연구자들은 자동적으로 그리고 인위적으로 이야기에 대한 특정 댓글에 즉시 좋아요 투표를 부여했는데, 이는 어떤 댓글이 가장 먼저 받게 되는 투표였다. 여러분은 수백 또는 수천 번의 방문자와 평가가 이루어진 후라면 댓글에 대한 단 하나의 첫 투표가 중요할 리가 없다고 생각할지도 모른다. 그것은 합리적인 생각이지만, 잘못된 것이다. 첫 번째 좋아요 투표를 본 후 (그리고 그것이 전적으로 인위적이라는 것을 기억하라) 다음 독자는 좋아요 투표를 할 가능성이 32% 더 커졌다.

문제 해설

댓글에 대한 첫 번째 반응이 다음 사람의 반응에 많은 영향을 미친다는 내용이므로, 글의 주제로 가장 적절한 것은 ④ '최초의 투표가 온라인 의견 형성에 미치는 영향'이다.
① 평가와 댓글이 자존감에 미치는 영향
② 부정적인 댓글 이면의 심리적 이유
③ 웹사이트상의 개인 이야기에 대한 일반적인 반응
⑤ 소셜 미디어에서 의견을 공유하는 것의 단점

구조 해설

- [After seeing an initial upvote] (and recall that it was entirely artificial), the next viewer became 32% more likely to give an upvote.
 [ ]는 접속사 After 뒤에 분사구 -ing가 사용된 분사구문이다.

어휘 및 어구

dynamic 역학 관계, 역학
interaction 상호 작용
psychology 심리
process 과정
comment 댓글
colleague 동료
carry out ~을 수행하다
diverse 다양한
post 게시하다
vote up[down] 좋아요[싫어요] 투표를 하다
automatically 자동적으로
artificially 인위적으로
rating 평가
sensible 합리적인

Word Search

| 정답 | 1. comments 2. diverse 3. sensible

1. 온라인에서 무례하거나 공격적인 댓글을 다는 것은 부적절하다.

2. 그 대학은 전 세계에서 온 다양한 학생 집단[재학생]으로 알려져 있다.

3. 그녀는 불필요한 갈등을 일으키지 않고 문제를 해결하기 위한 합리적인 제안을 했다.

Exercise 3 ────── 정답 ⑤

소재 이야기하기를 통한 자신에 대한 심도 있는 이해

해석 우리는 이야기를 통해 우리의 세상을 설명한다. 우리의 이야기 속에서 우리는 이야기꾼이자 주인공이다. 우리의 인생 이야기를 할 수 있다는 것은 우리가 그것을 이해하는 데 도움이 된다. 어쩌면 우리는 조용히 숙고하면서 스스로에게 그것을 말할 수 있다. 어쩌면 우리는 그것을 적어서, 그 당시에는 깨닫지 못했던 것을 다시 읽으면서 깨달을 수 있다. 그러나 우리 대다수에게는 친구와 대화를 나누거나 절친한 친구와 함께 곰곰이 생각하는 것이 자신의 이야기를 하는 방식이다. 그리고 우리의 이야기를 하면서 우리는 그것을 새로이 듣는다. 이야기를 하는 것은 우리가 세부 사항을 알아차리고 해석하거나, 더 큰 그림을 인식하거나, 혹은 이전에 간과하거나 부정했던 측면을 식별하는 데 도움이 된다. 우리의 이야기에 온전히 집중해 줄, 그 이야기에 완전히 빠져들 준비가 되어 있는 청자를 찾는 것은 고귀한 희망과 비참한 실패를 모두 안고 있는 우리 자신을 만나고, 우리 자신과 주변 세계를 더 진실하고 더 도움이 되는 방식으로 이해할 수 있는 기회이다.

문제 해설

다른 사람에게 자신의 이야기를 하면서 미처 알지 못했던 것을 깨닫게 되고, 자신의 진정한 모습을 마주하게 되며, 스스로를 깊이 이해할 수 있게 된다는 내용이므로, 글의 제목으로 가장 적절한 것은 ⑤ '자신의 이야기를 나누면 자신에 대한 더 깊은 이해의 문이 열린다'이다.

① 백일몽: 자기 발견의 여정

② 다른 사람의 관점을 통해 진실 발견하기

③ 개인적인 이야기를 나누며 더 강한 관계 형성하기

④ 스토리텔링의 기술: 성공을 달성하기 위한 필수 기술

구조 해설

▪ [Finding a listener {who will give our story their full attention}, {who is prepared to become completely caught up in the tale}], is an opportunity [to meet ourselves, {complete with our noble hopes and miserable failings}], and [to understand ourselves and

the world around us in a truer, more helpful way].

첫 번째 [ ]는 주어 역할을 하는 동명사구이며, 그 안에 있는 두 개의 { }는 a listener를 수식하는 관계절이다. 두 번째와 세 번째 [ ]는 an opportunity의 구체적 내용을 설명하고 있다. 두 번째 [ ] 안의 { }는 ourselves를 부가적으로 설명하는 어구이다.

어휘 및 어구

anew 새로이
interpret 해석하다
discern 식별하다
overlook 간과하다
deny 부정하다, 부인하다
tale 이야기
noble 고귀한
miserable 비참한

Word Search

| 정답 | 1. interpret 2. overlook 3. miserable

1. 그 회사는 계약서의 법적 용어를 해석하는 데 도움을 줄 전문가를 고용했다.

2. 친절이 담긴 작은 표현의 중요성을 간과하기 쉽다.

3. 독감으로 집에 갇히자 그는 비참함을 느꼈다.

Exercise 4 ────── 정답 ⑤

소재 상품 배치가 선택에 미치는 영향

해석 한 연구에서, 가로 진열에서 중앙의 왼편에 놓인 제품이 선택되는 경우는 24%에 불과했다. 왜일까? *Retailer's Edge*의 저자 Bruce D. Sanders는 뇌의 좌반구, 즉 수 계산을 하는 부위가 활성화될 때 눈은 오른쪽으로 움직인다는 것에 대해 언급한다. 따라서 우리가 좋은 값으로 사게 되었다고 생각할 때 행복감을 느끼는 좌반구가 우세해지고, 눈은 오른쪽으로 이동한다. 그것이 중앙의 왼쪽에 있는 제품보다 중앙의 오른쪽에 있는 제품이 더 자주 구매되는 이유에 대한 설명이다. 이처럼 왼쪽보다 오른쪽을 선호하는 것은 의식적이거나 이성적인 의사 결정과는 관련이 없다. '동일한' 스타킹 네 켤레를 평가하도록 요청했을 때, 맨 오른쪽에 위치한 스타킹이 맨 왼쪽에 위치한 스타킹보다 4 대 1로 더 많이 선택되었다. 심지어 실험자들이 스타킹의 위치가 선택에 영향을 미쳤을 수도 있다고 넌지시 말했을 때에도 참가자들의 반응은 혼란스러워하는 것부터 무시하는 것까지 다양했다.

문제 해설

(A) when절의 주어의 핵심어구인 the left hemisphere가 단

수이므로 gets가 어법상 적절하다.

(B) 전치사 for의 목적어로「의문사＋주어＋동사」 구조의 명사절이 와야 하는데 '~인 이유'라는 의미의 의문사 why가 어법상 적절하다.

(C) 술어동사 was chosen이 존재하므로 주어인 the pair를 수식하는 분사구를 이끄는 과거분사 positioned가 어법상 적절하다.

구조 해설

- Bruce D. Sanders, [author of *Retailer's Edge*], notes [that our eyes move to the right {when the left hemisphere of the brain, ⟨where we do the math⟩, gets active}].

  첫 번째 [ ]는 Bruce D. Sanders와 동격 관계에 있는 명사구이다. 두 번째 [ ]는 notes의 목적어 역할을 하는 명사절이다. { }는 시간의 부사절이고, ⟨ ⟩는 the left hemisphere of the brain을 추가적으로 설명하는 관계절이다.

어휘 및 어구

horizontal 가로의, 수평의
display 진열
hemisphere 반구
get a good deal 좋은 값으로 사다
dominate 우세하다, 지배하다
shift 이동하다
purchase 구매하다, 구입하다
conscious 의식적인
rational 이성적인
rate 평가하다
identical 동일한
dismissive 무시하는

Word Search

| 정답 | **1.** hemisphere **2.** dominate **3.** conscious
1. 뇌에서 좌반구는 일반적으로 언어 및 분석적 사고와 관련이 있다.
2. 그 대기업이 계속해서 세계 시장을 지배하고 있다.
3. 그녀는 대중 연설 수업에 참석하며, 의사소통 기술을 향상시키기 위해 의식적으로 노력했다.

Exercise 5
정답 ③

소재 인터넷 시대 의료 소비자의 요구

해석 전 세계적으로 인터넷에 대한 접근성이 높아지면서 의료 소비자(환자)가 더 잘 교육되어 있다. 이들은 역사상 이전의 어느 때보다 전 세계에서 일어나는 일, 시기적절한 뉴스 및 의료 분야

의 최신 경향에 대한 정보를 더 잘 알 수 있다. 오늘날의 소비자들은 자신의 바쁜 일정에 맞추어 주고, 빠르고 쉽게 획득할 수 있는 유용한 정보를 제공하며, 그들이 의사 결정에 참여할 수 있게 하는 의료 시스템과 경험을 요구하고 있다. (최고의 교육, 최고의 헌신, 최고의 동기부여에도 의학적인 처치에 오류가 여전히 발생할 수 있지만, 의료 기기의 설계를 통해 오류의 수를 줄일 수 있다.) 이런 소비자들은 치료에 대한 더 나은 접근성, 그리고 의료 제공자들과의 더 나은 소통 및 새로운 수준의 참여 또한 요구하므로, 과거의 소비자들과는 다르다. 많은 경우 치료 계획이 공동으로 세워지고 창의적인 방법을 통해 실행된다.

문제 해설

인터넷의 발달로 의료 정보에 대한 접근성이 높아진 의료 소비자들은 과거와는 다른 더 나은 의료 시스템, 의사 결정 참여, 의료 제공자와의 소통을 원한다는 내용의 글인데, ③은 다양한 노력에도 의학적 처치에 오류가 발생할 수 있지만, 의료 기기의 설계를 통해 오류를 줄일 수 있다는 내용이므로, 글의 전체 흐름과 관계가 없는 문장은 ③이다.

구조 해설

- These consumers are different from **those** in the past [because they also demand {better access to care} and {better communication and a new level of participation with their healthcare providers}].

  those는 consumers를 대신하여 사용되었다. [ ]는 이유를 나타내는 부사절이고, 두 개의 { }는 and로 연결되어 demand에 이어진다.

어휘 및 어구

access 접근성
consumer 소비자
timely 시기적절한
previous 이전의
demand 요구하다
accommodate (요구 등에) 맞추다, 부응하다
obtain 획득하다, 얻다
involve 참여시키다
procedure (의학적인) 처치
treatment 치료
devise (계획 등을) 세우다
jointly 공동으로
execute 실행하다

Word Search

| 정답 | **1.** access **2.** obtain **3.** Consumers
1. 의료 접근성을 개선하는 것이 정부의 우선 과제이다.
2. 공식 웹사이트에서 그 행사에 대한 더 많은 정보를 얻을 수 있다.

3. 소비자들은 자신들의 구매가 환경에 미치는 영향에 대해 점점 더 인식하고 있다.

Exercise 6

정답 ⑤

소재 데이터에 기반한 이해

해석 보통 직감은 세상이 어떻게 돌아가는지에 대한 적당한 일반적 이해를 제공하지만, 정확하지 않은 경우가 많다. 그림을 선명하게 그리려면 데이터가 필요하다. (C) 예를 들어 날씨가 기분에 미치는 영향을 생각해 보라. 아마도 여러분은 사람들이 70도인 날보다 10도인 날에 더 우울한 기분을 느낄 가능성이 더 높다고 추측할 것이다. 실제로 이것은 맞다. (B) 하지만 여러분은 이 온도 차이가 얼마나 큰 영향을 미칠 수 있는지 짐작하지 못할 수도 있다. 나는 한 지역의 우울증에 대한 인터넷 검색과 경제 상황, 교육 수준 및 교회 출석률을 포함하여 다양한 요인 간의 상관관계를 조사했다. 겨울철 기후가 나머지 모든 요인을 압도했다. 겨울철에 호놀룰루의 기후와 같은 따뜻한 기후에서는 시카고의 기후와 같은 추운 기후보다 우울증에 대한 검색이 40퍼센트 더 적었다. (A) 이 효과는 얼마나 의미 있는가? 항우울제의 효능에 대해 낙관적인 해석을 하더라도, 가장 효과적인 약물이 우울증 발생률을 약 20퍼센트 정도만 줄인다는 것을 알게 될 것이다. 검색 결과 수치로 판단하면 시카고에서 호놀룰루로 이사하는 것이 겨울철 우울증에 대한 약물 치료보다 최소 두 배는 더 효과적일 것이다.

문제 해설
세상에 대한 정확한 이해를 위해서는 직감이 아닌 데이터가 필요하다는 주어진 글 다음에는, 날씨와 기분의 관계에 대한 직감에 의한 설명이 언급된 (C)가 이어지고, 이와 관련해 다양한 요인을 조사한 내용이 언급된 (B)가 이어져야 한다. 다음으로 데이터로 파악된 우울증에 대한 날씨 효과의 중요성을 약물 치료와 비교해 구체적으로 설명하는 (A)가 이어지는 것이 자연스럽다. 따라서 주어진 글 다음에 이어질 글의 순서로 가장 적절한 것은 ⑤ '(C)−(B)−(A)'이다.

구조 해설

- [While our gut may usually give us a good general sense of {how the world works}], it is frequently not precise.
  [ ]는 양보의 의미를 가진 부사절이고, { }는 of의 목적어 역할을 하는 명사절이다.

어휘 및 어구
precise 정확한
sharpen 선명하게 하다

significant 의미 있는
optimistic 낙관적인
effectiveness 효능
antidepressant 항우울제
incidence 발생률
depression 우울증
correlation 상관관계
factor 요인
conditions 상황
gloomy 우울한

Word Search

| 정답 | **1.** precise **2.** optimistic **3.** factor
1. 그 건축가는 건물 배치도의 정확한 청사진을 스케치했다.
2. 그녀는 매일을 낙관적인 마음가짐으로 살며, 도전을 기회로 여긴다.
3. 교육에 대한 접근성은 불평등과 빈곤을 줄이는 데 중요한 요인이다.

Exercise 7

정답 ⑤

소재 디자인에 대한 시험의 필요성

해석 디자이너들은 기계의 전원을 켜거나 끄는 방법이든 DNA 합성이나 심지어 세탁기를 사용하는 방법이든, 자신들의 작업을 설명하기 위해 도표, 그림 및 간단한 도해를 사용한다. 디자이너들은 자신의 노력이 대상으로 삼고 있는 사람들에게 그것을 시험할 필요성을 이해한다. 항상 시험은 문제점, 즉 사람들이 이해하지 못하거나 혼란스러워하거나, 혹은 더 나쁜 경우에는, 고려 중인 기기를 사용할 때 심각한 오류를 범할 수도 있어서, 어쩌면 그들이 한 작업 전체를 잃게 되고 기계나 의료 기기의 경우 어쩌면 사고나 심각한 부상으로 이어질 수 있는 부분들을 드러낸다. 이러한 초기 시험의 이유는 제품이 출시되기 전에 디자인 결함을 찾아서 그 디자인을 수정할 수 있도록 하기 위함이다. 디자인, 특히 '인간 중심 디자인'이라는 일반적인 명칭으로 통하는 절차의 집합에서, 디자인의 많은 수정판이 구성되고 시험된다. 각각의 새로운 수정판, 즉 디자인에 대한 각각의 수정은 시험 결과에 따라 이끌어진다.

문제 해설
⑤ multiple iterations of the design이 주어이고 주어의 핵심 어구인 multiple iterations가 복수형이므로, is는 복수형 동사 are로 고쳐야 한다.
① 뒤에 이어지는 절이 필요한 요소를 모두 갖춘 완전한 절이고, 문맥상 '~이든지 (간에)'의 의미를 나타내야 하므로 접속사 whether는 어법상 적절하다.

② they are intended for에서 뒤에 있던 전치사 for가 관계사 whom 앞으로 이동하여 선행사 the people을 수식하는 관계절을 이끌고 있으므로, for whom은 어법상 적절하다.

③ places를 수식하는 관계절을 이끌어야 하는데 뒤에 필요한 요소를 모두 갖춘 완전한 절이 이어지므로, 관계부사 where는 어법상 적절하다.

④ 술어동사 is 뒤에 주격 보어로 쓰인 to부정사구를 유도하는 to find가 쓰인 것은 어법상 적절하다.

구조 해설

▪ Designers use diagrams, pictures, and simple illustrations [to explain their work], whether it is [how to turn the power on or off in a machine] or [how to use a DNA synthesizer or even a washing machine].

첫 번째 [ ]는 목적의 의미를 나타내는 to부정사구이고, 두 번째와 세 번째 [ ]는 모두 「의문사+to부정사구」의 형태로서 or로 대등하게 연결되어 있다.

어휘 및 어구

diagram 도표, 도해
illustration 도해, 삽화
synthesizer 합성기
intend 대상으로 삼다, 의도하다
invariably 항상, 변함없이
confused 혼란스러워하는
device 기기, 장치
injury 부상
flaw 결함
release 출시하다
procedure 절차
modification 수정

Word Search

| 정답 | **1.** intend **2.** confused **3.** flaws
1. 나는 어떤 무례함도 의도하지 않았다.
2. 그는 외국 도로 표지판 때문에 잠시 혼란스러웠다.
3. 그 주장은 근본적인 결함으로 가득 차 있다.

Exercise 8   정답 ③

소재 과학의 철학으로부터 분리

해석 모든 탐구가 한때 과학의 그 위대한 어머니(*mater scientiarum*)인 철학의 일부였으며, 철학이 그것들 모두를 구분되지 않고 무정형의 방식으로 포용했다는 사실에 주목하는 것은 흥미롭다. 그러

나 서양 문명이 발전하면서, 다양한 과학이 분리되고 독립적인 경로를 추구하기 시작했다. 천문학과 물리학이 분리된 최초의 것에 속했고, 이후 화학, 생물학, 지질학이 그 뒤를 이었다. 19세기에는 두 개의 새로운 과학이 등장했는데, 심리학(인간 행동에 관한 과학)과 사회학(인간 사회에 관한 과학)이었다. 따라서 한때 자연 철학이었던 것은 물리 과학이 되었고, 정신 철학, 즉 마음의 철학이었던 것은 심리 과학이 되었으며, 한때 사회 철학, 즉 역사 철학이었던 것은 사회 과학이 되었다. 고대의 어머니, 즉 철학에 몇 종류의 중요한 탐구 분야, 특히 형이상학, 논리학, 윤리학 및 미학이 여전히 속해 있지만, 과학 자체는 더 이상 철학의 하위 분야로 연구되지 않는다.

문제 해설

과학을 포함하여 모든 탐구가 한때 철학의 일부였으나 서양 문명이 발전하면서 다양한 과학이 분리되고 독립적인 경로를 추구하기 시작했고 과학 자체는 더 이상 철학의 하위 분야로 연구되지 않는다는 내용의 글이다. 따라서 글의 주제로 가장 적절한 것은 ③ '철학적 뿌리로부터의 과학의 분리'이다.

① 사람들이 철학 공부를 회피하는 이유
② 과학적 그리고 철학적 사고의 균형을 맞추는 것의 이점
④ 과학적 사고가 철학을 이해하는 것에 미치는 영향
⑤ 철학의 새로운 분야를 형성하는 데 있어 과학의 기여

구조 해설

▪ [To the ancient mother, **philosophy**], still belong [several important kinds of enquiries — notably metaphysics, logic, ethics and aesthetics —] but the sciences themselves are no longer studied as subdivisions of philosophy.

전치사구인 첫 번째 [ ]가 문장의 맨 앞에 위치하면서 동사 부분인 still belong과 주어인 두 번째 [ ]가 도치되었다. 첫 번째 [ ] 안의 philosophy는 the ancient mother와 동격 관계이다.

어휘 및 어구

inquiry(= enquiry) 탐구, 조사
embrace 포용하다, 받아들이다
undifferentiated 구분되지 않는, 획일적인
civilisation 문명
separate 분리된
physics 물리학
chemistry 화학
biology 생물학
geology 지질학
notably 특히, 뚜렷하게
ethics 윤리학
aesthetics 미학

subdivision 하위 분야

Word Search

| 정답 | **1.** inquiry **2.** embraced **3.** undifferentiated
1. 과학적 탐구는 우주에 대한 진실을 밝히기 위해 노력한다.
2. 이 아이디어는 과학계에서 널리 받아들여져 왔다.
3. 그 치료는 모든 환자들에게 획일적인 방식으로 적용되었다.

Exercise 9

정답 ⑤

소재 인간의 끝없는 열망이 행복에 미치는 해로운 효과

해석 사람들의 열망이 돈이 많아질수록 높아진다고 가정해 보자. 여러분이 학생 기숙사에 살면 비록 작을지라도 여러분만의 아파트를 갖기를 열망할지도 모른다. 여러분이 작은 아파트에 살면 더 큰 곳에 살기를 열망할지도 모른다. 여러분이 더 큰 아파트에 살면 여러분 자신의 건물에서 살기를 열망할지도 모른다. 그리고 기타 등등의 상황이 펼쳐질지도 모르는데, 여러분이 더 많이 가질수록 더 많이 기대하고, 아마도 더 많이 누릴 자격이 있다고 생각할지도 모른다. 만약 그렇다면, 돈을 더 많이 버는 것의 유익한 행복 효과는 상승하는 열망이 행복에 미치는 해로운 효과에 의해 (적어도 부분적으로는) 상쇄될 것이다. 행동 경제학자들은 이 현상을 '열망의 쳇바퀴'라고 부른다. 기본적인 개념은 아주 오래됐다. 스토아 철학자 Seneca는 약 2,000년 전에 이 문제를 진단했다. 그는 '과도한 번영은 정말 실제로 인간에게 탐욕이 생기게 하며, 욕망은 일단 충족되더라도 사라질 만큼 결코 잘 통제되지 않는다.'라고 썼다.

문제 해설
사람들의 열망은 돈이 많아질수록 높아지고, 돈을 더 많이 버는 것이 주는 유익한 행복 효과는 상승하는 열망이 행복에 미치는 해로운 효과에 의해 상쇄된다는 내용이므로, 글의 제목으로 가장 적절한 것은 ⑤ '점점 더 원하는 것: 그것이 어떻게 여러분의 행복을 해칠 수 있는가'이다.
① 탐욕: 사회 내 내부 분열의 원인
② 부의 축적을 추구하는 데 있어 도덕적 딜레마
③ 왜 현대 사회에서 행복을 달성하는 것이 중요한가?
④ 재정적 자유를 위해 여러분의 초점을 어떻게에서 언제로 바꿔라

구조 해설
- Suppose [that people's aspirations rise **as** they get more money].
  명령문으로서 [ ]가 Suppose의 목적어 역할을 하는 명사절이다. as는 '~할수록'이라는 의미의 접속사이다.

어휘 및 어구
suppose 가정하다
aspiration 열망
dorm(= dormitory) 기숙사
property 건물, 부동산
deserve (~을 누릴) 자격이 있다
ancient 아주 오래된
diagnose 진단하다
excessive 과도한
prosperity 번영
greed 탐욕
desire 욕망
vanish 사라지다

Word Search

| 정답 | **1.** aspired **2.** property **3.** deserved
1. 그는 유명해지기를 열망한 적이 없다고 말했다.
2. 부동산 가격이 엄청나게 상승했다.
3. 나는 그보다 더 나은 대우를 받을 자격이 있다고 생각했다.

Exercise 10

정답 ③

소재 세상을 변화시키는 예측 알고리즘

해석 인류는 새로운 시대에 접어들었다. 우리는 이제 수십억 개의 예측 알고리즘에 의해 점점 더 연결되는 세상, 즉 '거의 모든 것'이 예측 가능하고 거의 모든 삶의 영역에서 위험과 불확실성이 줄어들고 있는 것처럼 보이는 세상에서 살고 있다. 우리는 의료 서비스와 정밀 의학의 발전 덕분에 더 오래 살고 있다. 우리는 우리에게 다른 행성을 탐험하고 수십억 개의 은하를 시각화할 수 있게 하는 새로운 기술을 꿈꾸고 구축할 수 있게 하는 물리적 세계에 대한 더 큰 장악력을 가지고 있다. 우리는 시장, 질병, 그리고 교통을 점점 더 높은 정확성으로 모형화할 수 있으며, 자동차가 스스로 주행할 수 있도록 자동차에게 열쇠를 넘겨주는 것에 매우 가까워지고 있다. 더욱 더 놀라운 것은, 우리의 도구들이 인간 행동의 일부 이해하기 어렵고 다루기 힘든 복잡성의 기원을 밝혀내고 있을지도 모른다는 점이며, 알고리즘이 심지어 사람들의 행동을 바꾸는 데 사용되고 있다는 점이다. 예측 알고리즘은 세상을, 그리고 앞으로 올 모든 세상을 변화시켰으며, 되돌아갈 수 없다.

문제 해설
③ explore와 and로 대등하게 연결되어 to에 이어져 to부정사구를 이루어야 하므로 visualizes를 visualize로 고쳐야 한다.
① 「전치사+관계사」의 형태로 선행사 a world를 수식하는 관계

절을 이끌고 있으므로 어법상 적절하다.

② 부사인 long의 비교급으로서 동사구 are living을 수식하므로 어법상 적절하다.

④ 주어 it(the car)과 동사 drive의 목적어가 가리키는 대상이 같으므로, 재귀대명사 itself는 어법상 적절하다.

⑤ 「be used to+-ing」는 '~하는 데 익숙하다'라는 뜻이며, 「be used to+동사원형」은 '~하는 데 사용되다'라는 뜻이다. 주어진 문장에서는 사람들의 행동을 바꾸는 데 알고리즘이 사용된다는 내용이므로, to 다음에 동사원형이 쓰인 것은 어법상 적절하다.

구조 해설

- We have a greater mastery of the physical world [that allows us to {dream of}, and {build}, new technologies {that allow us to ⟨explore other planets⟩ and ⟨visualize billions of galaxies⟩}].

[ ]는 a greater mastery of the physical world를 수식하는 관계절이며, 그 안의 첫 번째와 두 번째 { }는 and에 의해 대등하게 to에 연결되고 new technologies를 공통의 목적어로 취한다. 세 번째 { }는 new technologies를 수식하는 관계절이고, 그 안의 두 개의 ⟨ ⟩는 and에 의해 대등하게 to에 연결되어 allow의 목적격 보어 역할을 한다.

어휘 및 어구

humanity 인류
era 시대
wired (컴퓨터 시스템에) 연결된
predictive 예측의
diminish 줄어들다
precision 정밀, 정확성
mastery 장악력
physical world 물리적 세계
galaxy 은하
alter 바꾸다, 고치다

Word Search

| 정답 | 1. predictive 2. diminish 3. precision

1. 예측 모델링을 사용하여 기상 전문가들은 심각한 기상 현상을 더 잘 예측할 수 있다.
2. 경제 침체 이후 그 회사의 수익은 줄어들기 시작했다.
3. 그 의사는 복잡한 뇌 수술을 정확하게 집도했다.

Tips

precision medicine(정밀 의학): 개인의 유전적, 환경적, 생활 습관적 특성을 고려하여 맞춤형 치료와 예방 전략을 개발하는 접근 방식을 의미한다. 정밀 의학은 질병의 예방과 치료를 개인화하여 보다 정확하고 효과적인 의료 서비스를 제공하는 것을 목표로 한다.

Exercise 11 정답 ③

소재 환경의 물리적, 사회적 측면에 대한 인간의 이해 진보 차이

해석 환경의 물리적 측면에 대한 인간의 이해 진보가 사회적 측면에 비해 훨씬 더 먼저 발생했다는 점은 주목할 만한 가치가 있다. 물리적 환경과 사회적 환경이라는 두 가지 다른 환경에 대한 인간 이해의 발전에서 이러한 차이의 주된 이유는 아마도 물리적 현상이 제공하는 특정 개인과 상관없는 접근 방식뿐만 아니라 그것(물리적 현상)에 대한 더 큰 관찰 가능성과 통제에 있을 것이다. 물리적 현상은 보통 사회적 현상보다 더 구체적이어서 더 관찰 가능하다. (물리적 환경과 사회적 환경은 모두 개인의 행복과 전반적인 공동체의 건강을 형성하는 데 중요한 역할을 한다.) Samuel Koenig의 말에 따르면, 물리적 현상을 관찰하려고 시도하면서 인간은 상당히 일찍 어느 정도의 객관성을 발전시킬 수 있었으나 사회적 현상과 관련해서는 그렇게 하기가 매우 어렵다는 것을 알게 되었다. 후자의 경우 인간은 자신이 연구 대상에 지나치게 가까이 있고, 그것에 지나치게 관여되어 있어 모든 과학에 필수적인 객관성을 달성하기가 어렵다는 것을 알게 되었다.

문제 해설

환경의 물리적 측면에 대한 인간의 이해 진보가 사회적 측면에 비해 훨씬 먼저 발생했으며 그 이유는 물리적 현상은 보통 사회적 현상보다 더 관찰 가능한 반면, 사회적 현상에는 인간이 지나치게 관여되어 있어 객관성을 달성하기가 어렵기 때문이라는 내용의 글이다. 물리적 환경과 사회적 환경은 모두 개인의 행복과 전반적인 공동체의 건강을 형성하는 데 중요한 역할을 한다는 내용인 ③은 글의 전체 흐름과 관계가 없는 문장이다.

구조 해설

- [The main reason for this discrepancy in the development of man's understanding of the two distinct environments—the physical and the social—] probably **lies** [in the greater observability and control of physical phenomena], **as well as** [in the impersonal approach {that **they** afford}].

첫 번째 [ ]는 문장의 주어 역할을 하는 명사구이고, 주어의 핵심어구가 The main reason으로 단수이므로 단수 동사 lies를 사용하였다. 두 번째와 세 번째 [ ]는 '…뿐만 아니라 ~(도)'라는 의미의 「~ as well as ...」에 의해 연결되어 있다. { }는 the impersonal approach를 수식하는 관계절이고, they는 physical phenomena를 가리킨다.

어휘 및 어구

progress 진보
physical 물리적인

aspect 측면
in comparison to ～에 비해
distinct 다른, 별개의
lie in ～에 (놓여) 있다
phenomenon 현상(pl. phenomena)
impersonal 특정 개인과 상관없는, 개인적인 것이 개입되지 않은
afford 제공하다
concrete 구체적인
a measure of 어느 정도의
detachment 객관성, 초연함, 떨어져 있음
investigation 연구, 조사
involved in ～에 관여된
objectivity 객관성

Word Search

| 정답 | **1.** progress **2.** comparison **3.** aspect
1. 그 프로젝트는 느리지만 꾸준한 진보를 보였다.
2. 그들은 다른 나라들의 식습관을 비교했다.
3. 이것은 그가 이전에 보지 못했던 그녀의 성격 중 한 측면이었다.

Exercise 12   정답 ②

소재 상황에 따른 음악 효과 처리의 시간적 측면

해석 컴퓨터가 시간에 맞게 처리할 때, 우리는 일반적으로 그것이 실시간으로도 작동하기를 원한다. 예를 들어 여러분이 공연에서 마이크를 에코 효과와 함께 설치하면, 여러분은 그것(에코 효과)이 시간에 맞게 그리고 실시간으로 작동하기를 원할 것이다. (B) 하지만 미리 녹음된 노래에 그저 에코만 추가하기를 원한다면, 그것이 언제 또는 얼마나 빨리 일어나는지는 중요하지 않다. 여러분은 아마도 컴퓨터가 '실시간보다 더 빠르게' 처리하는 것을 선호할지도 모른다. (A) 컴퓨터가 3분짜리 소리 파일을 7초 안에 처리할 수 있다면, 그것의 처리를 마치기 위해 3분을 기다릴 필요가 없다. 여러분은 또한 '실시간보다 더 느리게' 실행되는 연산량이 많은 에코 모델을 갖게 될지도 모른다. 스튜디오에서 작업하는 프로듀서에게는 처리가 실시간으로 실행되는지 여부가 중요하지 않다. (C) 하지만 무대 위의 공연자에게는 실시간 모드로 실행하는 것 외에 다른 선택지가 없다. 콘서트는 '지금' 일어나고 있으므로 사용되는 도구는 시간에 맞게 그리고 실시간으로 작동할 필요가 있다.

문제해설

공연에서 마이크를 에코 효과와 함께 설치하면 에코 효과가 시간에 맞게 그리고 실시간으로 작동하기를 원한다는 내용의 주어진 글 다음에, 역접의 연결사(However)와 함께 이와 달리 미리 녹음된 노래에 에코만 추가하는 상황이라면 그것의 실시간 여부는 중요하지 않다는 내용의 (B)가 이어지고, 이에 대한 부연 설명으로 스튜디오 작업에서는 컴퓨터가 소리 파일을 실시간보다 더 빠르게 혹은 더 느리게 처리하는 것이 가능하다는 내용의 (A)가 이어지며, 이와 대조하여 (C)에서 역접의 연결사(however)로 이어져, 무대 위의 공연자에게는 사용되는 도구가 시간에 맞게 그리고 실시간으로 작동해야 한다고 말하고 있다. 따라서 주어진 글 다음에 이어질 글의 순서로 가장 적절한 것은 ② '(B)-(A)-(C)'이다.

구조해설

- For a producer [working in a studio], **it** is not crucial [whether a process runs in real time].

  첫 번째 [ ]는 a producer를 수식하는 분사구이다. it은 형식상의 주어이고, 두 번째 [ ]가 내용상의 주어이다.

어휘 및 어구

process 처리하다; 처리
in time 시간에 맞게, 시간 내에
in real time 실시간으로
set up ～을 설치하다
performance 공연
crucial 중요한
prefer 선호하다

Word Search

| 정답 | **1.** performance **2.** processed **3.** prefer
1. 오늘 저녁 공연은 7시에 시작될 것이다.
2. 망원경으로 수집한 정보는 컴퓨터로 처리될 것이다.
3. 어떤 사람들은 초콜릿 아이스크림을 좋아하지만, 나는 바닐라를 선호한다.

1 ③	2 ③	3 ④	4 ①	5 ④	6 ③
7 ③	8 ⑤	9 ②	10 ③	11 ④	12 ⑤

Exercise 1   정답 ③

소재 밝기 대비의 중요성과 시각적 인식에 미치는 영향

해석 가장 근본적으로, 상대적 밝기는 우리가 살아가는 시각적 대비를 알아보게 한다. 대비가 없으면 시각의 본질적 요소가 작동하지 않아, 뇌는 경계와 형태를 식별하지 못하고 깊이나 거리도 인식하지 못한다. 대비가 적을수록 우리가 세부 사항을 식별하는 능력이 더 떨어지고, 환경을 탐색하는 능력도 더 떨어진다. 이상적인 시각 상황은 상대적 밝기의 대비가 가장 쉽게 식별되고, 광원이 강하며 선명하게 윤곽이 드러나는 그림자를 드리우는 경우이다. 따라서 이상적인 시각 상황은 긍정적인 감정 반응을 유발할 가능성이 매우 높다. 안전/위험 스펙트럼의 반대쪽 끝에서, 가장 안전한(→ 위험한) 시각 상황은 다양한 상대적 밝기에 의해 제공되는 대비가 더 이상 존재하지 않는 경우이다. 완전한 어둠(지하 동굴에서처럼)과 완전한 밝음(눈보라 속 화이트아웃에서처럼)은 똑같이 위험으로 가득 차 있다. 이러한 극단적인 밝기 상황에서는 세부 사항이 지워지고 길을 찾는 것이 위험해진다. 이와 같은 상황에서 유발되는 감정적 느낌은 부정적일 가능성이 매우 크다.

문제 해설
상대적 밝기는 시각적 대비를 식별하는 데 필수적인데, 이상적인 시각 상황은 강한 광원과 선명한 그림자를 통해 대비가 쉽게 식별되는 경우이고, 이는 긍정적인 감정을 유발하는 반면, 완전한 어둠이나 완전한 밝음은 대비를 없애 위험을 초래하고 부정적인 감정을 유발할 가능성이 매우 크다고 했으므로 상대적 밝기에 의한 대비가 존재하지 않는 시각 상황은 안전하지 않을 것이다. 따라서 ③의 secure를 hazardous와 같은 낱말로 바꿔야 한다.

구조 해설
- **The fewer** the contrasts, [**the less able** we are to identify details, and **the less able** we are to navigate the environment].
「the+비교급(~), the+비교급(...)」 구문으로 '~할수록 더 …하다'의 의미를 나타낸다. [ ]는 「the+비교급(...)」에 해당하는 부분으로 「the+비교급(...)」의 구조에 해당하는 두 개의 절이 and로 연결되어 있다.

어휘 및 어구
fundamental 근본적인
relative 상대적인
identify 알아보게 하다, 식별[감별]하다, 발견하다
contrast 대비, 대조
operate 작동하다, 작용하다
edge 경계, 테두리
perceive 인지[감지]하다
complete 완전한, 전적인
extreme 극단적인
hazardous 위험한, 모험적인
arouse (느낌·태도를) 유발하다

Word Search

| 정답 | **1.** operate   **2.** identify   **3.** complete
1. 리모컨을 사용하면 TV의 모든 기능을 작동할 수 있다.
2. 과학자들은 열대 우림의 외딴 지역에서 새로운 종들을 발견하고자 연구한다.
3. 난민들은 그들이 원하는 것은 무엇이든 할 수 있는 완전한 자유를 요구했다.

Exercise 2   정답 ③

소재 특허 가능성에서 발견과 발명의 차이

해석 특허를 받을 수 있는 대상에 대한 가장 중요한 내재적인 제약은 '발명'과 '발견' 사이의 때때로 모호한 구분을 중심으로 전개된다. 연구자가 '발견할' 수도 있는 자연법칙이나 과학적 원리는 특허를 받을 수 없다. 예를 들어, 과학자가 어떤 이전에 알려지지 않은 형태의 자기파가 먼 우주 공간에서 지구의 대기를 강타하며, 이런 종류의 파동이 지구와 통신 위성 간의 데이터 전송에 부정적인 영향을 미친다는 것을 발견한다고 가정해 보자. 그러한 발견은 엄청난 상업적 영향을 미쳐, 어쩌면 그 파동의 유해한 영향을 걸러내는 장치의 발명으로 이어질 수도 있다. 그 여과 장치의 발명가는 특허를 받을 자격이 있을 수도 있다. 자기파의 발견자는 자신의 발견에 대해 특허를 받을 수 없다. 오늘날, 유전공학 및 생명공학 분야에서 발견과 발명의 구분은 점점 더 중요해지고 있다.

문제 해설
주어진 문장의 Such a discovery는 ③의 앞 문장에서 언급한 과학자의 발견을 가리키고, ③의 다음 문장에서 The inventor of the filtering device는 주어진 문장에서 등장한 the invention of a device의 주체라고 할 수 있다. 따라서 주어진 문장이 들어가기에 가장 적절한 곳은 ③이다.

구조 해설
- For example, imagine that a scientist discovers [that a certain previously unknown form of magnetic wave

strikes the earth's atmosphere from deep space], and [that this kind of wave has an adverse effect on the transmission of data between the earth and communications satellites].

두 개의 [ ]는 각각 that이 이끄는 명사절로 and로 연결되어 discovers의 목적어 역할을 한다.

Word Search

| 정답 | **1.** inherent **2.** principle **3.** entitled
1. 그 예술가는 창의성이 모든 개인에게 내재해 있다고 믿었다.
2. 평등의 원칙은 법체계의 근간이다.
3. 모든 시민은 그들의 사회적 지위에 상관없이 기본적 인권에 대한 자격이 있다.

Exercise 3   정답 ④

소재 동물을 언급하는 언어와 그 의미

해석 '농장 동물, 실험실 동물, 스포츠 말…'과 같이 일부 동물을 언급할 때 쓰는 우리의 언어는 예를 들어, 식량 공급원으로서의 동물, 생물 의학 연구에 이용되거나 오락 활동을 위해 이용되는 동물처럼, 동물의 이용과 인간이 동물로부터 얻는 이익에 초점을 맞추는 경우가 많다. 우리가 일부 동물을 '애완동물'이라고 부를 때, 우리는 그 동물이 어떤 목적상 이용되고 있음을 암시하고 있는 것일까? 애완동물은 중립적으로, 인간과 함께 사는 동물로 정의될 수 있다. 또 다른 널리 사용되는 용어는 '반려동물'인데, 이는 마찬가지로, 우리가 결국 반려 관계를 얻기 위해 그 동물을 집에

서 키운다는 것을 암시한다. 일부 사람들은 어항 속의 물고기는 실제로 누구의 동반자도 아니라고 주장할 수도 있다. 그러나 이 용어는 우리가 친절하게 대하고, 가족의 일원이자 '가장 친한 친구'로 여기며, 시간과 재정적 자원을 공유하는 개와 고양이와 같은 가정에서 키우는 종에는 잘 어울린다. 따라서 다른 종과 달리 반려동물은 경제적 자원보다는 인간에게 지지, 위안, 동반자 관계를 제공하는 정서적 자원에 해당하는 것으로 보인다.
→ 동물에 대한 담론에 관한 한, 우리의 언어는 일반적으로 일부 동물의 유용성을 강조하는 반면, 가정에서 키우는 동물에 대한 용어는 그 동물의 관계적인 중요성을 암시한다.

문제 해설
우리가 동물을 언급할 때 사용하는 언어는 자주 그 동물의 이용과 인간이 얻는 이익에 초점을 맞추고 있는 데 반해, 반려동물처럼 가정에서 함께 생활하는 동물에 대한 용어는 동물이 인간과 함께 사는 동반자로 여겨짐을 암시하고 그러한 동물은 인간에게 정서적 자원을 제공한다는 내용이다. 따라서 요약문의 빈칸 (A), (B)에 들어갈 말로 가장 적절한 것은 ④ '유용성(utility) – 관계적인 (relational)'이다.
① 행동 – 인지적인
② 수익성 – 실용적인
③ 감정 – 오락적인
⑤ 지능 – 문화적인

구조 해설
■ "Farm animals, lab animals, sport horses…": our language [when we refer to some animals] often focuses [on their exploitation] and [on the benefits {humans get from them}], for example, animals as a source of food, used in biomedical research or for recreational activities.
첫 번째 [ ]는 시간을 나타내는 부사절이고, 두 번째와 세 번째 [ ]는 on이 이끄는 전치사구로 and로 연결되어 동사 focuses에 이어진다. { }는 관계사가 생략된 관계절로 the benefits를 수식한다.

Word Search

| 정답 | **1.** benefits **2.** research **3.** companionship
1. 규칙적인 운동은 더 나은 기분 조절을 포함하여 많은 이점을 제공한다.
2. 과학자들은 지식을 발전시키고 문제를 해결하기 위해 연구를 수행한다.
3. 반려동물은 주인에게 소중한 동반자 관계를 제공한다.

Exercise 4 ——— 정답 ①

소재 도시 문제의 전략적 개입을 위해 만들어진 축제들

해석 일부 이해 당사자들이 예술과 문화 축제가 도시 정책 도구로 재구성되는 것으로부터 예술과 문화 축제의 무결성을 보호하려 할 수도 있지만, 우리는 정책 지향적인 축제가 여전히 매우 긍정적인 사회적, 문화적 효과를 가져올 수 있다는 점을 인식해야 한다. 그리고 우리는 일부 축제는 도시 지역을 전략적으로 지원하기 위해 만들어졌다는 사실을 <u>무시할</u> 수 없다. 즉, 그것들은 예술적, 사회적 또는 문화적 현상이라기보다는 언제나 전략적 개입이었다. 영화제가 좋은 예이며, 이러한 행사 중 많은 수가 <u>경제적인</u> 이유로 만들어졌는데, 예를 들어, Cannes Film Festival은 관광철을 연장하기 위해 시작되었다. Brighton Festival도 유사한 이유로 만들어졌다. 오랜 휴지기를 거친 후 1979년 Venice Carnival을 다시 시작한 것 또한 젊은이들을 위한 (프로그램 등에 대한) 공급 부족을 포함하여, 그 당시 도시가 직면하고 있던 일부 문제를 해결하기 위한 <u>의도적인</u> 시도였다. 이러한 축제들은 도시 정책 도구로 전용된 것이 아니라 항상 더 광범위한 목표를 염두에 두고 개최되었다.

문제 해설

(A) 이어지는 문장에서 일부 축제들은 예술적, 사회적 또는 문화적 현상이라기보다는 전략적 개입이라고 했으므로, 축제가 도시 지역을 지원하기 위해 전략적으로 만들어졌다는 측면을 무시할 수 없다는 문맥이 되어야 한다. 따라서 ignore(무시하다)가 문맥상 적절하다. acknowledge는 '인정하다'라는 뜻이다.
(B) 예로 제시된 문장에서 관광철을 연장하기 위해 축제가 시작되었다고 했으므로, 경제적인 이유로 축제가 만들어졌다는 문맥이 되어야 한다. 따라서 economic(경제적인)이 문맥상 적절하다. recreational은 '오락적인'이라는 뜻이다.
(C) 당시 도시가 직면한 문제를 해결하려 했다는 내용이 이어지고 있으므로, 의도적인 시도로 해당 축제가 다시 시작되었다는 문맥이 되어야 한다. 따라서 deliberate(의도적인)가 문맥상 적절하다. accidental은 '우연한'이라는 뜻이다.

구조 해설

■ And we cannot ignore [the fact {that some festivals were established to strategically assist urban areas}].
[ ]는 ignore의 목적어이며, 그 안에서 { }는 the fact와 동격 관계를 이룬다.

어휘 및 어구
integrity 무결성, 온전함, 청렴
urban 도시의
policy 정책
positive 긍정적인
intervention 개입, 조정
phenomenon 현상(*pl.* phenomena)
launch 시작하다
prolong 연장하다
address 해결하다, 다루다

Word Search

| 정답 | **1.** policy **2.** Positive **3.** phenomena
1. 그 정책은 모든 직원이 매년 데이터 개인 정보 보호에 대한 교육을 받을 것을 요구한다.
2. 긍정적인 사고는 정신 건강과 전반적인 행복의 개선으로 이어질 수 있다.
3. 과학자들은 토네이도와 허리케인과 같은 다양한 기상 현상을 연구한다.

Exercise 5~6 ——— 정답 5 ④ 6 ③

소재 영양소의 원천에 따른 영향의 차이

해석 '칼로리는 칼로리일 뿐'이라는 말을 들어 봤을 것이다. 각 칼로리가 동일한 양의 에너지를 산출한다는 것은 사실이지만, 음식은 다르게 대사되고 사용될 수 있으며, 어떤 음식이 다른 음식보다 더 효율적일 수 있다. 예를 들어, 칼로리를 제공하는 다량 영양소인 탄수화물, 단백질, 지방은 배고픔과 섭식 행동을 조절하는 호르몬과 뇌 중추에 각각 다른 영향을 미칠 수 있다. 다량 영양소는 표준 칼로리 양을 가지고 있다. 단백질 1그램에는 4칼로리가 있다. 탄수화물 1그램에는 4칼로리가 있다. 지방 1그램에는 9칼로리가 있다.
그러나 원천은 무관하다(→ 중요하다). 다량 영양소 외에도 음식에는 신체의 과정과 반응 기제에 영향을 미칠 수 있는 많은 화합물이 포함되어 있다. 탄수화물 1그램이 그램당 4칼로리를 제공한다고 할지라도, 모든 탄수화물이 동일하게 만들어지지는 않는다. 예를 들어, 흰 빵은 정제되고 섬유질이 제거되었다. 이 때문에 흰 빵은 혈당의 급격한 상승을 야기할 수도 있는 반면에, 통곡물빵은 덜 가공되어 섬유질이 온전하게 남아 있다. 이로 인해 당이 혈액으로 더 천천히 방출되는데, 이는 흰 빵과 통곡물빵이 가지고 있

는 탄수화물의 총'량'은 동일할 수도 있지만, 그것이 혈당에 '영향을 미치는' 방식은 매우 다르다는 것을 의미한다.

(문제 해설)

5 칼로리는 같은 양의 에너지를 산출하지만, 음식에 따라 다르게 대사되고 사용될 수 있는데, 예를 들면 흰 빵은 정제되어 혈당을 급격히 상승시키는 반면, 통곡물빵은 섬유질이 남아 있어 혈당 상승이 더 천천히 이루어지는 것처럼 같은 칼로리를 제공하더라도 음식의 원천과 가공 방식에 따라 신체에 미치는 영향은 달라진다는 내용의 글이다. 따라서 글의 제목으로 가장 적절한 것은 ④ '칼로리를 넘어서: 영양소 원천의 영향을 밝히기'이다.
① 여러분의 음식이 충분한 칼로리를 함유하고 있는가?
② 숨겨진 사실: 당분 섭취가 여러분을 건강하게 만든다
③ 칼로리 표시를 보지 말고, 그냥 음식을 즐기라
⑤ 요리의 비밀: 조리법에 따라 달라지는 맛

6 해당 문장에 이어서 같은 칼로리를 제공하지만 서로 다른 칼로리 원천인 흰 빵과 통곡물빵이 혈액으로 당을 방출하는 방식이 서로 다르고 각각 신체에 미치는 영향이 서로 다르다는 내용이 제시되고 있으므로, (c)의 irrelevant를 important와 같은 낱말로 바꿔야 한다.

(구조 해설)

- Because of this, white bread may cause a spike in blood sugar, [**whereas** whole grain bread {is less processed} and {has its fiber intact}].
  [ ]는 '~인 반면에'의 의미를 가진 whereas가 이끄는 부사절인데, 그 안에서 두 개의 { }가 and로 연결되어 술어를 이루고 있다.

(어휘 및 어구)

yield 산출하다, 생산하다
carbohydrate 탄수화물
protein 단백질
compound 화합물
influence 영향을 미치다[주다]
refine 정제하다, 다듬다
strip ~ of ... ~에게서 …을 제거하다
fiber 섬유질
spike (급격한) 상승
release 방출, 방면, 출시

(Word Search)

| 정답 | **1.** influence **2.** refine **3.** release
1. 우리가 지금 취하는 결정은 미래의 사건 흐름에 영향을 미칠 수도 있다.
2. 그 주방장은 완벽한 맛의 균형을 이루기 위해 계속해서 조리법을 다듬었다.
3. 그 회사는 최신 제품의 출시를 발표하여 소비자 사이에서 흥분을 불러일으켰다.

Exercise 7 정답 ③

(소재) 정서와 감정의 차이

(해석) 정서는 의미상 감정과 매우 가깝지만, 사회 심리학자들이 정서라는 용어를 사용할 때 그들은 감정의 사회적 측면을 강조한다. 초기 사회 심리학자들은 동물이 가질 수 있는 유사한 반응과 인간 반응을 구분 짓는 인간 반응의 요소들을 지칭하기 위해 정서를 사용했다. 예를 들어, Cooley는 사랑과 욕망을 대조한다. (비록) 욕망이 본능적이지만, 우리는 사회적 상호 작용을 통해 사랑이 무엇인지 배운다. 다시 말해, 정서는 자극에 대한 개인의 반응뿐만 아니라 다른 사람들이 그 자극을 어떻게 이해하는지에도 의존한다. 이후에 사회 심리학자들이 사회적 요소가 감정의 중요 요소임을 점점 더 받아들이게 되면서, 정서의 개념은 감정의 개념과 더(→ 덜) 구별될 수 있게 되었다. 현대의 연구에서 사회 심리학자들은 즉각적인 감정 반응과 사랑, 슬픔, 그리고 질투와 같은 장기적인 감정 상태를 구별하기 위해 정서라는 용어를 자주 사용한다. 이러한 정서는 그것을 유발한 최초 사건 이후에도 며칠, 몇 주, 심지어 몇 년 동안 지속될 수 있다.

(문제 해설)

초기 사회 심리학자들은 정서라는 용어를 사용해서 동물의 반응과 구분 짓고, 감정의 사회적 측면을 강조했는데, 이후에는 그들이 사회적 요소가 정서와 마찬가지로 감정의 중요한 요소임을 점점 더 받아들이게 되었다는 맥락이므로, 이 둘을 구별하기가 쉽지 않아졌을 것이다. 따라서 ③의 more를 less와 같은 낱말로 바꿔야 한다.

(구조 해설)

- Sentiment is very close to emotion in meaning, but [when social psychologists use the term sentiment], **they** emphasize the social aspect of emotion.
  [ ]는 시간의 부사절이고, they는 social psychologists를 대신한다.

(어휘 및 어구)

sentiment 정서
emotion 감정
emphasize 강조하다
response 반응
contrast 대조하다
instinctive 본능적인
social interaction 사회적 상호 작용
individual 개인
stimulus 자극
contemporary 현대의
distinguish 구별하다
immediate 즉각적인

grief 슬픔
jealousy 질투
endure 지속되다
trigger 유발하다

Word Search

| 정답 | **1.** contrast **2.** individual **3.** contemporary
1. 그 연구는 도시와 농촌의 생활 환경을 대조하고자 한다.
2. 각 개인은 각자의 고유한 환경에 따라 스트레스에 다르게 반응한다.
3. 그 현대 미술 전시회에는 전 세계 예술가들의 작품이 특별 전시되었다.

Exercise 8 ——— 정답 ⑤

소재 화석이 되기 위한 조건

해석 화석이 되는 것은 길고 복잡한 과정이며, 대다수의 개체에 일어날 가능성이 없는 결과이다. 유기체가 결국 화석이 되기 위한 가장 분명한 필요조건은 일반적으로 매장에 의해 그것이 암석 내부에 묻힌 상태가 되어야 한다는 것이다. 이런 일을 막을 수 있는 많은 가능성이 있다. 만약 껍데기 두 개를 가진 연체동물이 불행히도 포식자에게 죽임을 당하면, 부드러운 부분은 먹힐 것이다. 포식자는 아마도 살에 접근하기 위해 적어도 껍데기 하나를 부숴야 할 것이기 때문에, 두 껍데기 모두가 온전하게 남을 가능성은 없다. 설사 그렇더라도, 두 껍데기는 아마도 분리될 것이다. 반면에, 그 유기체가 자연사할 경우, 몸의 부드러운 부분은 거의 확실히 썩어 없어질지라도, 어느 쪽의 껍데기도 손상되지 않을 가능성이 있다. 껍데기는 가장 마지막에 썩는 것 중 하나인 섬유 조직에 의해 이음매에 서로 붙어 있으므로, 두 껍데기가 묻히게 될 때까지 계속 함께 붙어 있을 상당한 가능성이 있다.

문제 해설

④ 다음 문장에서 두 개의 껍데기가 온전하더라도 분리되어 있을 것이라는 내용을 언급한 후, ⑤ 다음 문장에서 가장 마지막에 가서야 썩는 섬유 조직에 의해 이음매에 두 껍데기가 함께 붙어 있을 가능성이 꽤 있다고 하였으므로 논리적 단절이 있는데, 그 유기체가 자연사할 경우, 두 껍데기는 손상되지 않을 가능성이 있다는 주어진 문장을 넣으면 논리적 단절이 해소된다. 그러므로 주어진 문장이 들어가기에 가장 적절한 곳은 ⑤이다.

구조 해설

- [The most obvious requirement for an organism {to end up as a fossil}] is [that it must become entombed within rock, usually by burial].

첫 번째 [ ]는 주어이며, 그 안의 { }는 an organism을 의미

상의 주어로 하는 to부정사구이다. 두 번째 [ ]는 주격 보어의 역할을 하는 명사절이다.

어휘 및 어구

rot away 썩어 없어지다
shell 껍데기
fossil 화석
involved 복잡한, 관련된
unlikely 일어날 가능성이 없는
outcome 결과
entomb 묻다
rock 암석
burial 매장
prevent 막다, 방해하다, 저지하다
misfortune 불행
predator 포식자
flesh 살
fibrous tissue 섬유 조직
reasonable 상당한, 꽤 많은

Word Search

| 정답 | **1.** unlikely **2.** outcome **3.** reasonable
1. 그 연약한 꽃이 보호 없이 혹독한 겨울을 살아남을 가능성은 없다.
2. 그 실험 결과에 따라 연구 과정의 다음 단계가 결정될 것이다.
3. 그녀는 마감일 전에 그 프로젝트를 완료할 수 있는 상당한 시간을 가졌다.

Exercise 9 ——— 정답 ②

소재 사회적 보상과 신경 경로

해석 때때로 원숭이는 다른 원숭이를 손질해 주고 아무런 보상을 받지 못한다. 때때로 유인원은 짝에게 구애하지만, 완전히 무시당한다. 코르티솔은 포유류에게 다른 것을 시도해 보도록 유도하지만, 몇 번의 실망 후에 그 포유류는 자신의 에너지를 어디에 투자할지 예측하기 어려울 수 있다. 이것이 우리가 젊었을 때 미엘린을 형성한 신경 초고속도로에 자주 의지하는 이유이다. 여러분의 전기는 과거에 믿을 만하게 보상받았던 행동에 의해 형성된 경로를 따라 쉽게 흐른다. 아마도 그것은 터치다운을 하거나, 친구들과 함께 가장 좋아하는 쿼터백의 득점을 지켜보는 것이었을 것이다. 물론 공을 들고 선을 넘는 것이 실제 생존 욕구를 충족시키지는 않지만, 사회적 보상을 기대할 때 도파민이 급증한다. 각자의 뇌는 자신의 삶의 경험을 통해 사회적 보상을 예측한다. 아마도 여러분은 푸짐한 식사를 요리하거나, 어려운 방정식을 풀거나, 영업 시간이 끝난 후에도 문을 연 술집을 찾아낸 사람에게 사

회적 보상이 주어지는 세상에 살았을지도 모른다. 사회적 보상을 얻는 방법은 무한하지만, 우리가 젊었을 때 관찰하고 즐기는 것들이 오래 지속되는 기대를 형성한다.
→ 반복되는 사회적 실망이 개인이 젊었을 때 강화된 <u>익숙한</u> 행동으로 돌아가게 만들 수 있는 것처럼, 뇌는 <u>보상을 주는</u> 경험을 예측하고 찾기 위해 과거 경험과 신경 경로에 의존한다.

문제 해설
포유류는 젊었을 때 형성된 익숙한 보상 행동 경로를 따르기 쉽고, 사회적 보상을 기대할 때 도파민이 급증하며, 젊을 때 경험한 사회적 보상이 지속되는 기대를 형성한다는 내용의 글이다. 따라서 요약문의 빈칸 (A), (B)에 들어갈 말로 가장 적절한 것은 ② '익숙한(familiar) − 보상을 주는(rewarding)'이다.
① 충동적인 − 보상을 주는
③ 혁신적인 − 지치는
④ 파괴적인 − 좌절감을 주는
⑤ 습관적인 − 좌절감을 주는

구조 해설
- Cortisol prompts a mammal to try something different, but [after a few disappointments] **it** can be hard [{for the mammal} to predict where to invest its energy].
  첫 번째 [ ]는 시간의 부사구이고, it은 형식상의 주어이며, 두 번째 [ ]가 내용상의 주어인데, 그 안에서 { }는 to부정사구의 의미상의 주어를 나타낸다.

어휘 및 어구
groom 손질하다
prompt 유도하다, 촉발하다
mammal 포유류
disappointment 실망
predict 예측하다
invest 투자하다
neural superhighway 신경 초고속도로
electricity 전기
pathway 경로
behavior 행동
reliably 믿을 만하게
reward 보상하다
score a touchdown 터치다운(미식축구의 주요 득점 수단)을 하다
quarterback 쿼터백(미식축구에서 팀의 공격을 지휘하는 역할을 맡고 있는 선수)
survival 생존
surge 급증하다
social reward 사회적 보상

equation 방정식
limitless 무한한
expectation 기대

Word Search

| 정답 | 1. disappointment   2. reliably   3. survival
1. 결승전에서 그 팀의 패배는 팬들에게 큰 실망을 안겨 주었다.
2. 그 기계는 고품질의 결과를 믿을 만하게 만들어 낸다.
3. 깨끗한 물에 대한 접근은 모든 살아 있는 유기체의 생존에 필수적이다.

Tips

myelinate(미엘린을 형성하다): 미엘린을 형성하는 것은 뇌와 척수를 포함한 신경 주위에 미엘린(수초)이라는 지방층이 형성되는 과정이다. 이 미엘린은 신경 세포를 따라 전기 자극이 전달되는 속도를 높이고, 효율적인 뇌 기능을 촉진하는 것으로 알려져 있으며, 중요한 신경 회로들을 더 강화하여 정보 전달 속도를 높이고, 신경 신호의 효율성을 증대시킨다.

Exercise 10 ——— 정답 ③

소재 거짓말 탐지와 직관의 한계

해석 누군가가 거짓말을 하고 있는지를 알아내기 위해, 우리는 직관에 크게 의존하는 경향이 있다. 사실에 기반한 근거에 의해 거짓말을 포착하지 않는 한, 어떤 사람의 부정직함을 나타내는 유일한 지표는 목소리 톤, 신체 언어, 그리고 표정처럼 우리가 의식적으로 인식하기에는 너무 미묘할 수도 있지만, 여전히 강한 직감을 불러일으킬 수 있는 신호들뿐이다. 문제는 우리가 사회적 상호작용에서 다른 사람들의 진실성을 평가하는 기술을 연습할 수 있지만, 우리의 판단이 올바른지에 대한 명확한 피드백이 없으면, 우리가 지나치다 싶을 정도로 쉽게 속고 있는 것인지, 혹은 불신하고 있는 것인지 우리는 알 수 없다는 것이다. 이는 우리가 시간이 지남에 따라 향상될 수 있다는(→ 될 수 없다는) 것을 의미한다. 많은 사람들이 진실과 거짓을 구별하는 데 자신이 꽤 능숙하다고 믿지만, 일반 대중 중 우연보다 더 높은 정확도로 (이를) 수행하는 사람은 거의 없다. 평균적으로, 심지어 경찰관, 변호사, 판사, 정신과 의사, 그리고 보통의 사람들보다 더 빈번하고 심각한 거짓말을 접하는 다른 집단의 구성원들조차도 일반 대중보다 더 잘 수행하지 못한다.

문제 해설
사람들은 거짓말을 판단할 때 직관에 의존하지만, 명확한 피드백이 없으면, 쉽게 속는 것인지, 아니면 불신하는 것인지 알 수 없다는 맥락이므로, 거짓말을 하는지 판단하는 능력은 시간이 흘러도

향상될 수 없다는 내용의 글이다. 따라서 ③의 able을 unable과 같은 낱말로 바꿔야 한다.

(구조 해설)

- On average, [even police officers, lawyers, judges, psychiatrists, and members of other groups {that encounter more frequent and serious lies than ordinary people}] perform no better.

[ ]는 문장의 주어이고, 그 안의 { }는 members of other groups를 수식하는 관계절이다.

(어휘 및 어구)

determine 알아내다
intuition 직관
factual 사실에 기반한
ground (보통 복수로) 근거
indicator 지표
tone of voice 목소리 톤
body language 신체 언어, 몸짓 언어
facial expression 표정
subtle 미묘한
consciously 의식적으로
evoke 불러일으키다
feedback 피드백
judgment 판단
distrustfulness 불신
distinguish 구별하다
psychiatrist 정신과 의사

Word Search

| 정답 | 1. determine 2. subtle 3. distinguish
1. 철저한 조사 없는 문제의 원인을 알아내기 어려울 수 있다.
2. 그녀의 어조에 그녀가 화가 났음을 나타내는 미묘한 변화가 있었다.
3. 진품과 가품을 구별하는 것은 어려울 수 있다.

Exercise 11~12 — 정답 11 ④ 12 ⑤

(소재) 연민의 개념과 변화

(해석) 고대에는 모든 종류의 고통이 연민을 받을 자격이 있는 것은 아니었다. 아무도 포로로 잡히고 고문당하고 죽임을 당한 적에게 연민을 느껴서는 안 되었다. 노예나 기독교 순교자도 모두 연민을 받을 자격이 없었다. 이후, 형제애와 이웃 사랑의 복음을 전파하는 종교 공동체들은 (자신들에게) 소속되지 않고 믿지 않는 사람들에게 그것(연민)을 거부하는 데 어려움을 느끼지 않았다.

이러한 태도는 18세기에서야 변화하기 시작했다. 아리스토텔레스의 카타르시스 개념을 되살린 극작가들은 연민을 연극, 문학, 그리고 음악에 의해 길러야 할 가장 자연스럽고 가장 도덕적인 인간 능력임을 발견했다. 근대 시민 사회의 도덕적 토대를 모색한 철학자들은 연민과 동정을 자기애와 이기심의 대항력으로 칭송했다. 소설가들은 독자들의 연민을 이끌어 낼 줄거리와 이야기를 열심히 고안하여, 세심하고 분별 있는 시민이 되도록 그들에게 영감을 주려고 애썼다. 동시에, 수십만 명의 유럽 및 북미 남녀가 노예제 종식과 노예 해방을 위한 캠페인을 벌였다. 그들은 '상상된 공감'에 의해 고무되고, 자유와 인간 존엄성이 존중된(→ 침해된) 그 '형제'와 '자매'에 대한 사랑의 언어를 사용하여 노예 무역과 노예 제도에 맞서 전례 없는, 그리고 궁극적으로는 성공한 투쟁을 벌였다.

(문제 해설)

11 고대와 그 다음 시대에는 적군, 노예, 순교자 및 이교도에 대한 연민을 거부했으나, 18세기에 연극, 문학, 철학이 연민을 중요한 인간 능력으로 강조하기 시작했으며, 이는 노예제 폐지 운동에도 영향을 미쳐, 많은 사람들이 공감하여 노예 해방을 위해 싸웠다는 내용의 글이다. 따라서 글의 제목으로 가장 적절한 것은 ④ '연민의 진화: 고대의 태도에서 근대의 태도까지'이다.

① 공감과 친절함을 키우기 위한 조언
② 연민의 역설: 가엾게 여기는 것이 진정한 도움으로 이어질 수 있을까?
③ 지금 그 어느 때보다 공감이 중요한 이유
⑤ 양날의 검: 연민의 힘과 한계 조사하기

12 유럽인과 북미의 남녀가 노예제를 끝내기 위한 캠페인을 벌일 때, 노예 무역과 노예 제도에 의해 자유와 인간 존엄성이 침해된 형제와 자매에 대한 사랑의 언어를 사용했다는 내용이 되어야 하므로 (e)의 respected를 violated와 같은 낱말로 바꿔야 한다.

(구조 해설)

- At the same time, [hundreds of thousands of European and North American men and women] campaigned for [the ending of slavery] and [the liberation of slaves].

첫 번째 [ ]는 문장의 주어이고, 두 번째와 세 번째 [ ]는 and로 대등하게 연결되어 전치사 for에 이어진다.

(어휘 및 어구)

warrant (~을 받을) 자격이 있다, ~을 정당화하다
pity 연민
capture 포로로 잡다
torture 고문하다
deserve (~을 받을) 자격이 있다
religious 종교의, 종교적인
preach 전파하다, 설교하다

deny 거부하다
attitude 태도
playwright 극작가
revive 되살리다
faculty 능력
cultivate (말·행동 방식 등을) 기르다
seek 모색하다, 찾다
foundation 토대, 기초
civil society 시민 사회
praise 칭송하다
sympathy 동정
counter-force 대항력
devise 고안하다
elicit 이끌어 내다
inspire 영감을 주다
liberation 해방
empathy 공감
dignity 존엄성
unprecedented 전례 없는
struggle 투쟁

Word Search

| 정답 | **1.** pity **2.** religious **3.** attitude

1. 고통받고 있는 사람들에 대해 연민을 느끼고, 가능할 때마다 도움을 주는 것이 중요하다.

2. 종교적 믿음이 강하더라도, 의심을 느끼는 것은 정상이다.

3. 혁신에 대한 그 회사의 긍정적인 태도는 많은 성공적인 제품으로 이어졌다.

Week 5 10강   본문 114~125쪽

| **1** ⑤ | **2** ⑤ | **3** ① | **4** ③ | **5** ① | **6** ③ |
| **7** ② | **8** ⑤ | **9** ② | **10** ④ | **11** ⑤ | **12** ⑤ |

Exercise 1 ────────────── 정답 ⑤

소재 외래종의 침입에 대한 토착종의 반응

해석 많은 침입의 두드러진 특징 중 하나는 외래종의 형태를 눈에 띄고 보통 매우 파괴적으로 만드는 대규모의 초기 개체군 급증이다. 일부 종에서는 이러한 급증이 최초 도입 직후에 발생할 수도 있고, 다른 종에서는 수 년, 수십 년 또는 심지어 한 세기 동안의 유예 기간 이후에야 발생한다. 그러나 급증 이후에는 외래종의 개체 수가, 그것이 생태계에 미치는 영향과 더불어 크게 감소할 수도 있다. 많은 경우, 토착종도 외래종의 영향으로부터 그들을 보호해 주는 생태적, 진화적 적응을 한다. 또한, 외래종의 개체 수급증은 토착 군락의 구성원들에게 엄청난 생태적, 진화적 이용의 기회를 창출한다. 생태계의 반응이 빠르게 일어날 수도 있는데, 토착 군락 구성원이 새로운 군락 구성원을 식량 공급원으로 이용하고 그것의 직접적인 해로운 영향을 피하는 법을 알게 되기 때문이다. 나중에 진화적 적응은 토착 군락 구성원이 외래종을 자원으로 이용하거나 외래종의 부정적인 영향을 피할 수 있는 능력을 <u>감소시킨다(→ 향상한다)</u>.

문제 해설

외래종이 침입한 경우 토착종도 자신을 보호하는 빠른 생태적 반응을 보일 수 있으며 시간이 지남에 따라 진화적 적응을 통해 외래종을 자원으로 이용하거나 외래종의 부정적인 영향을 피할 수 있는 능력을 향상하게 된다는 내용이 되어야 문맥상 적절할 것이므로, ⑤의 reduce를 improve와 같은 낱말로 바꿔야 한다.

구조 해설

▪ [Following the outbreak], however, the population of the alien, **as well as** its ecological impact, may decline substantially.

[　]는 때를 나타내는 전치사구이며, 「as well as ~」은 '~과 더불어'라는 뜻이다.

어휘 및 어구

striking 두드러진
feature 특징
invasion 침입
massive 대규모의
outbreak 급증, 창궐
alien 외래(종)의, 이질적인; 외래종

destructive 파괴적인

initial 최초의

ecological 생태계의, 생태적인

substantially 크게, 상당히

adjustment 적응

shield 보호하다

community (식물의) 군락, (생물의) 군집

Word Search

| 정답 | **1.** invasion **2.** massive **3.** shield

1. 적이 마을로 급작스럽게 침입하여 모두를 놀라게 했다.

2. 그 도시는 전국 각지에서 수천 명의 사람들이 모인 대규모 축제를 개최했다.

3. 그 나무들은 동물들을 강한 바람과 많은 비로부터 보호할 수 있다.

Exercise 2

정답 ⑤

소재 스포츠 마케팅

해석 스포츠 마케팅에는 두 가지 측면이 있다는 점에 주목하는 것이 유용하다. 첫 번째는 스포츠 상품과 서비스를 소비자에게 직접 마케팅할 수 있다는 점이다. 두 번째는 스포츠를 활용하여 다른, 스포츠가 아닌 상품과 서비스를 마케팅할 수 있다는 것이다. 다시 말해, 스포츠 마케팅은 스포츠의 마케팅과 스포츠를 통한 마케팅을 포함한다. 예를 들어, 스포츠 소비자에게 직접 스포츠 상품과 서비스를 마케팅하는 것에는 스포츠 장비, 프로 대회, 스포츠 이벤트 및 지역 클럽이 포함될 수 있다. 다른 간단한 예에는 팀 광고, 운동선수를 홍보하기 위한 떠들썩한 선전 기획, 시즌권 판매, 판매용 라이선스 의류 개발이 포함된다. 이와는 대조적으로 스포츠를 통한 마케팅은 스포츠와의 연관을 통해 스포츠가 아닌 상품을 마케팅하는 경우 일어난다. 몇몇 예에는 프로 운동선수가 아침 식사 시리얼을 홍보하거나, 기업이 스포츠 이벤트를 후원하거나, 심지어 맥주 회사가 스포츠 행사장이나 이벤트에서 맥주를 제공하기 위한 독점권을 가지도록 합의하는 것이 포함될 수 있다.

문제 해설

스포츠 상품과 서비스를 소비자에게 직접 마케팅하는 경우와 스포츠가 아닌 상품을 스포츠를 활용하여 마케팅하는 스포츠 마케팅의 두 가지 측면을 설명하는 글이다. 주어진 문장은 대조의 의미를 나타내는 In contrast로 시작하여 스포츠를 통한 마케팅의 개념을 소개하면서, 이는 스포츠와의 연관을 통해 스포츠가 아닌 상품을 마케팅하는 것이라고 설명하고 있는데, ⑤의 앞에서는 스포츠 상품과 서비스를 직접 마케팅하는 경우의 예시를 들었고, ⑤의 뒤에서는 스포츠가 아닌 상품을 마케팅하는 경우의 예시가

나오므로, 주어진 문장이 들어가기에 가장 적절한 곳은 ⑤이다.

구조 해설

■ **It** is useful [to note {that there are two angles to sport marketing}].

It은 형식상의 주어이고, [ ]는 내용상의 주어이다. [ ] 안의 { }는 note의 목적어 역할을 하는 명사절이다.

어휘 및 어구

association 연관

note 주목하다

involve 포함하다

equipment 장비

competition 대회

athlete 운동선수

corporation 기업, 회사

Word Search

| 정답 | **1.** association **2.** equipment **3.** athlete

1. 그 로고의 디자인은 그 회사의 브랜드 정체성과 강한 연관을 가지고 있다.

2. 이 사이클링 프로그램은 자전거와 헬멧과 같은 필요 장비를 제공한다.

3. 그 운동선수는 큰 경주를 준비하기 위해 매일 아침 훈련한다.

Exercise 3

정답 ①

소재 홍수와 가뭄 예측

해석 특정 사건이 특정 시간 범위 내에서 특정 위치에서 발생할 확률을 알아내는 것은 재해 평가의 필수적인 목표이다. 많은 큰 강의 경우, 우리는 특정 기간 동안 발생할 특정 규모의 평균 홍수 횟수를 합리적으로 예측할 수 있는 확률 모델을 개발하기에 충분한 범람 기록을 가지고 있다. 마찬가지로 가뭄도 그 지역의 과거 강우 발생에 기반하여 확률이 정해질 수 있다. 그러나 이러한 확률은 주사위를 던져 특정 숫자가 나올 확률이나 포커에서 인사이드 스트레이트 패를 완성할 확률과 유사한데, 항상 우연의 요소가 존재한다. 예를 들어, 10년 주기 홍수는 평균적으로 10년마다 발생할 수도 있지만, 주사위를 던져 6이 두 번 연속으로 나올 수 있는 것과 꼭 마찬가지로 어떤 한 해에 이 정도 규모의 홍수가 여러 번 발생하는 것이 가능하다.

→ 재해 평가에는 기록을 기반으로 특정 시간과 장소에서 홍수나 가뭄과 같은 사건의 가능성을 추정하는 것이 포함되는데, 항상 자체에 무작위성의 요소를 지닌다.

문제 해설

재해 평가는 과거의 기록을 기반으로 특정 사건이 발생할 확률을

알아내는 것인데, 이 확률은 주사위를 던지거나 포커를 할 때 특정 숫자가 나올 확률과 유사하여 우연이라는 요소가 항상 존재하기 때문에, 10년 주기로 홍수가 일어난 과거 기록이 있다고 하더라도 실제로는 그 정도 규모의 홍수가 한 해에 여러 번 발생하는 것도 가능하다는 내용의 글이다. 따라서 요약문의 빈칸 (A), (B)에 들어갈 말로 가장 적절한 것은 ① '가능성(likelihood) – 무작위성(randomness)'이다.

② 가능성 – 특정 지역
③ 강도 – 특정 지역
④ 기간 – 무작위성
⑤ 기간 – 표준화

구조 해설

■ For example, the 10-year flood may occur on the average of every 10 years, but **it** is possible [{for several floods of this magnitude} to occur in any one year], just as **it** is possible [to throw two straight sixes with a die].

두 개의 it은 각각 형식상의 주어이고, 두 개의 [ ]는 각각 it에 대한 내용상의 주어이다. 첫 번째 [ ] 안의 { }는 to occur in any one year의 의미상의 주어를 나타낸다.

어휘 및 어구

determine 알아내다, 결정하다
probability 확률
particular 특정한
hazard evaluation 재해 평가
reasonably 합리적으로
drought 가뭄
assign (날짜나 시간 등을) 정하다, 배정하다
on the basis of ~에 기반하여
element 요소

Word Search

| 정답 | **1.** probability **2.** reasonably **3.** element

1. 간단히 말해서, 확률은 어떤 사건이 발생할 가능성을 숫자로 표현한 것이다.

2. 그 가족은 모두의 선호도를 고려하면서 휴가 계획에 대해 합리적으로 논의했다.

3. 팀 협업은 그 업무의 중요한 요소이다.

Tips

인사이드 스트레이트: 예컨대 포커 게임에서 손에 5, 3, 7, 8, 9 카드 다섯 장을 쥐고 있을 때 3카드를 버리고 6카드를 받으면 스트레이트 (5, 6, 7, 8, 9)를 만들 수 있다. 이때 손에 쥔 카드 가운데 3카드를 제외한 넉 장의 패를 인사이드 스트레이트라고 한다.

Exercise 4
정답 ③

소재 시각적 코드에 의존한 기억

해석 언어적인 자극에 대해, 말에 의한, 즉 음운론에 기반한 일시적인 기억과 대조하여 명백히 시각적인 기억의 사용을 시사한 최초의 연구 중에는 심리학자 Posner와 그의 동료들에 의해 수행된 연구가 있는데, 이들은 쌍을 이룬 글자가 대문자(예를 들어 AB) 또는 소문자(예를 들어 aa) 버전 또는 두 가지가 혼합된 버전(예를 들어 Bb)으로 제시되는 시각적 글자 조합 과제를 개발했다. 피험자들의 과제는 쌍에 있는 글자가 같은 이름(예를 들어 Aa)인지 다른 이름(예를 들어 Ab)인지에 기반하여 응답하는 것이다. 글자들의 대소문자 여부가 같고 형태가 동일한 경우(예를 들어 AA), 피험자들은 대문자와 소문자 버전이 다른 경우보다 훨씬 더 빠르게 응답했다. 형태가 동일한 글자에 대한 이점은 각 쌍에 있는 글자가 최대 2초의 글자 간 지연을 두고 차례로 제시될 때도 유지되었다. 이는 피험자들이 지연 시간 동안은 글자에 대해 시각적 코드에 의존하고 있었으며, 그 후에는 이름 코드가 판단을 위해 사용되고 있었음을 시사했다.

문제 해설

(A) 실험 내용상 사람들이 시각적 글자 조합 과제에서 글자 형태에 의존했다는 면에서 시각을 기반으로 한 일시적 기억을 사용했음을 시사하는 연구임을 알 수 있으므로, use(사용)를 써야 한다. avoidance는 '회피'라는 뜻이다.

(B) 제시된 쌍의 글자가 같은 이름인지 다른 이름인지에 기반하여 응답하는 과제에서 형태가 동일한 글자에 대해 이점이 있었다고 했고, 따라서 대소문자 여부가 같고 형태가 동일한 경우 훨씬 더 빠르게 응답했을 것이므로, quickly(빠르게)를 써야 한다. slowly는 '느리게'라는 뜻이다.

(C) 각 쌍의 글자가 지연을 두고 차례로 제시될 때에도 형태가 동일한 글자에 대한 이점이 유지되었다고 한 것으로 미루어 볼 때, 그 시간 동안에는 피험자들이 시각적 코드에 의존하고 있었음을 알 수 있으므로, relying(의존하는)을 써야 한다. pass는 '지나치다, 전달하다'라는 뜻이다.

구조 해설

■ [Among the first studies to suggest the use of specifically a visual in contrast to verbal or phonologically based temporary memory for verbal stimuli] **were** those conducted by psychologist Posner and his colleagues, [who developed a visual letter-matching task {in which pairs of letters were shown in their upper case (e.g., AB) or lower case (e.g., aa) versions or a mixture of both (e.g., Bb)}].

전치사구인 첫 번째 [ ]가 맨 앞으로 나오면서 동사(were)와 주어가 도치되었다. 두 번째 [ ]는 psychologist Posner and his colleagues를 부연 설명하는 관계절이고, 그 안의 { }는 a visual letter-matching task를 수식하는 관계절이다.

어휘 및 어구

specifically 명백히, 특히
verbal 말에 의한, 언어의
stimulus 자극(*pl.* stimuli)
identical 동일한
advantage 이점, 장점

Word Search

| 정답 | **1.** stimuli **2.** identical **3.** advantage
1. 식물은 빛과 온도 변화와 같은 환경 자극에 반응한다.
2. 그녀의 필체가 그녀 어머니의 필체와 거의 동일해서 둘을 구별하기 어려웠다.
3. 여러 언어에 능통하다는 것은 여행할 때 상당한 이점일 수 있다.

Exercise 5~6
정답 5 ① 6 ③

소재 인플루언서를 이용한 마케팅 전략

해석 점점 더 많은 브랜드와 조직이 영향력을 가진 사람들과 협력하여 작업하기 시작하면서 인플루언서에 대한 수익률(네, 언어 유희를 의도했습니다)이 감소했다. 사람들은 자신의 영웅이 소셜 미디어에서 브랜드를 홍보하도록 돈을 받고 있다는 것을 깨닫는다. 유명인이 브랜드 홍보를 위해 단 하나의 인스타그램 게시물에 한 번 언급하는 것으로 받을 수 있는 금액은 최대 50만 달러까지 올라간다. 따라서 지난 몇 년 동안 브랜드는 인플루언서 마케팅 전략을 소위 마이크로 인플루언서에게까지 서서히 확장해 왔다. 이 사람들은 팔로워 수가 10,000명에서 50,000명에 이르는 사람들로, 더 소규모인 사람들 집단에게 전문가 또는 영웅으로 여겨지는 사람들이다. 팔로잉의 규모 때문에 소셜 미디어에서 이들의 메시지는 진정성 있고 그 결과 더 신뢰할 만하다고 인식된다. 이 전략이 또한 (너무 많은) 인기를 얻으면서 몇몇 전문가들은 2020년대가 나노 영향력을 가진 사람들의 시대가 될 것이라고 예측했다. 이들은 1,000~10,000명의 팔로워를 가지고 있으며 여러분 자신이 가장 좋아하는 친구나 가족과 꼭 같은 사람들인데, 그들이 어떠한 실제적인 명성도 가지고 있지 않다는 것이 그들을 <u>신뢰할 수 없음(→ 신뢰할 수 있음)</u>을 나타내는 표시일 것이다. 이 나노 카테고리도 결국에는 진정성을 잃고 신뢰가 마지막 카테고리인 피코 영향력을 가진 사람들로 옮겨갈 것으로 예상된다. 이것은 1,000명 미만의 인맥을 가진 사람들, 즉 여러분, 나, 그리고 개를 키우는 내 이웃의 수준이다. 이들은 우리가 실생활에서 알고 있고 관계를 맺고 있는 사람들이다. 흥미롭게도, 그렇다면, 우리가 완전히 한 바퀴 돌아서 우리가 역사상 존재했던 곳으로, 즉 우리가 실제로 매우 잘 아는 사람들을 신뢰했던 때로 되돌아갔다는 것이다.

문제 해설

5 돈을 받고 인플루언서들이 제품 홍보를 한다는 것을 알게 되면서 점차 팔로워가 적은 인플루언서로 마케팅 전략이 이동하고 있으며 앞으로 이동하게 될 것이라는 내용의 글이므로, 글의 제목으로 가장 적절한 것은 ① '인플루언서 마케팅의 트렌드와 예측'이다.
② 디지털 마케팅이 비즈니스 성장에 미치는 영향
③ 고객 후기를 활용한 브랜드 이미지 개선하기
④ 더 젊은 세대에게 호소하기 위해 소셜 미디어 활용하기
⑤ 긍정적인 브랜드 경험을 통해 고객 충성도 높이기

6 인플루언서가 돈을 받고 제품 홍보를 한다는 사실이 알려져, 브랜드는 제품 홍보를 위해 더 진정성이 있고 신뢰성이 있다고 여겨지는 팔로워 수가 더 적은 인플루언서를 활용하는 방향으로 나아가고 있다고 했다. 따라서 1,000명에서 10,000명에 해당하는 팔로워를 가진 인플루언서가 실제 명성이 부족한 것은 신뢰성이 있음을 나타낸다는 내용이 되도록 (c)의 untrustworthiness를 trustworthiness와 같은 낱말로 바꿔야 한다.

구조 해설

■ These are the people [that we know in real life] and [with whom we have a relationship].
두 개의 [ ]는 모두 관계절로 and로 연결되어 the people을 수식한다.

어휘 및 어구

influence 영향력을 가진 사물이나 사람, 영향(력)
pun (다의어, 동음이의어를 이용한) 언어 유희, 말장난
promote 홍보하다
celebrity 유명인
strategy 전략
perceive 인식하다
authentic 진정성 있는
trustworthy 신뢰할 수 있는
lack 가지고 있지 않음, 부재

Word Search

| 정답 | **1.** perceive **2.** strategy **3.** trustworthy
1. 예술가들은 자주 세상을 고유하고 상상력 넘치는 방식으로 인식한다.
2. 우리 팀의 승리 전략은 강력한 수비와 빠른 역습의 조합이다.
3. 그 웹사이트는 정확하고 신뢰할 수 있는 정보를 제공하기 때문에 신뢰할 수 있다고 여겨진다.

Exercise 7

정답 ②

소재 과대광고의 부정적인 측면

해석 인식론적 관점에서, 핵심 질문은 광고가 진실하도록 기대되는 정도와 관련되어 있다. 광고는 흔히 몹시 과장되거나 비유적인 주장을 사용하는데, 이는 '과대광고'라고 불리는 관행이며, 이것은 오도하는 광고와는 달리 예를 들어 미국과 같은 일부 국가에서 합법적으로 허용된다. 이것의 합법성은 사람들이 그러한 주장을 문자 그대로 받아들이거나 그에 따라 행동하지 않을 것이라는 가정에 기반한다. 하지만 이러한 가정은 소비자들이 사실 부풀려진 문구에 '실제로' 반응한다는 것을 보여 주는 실증적 증거에 의해 확인된다(→ 반박된다). 하기야 일부 소비자들이 과대광고를 있는 그대로의 것으로 인식한다 할지라도, 다른 소비자들은 더 취약하고 그것을 액면 그대로 받아들인다. 따라서 이러한 관행은 '기업이 더 견문이 넓거나 더 숙고하는 소비자들은 믿지 않겠지만, 다른 소비자들은 희생자가 될 수 있는 기만적인 진술을 하는 것이 정당한가?'라는 공정성의 문제를 또한 발생시킨다. 그러나 세상사에 밝은 소비자들조차도 이러한 전략에 시달리는데, 그것은 어떤 광고를 진지하게 받아들여야 하는지 재확인해야 하기 때문이다. 미국 법에서의 과대광고의 처리를 분석하면서, 법학자 David Hoffman은 그것이 부정적인 외부 효과를 유발하는 것으로 이해되어야 한다고, 즉 그것은 '판매자로부터의 보상 없이 구매자가 현재 감당하는 정보 부담'을 만들어 낸다고 주장한다.

문제해설
과대광고의 합법성은 사람들이 과대광고의 주장을 문자 그대로 받아들이거나 그에 따라서 행동하지 않을 것이라는 가정에 기반하지만, 일부 소비자들은 더 취약하고 그것을 액면 그대로 받아들인다는 증거가 존재하기 때문에 이러한 가정이 실증적 증거에 의해 반박된다고 해야 문맥상 자연스럽다. 따라서 ②의 confirmed를 contradicted와 같은 낱말로 바꿔야 한다.

구조해설

• Advertisements often use widely exaggerated or metaphorical claims, [a practice called "puffery," {which, in contrast to misleading advertising, is legally permitted in some countries, for example, the United States}].
[ ]는 앞 절과 의미상 동격 관계를 이룬다. { }는 a practice called "puffery"를 부가적으로 설명하는 관계절이다.

어휘 및 어구
extent 정도
metaphorical 비유적인, 은유적인
misleading 오도하는
legality 합법성, 적법성
literally 문자 그대로
empirical 실증적인
vulnerable 취약한
take ~ at face value ~을 액면 그대로 받아들이다
legitimate 정당한
reflective 숙고하는
fall prey to ~의 희생자가 되다
sophisticated 세상사에 밝은, 지적 수준이 높은
bear (책임 등을) 감당하다
compensation 보상

Word Search

| 정답 | 1. legality   2. vulnerable   3. sophisticated

1. 그 변호사는 계약의 불명확한 조항 때문에 그 합법성에 의문을 제기했다.

2. 둥지 속의 어린 새들은 어미의 보호 없이 취약했다.

3. 그 지적 수준이 높은 신사는 그의 폭넓은 예술 지식으로 모두에게 깊은 인상을 주었다.

Exercise 8

정답 ⑤

소재 생태계 서비스에 부여하는 경제적 가치

해석 환경 경제학자들은 흔히 생태계 서비스에 경제적 가치 평가액을 매기고자 한다. 이렇게 하는 이유는 생태계 서비스가 역사적으로 환경 관련 의사 결정과 정책에서 덜 고려되었고 저평가되었기 때문이다. 생태계 서비스에 금전적 가치를 부여하는 것은 생태계 서비스의 중요성을 높인다. 전 세계 생태계 서비스의 경제적 가치는 연간 125조 달러로 추정된다. 이것은 잘 작동하는 생태계가 인간의 행복에 정말로 얼마나 대단히 중요한지, 그리고 아울러 인간의 시스템이 잘 작동하는 생태계와 밀접하게 관련되고 그것에 의존하는 정도를 분명하게 알 수 있도록 한다. 하지만, 환경 관련 의사 결정에는 흔히 전 세계적 또는 거시적 평가보다 장소나 (생태) 시스템에 국한된 경제적 가치 평가가 더 중요하다. 모든 생태계와 공간이 동일한 생태계 서비스와 천연자원 가치를 갖는 것은 아니다. 예를 들어, 평균 1헥타르의 외해(外海)는 평균 1헥타르의 암초보다 더 적은 (생태계) 서비스를 제공하며, 평균 1헥타르의 사막은 평균 1헥타르의 열대 우림보다 더 적은 생태계 서비스를 제공한다.

문제해설
주어진 문장은 모든 생태계와 공간이 동일한 생태계 서비스와 천연자원 가치를 갖는 것은 아니라는 내용으로, 이 문장 앞에는 이러한 경제적 가치 평가는 거시적인 평가보다 장소 또는 생태 시스템에 국한된 평가가 더 중요하다는 문장이 있어야 하고, 바로 뒤

에는 주어진 문장의 예시로 같은 면적이라 할지라도 외해와 암초, 사막과 열대 우림이라는 서로 다른 생태계와 공간이 제공하는 생태계 서비스의 양이 다를 수 있다는 내용의 문장이 이어져야 한다. 따라서 주어진 문장이 들어가기에 가장 적절한 곳은 ⑤이다.

구조해설

- For example, an average hectare of open ocean provides fewer services [than **does** {an average hectare of reef}], and an average hectare of desert provides fewer [than **does** {an average hectare of tropical rainforest}].

두 개의 [ ]에서는 모두 than 다음에 대동사 does와 주어인 { }가 도치되었다.

어휘 및 어구

valuation 가치 평가(액)
underappreciate 저평가하다
monetary 금전적인
trillion 조(兆)
put ~ into perspective ~을 분명하게 알 수 있게 하다
crucial 대단히 중요한, 결정적인
dependent 의존하는
macro 거시적인
hectare 헥타르(1만 제곱미터)
open ocean 외해(육지에서 멀리 떨어진 바다)
tropical rainforest 열대 우림

Word Search

| 정답 | 1. monetary 2. crucial 3. dependent
1. 한 젊은 사업가가 투자자들로부터 새로운 사업을 시작하기 위한 금전적 지원을 받았다.
2. 균형 잡힌 식사를 하는 것은 좋은 건강을 유지하고 질병을 예방하는 데 대단히 중요하다.
3. 그 작은 섬의 경제는 생존을 위해 관광업에 크게 의존하고 있다.

Tips

ecosystem service(생태계 서비스): 생태계의 존재와 기능이 인간에게 제공하는 여러 혜택. 예컨대 깨끗한 물과 공기, 식량 생산, 기후 조절, 홍수 예방, 토양 형성 등이 있다.

Exercise 9 ── 정답 ②

소재 Hume의 자연법칙에 대한 견해

해석 자연법칙은 우리가 보거나 만지거나 들을 수 있는 것이 아

니라, 그것은 스코틀랜드 출신 철학자 David Hume에 따르면, 단지 과거 사건의 집합에 관한 기술일 뿐이다. 예를 들어, 일련의 사건들을 생각해 보자. 나는 펜을 놓는다. 펜이 떨어진다. 나는 펜을 놓는다. 펜이 떨어진다. 나는 펜을 놓는다. 그리고 계속해서 반복된다. 이러한 (사건의) 연속은 중력과 관성과 같은 일군의 법칙으로 설명될 수도 있다. 그러나 이 일련의 사건은 그 자체만으로는 과거에 일어난 사건들을 단지 '기술'할 뿐이고, 그것은 '왜' 펜이 그런 식으로 반응하는지를 우리에게 알려 주지 않으며, 또 펜이 항상 그런 식으로 반응할 것이라는 점도 우리에게 알려 주지 않는다. 그리고 설령 과거의 그 사건들이 법칙 때문에 일어났다고 해도, 왜 그 법칙들이 미래에도 사건들에 계속 영향을 미칠 것이라고 믿어야 하는가? 만약 우리가 과거에 법칙이 존재했다고 말한다면, 우리는 그 법칙들이 미래에 존재할 것이라고 우리에게 알려 주는 추가적인 어떤 법칙이 필요하다. 그러나 우리는 단지 과거에 그 법칙들이 작용한 방식을 기술할 수 있을 뿐인 듯하다.
→ David Hume이 단지 과거 사건의 집합을 나타낼 뿐이라고 주장하는 자연법칙은, 그것들의 지속적인 유효성에 대한 설명이 없으므로 미래의 반응을 보장하지 않는다.

문제해설

David Hume에 따르면 자연법칙은 단지 과거 사건의 집합에 관한 기술일 뿐이며, 그 법칙이 미래에도 존재할 것이라고 말하려면 추가적인 법칙이 필요하다는 내용의 글이다. 따라서 요약문의 빈칸 (A), (B)에 들어갈 말로 가장 적절한 것은 ② '나타내다(represent) – 지속적인(continuing)'이다.
① 왜곡하다 – 보편적인
③ 추적하다 – 통계적인
④ 지나치게 단순화하다 – 지속적인
⑤ 미화하다 – 통계적인

구조해설

- And [even if those past events happened {because of laws}], why believe [those laws *will* continue to govern events in the future]?

첫 번째 [ ]는 양보의 의미를 나타내는 부사절이고, { }는 이유의 부사구이다. 두 번째 [ ]는 believe의 목적어 역할을 하는 명사절이다.

어휘 및 어구

natural law 자연법칙
grouping 집합, 집단
gravity 중력
govern 영향을 미치다, 규정하다
additional 추가적인

Word Search

| 정답 | 1. gravity 2. govern 3. additional

1. 그 공은 중력, 즉 사물을 아래로 당기는 힘 때문에 땅에 떨어졌다.
2. 새로운 법률은 이 나라에서 기업이 운영되는 방식을 규정할 것이다.
3. 숙제에 추가 도움이 필요하면 수업 후에 여러분의 선생님께 요청하세요.

Exercise 10
정답 ④

소재 고전 물리학과 양자 원리

해석 우주는 지속적으로 팽창하고 있음에도 불구하고 여전히 질서 정연하다. 관찰하고 측정할 수 있는 일관된 규칙들이 존재한다. 다양한 규모에서 한 입자와 다른 입자의 관계에 영향을 미치는 신뢰할 수 있는 물리적 원리들이 있다. 300년 전, 뛰어난 수학자였던 Isaac Newton 경은 물리적 공간에서의 큰 물체들의 움직임을 정의하는 운동 법칙들을 체계적으로 나타냈다. 이것은 우리가 포탄이나 로켓 항공기의 정확한 궤도를 계산할 수 있게 해주는 신뢰할 수 있고 재현할 수 있는 물리적 규칙들이다. 이 규칙들은 우리가 직접 관찰할 수 있고 크기가 미시적이지 않은 물체들의 규모에 적용된다. 하지만 지난 한 세기 동안, 원자 규모가 어떻게 작동하는지에 영향을 미치는 새로운 일련의 양자 규칙들이 발견되었다. 이러한 양자 원리 중 일부는 우리의 일반적인 감각과 모순된다. 이 미시적인 규모에서 입자 간의 작용은 물리적 원인과 결과 사이의 명확한 관계에 대한 우리의 확립된 기대를 확증한다(→ 에 따라 일어나지 않는다). 대신, 양자 측정은 한 물체와 다른 물체 사이의 관계에 대한 내재하는 불확실성을 드러낸다. 그러나 이러한 내재된 불확실성은 두 자동차가 서로 충돌할 때 일어나는 일에 영향을 미치는 그러한 고전 물리학에서의 작용력만큼이나 우리의 삶과 정말 관련이 있다.

문제 해설
양자 원리 중 일부는 우리의 일반적인 감각과 모순되며 한 물체와 다른 물체 사이의 관계에 대한 불확실성을 드러낸다고 했으므로, 미시적 규모에서 입자 간의 작용은 물리적 원인과 결과 사이의 명확한 관계에 대한 우리의 확립된 기대에 따라 일어나지 않는다는 맥락이 되어야 한다. 따라서 ④의 confirm을 defy와 같은 낱말로 바꿔야 한다.

구조 해설
- Yet, these built-in uncertainties are just **as relevant** to our lives **as** those Newtonian forces [that govern {what happens when two cars crash into each other}].
  '…만큼 ~한'이라는 의미의 「as ~ as ...」의 원급 비교가 사용되었다. [ ]는 those Newtonian forces를 수식하는 관계절이고, { }는 govern의 목적어 역할을 하는 명사절이다.

어휘 및 어구
inflate 팽창하다

consistent 일관된
particle 입자
formulate 체계적으로 나타내다
law of motion 운동 법칙
reproducible 재현할 수 있는
microscopic 미시적인, 극히 작은
contradict ~과 모순되다
clear-cut 명확한
reveal 드러내다
inherent 내재하는
Newtonian 고전 물리학의, 뉴턴 학설의

Word Search

| 정답 | **1.** consistent **2.** reproducible **3.** contradict
1. Kate의 영어를 연습하는 일관된 노력은 그녀의 말하기 능력을 크게 향상시켰다.
2. 실험은 다른 연구자들이 결과를 재현할 수 있도록 명확한 지침을 가져야 한다.
3. 그 정치인의 세금 인하 약속은 세금 인상을 지지한 그의 투표 기록과 모순된다.

Exercise 11~12
정답 11 ⑤ 12 ⑤

소재 인간 뇌 발달의 과정

해석 인간은 매우 장기간에 걸친 뇌 발달 기간을 가지고 있다. 이는 사람의 20대 초반까지 완전히 성숙되지 않은 전전두엽 피질과 같이 행동 제어를 조정하는 회로에 특히 해당된다. 그때까지 시냅스는 여전히 대규모로 바뀌고 있다. 이는 이러한 회로가 경험에 의해 형성될 수 있는 풍부한 기회를 제공하며, 아마도 '인지적 적소'라고 불리는 것을 성공적으로 차지할 수 있는 우리 능력의 핵심 요인 중 하나일 것이다. 특정 환경에 적응하는 대신, 제한된 일련의 타고난 본능적 행동으로, 우리는 인지적 유연성과 반응성을 진화시켰으며, 그것은 우리가 각자 개인이 처한 환경에 적응할 수 있도록 했다. 반복되는 패턴은 강화되고 습관적인 행동 방식이 나타난다. 우리는 점차 우리 자신이 된다.
하지만 어느 시점에 우리는 끊임없이 변하기를 멈추고 경력을 쌓거나 짝을 찾는 것과 같은 중요한 일을 그냥 해 나가야 한다. 그것은 우리가 만들어 온 적응 형태를 공고히 하고 더 이상의 변화를 제한해야 한다는 것을 의미한다. 우리는 제어되지 않는 양성 피드백 고리를 영원히 가질 수 없다. 즉 우리는 계속하여 우리 자신으로 남기 위해 이러한 신경 구성을 유지해야 하는 것이다. 대대적인 가소성(뇌나 신경 회로가 환경 요인에 따라 특정 방향으로 변화하는 성질) 기간은 감각 회로보다 행동 및 인지 회로에서 상당히 더 오래 지속되지만, 그럼에도 우리가 성인이 되면 이 기간도 끝난다. 가소성 과정 자체는 발달하는 뇌의 '자유도'를 계속하여 팽창시켜(→ 좁혀) 놓아, 정적 강화와 덜 선호되는 상태를 매개하는 연결의 계속된 제거를 통해서 초기의 편향을 강화할 것이다. 하지만 뇌의 생화학적 성질 또한 성숙과 함께 변화하며, 그리하여 가소성과 유연성의 기제가 안정성과 유지의 기제로 대체된다.

(문제 해설)

11 인간의 뇌 발달 과정은 처음에는 인지적 유연성과 반응성을 바탕으로 개별 환경에 적응하는 방향으로 진행되다가 어느 시점에 가서는 변화를 제한하고 안정성을 추구하는 방향으로 진행된다는 내용의 글이므로, 글의 제목으로 가장 적절한 것은 ⑤ '유연한 상태에서 안정된 상태로: 인간 뇌의 점진적 성숙'이다.
① 뇌를 속여서 지속적인 습관 형성을 하게 하는 방법
② 유연한 사고방식이 뇌 발달에 기여하는 이유
③ 노화하는 뇌: 인지 저하에 대한 근거 없는 믿음과 진실
④ 물리적 환경 변화에 우리 뇌가 반응하는 방법

12 성인이 되면 뇌가 유연하게 발달할 수 있는 가소성 기간이 끝나게 되고, 뇌 발달 과정에서 정적 강화와 덜 선호되는 상태를 매개하는 연결이 계속 제거되는 과정을 통해 초기의 편향이 강화된다고 했으므로, 가소성 과정이 결국 뇌의 자유도를 좁힐 것임을 추론할 수 있다. 따라서 (e)의 expanded를 narrowed와 같은 낱말로 바꿔야 한다.

(구조 해설)

- [**Rather than being adapted** for specific environments], with a limited set of hardwired, instinctive behaviors, we have evolved cognitive flexibility and responsiveness, [allowing us to adapt ourselves to our individual environments].

  첫 번째 [ ]에서는 Rather than 다음에 동명사의 수동형 (being p.p.)이 쓰였다. 두 번째 [ ]는 앞 절의 내용의 결과를 나타내는 분사구문이다.

(어휘 및 어구)

circuit (뇌) 회로
mediate 조정하다, 매개하다
mature 성숙하다
synapse 시냅스(신경 세포 간의 연결 부위)
modify 바꾸다, 수정하다
populate (장소를) 차지하다
hardwired 타고난
instinctive 본능적인
reinforce 강화하다
adaptation 적응 (형태)
restrict 제한하다
runaway 제어되지 않는
neural 신경의
wholesale 대대적인, 대규모의
plasticity 가소성
sensory 감각의
progressively 계속해서
magnify 강화하다, 증대하다
bias 편향
positive reinforcement 정적 강화
biochemistry 생화학적 성질[작용]

Word Search

| 정답 | **1.** instinctive   **2.** adaptation   **3.** sensory
1. 그 노인은 낯선 사람에 대한 본능적인 불신을 가지고 있었다.
2. 카멜레온이 피부색을 바꿀 수 있는 능력은 포식자로부터 숨는 데 도움이 되는 적응 형태이다.
3. 우리의 눈과 귀 같은 감각 기관은 주변 세상을 인식하는 데 도움을 준다.

(Tips)

positive feedback loop(양성 피드백 고리): 반응의 결과물이 그 반응의 시작을 증폭하는 현상. 이것은 빠른 반응과 극적인 변화를 가능하게 하여 생명체가 특정 상황에 효과적으로 대응할 수 있게 한다.

| 1 ⑤ | 2 ④ | 3 ① | 4 ③ | 5 ③ | 6 ④ |
| 7 ④ | 8 ② | 9 ③ | 10 ④ | 11 ② | 12 ⑤ |

Exercise 1
정답 ⑤

소재 이상 행동의 유형

해석 심리학자인 Daniel Kahneman과 고(故) Amos Tversky의 선구적인 연구에서 영감을 받아, 심리학 분야에서 사람들이 표준 경제 모델의 예측과 규칙을 명백히 위반하는 이상 행동에 대한 대규모 목록을 작성했다. 예를 들어, 어떤 사람이 20달러짜리 시계 라디오에 10달러를 절약하기 위해 기꺼이 시내를 가로질러 차를 몰고 가지만, 1,000달러짜리 텔레비전에 10달러를 절약하기 위해서는 그렇게 하고 싶어 하지 않는 것이 보통이다. 그러나 그 운전을 한 이익은 각각의 경우 10달러이다. 따라서 운전의 암묵적 비용이 10달러보다 더 적다면, 합리적인 사람이라면 두 경우 모두 시내를 가로질러 운전할 것이다. 사람들은 절약한 10달러가 그 가격에 비해 매우 작은 비율에 불과하다고 말함으로써 텔레비전을 사기 위한 운전을 주저하는 것을 흔히 설명한다. 하지만 합리적인 사람은 이익과 비용을 백분율로서가 아니라 상대적인 (→ 절대적인) 면에서 생각한다.

문제 해설
표준 경제 모델의 예측과 규칙을 위반하는 이상 행동의 유형을 설명하는 글이다. 20달러짜리 시계 라디오를 사면서 10달러를 절약할 수 있으면 기꺼이 시내를 가로질러 운전해 가지만, 1,000달러짜리 텔레비전 세트를 사면서 10달러를 절약할 수 있으면 운전해 가지 않는다는 사례를 통해 이익과 비용을 계산할 때 사람들이 보통 절대적인 금액이 아니라 백분율로 이익을 계산해 행동함을 보여 주고 이와 반대로 합리적인 사람은 금액으로 이익을 계산해 두 경우 모두 운전했을 것이라고 했으므로, ⑤의 relative를 absolute와 같은 낱말로 바꿔야 한다.

구조 해설

■ People often explain [their reluctance {to make the drive for the television}] **by saying** [the $10 savings is such a small percentage of its price].
첫 번째 [ ]는 explain의 목적어 역할을 한다. { }는 their reluctance의 구체적인 내용을 설명하는 to부정사구이다. '~함으로써'의 의미를 나타내는 「by+-ing」가 쓰였고, 두 번째 [ ]는 saying의 목적어 역할을 한다.

어휘 및 어구
inspire 영감을 주다

pioneering 선구적인
inventory 목록
violate 위반하다
prediction 예측
prescription 규칙, 규정, 처방
implicit cost 암묵적 비용(자신이 선택하지 않고 포기하는 다른 기회의 잠재적 비용)
rational 합리적인
reluctance 주저, 머뭇거림

Word Search

| 정답 | 1. inspired 2. reluctance 3. rational
1. 그 디자이너의 최신 컬렉션은 빈티지 패션 트렌드에서 영감을 받았다.
2. 새로운 정책을 채택하는 것에 대해 직원들 사이에 눈에 띄는 주저함이 있었다.
3. 복잡한 문제를 해결하기 위해서, 우리는 합리적이고 체계적인 접근이 필요하다.

Exercise 2
정답 ④

소재 다양한 세대와의 소통 방식

해석 세대에 상관없이 모든 사람은 소중하게 여겨지고 존중받으며 경청되기를 바란다. 하지만 우리 각자는 자신의 생각을 표현하고 정보를 받아들이는 방식이 다를 수도 있을 것이다. 내 고객 중 한 사람은 베이비붐 세대 직원 중 일부가 밀레니얼 세대 직원들과 개인적인 관계를 형성하려고 애쓰고 있는 모습에 고심하고 있었다. 이들 직원 중 몇 명은 각 세대와 소통하는 구체적인 방법을 추천하는 기사를 읽었다. 그런 다음 베이비붐 세대 직원들은 이어서 다양한 세대의 모든 동료들과 소통하는 방식에 영향을 주기 위해 이 정보를 사용했다. 그들은 베이비붐 세대 동료들의 사무실을 방문했고, X세대 동료들에게 이메일을 보냈고, 밀레니얼 세대에게 문자를 보냈다. 이것은 표면적으로는 괜찮아 보였으나 곧 약간의 논란을 불러일으키기 시작했는데, 왜냐하면 사람들이 다른 사람은 대면 대화를 하고 있는데 왜 자신은 이메일을 받고 있는지 이해하지 못했기 때문이다. 대면 대화를 나누는 것을 특정 동료로만 제한함으로써, 이 직원들은 의도치 않게 다른 직원들이 소외되고 경시되는 느낌을 가지도록 만들었다.

문제 해설
④의 앞 문장에는 베이비붐 세대 직원들이 자신들이 읽은 기사에서 추천한 방식으로 각 세대 동료들과 소통하려 했다는 내용이 언급되어 있고, ④의 뒤에는 이것이 표면적으로는 괜찮아 보였으나 사람에 따라 차별 대우를 받는 것 같은 느낌을 주게 되어 논란이

생겨났다는 내용이 언급되어, 두 문장 사이에 정보의 단절이 있다. 따라서 베이비붐 세대가 각 세대의 동료들과 어떤 방식으로 소통하려고 시도했는지를 설명하는 주어진 문장이 들어가기에 가장 적절한 곳은 ④이다.

구조 해설

- On the surface, this seemed fine, but it soon started to stir up some controversy [because people did not understand {why they were getting an email, ⟨when someone else was getting a face-to-face conversation⟩}].

  [ ]는 이유를 나타내는 부사절이다. { }는 understand의 목적어 역할을 한다. 그 안의 ⟨ ⟩에서 when은 '~하는데도, ~함에도 불구하고'라는 뜻이다.

어휘 및 어구

regardless of ~에 상관없이
generation 세대
value 소중하게[가치 있게] 여기다[생각하다]
express 표현하다
receive 받아들이다
client 고객
struggle with ~로 고심하다
inform ~에 영향을 주다
stir up ~을 불러일으키다
controversy 논란
limit 제한하다
unintentionally 의도치 않게

Word Search

| 정답 | **1.** generation(s) **2.** unintentionally **3.** controversy
1. 젊은 세대의 많은 수가 사회 변화를 촉구하고 있다.
2. 그녀는 의도치 않게 자신의 컴퓨터에서 중요한 파일을 삭제했다.
3. 도서관에서 특정 책을 금지하기로 한 결정은 학생들 사이에서 논란을 불러일으켰다.

Tips

베이비붐 세대: 나라마다 정확한 연도의 차이는 있으나 보통 제2차 세계 대전 직후 출생률이 급격히 증가한 시기에 태어난 사람들을 말한다.
X세대: 1965년에서 1980년대 초반에 태어난 세대로 대체적으로 개성을 중시하고 독립적인 성향을 보인다.
밀레니얼 세대: X세대 이후의 1990년대 중반까지 출생한 세대로, 인터넷과 디지털 기기를 접하며 자라온 세대이다.

Exercise 3
정답 ①

소재 인류에게 있어서 음악의 중요성

해석 우리 종(種)은 아마도 인류로 존재해 온 만큼 오랫동안 음악을 만들어 왔을 것이다. 음악은 우리의 영원한 일부인 것 같다. 알려진 가장 오래된 악기는 약 4만 년 전으로 거슬러 올라간다. 이 중에서 가장 흥미로운 것들 중 하나는 섬세한 뼈로 만든 피리인데, 오늘날 남부 독일인 곳에서 발견되었다. 이러한 발견물들은 우리 조상들이 음악을 창조하기 위해 적용한 고도의 기술을 증명하는데, 손가락 구멍은 연주자의 손가락으로 단단히 밀폐될 수 있도록 세심하게 비스듬히 잘라져 있으며, 구멍들 간의 거리는 아마도 특정 음계에 맞추기 위해 정밀하게 측정된 것으로 보인다. 이 시기는 북반구의 마지막 빙하기에 해당하는데, 그 당시 살던 사람들에게는 삶이 쉽지 않았을 것이다. 그럼에도 불구하고 시간, 에너지, 그리고 장인들의 기술이 '삶의 일상 행위에 가장 쓸모가 적은' 추상적인 소리를 만들기 위해 사용되었다. 따라서 그들에게 음악은 매우 의미 있고 중요했음에 틀림없다.

→ 40,000년 전에 살던 인간들이 혹독한 생활 환경에도 불구하고 정교한 방법을 사용해 세심하게 만든 악기를 가지고 있었다는 사실은 그들이 음악에 둔 의의를 보여 준다.

문제 해설

일상생활을 영위하기도 혹독한 빙하기에 고도의 기술을 사용해 악기를 만들었다는 것으로 보아, 고대 인류는 음악에 의의를 두었을 것이라고 추정하는 글이다. 따라서 요약문의 빈칸 (A), (B)에 들어갈 말로 가장 적절한 것은 ① '정교한(sophisticated) – 의의(significance)'이다.
② 정교한 – 복잡성
③ 유연한 – 복잡성
④ 구식의 – 본질
⑤ 구식의 – 의의

구조 해설

- Our species has been making music most likely for **as long as** we've been human.

  for long(오랫동안)에 「as ~ as ...」이 사용되어 '…만큼 오랫동안'이라는 의미를 나타낸다.

어휘 및 어구

permanent 영원한, 영구적인
musical instrument 악기
intriguing 흥미로운
delicate 섬세한
testify to ~을 증명하다
ancestor 조상
seal 밀폐

correspond to ~에 맞추다, ~에 해당하다
musical scale 음계
hemisphere 반구
craftworker 장인
abstract 추상적인

| 정답 | **1.** testify **2.** intriguing **3.** abstract
1. 전문가들이 시험 기간 중에 새로운 약물의 안전성과 효능을 증명할 것이다.
2. 그 예술 전시회는 많은 대화를 불러일으킨 매우 흥미로운 작품들을 선보였다.
3. 그 교수의 강의는 수학의 복잡한 추상적 개념을 다루었다.

Exercise 4 ──────── 정답 ③

소재 전화위복의 상황

해석 때로는 겉보기에 운이 없어 보이는 시작이 알고 보면 장기적인 성공을 더 유망하게 만들기도 한다. Gladwell은 20세기 초 뉴욕으로 이주하고 이어서 의류 산업에서 성공한 유대인들의 경험을 예로 든다. 많은 사람들이 로스쿨을 졸업했지만 당시 부유한 개신교 가정 출신의 변호사들을 주로 고용했던 뉴욕의 주요 법률 회사들로부터 결국 (채용이) 거절되었던 자녀들을 길렀다. 유대인 졸업생들은 작은 법률 회사를 직접 설립하는 것보다 더 나은 선택지가 남아 있는 경우가 거의 없는 경우가 많았다. 그러한 법률 회사는 적대적 기업 인수 소송과 같이 엘리트 법률 회사가 자기네 수준보다 못하다고 생각하는 소송 사건을 전문으로 다루는 경우가 흔했다. 따라서 의류 노동자가 키운 변호사들은 1970년대와 1980년대에 발생한 적대적 인수 소송의 폭발적인 증가를 활용할 수 있는 전문 지식을 갖춘 유일한 변호사였던 경우가 많았다. 그 새로운 시장을 장악함으로써 그들은 이어서 이전에 그들을 기피했던 법률 회사의 변호사들보다 엄청나게 더 많은 수입을 올렸다.

문제 해설
(A) 뉴욕의 주요 법률 회사에서는 부유한 개신교도 출신을 고용했다고 했으므로, 유대인 출신 변호사에 대해서는 rejected(거절되었다)를 써야 한다. hired는 '고용되었다'라는 뜻이다.
(B) 엘리트 법률 회사가 적대적 기업 인수 소송을 다루지 않아 유대인 법률 회사가 맡았다는 내용이 되어야 하므로, beneath((수준 등이) ~보다 못한)를 써야 한다. above는 '~보다 상위에'라는 뜻이다.
(C) 유대인 변호사가 자신들을 기피했던 법률 회사의 변호사보다

훨씬 더 많은 수입을 올렸다고 했으므로, dominating(장악함)을 써야 한다. sharing은 '공유함'이라는 뜻이다.

구조 해설
■ Gladwell cites the experience of Jews [who {immigrated to New York in the early twentieth century} and {went on to prosper in the garment industry}].
[ ]는 Jews를 수식하는 관계절이다. 두 개의 { }는 관계절에서 and로 연결되어 who에 이어진다.

어휘 및 어구
seemingly 겉보기에는
Jew 유대인
immigrate 이주하다
prosper 성공하다, 번영하다
garment 의류
raise 기르다, 양육하다
hire 고용하다
Protestant 개신교의; 개신교도
option 선택지
case 소송 사건, 소송
hostile 적대적인
corporate 기업의
takeover 인수
expertise 전문 지식
capitalize on ~을 활용하다
explosive 폭발적인
vastly 엄청나게, 대단히

| 정답 | **1.** explosive **2.** expertise **3.** vastly
1. 그 가수의 최신 앨범은 팬들로부터 폭발적인 반응을 받았다.
2. 그의 공학 분야의 전문 지식은 그를 그 프로젝트의 이상적인 후보자로 만들었다.
3. 그 새로운 기술은 직장에서의 효율성을 엄청나게 향상시켰다.

Exercise 5~6 ──────── 정답 5 ③ 6 ④

소재 의미적 언어 처리가 기억에 미치는 영향

해석 단어 처리 깊이의 효과라는 전형적인 예를 인지 심리학에서 들어 보자. 내가 세 집단의 학생들에게 60개의 단어로 구성된 목록을 제시한다고 생각해 보라. 첫 번째 집단에게는 단어의 글자가 대문자인지 소문자인지, 두 번째 집단에게는 단어가 'chair'와

운율이 맞는지, 세 번째 집단에게는 단어가 동물 이름인지 아닌지를 결정하도록 요청한다. 일단 학생들이 과제를 마친 후에, 나는 그들에게 기억력 테스트를 실시한다. 어느 집단이 단어를 가장 잘 기억할까? 단어의 더 피상적인 감각적 측면을 문자 수준(33% 성공)이나 운율 수준(52% 성공)에서 처리한 다른 두 집단에서보다 단어를 의미 수준(75% 성공)에서 깊이 있게 처리한 세 번째 집단에서 기억력이 훨씬 더 뛰어난 것으로 밝혀진다. 우리는 모든 집단에서 단어에 대한 약한 암시적이고 무의식적인 흔적을 실제로 발견하는데, 즉 학습이 철자와 음운 체계에 잠재의식적인 흔적을 남긴다는 것이다. 그러나 심층적인 의미적 처리만이 단어에 대한 명시적이고 상세한 기억을 보장한다. 동일한 현상이 문장 수준에서도 사라져서(→ 일어나서), 교사의 시도 없이 스스로 문장을 이해하려고 노력하는 학생들이 훨씬 더 나은 정보 기억을 보인다. 이는 일반적인 법칙인데, 이를 미국의 심리학자 Henry Roediger는 다음과 같이 말한다. "학습 상황을 더 어렵게 만들어 학생들이 더 많은 인지적 노력을 기울이도록 요구하면 많은 경우 향상된 기억으로 이어진다."

(문제 해설)

5 단어를 의미 수준에서 깊이 있게 처리하면 그 단어를 훨씬 더 잘 기억하고, 단어뿐 아니라 문장에서도 심층적인 의미적 처리만이 명시적이고 상세한 기억을 보장한다는 내용의 글이다. 따라서 글의 제목으로 가장 적절한 것은 ③ '의미적 처리의 깊이가 기억에 영향을 미치는 방식'이다.
① 학습 장애 아동을 지원하는 방법
② 동기와 그것이 기억 유지율에 미치는 영향
④ 잠재의식적인 언어 학습: 현실인가 아니면 공상 과학인가?
⑤ 우리가 사용하는 언어가 우리의 사고방식에 영향을 미칠 수 있다

6 심층적인 의미적 처리를 하면 단어에 대한 명시적이고 상세한 기억이 보장된다고 하였는데, 이는 교사의 지도 없이 스스로 문장을 이해하려고 노력하는 학생이 훨씬 더 나은 기억을 보이는 것과 같은 맥락이므로, 동일한 현상이 문장 수준에서도 일어남을 알 수 있다. 따라서 (d)의 disappears를 occurs와 같은 낱말로 바꿔야 한다.

(구조 해설)

- This is a general rule, [which the American psychologist Henry Roediger states as follows: "{Making learning conditions more difficult, thus requiring students to engage more cognitive effort}, often leads to enhanced retention]."
  [ ]는 a general rule을 부가적으로 설명하는 관계절이고, { }는 주어 역할을 하는 동명사구이다.

(어휘 및 어구)
cognitive 인지의

rhyme 운율이 맞다; 운율
superficial 피상적인
sensory 감각적인
implicit 암시적인, 내재하는
unconscious 무의식적인
trace 흔적
explicit 명시적인
phenomenon 현상
guidance 지도
retention 기억, 유지
enhanced 향상된

Word Search

| 정답 | **1.** superficial **2.** trace **3.** guidance
1. 주제에 대한 그의 이해는 피상적일 뿐이어서, 그는 더 깊은 개념을 이해하지 못했다.
2. 형사는 용의자에게로 이어지는 증거의 흔적을 발견했다.
3. 그 교사의 지도는 학생들이 글쓰기 기술을 향상하는 데 도움을 주었다.

Exercise 7 — 정답 ④

(소재) 수렵채집에서 농업으로 변화하며 얻게 된 단점

(해석) 성공적인 농업 공동체는 소규모 지역에서도 수렵채집인들이 할 수 있었던 것보다 훨씬 더 많은 인구를 부양할 수 있었다. 그러나 이 새로운 생활 방식에는 매우 커다란 단점이 있었다. 우선, 폭풍, 가뭄, 홍수, 폭우 및 기타 심각한 기상 이변은 농업 공동체에는 재앙이 될 수 있었지만, 수렵채집인들에게는 단지 큰 골칫거리를 야기할 뿐이었다. 후자(수렵채집인들)는 자신들의 최소한의 가벼운 소지품을 가지고 피해가 그다지 크지 않은 어떤 다른 지역으로 꽤 쉽게 갑자기 이동할 수 있었다. 그러나 더 중요한 것은 그들이 의존했던 자연의 집은 농업 공동체가 의존했던 농작물과 구조물이 그러했던 것보다 심각하게 훼손될 가능성이 훨씬 더 많았(→ 더 적었)다는 점이다. 자신들의 자연의 집을 인공적인 건축물로 대체함으로써, 농부들은 충분히 역설적이게도 스스로를 자연재해에 더 무방비 상태로 만들었고, 실제로는 수렵채집인들에게 전혀 재앙이 아니었던 자연 사건이 그들에게는 재앙이 되었다.

(문제 해설)
이 글은 수렵채집에서 농업 공동체로 변화하면서 얻게 된 단점에 대한 내용이다. 자연재해가 닥쳤을 때 수렵채집인들이 의존했던 자연의 집은 농업 공동체가 의존했던 농작물과 구조물보다 심각하게 훼손될 가능성이 훨씬 더 적었다고 해야 문맥상 적절하므로, ④의 more를 less와 같은 낱말로 바꿔야 한다.

구조 해설

■ But, more importantly, the natural home [they depended on] was much less likely to be severely damaged **than were** [the crops and structures {the agricultural community relied on}].

첫 번째 [ ]는 관계사가 생략된 관계절로 the natural home을 수식한다. than의 뒤에는 긴 주어인 두 번째 [ ]와 동사 were가 도치되어 있다. { }는 관계사가 생략된 관계절로 the crops and structures를 수식한다.

어휘 및 어구

region 지역
significant 커다란, 중요한
storm 폭풍
drought 가뭄
flood 홍수
torrential rain 폭우
severe 심각한, 엄격한
pose 야기하다, 제기하다
possession (보통 복수로) 소지품, 소유물
up and 갑자기 (어떤 일을 하다)
crop 농작물
replace 대체하다
construction 건축물
paradoxically 역설적이게도
defenseless 무방비 상태의

Word Search

| 정답 | 1. pose 2. possessions 3. crops

1. 핵무기는 모든 사람에게 위협을 제기한다.
2. 그 목걸이는 그녀가 가장 소중히 여기는 소유물 중 하나이다.
3. 이곳의 농부들은 여전히 손으로 자신들의 농작물을 심고 수확한다.

Exercise 8
정답 ②

소재 인간의 지각 경험에 미치지 못하는 언어

해석 사람들이 색에 대해 수백 개는 아닐지라도 수십 개의 고유한 단어를 알 수도 있는 특별한 상황이 있다. 한 유명 가정용 페인트 브랜드는 자신의 상업용 색채의 범위에 2천 가지가 넘는 색을 보유하고 있다. 이 중 다수는 *violet posy*, *wing commander*, *Aztec tan*과 같은 독특한 (그리고 분명히 기본적이지 않은) 영어 이름의 라벨이 붙어 있다. 그러나 심지어 기본적이지 않은 용어를 허용하고 심지어는 페인트 산업의 고도로 전문화된 어휘를 허용

하더라도, 언어가 색 공간에서 만드는 구분은 인간의 눈이 식별할 수 있는 색에서의 2'백만' 개가 훨씬 넘는 구분에 비하면 어마어마하게 적다. 심지어 고도로 전문화되고 기술적인 어휘에서도 사용 중인 색 용어의 총 목록은 인간의 시각 체계가 할 수 있는 식별 가능한 색 구분 중 1퍼센트의 10분의 1보다 더 적다. 대부분의 사람들이 사용하는 일상적인 어휘는 그것보다 또 100배 더 적을 것이다. 우리의 지각 경험은 풍부하지만, 우리의 언어는 그 풍부함 가까이 그 어디에도 미치지 못한다. 여러분이 언어가 지각 경험을 담아내는 데 유용하다고 생각한다면, 다시 생각해 보라.

문제 해설

주어진 문장은 언어가 색 공간에서 만드는 구분은 인간의 눈이 식별할 수 있는 구분에 비하면 어마어마하게 적다는 내용이다. ②의 앞에서는 한 가정용 페인트 브랜드가 2천 가지가 넘는 색 목록을 가지고 있다고 언급한 다음 고유한 영어 이름의 구체적 예시를 언급하는 반면, ②의 다음 문장에서는 고도로 전문화되고 기술적인 어휘에서 사용되는 색 용어의 총 목록은 인간의 시각 체계가 할 수 있는 식별 가능한 색 구분 중 1퍼센트의 10분의 1보다 더 적다고 말하고 있다. 따라서 주어진 문장이 들어가기에 가장 적절한 곳은 ②이다.

구조 해설

■ But [even allowing nonbasic terms] and [even allowing the highly specialized vocabulary of the paint industry], the distinctions [that languages make in the color space] are astronomically small in comparison to well over 2 *million* distinctions in color [that the human eye can discern].

첫 번째와 두 번째 [ ]는 일반인을 의미상의 주어로 하며 양보의 의미를 내포한 분사구문으로 and로 대등하게 연결되어 있다. 세 번째 [ ]는 the distinctions를 수식하는 관계절이고, 네 번째 [ ]는 well over 2 *million* distinctions in color를 수식하는 관계절이다.

어휘 및 어구

term 용어
distinction 구분, 구별, 차이
distinct 고유한
commercial 상업용의
palette 색채의 범위, 팔레트
label 라벨을 붙이다
visual 시각의
perceptual 지각의
capture 담아내다, 포착하다

Word Search

| 정답 | 1. distinctions 2. labeled 3. visual

1. 두 디자인 사이에는 뚜렷한 차이가 없다.
2. 창고에 있는 각 상자에는 내용물이 표시된 라벨이 붙어 있었다.
3. 지도는 학습을 위한 시각적 도구이다.

Exercise 9

정답 ③

소재 과거 영국의 공유지와 현대 인터넷 디지털 공유지

해석 영국에서, '공유지'로 알려진 넓은 개방된 토지 영역은 토지를 소유한 귀족들의 소유지였지만, 농민들이 사용하도록 접근이 가능했다. 그들은 소규모 농업을 하고, 연료와 요리에 필요한 나무를 모으며, 가족을 위한 소규모 가축을 키우기 위해 공유지를 사용했다. 의류와 기타 상품에 사용될 양모에 대한 세계적인 수요가 증가하면서, 지주들은 양모 생산을 늘리기 위해 더 많은 목초지를 얻고자 했고, 그래서 그들은 공유지를 사유화하여 그것에 울타리를 치고 따라서 농민들의 접근을 차단했다. 공유지에 대한 이러한 '울타리 치기' 과정으로 인해 농민들은 수 세기 동안 생존을 위한 주요 자원이었던 토지로부터 차단되었다. 오늘날, 영국 목초지의 물리적 공유지와 인터넷의 디지털 공유지 사이에는 흥미로운 상관관계가 있다. 둘 다 자립, 표현, 번영의 기회가 풍부하지만, 둘 다 '울타리 치기'와 '지주'(예를 들어 접근이나 사용에 요금을 부과하는 플랫폼 소유자)로부터의 문제들로 가득 차 있다.
→ 영국에서 양모에 대한 수요가 증가하면서 공유지에 대한 농민들의 접근이 <u>차단되었고</u>, 이는 우리가 디지털 세계에서 직면하고 있는 현대의 상황과 <u>유사하다</u>.

문제 해설

영국에서 양모에 대한 수요가 증가하면서 지주들이 공유지를 사유화하여 울타리를 두르게 되었는데, 오늘날 인터넷의 디지털 공유지가 영국 목초지의 물리적 공유지와 유사하다는 내용의 글이다. 따라서 요약문의 빈칸 (A), (B)에 들어갈 말로 가장 적절한 것은 ③ '차단되었고(blocked) - 유사하다(resembles)'이다.
① 방해받았고 - 차이를 보인다
② 보장받았고 - 반영한다
④ 확장되었고 - 유사하다
⑤ 제한되었고 - 해결한다

구조 해설

- They used the commons [to conduct small-scale agriculture], [to collect wood for fuel and cooking], and [to raise small amounts of livestock for their families].
  세 개의 [ ]는 모두 목적을 나타내는 to부정사구로서 콤마와 and로 대등하게 연결되어 있다.

Word Search

| 정답 | **1.** conducted **2.** livestock **3.** privatized
1. 그 실험들은 과학자들에 의해 행해졌다.
2. 농부는 매일 아침 일찍 자신의 가축을 돌보았다.
3. 항공 교통 관제가 민영화되었다.

Exercise 10

정답 ④

소재 자기도취적 성향으로 인한 경청의 어려움

해석 경청이 매우 어려울 수 있는 이유는 우리의 자기도취적 성향 때문인 것 같다. 너무 자주, 재치 있는 무언가를 생각해 내려고 애쓰며 우리의 머리가 정신없이 돌아가는 동안 우리는 경청하고 있는 척한다. 그러나 재치 있는 것이 지혜로운 것은 아니다. 아울러 우리의 자기도취적 경향을 악화시키는 것으로, 상대방이 무엇을 말할지 우리가 이미 알고 있다고 가정하는 한쪽 귀로만 듣는 유형도 있다. 나는 부주의한 듣기, 즉 그저 말할 기회를 기다리며, 심지어는 조급해져서 상대방(의 말)을 끝내고 싶어 하는 것을 말하는 것이다. 철학자이자 시인인 Ralph Waldo Emerson이 한때 말했듯이, '진정으로 경청하는 것과 자신의 말할 차례를 기다리는 것에는 차이가 있다.' 말을 가로막고 대화에 끼어들어 한마

디 하고자 하는 이러한 꺼림(→ 충동)은 꽤 강력할 수 있다. 어떤 사람들은 단지 자신의 존재를 확인하고 입증하기 위해 그저 자기 말을 하고 싶어 한다. 자아도취된 사람은 작은 귀를 가졌다는 말이 있다.

(문제 해설)
우리의 자기도취적 성향 때문에 경청이 어려울 수 있다는 내용으로, 자신이 말할 기회를 기다리기만 하면서 상대방의 말을 끝내고 싶어 하는 유형에 대해 이야기하고 있다. 따라서 ④의 reluctance를 urge와 같은 낱말로 바꿔야 한다.

(구조 해설)
- [This urge to {interrupt} and {get a word in}] can be quite powerful.
  [ ]는 문장의 주어 역할을 하고, 그 안의 두 개의 { }는 and로 대등하게 연결되어 to에 이어지며, to부정사구를 만들어 This urge를 수식한다.

(어휘 및 어구)
narcissistic 자기도취적인
race (머리가) 정신없이 돌아가다
tendency 성향
presume 가정하다
refer to ~을 말하다
impatient 조급한
philosopher 철학자
poet 시인
interrupt (말·행동을) 가로막다, 중단하다
get a word in 대화에 끼어들어 한마디 하다
confirm 확인하다
validate 입증하다
existence 존재
big ego 자아도취된 사람

(Word Search)

| 정답 | 1. poet 2. interrupt 3. validate
1. 그 시인은 식당에서 자신의 최신 시를 낭송했다.
2. 내가 말하는 동안 가로막지 말아 주십시오.
3. 그녀는 자신을 한 인간으로 입증하기 위해 그의 찬사가 필요한 것 같았다.

Exercise 11~12
정답 11 ② 12 ⑤

(소재) 질소 비료의 필요성
(해석) 은행 계좌 비유는 농업 생산량 증가에 왜 비료 사용이 동반되어야 하는지를 설명하는 데 도움이 된다. 여러분이 작물을 수확하고 그것을 먹을 때마다, 그 작물을 좋은 식량으로 만드는 바로 그 영양소인 그 식물에 있는 질소(그리고 기타 영양소)가 토양에서 제거되어 여러분이 있는 곳이 어디든지 그곳으로 이동한다. 그러한 영양소 중 일부는 (여러분이 성장 중이라면) 여러분의 체내에 축적되지만 대부분은 빠져나간다. 어느 쪽이든, 여러분과 여러분의 배설물이 농장으로 돌아가지 않는 한, 토양으로부터 영양소의 순 손실, 즉 영양소 은행으로부터의 순 인출이 있다. 사람이 비교적 적고, 그들 대부분이 농장이나 그 근처에서 살며, 배변하고, 죽던 때에는 밭을 경작하지 않은 채로 두거나 콩과(科) 식물을 심고 그것들을 갈아엎는 것이 식량을 생산하는 꽤 지속 가능한 방식이었다. 하지만 80억 명의 사람들을 위해서라면? 혹은 100억 명을 위해서라면? 우리 모두를 먹이기 위해 필요한 식량은 많은 질소를 농장으로부터 제거하는 것을 필요로 하고, 이것(질소)은 되돌려져야 하는데, 그렇지 않다면 질소의 토양 은행 계좌는 고갈될 것이다. 이는 적어도 예측 가능한 미래에, 인간을 거쳐 빠져나가는 영양소를 원래 있던 농장 토양으로 돌려보낼 안전하고 효과적인 방법을 우리가 생각해 낼 때까지는 세계를 먹이는 데 산업적으로 생산된 질소 비료가 필요하다는 것을 의미한다. 다시 말해, 우리는 질소를 제거하는(→ 획득하는) 우리의 혁신 덕분에 번영한다.

(문제 해설)
11 왜 질소 비료를 사용해야 하는지를 은행 계좌 비유를 통해 설명하는 글로, 전 세계의 인구를 먹이기 위해 필요한 식량은 많은 질소를 농장에서 제거하게 되고 질소의 토양 은행 계좌의 고갈을 야기하는데, 이를 막기 위해 산업적으로 생산된 질소 비료가 필요하다는 내용이다. 따라서 글의 제목으로 가장 적절한 것은 ② '왜 질소 비료는 오늘날 세계에서 필수적인가'이다.
① 질소는 생물 다양성의 핵심 요소인가?
③ 은행 계좌의 총량: 탐욕의 거울
④ 화학물질 사용으로부터 토양을 보호해야 한다는 증가하는 요구
⑤ 어느 것이 기후 위기의 원인인가, 질소인가 아니면 사람인가?
12 인간을 거쳐 빠져나가는 질소를 원래 있던 농장 토양으로 돌려보낼 안전하고 효과적인 방법이 나올 때까지는 세계를 먹이는 데 산업적으로 생산된 질소 비료가 필요하므로, 질소를 획득하는 우리의 혁신, 즉 질소 비료 덕분에 우리가 번영한다는 내용이다. 따라서 (e)의 eliminating을 capturing과 같은 낱말로 바꿔야 한다.

(구조 해설)
- [When there were relatively few people, {most of whom lived, went to the bathroom, and died on or near the farm}], [leaving fields uncultivated] or [planting legumes and plowing them under] was a reasonably sustainable way to produce food.
  첫 번째 [ ]는 접속사 When이 이끄는 부사절이고, 그 안의

{ }는 relatively few people을 부가적으로 설명하는 관계절이다. 두 번째와 세 번째 [ ]는 or로 대등하게 연결되어 문장의 주어 역할을 하는 동명사구이다.

어휘 및 어구

account 계좌
agricultural 농업의
accompany 동반하다
fertilizer 비료
harvest 수확하다
crop 작물
nitrogen 질소
nutrient 영양소
accumulate 축적되다
pass through ~을 (거쳐) 빠져나가다
waste 배설물
net 순
withdrawal 인출
relatively 비교적, 상대적으로
uncultivated 경작하지 않은
reasonably 꽤, 합리적으로
sustainable 지속 가능한
replace (원위치로) 되돌리다, 대체하다
run out 고갈되다
for the foreseeable future 예측 가능한[가까운] 미래에
figure out ~을 생각해 내다

Word Search

| 정답 | **1.** accompanied **2.** harvest **3.** withdrawals
1. 뇌우에 강풍이 동반되었다.
2. 농부들은 수확할 때 채소를 분류한다.
3. 당신은 하루에 500달러까지 인출할 수 있습니다.

Week 6 12강 본문 138~149쪽

| 1 ④ | 2 ③ | 3 ④ | 4 ③ | 5 ③ | 6 ⑤ |
| 7 ④ | 8 ② | 9 ② | 10 ④ | 11 ① | 12 ⑤ |

Exercise 1 정답 ④

소재 영토의 중요성

해석 사회는 영토를 '재생산해'야 하는가? 한편으로, 역사는 영토 갈등으로 가득 차 있으며, 영토가 제거된 사회 체제는 실패할 운명이다. 사회 체제의 인구의 대사적 재생산에서 영토는 어떤 역할을 하는가? 가장 중요한 것은 영토가 인구에 합법적인 물리적 '공동의 생활 공간'을 제공한다는 사실인데, 바꾸어 말하면, 영토는 '인간과 그들의 하부 구조를 위한 저장소' 역할을 한다는 사실이다. 이러한 '저장소 기능'은 이른바 '자유재', 즉 (예컨대 깨끗한 공기와 물과 같은) 영토 내의 생태계 서비스의 소비에 참여할 수 있는 기회를 제공한다. 그러나 대부분의 경우에, 그것은 이 이상의 의미를 갖는다. 국가는 어떤 의미에서 '공익'에 대한 책임이 있고, 따라서 자국 국민의 재생산을 지원하는 환경을 보장할 책임이 있다. 어떤 경우든 적어도 자국 영토 내에서, 국가는 그들의 대사적 재생산이 달성할 수 없는(→ 가능한) 것을 보장해야 한다. 그러므로 영토는 경제 과정에서 사용할 수 있는 천연자원을 포함한다는 점에서 의미가 있다. 그것은 또한 이러한 과정에서 나오는 폐기물을 둘 수 있는 배출구를 제공하며, 다양한 비공급 생태계 서비스의 원천이다.

문제 해설
영토는 인구에 합법적인 물리적 '공동의 생활 공간'을 제공하고 국민의 대사적 재생산을 지원하는 환경을 보장할 책임이 있다고 했다. 따라서 ④의 unattainable을 possible과 같은 낱말로 바꿔야 한다.

구조 해설
- The territory is therefore meaningful in containing natural resources [which economic processes can appropriate].
[ ]는 natural resources를 수식하는 관계절이다.

어휘 및 어구
reproduce 재생산하다, 번식시키다
territory 영토
conflict 갈등
be doomed to ~할 운명이다
significance 중요한 것, 중요성
legitimate 합법적인

consumption 소비
ecosystem 생태계
answerable to ~에 대한 책임이 있는
subject 국민, 백성
appropriate 사용하다, (마음대로) 가져가다
deposit 놓아 두다, 퇴적하다, 침전하다
provision 공급하다; 공급, 제공

Word Search

| 정답 | **1.** territory **2.** legitimate **3.** answerable
1. 그들은 군대가 그들의 영토에 주둔하는 것을 거부해 왔다.
2. 그녀의 여권은 합법적인 것처럼 보였지만, 위조된 것으로 밝혀졌다.
3. 의회는 자신을 선출한 국민에게 책임을 져야 한다는 것을 기억하라.

Tips

non-provisioning ecosystem services(비공급 생태계 서비스): 생태계가 제공하는 서비스는 일반적으로 공급 서비스(provisioning services, 식량, 물, 공기, 천연자원 등), 조절 서비스(regulating services, 오염 정화, 홍수 및 침식 방지, 서식지 제공, 병충해 예방 등), 문화 서비스(cultural services, 경관, 관광, 휴양등), 지지 서비스(supporting services, 생물 다양성, 영양 순환, 기초 생산 등)의 4가지로 구분되는데, '비공급 생태계 서비스'란 이 중 공급 서비스를 제외한 나머지 3가지 서비스를 말한다.

Exercise 2 ——————————————— 정답 ③

소재 직원 보호를 위한 물리적 장벽의 필요성과 문제점
해석 직원의 보호를 위해 물리적 장벽이 합리적이고도 필요한 몇몇 상황이 있다. 교정 시설 안에 이에 대한, 그리고 방문객의 안전을 위한 배치가 이루어질 수 있는 방식에 대한 전형적인 사례가 있다. 그렇지만 많은 공공 기관의 접수처에서 폭력이나 가학적 행위를 저지하기를 바라면서, 그리고 직원의 보호를 위해 때때로 안전 칸막이를 사용해 왔다. 그러나 이것들이 항상 효과가 있었던 것은 아니고, 때로는 폭력적인 행동이 예상되고 있다는 메시지를 주었다. 어떤 사람들은 그러한 칸막이가 너무 비인격적이어서 그것이 수상쩍은, 심지어 공격적인 반응을 일으킨다고 생각한다. 그들은 개인을 존중하여 결국 그들이 직원을 향해 존중하는 행동을 하도록 장려하는 개방되고, 따뜻하고, 인간적인 환경을 만드는 것을 훨씬 선호한다. 칸막이가 비밀을 위험에 빠뜨리는 것도 있을 수 있는 일인데, 만약 사람들이 자신의 목소리가 들리게 하기 위해서 더 큰 소리로 말해야 한다고 느낀다면 그렇다. 지역 보건 진료소의 칸막이가 이것의 좋은 예이다.

문제 해설
주어진 문장은 앞에서 나온 칸막이를 지칭하면서 그것이 너무 비인격적이어서 부정적인 반응을 일으킨다고 생각하는 사람들이 있다는 내용이다. 따라서 주어진 문장은 칸막이의 부정적인 효과에 대한 소개가 제시되고 그것을 싫어하는 사람들이 더 선호하는 환경에 대한 내용이 이어지는 ③에 들어가는 것이 가장 적절하다.

구조 해설
■ **It** is also possible [that screens can put confidentiality at risk {if people feel ⟨that they have to speak more loudly in order to make themselves heard⟩}].
It은 형식상의 주어이고 [ ]가 내용상의 주어이다. { }는 조건을 나타내는 부사절이고, ⟨ ⟩는 feel의 목적어 역할을 하는 명사절이다.

어휘 및 어구
impersonal 비인격적인, 비인간적인
provoke (감정 등을) 일으키다
suspicious 수상쩍은, 의심하는
aggressive 공격적인
barrier 장벽
sensible 합리적인
arrangement 배치
reception 접수
security 안전, 보안
discourage 저지하다
violence 폭력
abusive 가학적인, 모욕하는
put ~ at risk ~을 위험에 빠뜨리다

Word Search

| 정답 | **1.** impersonal **2.** suspicious **3.** aggressive
1. 칙칙한 회색 방에 똑같은 침대가 줄지어 있으면 병원이 비인간적으로 보인다.
2. 순찰 중이던 경찰관들이 차에 탄 남자를 의심하게 되었다.
3. 우리가 그녀를 비판하면, 그녀는 공격적이 되고 소리를 지르기 시작한다.

Exercise 3 ——————————————— 정답 ④

소재 암호 해독과 설명의 차이
해석 암호 해독은 그 대상을 신호의 집합으로 간주하는 언어 활동인데, 각각의 신호는 암호에 의해 결정되는 특정한 의미를 전달한다. 암호 해독의 대상은 언어적일 수도 있고 비언어적일 수도 있다. 그러나 비언어적 대상은 해독하는 과정에서 언어적 대상과

유사해지는데, 즉 신호들의 집합으로 여겨지는 것이다. 적의 메시지를 해독하는 것은 언뜻 보기에 무작위적인 신호들이 일관된 의미를 숨기고 있다는 믿음에서 시작된다. 설명의 경우와 마찬가지로, 암호 해독은 이질적인 사실들을 통합하는 것을 목표로 한다. 설명은 일반적인 법칙으로 다루어지는 특정한 것을 포함하는 반면, 암호 해독은 특정한 사례에 대해 별개일 수도 있는 규칙에 따라 한 요소를 다른 요소로 대체한다. 곧 적의 두 번째 메시지가 첫 번째와 같은 암호로 작성될 수도 있고 아닐 수도 있는 것이다. 암호를 해독하는 사람은 올바른 암호를 제공하기 위해 겉으로 비슷해 보이는 사례에 의존할 수 없는데, 외견상의 유사성은 오해를 일으킬 수도 있기 때문이다. 반면에, 유사한 사례는 유사하게 설명되어야 하는데, 왜냐하면 설명은 관련된 모든 사례에 공통되는 일반적이고 일관된 규칙을 가정하기 때문이다.

→ 유사한 사례에서 일관된 규칙에 의존하는 설명과는 대조적으로, 암호 해독은 신호를 해석하는 과정인데, 흔히 특정한 사례에 대한 특정한 규칙을 필요로 한다.

(문제 해설)

암호 해독은 특정한 사례에 대해 별개일 수 있는 규칙에 따라 한 요소를 다른 요소로 대체한다고 했으므로, (A)에는 specific(특정한)이 적절하다. 반면에 설명의 경우에는 관련된 모든 사례에 공통되는 일반적이고 일관된 규칙을 가정하기 때문에 유사한 사례는 유사하게 설명되어야 한다고 했으므로, (B)에는 similar(유사한)가 적절하다.

① 일반화된 – 무작위적인
② 일반화된 – 특유의
③ 특정한 – 우연의
⑤ 특정한 – 복잡한

(구조 해설)

- While explanation includes the particular [covered by a general law], decoding substitutes one element with another in accordance with a rule [that may be individual to the particular case]: ~.

  첫 번째 [ ]는 the particular를 수식하는 분사구이고, 두 번째 [ ]는 a rule을 수식하는 관계절이다.

(어휘 및 어구)

verbal 언어의, 언어적인, 구두의
signal 신호
determine 결정하다
apparently 언뜻 보기에, 외견상
random 무작위적인
conceal 숨기다, 감추다
substitute 대체하다
in accordance with ~에 따라, ~과 일치하여

seemingly 겉보기에
misleading 오해를 일으키는
relevant 관련된

(Word Search)

| 정답 | **1. conceal 2. apparently 3. verbal**

1. Mark는 실망감을 거의 감추지 못했다.
2. 언뜻 보기에 연관성이 없는 일련의 사건들이 그녀의 사임으로 이어졌다.
3. 그들은 그와 구두로 합의한 상태이다.

Exercise 4 ————————— 정답 ③

(소재) 이산화 탄소와 산소의 반대 작용

(해석) 이산화 탄소는 호흡의 강력한 조절자이며(그것은 뇌에서 발견되는 서로 다른 화학 수용체 무리에 작용한다), 그것의 혈중 농도가 떨어지면 호흡은 억제된다. 이를 스스로 시연해 보는 것이 가능하다. 미리 짧은 시간 동안 매우 빠르게 호흡하면, 숨을 더 오래 참을 수 있다는 것을 알게 될 것이다. (1분 넘게 이것을 하지 말아야 하는데, 그렇지 않으면 어지러워질 수도 있다.) 그 이유는 숨을 참는 것은 산소에 대한 요구가 아니라 오히려 혈중 이산화 탄소 농도 상승에 의해 종료되기 때문이다. 이것이 임계 수준에 도달하면 여러분이 숨을 쉬도록 자극한다. 숨을 참기 전의 과호흡은 몸에서 이산화 탄소를 없애고 이산화 탄소가 호흡을 자극할 정도로 충분한 수준까지 쌓이기 전에 더 오랜 시간이 지나는 것을 가능하게 한다. 산소와 이산화 탄소로부터 오는 상반되는 작용력은 왜 해발 3,000미터 미만에서 호흡의 변화가 일어나지 않는지 설명한다.

(문제 해설)

(A) 바로 다음 문장에서 빠르게 호흡하여 혈중 이산화 탄소 농도가 떨어지면 숨을 오래 참을 수 있다고 하였는데, 이 말은 곧 호흡이 억제될 수 있다는 뜻이다. 따라서 (A)에 적절한 말은 inhibited(억제된)이다. facilitated는 '촉진된'이라는 뜻이다.

(B) 숨을 쉬도록 자극하는 것은 숨을 참는 것과 반대되는 상황이고, 숨을 참기 전의 과호흡은 몸에서 이산화 탄소를 없앤다고 했다. 과호흡으로 이산화 탄소가 몸에서 없어져 숨을 더 오래 참을 수 있게 되는데, 이는 돌려 말하면 숨을 참는 것이 종료되려면 이산화 탄소가 몸에 많아져야 한다는 뜻이다. 따라서 (B)에 적절한 말은 rising(상승)이다. falling은 '하락'이라는 뜻이다.

(C) 과호흡으로 몸에서 이산화 탄소가 없어지면 숨을 참을 수 있고 반대로 이산화 탄소가 쌓이면 숨을 참을 수 없게 된다. 따라서 (C)에 적절한 말은 stimulate(자극하다)이다. obstruct는 '차단하다'라는 뜻이다.

- Hyperventilation before holding your breath [blows off carbon dioxide from the body] and [enables a longer period to go by {before carbon dioxide builds up to a level ⟨sufficient to stimulate breathing⟩}].

두 개의 [ ]가 and로 연결되어 문장의 술부를 구성한다. 두 번째 [ ] 안의 { }는 시간의 부사절이고, ⟨ ⟩는 a level을 수식하는 형용사구이다.

어휘 및 어구

carbon dioxide 이산화 탄소
regulator 조절자, 조정자
concentration 농도
terminate 종료하다, 해지하다
oxygen 산소
stimulate 자극하다, 촉진하다
sufficient 충분한

Word Search

| 정답 | 1. regulator   2. terminated   3. stimulate
1. 공정한 경쟁을 위해 독립된 조정자가 지정될 것이다.
2. 그들은 그의 계약을 12월에 해지했다.
3. 정부는 사업 발전을 촉진하기 위한 조치를 취했다.

Exercise 5~6 ── 정답 5 ③ 6 ⑤

소재 교육 분야에서 권력을 행사하는 정책

해석 정책이 개인들에게 그들이 하고 싶어 하지 않았을 방식으로 행동하도록 강요하는 경우, 정책의 수행이 권력의 행사로 드러나는 것은 아마도 자명한 일일 것이다. 정책은 권력 관계를 강조하기 위해 사용될 수 있지만, 그것은 또한 권력과 의사 결정의 몰개인화를 허용할 수도 있다. 즉 지도자는 자신이 특정한 행동 방침에 실제로 동의하지는 않지만, 정책이 그것을 요구하기 때문에 그것이 실행되어야 한다고 암시할 수도 있다. 마찬가지로 위계가 불필요할 정도로 정책이 권위의 원천 역할을 할 수도 있다. 즉, 한 무리의 개인이 그들의 방향을 정해 줄 확실한 권위자가 없을 때, 정책에 의존하여 의사 결정을 이끌어 갈 수도 있다. 결국 정책을 규정하는 사람들이 정책을 따르는 사람들에 대해 권력을 행사하는데, 적어도 교육 분야에서는 정책이 집단적으로 결정되는 경우가 드물어서, 기껏해야 유권자들을 위한 정책을 정할 수 있도록 지도자들이 민주적으로 선출된다. 따라서, 정치적 엘리트에 대한 불가피한 의존과 그 결과로 일어나는 권력의 위계가 존재한다. 학교에서 흔히 볼 수 있는 권력으로서의 정책의 한 예는 교복이다.

학교에 교복이 있는지의 여부, 그리고 있다면 그것이 무엇인지를 결정하는 것은 학교 '운영위원회'이지만, 학부모와 학생들은 따라야 할 의무가 있고 교직원은 그것을 시행할 의무가 있다. 수많은 아이들이 부적절한 교복이 그들의 학습에 영향을 주지 않는다는 논리에 호소할 수 없고, 자신의 주장을 호소하기 위해 어떤 개인에게도 의존할 수 없다는 것을 결국 알게 될 것인데, 이는 권력이 정책 그 자체 내에 있기 때문이다. 즉 교복을 입어야 하는 이유가 정책이 그렇게 시키기 때문이라고 말하는 것으로 충분하지 않기 (→ 충분하기) 때문이다.

문제 해설

5 권력을 행사할 수 있는 정책의 속성을 교육 분야에 적용하여 학교에서 흔히 보이는 권력으로서의 정책인 교복을 예로 들어 설명한 글이다. 따라서 글의 제목으로 가장 적절한 것은 ③ '권력으로서의 정책: 교육에서 그것의 영향 이해하기'이다.
① 교육 정책 탐색하기: 모두가 피해자다
② 다양성 교육: 교육 정책에 매우 중요하게 포함된 사람
④ 교육에 대한 정책적 참여를 통한 문화적 간극 메우기
⑤ 정책이 교육의 의사 결정에 미치는 부정적 영향 해소하기

6 교복을 입는 것에 대해 개인이 호소할 수도 어떤 개인에게 의존할 수도 없는 상황은 권력을 가지고 있는 정책이 시키는 것이기 때문이며, 이를 근거로 교복을 입어야 한다고 말할 수 있다. 따라서 (e)의 insufficient를 sufficient와 같은 낱말로 바꿔야 한다.

구조 해설

- Countless children will have discovered [that {they cannot appeal to the logic ⟨that inadequate uniform does not affect their learning⟩}, and {**nor** can they resort to any individual to plead their case}], for the power rests within the policy itself: ~.

[ ]는 discovered의 목적어 역할을 하는 명사절이다. 두 개의 { }는 and로 연결되어 명사절을 구성하고 있다. 첫 번째 { } 안의 ⟨ ⟩는 the logic과 동격 관계에 있고, 두 번째 { }는 부정어 nor가 앞으로 나와서 주어 they와 조동사 can이 도치되었다.

어휘 및 어구

self-evident 자명한
outworking 수행, 효과, 결과
exercise (권력·권리·역량 등의) 행사[발휘]; 행사하다, 발휘하다
compel 강요하다
depersonalisation 몰개인화
imply 암시하다, 넌지시 내비치다
authority 권위, 권한
resort to ~에 의존하다, ~에 의지하다
exert 행사하다

determine 결정하다
democratically 민주적으로
inevitable 불가피한
reliance 의존
consequent 결과로 일어나는
comply 따르다
enforce 시행하다
inadequate 부적절한

Word Search

| 정답 | **1.** authority **2.** exerted/exercised **3.** inevitable
1. 유엔은 그 지역의 평화를 회복하기 위해 자신의 권한을 사용해 왔다.
2. 그는 과학계에 상당한 영향력을 행사해 왔다.
3. 그 사고는 부주의로 인한 불가피한 결과였다.

Exercise 7
정답 ④

소재 자유와 통제권 사이의 역설

해석 전동 킥보드는 도시 교통에서 최신 유행이다. 그것들은 인도 위에 떼 지어 기다리고 있으며, 스마트폰과 신용 카드를 가진 누구든지 대여할 수 있게 준비가 되어 있다. 킥보드는 걷는 것보다 더 빠르고, 자전거를 타는 것보다 더 쉽고, 자동차보다 더 간단하다. 그것들은 도시 풍경을 놀이터로 만든다. 그것들을 타는 것은 근심 걱정 없는 경험이지만, (혹시 그렇게) 보일 수도 있는 것보다 더 제한적이다. 모든 여정은 시작부터 끝까지 추적된다. 아무리 조절판을 강하게 눌러도 킥보드는 특정 속도를 초과하지 않을 것이다. 그것들은 지정된 도시 지역을 벗어나지 않는다. 그리고 요금에 대해 흥정할 수 없는데, 앱이 여정의 길이에 따라 정확한 금액을 부과한다. 이 중 어느 것도 본질적으로 불쾌하지 않다. 그러나 킥보드는 디지털 기술의 역설을 보여 주는 유용한 예를 정말로 제공하는데, 즉 그것들은 자유를 제공하지만, 어느 정도 통제권을 일부 유지(→ 포기)하는 대가로만 그러하다. 이것은 결코 완전히 해결될 역설이 아니다. 자유와 통제권 사이의 균형이 올바르게 맞춰져 있는지의 여부가 항상 의문일 것이다.

문제 해설 전동 킥보드는 도시 교통에서 최신 유행으로, 누구나 스마트폰과 신용 카드로 대여할 수 있으며 걷는 것보다 빠르고 자동차보다 간단하지만, 모든 여정은 추적되고 속도 제한과 지정된 지역 내에서만 운행이 된다는 내용으로, 킥보드는 자유를 제공하지만 킥보드에 대한 통제권을 일부 포기해야 한다는 디지털 기술의 역설을 보여 준다. 따라서 ④의 maintenance를 surrender와 같은 낱말로 바꿔야 한다.

구조 해설
■ This is not a paradox [that will ever be fully resolved].
[ ]는 a paradox를 수식하는 관계절이다.

어휘 및 어구
urban 도시의
cluster 떼, 무리
pavement 인도
hire 대여(하다)
landscape 풍경
carefree 근심 걱정 없는
track 추적하다
designated 지정된
fare 요금
precise 정확한
sum 금액
inherently 본질적으로
objectionable 불쾌한, 반대할 만한

Word Search

| 정답 | **1.** urban **2.** designated **3.** inherently
1. 시 의회는 새로운 도시 공원 계획을 제안하고 있다.
2. 주차장은 명백히 허가증 소지자 전용으로 지정되어 있다.
3. 그 아이는 본질적으로 호기심이 많은 성격을 지니고 있어 끊임없이 질문을 했다.

Exercise 8
정답 ②

소재 언어 진화의 단계

해석 이론상으로, 언어가 인간 진화 과정에서 어떤 점진적이거나 중간 형태 없이 갑자기 완전한 형태로 나타났을 가능성이 있다. 그 개념, 즉 언어의 빅뱅은 Noam Chomsky와 같은 언어학자들에 의해 옹호되어 왔다. 그러나 생물학적인 면에서 말하자면, 우리의 말하기 능력과 같은 매우 복잡한 특성이 어디선지 모르게 그저 갑자기 생겨났다는 것은 극히 가능성이 낮다. 언어는, 예를 들어, 우리의 대뇌나 우리의 도구 제작 능력이 그랬던 것과 같은 방식으로 여러 단계를 거쳐 진화했을 가능성이 더 높다. 이 과정은 어린이가 겪는 과정과 유사했을 수도 있거나, 혹은 그 과정이 전혀 다르고 중간 형태를 포함했을 수도 있다. 그러나 말하지 못하는 유인원에서 말하는 인간으로 가는 여정에서 어떤 형태의 진화가 일어났음이 틀림없다. 어느 시점에 언어의 선행하는 형태, 곧 더 단순한 형태의 언어가 있었음에 틀림없다. 또한 어떤 조어(祖語), 즉 언어라고 불릴 수 있는 최초의 것이 있었음에 틀림없다.

주어진 문장은 언어도 우리의 대뇌가 진화했던 것처럼, 또는 우리의 도구 제작 능력과 같은 방식으로 여러 단계를 거쳐 진화했을 가능성이 더 높다는 내용이다. 글의 전반부에서는 언어가 인간 진화 과정에서 갑자기 완전한 형태로 나타났다는 언어의 빅뱅 이론을 소개하고 있으며, ②의 바로 앞 문장에서는 생물학적으로는 말하기 능력이 갑자기 생겨났을 가능성은 극히 낮다는 내용이 이어지고 있다. ②의 바로 다음 문장에서는 주어진 문장의 주장을 어린이와 유인원의 예를 들어 뒷받침하고 있다. 따라서 주어진 문장이 들어가기에 가장 적절한 곳은 ②이다.

구조 해설

■ It is more likely that language evolved in several stages in the same way, for instance, as [our large brains **did**] or [our tool-making ability].

두 개의 [ ]가 or에 의해 대등하게 연결되어 as에 이어진다. 첫 번째 [ ] 안의 did는 대동사로 앞서 언급된 evolved in several stages를 가리키며, 두 번째 [ ]의 마지막에는 did가 생략되어 있다.

어휘 및 어구

evolve 진화하다
in principle 이론상으로
intermediate 중간의
notion 개념
linguistic 언어의, 언어적인
champion 옹호하다
pop up 갑자기 나타나다
ape 유인원

Word Search

| 정답 | **1.** evolved **2.** linguistic **3.** intermediate
1. 시간이 지나면서 언어는 새로운 단어와 어구를 포함하도록 진화했다.
2. 그 시의 전체적인 의미를 이해하려면, 저자가 사용하는 언어적 장치를 고려해야 한다.
3. 이것은 최종 승인 전에 거치는 중간 단계이다.

Exercise 9
정답 ②

소재 정보의 생성과 배포
해석 디지털 시대 이전의 시대에 정보를 생성할 수 있는 권한은, 왕이 되었든 '뉴욕 타임즈'가 되었든 간에, 주로 지정된 권위자들의 수중에 있었다. 그러나 오늘날 정보 기술, 특히 컴퓨터와 원

거리 통신 전송기의 발전은 그 결과 정보가 누구에 의해서든지 생성될 수 있고 '누구나에서 다른 모든 사람에게'로 흘러갈 수 있는 상황이 되었다. 다음 놀라운 통계를 예로 들어보면, 이 문서를 작성하는 시점에 현재 온라인에 있는 웹사이트 수는 15억 개를 넘고, 가장 큰 소셜 미디어 플랫폼 중 일부의 활성 사용자 수는 20억 명을 넘으며, 인터넷 사용자 수는 40억 명을 넘는다. 소비자가 생성한 콘텐츠로 더 특정해 보면, 매 분마다 400시간 분량의 비디오가 업로드되고 있으며, 유명 비디오 사이트에서 '사용 방법' 비디오에 대한 검색은 매년 70%씩 증가하고 있다. 이 숫자들은 이 원고가 인쇄될 때쯤이면 아마 증가할 것이다. 오늘날 사람들은 정보의 단순한 소비자일 뿐만 아니라, 자신만의 콘텐츠 생성물을 통한 정보의 제공자이기도 한데, 그것은 곧 '사람들이 온라인 세계에 기여하는 자료'이다. 오늘날 정보의 생성과 배포는 거의 모든 사람이 참여할 수 있고, 실제로 참여하는 시장이다.

→ 디지털 시대에, 정보 기술의 발전은 콘텐츠를 생성하고 공유하는 권한을 지정된 권위자들로부터 개인들에게로 옮겼으며, 그 결과 거의 모든 사람이 기여하는 참여(의) 정보 시장이 형성되었다.

문제 해설

디지털 시대 이전에는 정보 생성이 소수에게만 한정되었지만, 디지털 기술의 발전으로 이제 누구나 정보를 생성하고 공유할 수 있게 되었으며, 오늘날 수십억 명의 인터넷 사용자가 매 분마다 방대한 양의 콘텐츠를 온라인에 업로드하고 정보 생성과 배포에 적극 참여하고 있다는 내용의 글이다. 따라서 요약문의 빈칸 (A), (B)에 들어갈 말로 가장 적절한 것은 ② '옮겼다(shifted) – 참여의(participatory)'이다.
① 옮겼다 – 폐쇄적인
③ 간직했다 – 경쟁적인
④ 유지했다 – 개인화된
⑤ 유지했다 – 소비자가 만든

구조 해설

■ People today are **not simply** [innocent consumers of information], **but also** [providers of information through their own content creation], [which is "the material people contribute to the online world."]

'~뿐만 아니라 …도'라는 의미의 「not simply ~, but also …」의 구조로 두 개의 [ ]가 대등하게 연결되어 있다. 세 번째 [ ]는 their own content creation을 부가적으로 설명하는 관계절이다.

어휘 및 어구

designated 지정된
authority 권위자
telecommunicator 원거리 통신 전송기[수신기]
statistics 통계

manuscript 원고
innocent 단순한, 순진한
distribution 배포

Word Search

| 정답 | **1.** statistics **2.** innocent **3.** distribution
1. 이 지역의 범죄율은 전국 통계보다 낮다.
2. 그 디자인은 단순하지만 매력적이며, 단순함과 순수함으로 가득 차 있었다.
3. 새로운 소프트웨어 업데이트의 배포는 다음 주에 시작될 것이다.

Exercise 10 ────────────── 정답 ④

소재 과학 연구

해석 과학 연구는 일반적으로 몇 가지 뚜렷한 단계를 포함하는 순환을 기반으로 한다. 그 순환은 보통 우연한 관찰의 수집과 이러한 관찰에 대한 가능한 설명의 체계적인 정리에서 시작된다. 이 설명은 이론으로 알려져 있다. 이 이론을 바탕으로, 우리는 그런 다음 우리가 아직 관찰하지 않은 상황들에 대해 예측할 수 있다. 다시 말해, 우리는 구체적인 가설을 세울 수 있다. 이제 우리는 우리의 이론이 틀렸음을 입증하는 것을 목표로 하는 실험을 설계할 수 있는데, 우리의 실험 결과가 그 이론과 일치할 때, 우리는 그것에 아마 어떤 문제가 있을 것이라고 생각할 이유가 없다. 그러나 결과가 일치하지 않을 때, 우리는 우리의 이론을 조정하거나 완전히 버려야 할 것이다. 이론을 '증명하는' 것은 불가능한데, 상충되는 증거를 찾지 못하는 한, 그 이론은 여전히 타당한 것 같다. 그러나 이것은 미래에 일치하는(→ 상충되는) 결과가 있을 수도 있다는 가능성을 배제하지는 않는다. 우리가 어떤 이론과 일치하는 결과를 더 많이 찾을수록 그것(그 이론)은 더 타당한 것처럼 보이겠지만, (적어도 원칙적으로는) 상충되는 결과를 내는 단 하나의 실험만으로 그것을 뒤집어엎기에 충분할 수 있다는 것은 여전히 사실이다.

문제 해설
과학 연구는 관찰에서 시작해 이론을 세우고, 이를 바탕으로 가설을 세운 후, 실험을 통해 이론을 검증하거나 조정 및 완전히 버리는 과정을 거친다. 이론에 상충되는 증거를 찾지 못하면 이론은 남아 있게 되지만, 상충되는 결과를 내는 실험이 단 하나만 있어도 이론은 뒤집어엎어질 수 있기 때문에 이론이 타당하게 남아 있다 할지라도, 미래에 그 이론을 뒤집어엎을 결과가 나올 가능성도 있다. 따라서 ④의 match를 conflict와 같은 낱말로 바꿔야 한다.

구조 해설
▪ Now, we can design an experiment [that aims to

disprove our theory]: [when the results of our experiment are consistent with the theory], we have no reason to suspect [that anything is wrong with **it**].
첫 번째 [ ]는 an experiment를 수식하는 관계절이다. 두 번째 [ ]는 시간의 부사절이며, 세 번째 [ ]는 suspect의 목적어 역할을 하는 명사절로, 그 안의 대명사 it은 앞에 언급된 the theory를 가리킨다.

어휘 및 어구
typically 일반적으로, 보통
incidental 우연한, 우연히 일어나는
observation 관찰
formulation 체계적인 정리, 공식화
explanation 설명
theory 이론
hypothesis 가설(*pl.* hypotheses)
disprove 틀렸음을 입증하다
consistent 일치하는
suspect 아마 ~일 것이라고 생각하다[여기다], 추측하다
exclude 배제하다
yield 내다, 생산하다

Word Search

| 정답 | **1.** incidental **2.** Observation **3.** suspect
1. 그들의 길은 순전히 우연한 기회로 커피숍에서 교차했다.
2. 그 팀의 상호 작용에 대한 관찰은 인내심과 열린 마음의 중요성을 보여주었다.
3. 나는 그녀가 우리에게 말하는 것보다 아마 더 많이 알고 있을 것이라고 생각한다.

Exercise 11~12 ────────── 정답 11 ① 12 ⑤

소재 과학 글쓰기의 장점

해석 과학에서 글쓰기의 진실 중 하나는 정확한 응답에 대해 전혀 모르는 학생의 머리에서 나올 수 있는 불성실한 허튼소리의 양을 상당히 제한한다는 것이다. 과학 이론에 대한 기본적인 이해가 없는 학생은 그것을 적용하는 데 엄청난 어려움을 겪을 것이다. 에세이를 통해 실제 과학 문제에 답할 때는 정답을 추측하여 '운 좋게' 답을 맞힐 수 있는 방법이 없으며 숨을 곳도 없다. 일반적으로 학생들에게 관련 있는 과학 원리와 이론을 언급하고 잠재적인 응용을 탐구하도록 요구하는 에세이는 이미 제공된 가능한 답과 짝을 이루는 짧은 질문으로 이루어진 객관식 시험보다 학생의 지식에 대한 더 진정성 있는 평가이다.
에세이 작성의 또 다른 장점은 교사가 학생의 이해도가 강한 부분

과 이해도가 떨어지기 시작하는 부분이 어디인지 파악할 수 있다는 점이다. 결국 글쓰기는 구체화된 생각의 한 형태이다. 따라서 개념에 대한 학생의 이해가 옆길로 새기 시작할 때 교사가 문제를 파악하고 도움을 제공할 수 있는 것은 생각이 가시화되는 오직 그 순간을 통해서이다. 글쓰기는 학생의 학업 과정 내내 그를 괴롭힐 수 있는 개념적 오해를 악화시킬(→ 방지할) 수 있는 일종의 개입을 위한 기회를 제공한다. 실제로 Heddy와 Sinatra(2013)와 Francek(2013)은 가르칠 때 가장 어려운 부분 중 하나가 학생들이 오개념을 버리게 하려고 애쓰는 것임을 발견했다.

문제 해설

11 과학에서 글쓰기는 학생의 진정한 지식을 평가하고, 개념적 오해를 방지하는 데 도움이 되며, 에세이를 통해 교사는 학생의 이해도를 파악하고, 필요한 도움을 제공할 수 있다는 내용이므로, 글의 제목으로 가장 적절한 것은 ① '과학에서 에세이 작성의 이점'이다.
② 과학에서의 에세이: 명확성보다 추측
③ 학술적 과학 글쓰기의 추악한 진실
④ 객관식 시험을 통한 잘못된 생각 버리기
⑤ 기억력 향상하기: 글쓰기가 지식을 강화하는 방법

12 학생들의 에세이 작성을 통해 교사가 학생의 이해도가 강한 부분과 이해도가 떨어지기 시작하는 부분을 파악할 수 있으며, 학생의 개념에 대한 이해가 옆길로 새기 시작하면 교사가 문제를 파악하고 도움을 제공할 수 있으므로, 개념적 오해를 방지할 수 있다는 것을 알 수 있다. 따라서 (e)의 worsen을 prevent와 같은 낱말로 바꿔야 한다.

구조 해설

■ In general, essays [that require students {to invoke relevant scientific principles and theories} and {to explore potential applications}] are more genuine assessments of student knowledge than objective tests of short questions [paired with possible answers already furnished].
첫 번째 [ ]는 essays를 수식하는 관계절로, 그 안의 두 개의 { }는 and에 의해 대등하게 연결되어 students에 이어진다. 두 번째 [ ]는 objective tests of short questions를 수식하는 분사구이다.

어휘 및 어구

insincere 불성실한
immense 엄청난
relevant 관련 있는
assessment 평가
furnish 제공하다
concretize 구체화하다

grasp 이해, 파악
get off-track 옆길로 새다, 벗어나다
intervention 개입
conceptual 개념적인, 개념의

Word Search

| 정답 | **1.** immense **2.** relevant **3.** intervention
1. 그 소방관들은 엄청난 산불과 며칠 동안 싸우며 위험에 직면했다.
2. 흥미롭기는 하지만, 그 유명인의 휴가에 관한 정보는 당면한 주제와 완전히 관련 있는 것은 아니었다.
3. 그 교사의 시기적절한 개입은 두 학생 간에 싸움이 발생하는 것을 막았다.

수능특강

영어영역 | **영어독해연습**

정답과 해설

Mini Test

01 ⑤	02 ④	03 ⑤	04 ⑤	05 ⑤	06 ⑤	07 ⑤
08 ③	09 ③	10 ⑤	11 ⑤	12 ④	13 ④	14 ②
15 ⑤	16 ①	17 ②	18 ③	19 ②	20 ③	21 ②
22 ④	23 ①	24 ⑤	25 ④	26 ④	27 ②	28 ③

01

정답 | ⑤

소재 자원봉사 일정 관리 방식 변경에 대한 항의

해석 Stevens 박사님께,

지난 4년 동안 Langford 과학 박물관에서 자원봉사를 한 것은 제가 대학교수로서 은퇴한 이후 자부심의 원천이었습니다. 그러나 최근의 자원봉사 일정 관리의 변화는 저와 많은 다른 사람들이 우리의 책무를 지속하기에 점점 더 힘이 드는 것으로 드러나고 있습니다. 이전에는 봉사자들이 (봉사) 일정표에서 가능한 시간대를 신청하도록 요청을 받았습니다. 이 방식은 저희 모두에게 매우 효과적이었는데 왜냐하면 그것이 저희 자원봉사 시간을 다른 다양한 시간적 요구에 맞출 융통성을 허용했기 때문입니다. 하지만 지난 6주 동안 자원봉사 서비스의 신임 관리자는 저희와의 상의 없이 저희에게 자원봉사 일정을 배정하기 시작했고, 이는 자원봉사자들 사이에 많은 혼란 상태를 불러왔고 수많은 교대 근무 취소를 야기했습니다. 저희는 무급 봉사자로서 저희의 자율성을 중요하게 생각하며, 관리자에 의해 근무 시간이 배정되는 급여가 지급되는 직원처럼 대우받아서는 안 됩니다. 이 문제가 너무 커져서 해결할 수 없어지기 전에 박사님께서 이 문제를 처리해 주실 수 있기를 바랍니다.

Kate Greenbaum 드림

문제 해설

자원봉사자들이 봉사 일정표에서 가능한 시간대를 선택하던 기존의 방식과는 달리 자원봉사자들에게 자원봉사 일정을 일방적으로 배정하는 것으로 일정 관리 방식이 변경된 것에 대해 항의하는 내용이므로, 글의 목적으로 가장 적절한 것은 ⑤이다.

구조 해설

■ In the last six weeks, though, [the new manager of volunteer services] has begun assigning us to volunteer schedules without consulting us, [{causing a great deal of upset among the volunteers} and {resulting in numerous cancellations of shifts}].

첫 번째 [ ]는 문장의 주어이다. 두 번째 [ ]는 앞 절의 내용에 대한 결과를 나타내는 분사구문으로 두 개의 { }가 and로 연결되어 있다.

어휘 및 어구

retirement 은퇴
sustain 지속하다
slot 시간대
flexibility 융통성
assign 배정하다
consult 상의하다
upset 혼란 상태
shift 교대 근무 (시간)
salaried 급여가 지급되는
address 처리하다, 대처하다

02

정답 | ④

소재 지진 상황에서 아들을 구하려는 절박한 상황

해석 강력한 지진이 한밤중에 우리 집을 강타하여, 갑자기 나를 거칠게 흔들어 깨우고 침대 밖으로 나를 거의 내던질 뻔했다. 나의 즉각적인 반응은 달려가서 우리 집의 반대편 쪽 끝에 있는 방에 있던 아들 Dustin을 데려오는 것이었다. 정전이 된 후 모든 것이 캄캄했다. 나는 두 손을 뻗어 나를 Dustin의 침실로 안내해 줄 벽과 가구를 앞이 안 보이는 채 더듬어 찾으면서 완전한 암흑 속에서 집을 가로질러 달려갔다. 격렬한 여진이 나를 넘어뜨려 무릎 꿇게 했지만, 나는 멈출 수 없었다. 무슨 일이 있어도 아들에게 가야 했다. 그때, 흔들림이 멈추었으며, 내가 아들의 침실을 자세히 들여다보았을 때 모든 것이 조용해졌다. 이제 어둠에 익숙해졌기 때문에 나는 Dustin의 책장이 방 중앙으로 넘어져 아들의 침대를 가까스로 비껴간 것을 알아볼 수 있었다. 잔해 위를 기어가 나는 마침내 Dustin에게 닿았다. 아들은 두려움에 떨고 있었지만 이불 속에 안전하게 웅크리고 있었다. 나는 Dustin을 꼭 껴안고 안도의 한숨을 쉬었다. 그 순간, 내 아들이 안전하다는 사실 외에는 다른 어떤 것도 중요하지 않았다.

문제 해설

강력한 지진이 'I'의 집을 강타하자 집 반대편 쪽 방에 있는 아들을 구하기 위해 무슨 일이 있어도 아들에게 가려는 'I'의 행동과 생각으로부터 'I'의 필사적인 심경을 알 수 있다. 이후 아들의 침실에 도착하여 아들이 이불 속에서 안전하게 웅크리고 있다는 것을 'I'가 확인하자 아들을 껴안고 안도의 한숨을 쉬었으므로, 'I'의 심경 변화로 가장 적절한 것은 ④ '필사적인 → 안도한'이다.

① 깜짝 놀란 → 궁금한
② 부끄러운 → 감사한
③ 좌절한 → 자신감 있는
⑤ 무서워하는 → 혼란스러운

Mini Test

구조 해설

- [By now accustomed to the darkness], I could make out [that Dustin's bookcases had fallen into the center of the room, {narrowly missing his bed}].

첫 번째 [ ]는 이유를 나타내는 분사구문이다. 두 번째 [ ]는 make out의 목적어 역할을 하는 명사절이고, 그 안의 { }는 앞 절의 부수적인 상황을 나타내는 분사구문이다.

어휘 및 어구

strike 강타하다
instant 즉각적인
outstretch 뻗다, 펴다
blindly 앞이 안 보이는 채
feel for ~을 더듬어 찾다
aftershock 여진
knock 넘어뜨리다
peer into ~을 자세히 들여다보다
accustomed to ~에 익숙해진
make out ~을 알아보다
crawl 기어가다
sigh 한숨
relief 안도
matter 중요하다

03

정답 | ⑤

소재 유의미한 생태 연구를 위한 직관의 계발

해석 좋은 직관은 유의미한 생태 연구를 설계하기 위한 첫 번째 요건이다. 그 직관을 계발하는 가장 좋은 방법은 현장에서 생물을 관찰하는 것에 의해서이다. 안타깝게도, 자연을 그저 관찰할 '시간이 있는' 사람은 우리들 중 거의 없다. 대학원 위원회와 교수 종신 재직권 심사위원들은 귀중한 시간을 이런 식으로 투자하는 것을 추천하지 않을 것이다. 하지만, 관찰은 여러분이 현실에 근거한 작업가설을 생성하는 데 절대적으로 필요하다. 그러니, 여러분의 생물을 알게 되도록 어느 정도의 시간을 할애하라. 만약 여러분이 수업이나 다른 맡은 일로 너무 바쁘다면, 어떠한 조작(또는 선입견)도 없이 생태계를 관찰하기 위해 실험을 시작하기 전에 이틀을 따로 잡아 두라. 이것을 실험실 친구나 동료와 함께 하면 많은 경우 재미가 있다. 반대의 방법도 효과적일 수 있다. 즉 주변에 다른 사람이나 방해 요소가 없는 상태로 그저 여러분의 생태계를 관찰하면서 하루를 꼬박 보내는 것도 생각해 보라.

문제 해설
유의미한 생태 연구를 설계하려면 좋은 직관이 필요한데 그 직관

을 계발하는 가장 좋은 방법은 현장에서 생태계를 어떤 조작과 선입견 없이 있는 그대로 관찰하는 것이라는 내용의 글이다. 따라서 필자가 주장하는 바로 가장 적절한 것은 ⑤이다.

구조 해설

- However, observations are absolutely essential [for you to generate working hypotheses {that are grounded in reality}].

[ ] 안에서 for you는 to부정사구의 의미상 주어를 나타내고, { }는 working hypotheses를 수식하는 관계절이다.

어휘 및 어구

intuition 직관
requirement 요건
organism 생물, 유기체
graduate committee 대학원 위원회
working hypothesis 작업가설(연구를 진행하기 위한 수단으로 세우는 가설)
ground 근거를 두다
reserve 따로 잡아 두다
manipulation 조작
preconceived notion 선입견
distraction 방해 요소, 주의를 산만하게 하는 것

04

정답 | ⑤

소재 지역별 기후 예측을 위한 기후 모델

해석 우리는 인간이 지구 기후 시스템에 미치는 영향을 더 잘 이해해 가고 있다. 지구가 더 더워짐에 따라, 고위도 지역은 열대 지역보다 더 빠르게 더워질 것이다. 지중해 지역은 더 건조해지고 열대 지역은 더 습해질 것이다. 하지만 이것은 베트남에 비가 올 것이기 때문에 스페인에서 우산을 챙겨야 한다고 말하는 것과 같다. 기후 변화에 대해 계획하고 효과적으로 적응하기 위해, 우리는 대기 대순환 모델(GCM)이 제공할 수 있는 것보다 훨씬 더 세밀한 규모의 미래 기후 정보가 필요하다. 제방을 쌓을지, 일부 주택을 옮길지, 작물을 바꿀지, 혹은 보험에 가입할지를 결정하기 위해, 우리는 100킬로미터 미만 규모의 데이터가 필요하다. 한 가지 접근 방식은 특정 관심 지역의 더 세밀한 규모의 모델을 더 큰 규모의 대기 대순환 모델에 끼워 넣는 것이다. 어떤 지역도 지구의 나머지 부분과 격리되어 있지 않으므로, 그 모델의 대기 대순환 모델 부분은 전 세계적으로 무슨 일이 일어나고 있는지를 추적하고 더 세밀한 규모의 지역 기후 모델과 정보를 교환할 수 있다.

문제 해설
기후 변화에 대해 효과적으로 계획하고 적응하려면 전 지구적인

규모의 대기 대순환 모델에만 의존하는 것이 아니라 훨씬 더 세밀한 규모의 지역 기후 모델을 같이 활용하는 것이 필요하다는 내용이므로, 밑줄 친 부분이 글에서 의미하는 바로 가장 적절한 것은 ⑤ '전 지구적 규모의 예측에만 의존하는 것은 지역 기후 변화 관련 문제에 효과적으로 대처하지 못하게 한다.'이다.
① 기후 위기에 대처하는 방법에 대한 합의에 도달하는 것은 어렵다.
② 한 지역의 기후에서 일어나는 일은 다른 지역의 기후에 영향을 미칠 수도 있다.
③ 주요 오염 유발 국가의 참여 없이는 기후 문제 해결을 위한 행동은 반드시 실패할 것이다.
④ 우리는 시대에 뒤진 척도를 사용하여 기후 변화의 영향을 측정하는 것을 멈춰야 한다.

구조 해설

- [No region is isolated from the rest of the planet], so [the GCM part of the model can {keep track of ⟨what is going on globally⟩} and {exchange information with the finer-scale regional climate model}].
  두 개의 [ ]가 so에 의해 대등하게 연결되었고, 두 개의 { }는 and로 연결되어 can에 이어진다. ⟨ ⟩는 keep track of의 목적어 역할을 하는 명사절이다.

어휘 및 어구

latitude 위도 (지역)
tropics 열대 지역
the Mediterranean 지중해 (지역)
fine 세밀한, 미세한
put up ~을 쌓다, ~을 세우다
insurance 보험
isolate 격리하다, 고립시키다
keep track of ~을 추적하다
regional 지역의

Tips

general circulation model(대기 대순환 모델): 대기학에서 대기 대순환 모델은 지구의 기후 시스템을 시뮬레이션하는 기후 모델이다. 이 모델은 대기, 해양, 육지, 해빙(海氷) 등 지구 시스템의 다양한 구성 요소 간의 상호 작용을 방정식으로 표현하여 날씨와 기후의 변화를 예측하는 데 사용된다.

05

정답 | ⑤

소재 삶에서 물러남과 복귀의 균형
해석 곰은 어디서 언제 먹이를 찾을지 알고 있을 뿐만 아니라, 어려운 시기를 잘 참고 견디기 위해 언제 자신의 굴로 물러나야

할지도 안다. 곰과 마찬가지로 우리는 때때로 세상으로부터 물러나고 싶은 욕구를 느끼는데, 특히 스트레스를 받은 시기 이후에 그렇다. '동면하기' 위한 회복의 시간을 할애하려는 이러한 곰과 같은 충동을 느낄 때, 그것을 우리의 내면에 있는 곰의 분별 있는 충동으로 생각해야 한다. 우리는 우리의 삶을 찬찬히 살펴보거나, 창의적인 프로젝트를 시작하거나, 여행을 계획하거나, 싹터서 미래에 결실을 맺길 바라는 생각의 씨앗을 심기 위해 일부 사회적 활동에서 물러나는 것을 고려할 수도 있을 것이다. 하지만, 우리는 봄이 오면 곰들이 굴 밖으로 나온다는 것을 또한 기억할 필요가 있다. 고립된 상태에서 너무 많은 시간을 보내는 것은 우리에게서 외부 세계와의 연결과 외부 세계로부터 오는 영감을 앗아갈 수 있다. 회복을 위해 물러나고 싶은 충동과 세상이 제공할 모든 것에 의해 새로운 활력을 얻을 기회의 균형을 맞추는 것이 가장 좋다.

문제 해설

스트레스가 많은 시기 후에 세상으로부터 물러나고 싶은 욕구를 느끼는데, 이는 내면의 회복과 미래를 위한 에너지 충전의 시간일 수 있지만, 너무 오래 고립된 상태로 지내면 외부와의 연결과 영감을 잃을 수 있기에 회복을 위해 물러나는 시간과 외부 세계로부터 활력을 얻을 수 있는 기회를 잡는 것의 균형을 맞추어야 한다는 내용의 글이다. 따라서 글의 요지로 가장 적절한 것은 ⑤이다.

구조 해설

- The bear **not only** knows [where and when to find food], he **also** knows [when to retreat to his den to ride out a challenging time].
  「not only ~ (but) also ...」 구문이 사용되어 '~뿐만 아니라 ...도'의 의미를 나타낸다. 두 개의 [ ]는 모두 「의문사+to부정사구」로 각각 앞에 있는 동사 knows의 목적어 역할을 하고 있다.

어휘 및 어구

retreat 물러나다, 후퇴하다
ride out ~을 잘 참고 견디다
urge 충동
carve out ~을 할애하다, ~을 떼어 내다
restorative (원기를) 회복시키는
sensible 분별 있는, 합리적인
impulse 충동
withdraw 물러나다, 철수하다
take stock of ~을 찬찬히 살펴보다
fruition 결실, 성과
isolation 고립
deprive ~ of ... ~에게서 ...을 앗아가다
inspiration 영감
revitalize 새로운 활력을 주다

06

정답 | ⑤

소재 식물을 인격체로서 대해야 하는 당위성

해석 식물은 우리 인간 드라마에서 무대 도구의 일부가 아니라, 명백한 경험과 욕구를 지닌 등장인물이다. 우리처럼, 그들은 애정, 보살핌, 그리고 고통을 느낄 수 있다. 우리가 식물의 경험이 가진 도덕적인 중요성을 묵살할 수 있지만(그리고 실제로 그렇게 하지만), 그것이 존재하지 않는다고 주장하는 것은 매우 위험한 무시의 형태이다. 기후 변화를 줄이기 위한 우리의 냉철하고 실용적인 접근법은 지금까지 대부분 효과가 없었다. 진정한 변화를 가져오려면, 우리는 식물을 단순히 우리 인간 자신의 번영을 위한 도구가 아니라 독립적이고 본질적으로 가치 있고 '지각력이 있는' 존재로 '보는' 법을 배워야 한다. 우리의 행복은 식물의 행복에 크게 의존하고 있으며, 우리가 이를 '이해'하는 가장 좋은 방법은 식물이 '인격체'임을 우리 자신에게 상기하는 것에 의해서이다. 그들도 나머지 우리 못지않게 깨끗한 물, 건강한 토양, 그리고 살기 좋은 대기에 대한 권리를 가지고 있으며, 이는 우리가 기후 변화의 윤리적 차원을 고려할 때 기억될 필요가 있다.

문제 해설
기후 변화를 줄이기 위한 기존의 접근법은 효과가 없었으며, 진정한 변화를 위해서는 식물을 독립적이고 본질적으로 가치 있으며 지각력이 있는 존재로 여겨야 하고, 기후 변화의 윤리적 차원을 고려할 때, 식물이 인격체임을 상기하는 것이 중요하다는 내용의 글이다. 따라서 글의 주제로 가장 적절한 것은 ⑤ '기후 변화를 다룸에 있어 식물을 지각 있는 존재로 인식할 필요성'이다.
① 지역 생태계에 가져온 비토착 식물의 파괴적인 결과
② 기후 변화로 인한 식물 종 다양성 감소
③ 인간의 윤리적 삶에 기준점을 제공하는 식물의 생태
④ 인간 본성을 회복하기 위해 인간이 식물로부터 배울 수 있는 윤리적 교훈

구조 해설
- [While we can (and do) dismiss the moral importance of plant experiences], [to pretend that they don't exist] is a very dangerous form of disregard.
  첫 번째 [ ]는 양보의 의미를 나타내는 부사절이다. 두 번째 [ ]는 to부정사구로 주절의 주어 역할을 하고 있다.

어휘 및 어구
set piece 무대 도구, 기성 형식
distinct 명백한, 분명한
affection 애정, 호의
dismiss 묵살하다
pretend (that) (사실이 아닌 것을) 주장하다
disregard 무시, 경시
utilitarian 실용적인, 공리적인
ineffectual 효과가 없는
intrinsically 본질적으로
flourish 번영하다, 풍요롭게 살다
welfare 행복, 복지
soil 토양
atmosphere 대기

07

정답 | ⑤

소재 현대 스포츠의 기원과 확산

해석 현대의 요리가 전 세계에서 발전된 영향력 있는 것들의 진정한 융합을 나타내는 반면, 현대의 스포츠는 그렇지 않다. 사실, 세계의 스포츠는 부리토보다는 흑사병에 더 가깝게 퍼졌다. 그것들은 거의 전적으로 서양 문명, 즉 유럽 국가와 유럽 정착민 사회에서 나타난 세계적인 현상이며 세계의 다른 지역으로 퍼져나갔다. 사커(축구)를 예로 들어 보자. 유럽인과의 접촉 이전에 손의 사용을 제한하거나 금지한 구기 경기가 메소아메리카에서 번창했다. 오늘날, 'fútbol'은 수백만 멕시코인의 열정에 불을 붙인다. Estadio Azteca는 두 번의 월드컵 결승전(1970년과 1986년)과 다수의 다른 국제 경기에서 넘칠 정도로 가득 찼다. 피자와 파스타가 구세계와 신세계 음식을 융합한 것과 꼭 마찬가지로 'fútbol'도 메소아메리카와 유럽의 스포츠 오락을 융합한 것이라고 여러분은 쉽게 결론지을 수 있을지도 모른다. (하지만) 그렇지 않다.

문제 해설
현대 요리는 전 세계적으로 영향을 받은 반면, 현대 스포츠는 주로 서구 문명에서 퍼져 나왔는데, 축구의 예를 들어 유럽과의 접촉 이전에 메소아메리카에서 손을 사용하지 않는 구기 경기가 존재했으며, 오늘날 월드컵과 같은 축구 경기가 큰 인기를 끌지만, 이는 두 문화의 스포츠 오락의 융합은 아니라는 내용의 글이다. 따라서 글의 제목으로 가장 적절한 것은 ⑤ '현대 스포츠: 서양의 뿌리와 세계적인 확산'이다.
① 새로운 맛, 서구의 동양화
② 유럽이 정말로 축구의 발상지인가?
③ 요리와 스포츠의 유사점
④ 스포츠가 요리보다 더 쉽게 퍼진 이유

구조 해설
- They are global phenomena [that {emerged almost exclusively from Western civilization — from European nations and European settler societies} — and {spread to other parts of the world}].
  [ ]는 global phenomena를 수식하는 관계절인데, 그 안에서 { }로 표시된 두 개의 동사구가 and로 연결되어 술어를 이루

고 있다.

어휘 및 어구

contemporary 현대의, 동시대의

cuisine 요리, 요리법

blending 혼합

phenomenon 현상(*pl.* phenomena)

emerge 나타나다, 출현하다

exclusively 전적으로, 독점적으로

civilization 문명

bar 금지하다

flourish 번창하다

ignite (~에) 불을 붙이다

filled to overflowing 넘칠 정도로 가득 찬

a multitude of 다수의

conclude 결론짓다

fuse 융합하다

Tips

Mesoamerica(메소아메리카): 현재의 멕시코 중부 및 남부, 과테말라, 벨리즈, 엘살바도르, 온두라스 서부, 니카라과 북부 및 코스타리카 북서부를 포함하는 지역으로, 고대 문명들이 번성했던 중요한 문화권이다. 이 지역은 다양한 고대 문명들이 상호작용하며 공통된 문화적 특징을 공유했던 곳으로, 아즈텍(Aztec), 마야(Maya), 올멕(Olmec), 테오티우아칸(Teotihuacan)과 같은 주요 문명들이 이 지역에 존재했다.

08

정답 | ③

소재 공립 초등학교와 중등학교의 교사 1인당 평균 학생 수

해석 위 도표는 2019년 OECD 7개국에 대한 공립 초등학교와 중등학교에서 국가 간 교사 1인당 평균 학생 수의 차이를 보여 준다. 노르웨이는 초등학교에서 교사 1인당 평균 학생 수가 가장 적었고, 반면에 벨기에가 중등학교에서 가장 적었다. 스페인에서는 초등학교와 중등학교 모두에서 교사 1인당 평균 학생 수가 벨기에보다 0.5명 더 많았다. 프랑스는 초등학교와 중등학교의 교사 1인당 평균 학생 수에서 두 번째로 큰 차이를 보였으며, 중등학교에서 (초등학교에 비해) 교사 1인당 학생 수가 5명보다 더 많았다. 미국은 유일무이하게 초등학교와 중등학교에서 동일한 학생 대 교사 비율을 기록했다. 한편, 멕시코는 도표에 포함된 국가들 중 중등학교의 학생 대 교사 비율이 초등학교보다 높은 유일한 국가였다.

문제 해설

프랑스는 초등학교에서 교사 1인당 학생 수가 중등학교보다 5.6명 많은 것으로 나타났으므로, 도표의 내용과 일치하지 않는 것은

③이다.

구조 해설

■ Meanwhile, Mexico was the only country among those listed in the chart [where the student-to-teacher ratio in secondary schools was higher than in elementary schools].

[ ]는 the only country among those listed in the chart를 수식하는 관계절이다.

어휘 및 어구

ratio 비율

public 공립(의), 공공(의)

uniquely 유일무이하게

09

정답 | ③

소재 삽화가 Helen Moore Sewell의 삶

해석 Helen Moore Sewell은 삽화로 알려진 미국의 화가이자 아동 도서 작가였다. Sewell은 어릴 때부터 그림을 그리기 시작했다. 12세에 그녀는 Pratt Institute에 다닌 역대 최연소자가 되었는데, 그곳은 예술과 디자인으로 특히 유명했다. 그녀는 또한 우크라이나계 미국인 화가 Alexander Archipenko 밑에서 공부했는데, 그는 그녀의 스타일에 극적으로 영향을 미쳤다. Sewell의 초기 작품은 작가이자 삽화가로서 둘 다였다. 1924년에 그녀는 자신의 첫 책, Susanne K. Langer의 *The Cruise of the Little Dipper, and Other Fairy Tales*에 삽화를 그렸다. 그녀는 1930년에 자신의 책 *ABC for Everyday*에 삽화를 그렸고, 1년 후에 자신의 여동생과 함께 *Building a House in Sweden*을 공동으로 작업했다. 그녀는 또한 미국 시인 Emily Dickinson과 Jane Austen과 같은 영국 작가들의 작품을 포함하여 고전 작품에도 삽화를 그렸다. 자신의 경력 동안 Sewell은 50권이 넘는 책에 삽화를 그렸다. 그녀는 1955년 미국 작가 Alice Dalgliesh가 쓴 *The Thanksgiving Story*의 삽화로 칼데콧(Caldecott) 상을 수상했다.

문제 해설

Sewell은 1930년에 자신의 책인 *ABC for Everyday*에 삽화를 그렸고, 그로부터 1년 후에 자신의 여동생과 함께 *Building a House in Sweden*을 공동 작업했다고 했으므로, 글의 내용과 일치하지 않는 것은 ③이다.

구조 해설

■ She also studied under the Ukrainian American artist Alexander Archipenko, [who dramatically influenced her style].

[   ]는 the Ukrainian American artist Alexander Archipenko를 부가적으로 설명하는 관계절이다.

(어휘 및 어구)

illustration 삽화
attend 다니다
dramatically 극적으로
collaborate 공동으로 작업하다
classic 고전의
poet 시인

10
정답 | ⑤

소재 신나는 언플러그드 게임의 밤

해석 신나는 언플러그드 게임의 밤

Windsor 시 청소년 서비스 센터는 신나는 언플러그드 게임의 밤을 주최합니다. 친구와 가족과 함께 보드게임, 카드 게임, 주사위 게임을 즐기러 나오세요!

일시: 4월 18일 금요일 오후 6시부터 오후 8시까지
장소: Windsor 청소년 서비스 센터
대상: Windsor 주민 모두 참여할 수 있습니다! 12세 미만의 어린이는 반드시 성인과 동행해야 합니다.
세부 사항:
– 무료 입장. (음료와 간식이 무료로 제공됩니다.)
– 게임은 제공되지만, 자신이 가지고 있는 고전적인 게임이나 비전자 게임을 가져오는 것도 환영합니다.
– 좌석은 100명으로 제한됩니다.
– 예약 없이 오는 사람을 수용할 수 없으므로 사전 등록이 필요합니다.
– 자리를 예약하기 위해서는 Sarah Bunyan에게 860-627-1482로 연락하십시오.

더 많은 정보가 필요하시면 unpluggn@townwindsorgu.co.uk로 저희에게 이메일을 보내 주세요.

(문제 해설)

예약 없이 오는 사람의 입장은 수용할 수 없고 사전 등록이 필요하다고 했으므로, 안내문의 내용과 일치하지 않는 것은 ⑤이다.

(구조 해설)

▪ Pre-registration is required [as we cannot accommodate walk-ins].
[   ]는 이유를 나타내는 부사절이다.

(어휘 및 어구)

accompany 동반하다

admission 입장
accommodate 수용하다
walk-in 예약 없이 오는 사람
reserve 예약하다

11
정답 | ⑤

소재 기부하는 금요일 행사

해석 기부하는 금요일 행사

기부하는 금요일은 나눔의 정신을 증진하는 데 헌신하는 연례 글로벌 행사입니다. 이 행사는 매년 미국 추수감사절 이후 금요일에 열립니다. 현지 시간 자정부터 시작해서 24시간 동안 진행됩니다. 비영리 단체들이 나눔과 공동체 의식을 촉진하는 활동에 참여하도록 장려합니다.

등록:
– 이 행사에 참여하기 위한 공식적인 등록 절차는 없습니다.
– 모든 비영리 단체의 참여를 환영합니다.

기부:
– 기부하는 금요일 웹사이트는 직접 기부금을 받거나 배분하지 않습니다.
– 비영리 단체들은 자신의 웹사이트를 통해서만 기부금을 받아야 합니다.

비용: 행사 참여는 무료입니다.

참여 방법: 단체들은 기부용 음식 모으기 운동을 주최하거나, 자원봉사자를 모집하거나, #GivingFriday 해시태그를 사용하여 소셜 미디어에서 자신들의 행동을 공유함으로써 참여할 수 있습니다.

※ 무료 자료는 기부하는 금요일 웹사이트에서 이용 가능합니다.

(문제 해설)

기부용 음식 모으기 운동을 주최하거나 자원봉사자를 모집하는 방식으로 행사에 참여할 수 있다고 했으므로, 안내문의 내용과 일치하는 것은 ⑤이다.

(구조 해설)

▪ GivingFriday is an annual global event [dedicated to promoting the spirit of giving].
[   ]는 an annual global event를 수식하는 분사구이다.

(어휘 및 어구)

dedicate 헌신하다
nonprofit 비영리의
foster 촉진하다

donation 기부(금)
distribute 배분하다
directly 직접으로, 직접적으로
host 주최하다
food drive 기부용 음식 모으기 운동
recruit 모집하다

12

소재 새로운 기술의 가능성과 잘 보이지 않는 제약

해석 뉴 미디어는 우리가 새로운 일을 하고, 새로운 종류의 의미를 만들며 우리 자신의 정체성을 구현할 수 있게 해 주지만, 또한 우리가 그것을 사용하고 있을 때 무엇을 하고 의미할 수 있는지, 어떻게 생각하고 관계를 맺을 수 있는지, 그리고 어떤 사람이 '될' 수 있는지에 대하여 예외 없이 제약을 부과한다. 예를 들어, 텔레비전 뉴스는 생생하고 극적인 이야기의 제시를 할 수 있게 하지만, 길고 면밀히 살피는 분석에는 신문이나 잡지보다 덜 적합할 수도 있다. 소셜 네트워킹 사이트는 우리가 친구들과 연결 상태를 유지하는 것을 더 쉽게 만들어 주지만, (특히 광고주로부터) 우리의 프라이버시를 유지하는 것을 더 어렵게 만든다. 대부분의 모바일 전화기에 기본으로 탑재된 발신자 번호 표시 서비스는 우리가 전화를 가려내는 것을 더 쉽게 만들어 주지만, 우리가 거는 전화 또한 다른 사람들이 가려내기 더 쉽게 만든다. 새로운 기술의 제약은 흔히 그것(그 기술)의 행동 유도성보다 우리에게 덜 가시적이다. 우리는 새로운 도구로 할 '수 있는' 새로운 일들에 너무 집중하는 경향이 있어서 그것으로 할 '수 없는' 일들에는 주의를 기울이지 않는다.

문제해설
④ makes it easier에서 대명사 it이 makes의 형식상의 목적어로 쓰였는데, 내용상의 목적어가 없으므로, is를 내용상의 목적어를 유도하는 to be로 고쳐야 한다.
① 문장에서 쓰인 대명사 they나 them이 가리키는 것은 new media로 media는 medium의 복수 형태이다. 그러므로 them을 쓴 것은 어법상 적절하다.
② Television news는 셀 수 없는 명사로 단수 취급해야 하므로, 술어동사 allows는 어법상 적절하다.
③ 의미상 Social networking sites를 통해서 우리가 친구들과 연결되는 것이므로, 과거분사 connected를 쓴 것은 어법상 적절하다.
⑤ 「so ~ that ...」 구문을 이루는 구조로 접속사 that은 어법상 적절하다.

구조해설
■ Social networking sites [make **it** easier {**for us** to stay connected to our friends}], but [make **it** more difficult {to maintain our privacy (especially from advertisers)}].
두 개의 [ ]가 but에 의해 병렬 구조로 연결되어 Social networking sites에 이어진다. 각 [ ] 안의 it은 형식상의 목적어이며, { }가 내용상의 목적어이다. 첫 번째 { } 안의 for us는 이어지는 to부정사구의 의미상 주어이다.

어휘 및 어구
enact 구현하다, 일어나다
invariably 예외 없이, 언제나
limitation 제약, 한계
vivid 생생한
suitable 적합한
lengthy 긴, 장황한
analysis 분석
privacy 프라이버시, 사생활
caller identification 발신자 번호 표시 서비스
screen 가려내다
constraint 제약

Tips

affordance(행동 유도성): 어떤 형태나 이미지가 사람의 행위를 유도하는 힘을 말한다. 예컨대 의자를 보면 앉고 싶은 것은 의자가 사람에게 앉는 행동을 유도하기 때문이다.

13

소재 시도 횟수에 따른 우연의 예측 가능성

해석 많은 기회가 주어지면 우연은 무작위적인 차이를 상당히 균등하게 분포시키는 경향이 있을 것이다. 그러나 많지 않은 기회가 주어진다면 그것(우연)이 무작위적인 차이를 매우 불균등하게 분포시킬 수도 있다. 따라서 동전 던지기에 의해 각 개인을 집단에 배정하는데 참가자가 많다면, 우연은 집단을 동등하게 만드는 것을 잘 해낼 것이다. 반대로 참가자가 적다면, 우연은 아마 집단 간 개인 차이의 영향의 균형을 잘 잡지 못할 것이다. 실제로 참가자가 너무 적으면 우연은 기회가 없다. 예를 들어, 연구에 참여한 사람이 4명이고 그중 한 명만 폭력적인 사람이라면 동전 던지기로는 동등한 집단을 얻을 수 없다. 참가자가 8명이고 그 중 4명이 폭력적인 사람이라고 해도 동전을 던지면 폭력적인 사람 4명이 모두 같은 집단에 속하는 결과가 나올 수도 있다. 왜 그럴까? 왜냐하면 단기적으로 우연은 예측이 불가능할 수 있기 때문이다. 예

를 들어, '앞면'이 4번 연달아 나오는 것은 그렇게 일반적이지(→ 특이하지) 않다. 우연이 단기적으로는 예측할 수 없지만, 장기적으로는 신뢰할 수 있다는 것을 이해하려면 카지노가 여러 번의 내기에서 연속으로 질 수도 있지만 결국에는 항상 카지노가 이긴다는 것을 인식하라.

문제 해설
주어진 기회가 적으면 무작위적인 차이가 불균등하게 분포된다고 했으므로, 동전 던지기를 통한 집단 배정 시 앞면과 뒷면이 나오는 무작위적인 차이가 균등해져서 앞면과 뒷면이 나오는 횟수가 거의 같아지는 장기적인 실행과는 달리, 8명의 방 배정을 위한 8번의 동전 던지기와 같은 단기적인 실행에서는 앞면이 연달아 4번 나오는 것도 있을 수 있는 일일 것이다. 따라서 ④의 usual을 unusual과 같은 낱말로 바꿔야 한다.

구조 해설
- [To appreciate {that chance can be ⟨unpredictable in the short run⟩ but ⟨dependable in the long run⟩}], realize [that {although a casino may lose several bets in a row}, the casino always wins in the end].

첫 번째 [ ]는 목적의 의미를 나타내는 to부정사구이고, 그 안의 { }는 appreciate의 목적어 역할을 하는 명사절이며, 그 안에서 두 개의 ⟨ ⟩는 but으로 대등하게 연결되어 주격 보어를 이룬다. 두 번째 [ ]는 realize의 목적어 역할을 하는 명사절이고, 그 안의 { }는 양보의 의미를 나타내는 접속사 although가 이끄는 부사절이다.

어휘 및 어구
chance 기회, 우연
distribute 분포시키다
random 무작위적인
assign 배정하다
flip 던지다, 뒤집다
conversely 반대로
unpredictable 예측할 수 없는
appreciate 이해하다
bet 내기

14 정답 | ②

소재 포식자 인식의 사회적 전파
해석 포식자 인식의 사회적 전파는 기능적으로 의미가 있는데, 포식자가 위험하다는 것을 배우기 위해 스스로 포식자를 경험해야 하는 개체는 그러한 경험에서 살아남지 못할 수도 있기 때문이다. 가장 잘 분석된 예는 원숭이의 뱀에 대한 두려움과 관련된다.

감금되어 길러진 원숭이는 살아있는 혹은 장난감 뱀을 처음 마주쳤을 때 두려움을 나타내지 않는다. 다른 원숭이가 뱀을 향해 두려워하는 행동을 하고 있는 것을 그것들이 보면 나중에 자신도 똑같이 행동한다. 학습을 시도하는 동안 경험이 없는 관찰자는 모델의 행동(이 경우에는 물러나기, 소리지르기 및 털 세우기 등과 같은 반응)과 같은 행동을 보인다. 경험이 없는 원숭이가 모델이 뱀에 대해서는 두려워하며 행동하고 꽃과 같은 다른 대상에 대해서는 중립적으로 행동하는 것을 관찰하면 동일한 구별을 습득한다. 예를 들어, 나중에 꽃이나 뱀 너머에 있는 손이 닿지 않는 건포도를 제공받으면, 그것(원숭이)은 꽃 너머로는 재빨리 손을 뻗지만, 뱀 너머로는 손을 뻗으려 하지 않는다.

문제 해설
원숭이가 처음 뱀을 접하면 두려움을 보이지 않다가, 다른 원숭이가 뱀에 대해 두려워하는 행동을 하는 것을 보면 그것을 배워서 똑같이 행동한다고 하였으므로, 이는 포식자에 대한 인식이 사회적으로 전달되는 것을 나타낸다. 따라서 빈칸에 들어갈 말로 가장 적절한 것은 ② '사회적 전파'이다.
① 유전적 근거
③ 오해하게 만드는 안내
④ 즉각적인 이해
⑤ 무의식적 억제

구조 해설
- Social transmission of predator recognition makes functional sense [because individuals {that must experience predators for themselves to learn they are dangerous} may not survive those experiences].

[ ]는 이유를 나타내는 부사절이고, 그 안의 { }는 individuals를 수식하는 관계절이다.

어휘 및 어구
predator 포식자
recognition 인식
functional 기능적인
involve 관련시키다
rear 기르다
in captivity 감금되어
exhibit 나타내다, 보이다
encounter 마주치다
observer 관찰자
withdrawal 물러나기, 철수
vocalization 소리지르기, 발성
neutrally 중립적으로
acquire 습득하다
discrimination 구별, 차별

15

소재 Ebbinghaus의 암기 기술

해석 1880년대에 Hermann Ebbinghaus라는 독일 심리학자는 기억이 어떻게 작동하는지 시험하기 위해 파리의 한 방에 틀어박혔다. 그는 스스로에게 구체적이며 시간이 정해져 있는 일정표에 따라 무의미한 단어를 학습하고, 복습하고, 기억해 내도록 강제했다. Ebbinghaus가 발견한 것은 망각률이 예측 가능하다는 것이었다. 그는 잊어버리는 데 정확히 얼마나 오래 걸리는지에 관한 패턴을 발견했다. 그가 무의미한 단어 중 하나를 곧 잊어버리게 될 것이라는 것을 알기 직전에 (그러나 그보다 더 이르지는 않게) 스스로에게 그것을 상기시키는 경우, 그는 공부 시간을 절약하면서도 여전히 그 정보를 정확히 기억할 수 있었다. 요령은 그가 그것을 곧 잊어버리게 될 시점을 아는 것이었다. Ebbinghaus의 암기 기술은 간격을 둔 반복으로 알려지게 되었다. 본질적으로, 이것은 여러분이 꿈꿀 수 있는 가장 고도로 구체적이고 과학에 기반한 학습 일정표였다. 100년이 넘는 시간이 흐른 후, 특별히 설계된 컴퓨터 프로그램들이 Ebbinghaus 일정표의 수정판을 따르는 것을 실현 가능하게 만들었다.

문제 해설
1880년대 Hermann Ebbinghaus는 실험을 통해 망각률이 예측 가능하다는 것을 발견했으며, 잊기 직전에 단어를 상기해서 학습 시간을 절약하면서도 기억을 유지할 수 있었다는 내용이므로, 빈칸에 들어갈 말로 가장 적절한 것은 ⑤ '그가 그것을 곧 잊어버리게 될 시점'이다.
① 연결하는 방법
② 그가 유의미한 질문을 할 수 있는 방법
③ 세부 사항이 아니라 전체 상황이 무엇인지
④ 어떤 동기 부여의 원천에 의존하는지

구조 해설
- [What Ebbinghaus discovered] was [that the rate of forgetting was predictable].
  첫 번째 [ ]는 주어 역할을 하는 명사절이며, 두 번째 [ ]는 주격 보어 역할을 하는 명사절이다.

어휘 및 어구
psychologist 심리학자
shut oneself up in ~에 틀어박히다
nonsense 무의미한
timed 시간이 정해져 있는
rate of forgetting 망각률
predictable 예측 가능한
remind 상기시키다
recall 기억해 내다, 회상하다
spaced repetition 간격을 둔 반복(간격을 두고 반복해 기억 효과를 늘

리는 것)

16

소재 자연 선택의 결과인 현재의 성도 상태

해석 언어학자 Philip Lieberman은 인류의 조상(예를 들어 *Homo erectus*)은 그들의 말이 현대인의 말처럼 세련되지는 않았을지라도 말할 수 있는 능력을 가지고 있었다고 주장한다. 이러한 결론은 인간의 성도가 현재의 모양을 가지는 이유에 대한 추론에 근거한다. Lieberman은 /i/(*meet*에서와 같이), /u/(*you*에서와 같이)와 같은 모음 소리를 내기 위해서는 목구멍의 후두 위 공간이 목구멍의 맨 윗부분과 입이 열리는 지점 사이의 수평 공간과 거의 같은 길이가 되어야 한다는 점을 언급한다. Lieberman의 주장에 따르면, 자연 선택이 이러한 배치를 만들어 유지하기 위해서는 틀림없이 몇 가지 기본적인 언어 능력이 사전에 존재했을 것이다. 자연 선택은 그런 다음 그들이 더욱 다양한 모음 소리를 낼 수 있도록 해 주는 신체적 특성을 가진 개체를 선호했을 수도 있다. *Homo sapiens*가 출현하기 전에 몇 가지 기본적인 언어 능력이 존재하지 않았다면, 후두가 낮아지고 그에 따라 더 많은 모음 소리를 생성하는 능력이 생긴 것은 자연 선택에 의한 점진적인 진화가 아니라 대규모의 그리고 믿을 수 없을 정도로 운이 좋은 돌연변이의 결과여야 할 것이다.

문제 해설
인간에게는 기본적인 언어 능력이 있었고, 자연 선택에 의해 후두가 낮아지고 그에 따라 더 많은 모음 소리를 생성하는 능력이 생겼다고 하였으므로, 빈칸에 들어갈 말로 가장 적절한 것은 ① '그들이 더욱 다양한 모음 소리를 낼 수 있도록 해 주는'이다.
② 그들의 언어 능력을 개선하기 위해 모음의 사용을 제한하는
③ 번식을 하고 유전자를 전달하는 그들의 능력을 향상하는
④ 협력을 위해 비언어적 의사소통 능력을 높이는
⑤ 그들의 언어 능력에 영향을 미치는 돌연변이에 더 저항하게 하는

구조 해설
- [{For natural selection} to produce and maintain this arrangement], [Lieberman argues], some basic speech abilities must have been present beforehand.
  첫 번째 [ ]는 '~하기 위해서'라는 뜻인데, 그 안의 { }는 to produce and maintain this arrangement의 의미상의 주어를 나타낸다. 두 번째 [ ]는 삽입절이다.

어휘 및 어구
linguist 언어학자
argue 주장하다
ancestor 조상
refined 세련된

110 EBS 수능특강 영어독해연습

reasoning 추론
horizontal 수평의
favor 선호하다
appearance 출현
massive 대규모의
incredibly 믿을 수 없을 정도로, 엄청나게
gradual 점진적인
evolution 진화

17
정답 | ②

소재 개념의 의미 파악

해석 개념의 의미는 무엇이며 그것은 문장의 의미에 어떻게 기여하는가? 철학자들은 이 질문 때문에 특히 고충을 느껴 왔으며 이에 대한 다양한 가능한 답을 발전시켜 왔다. 한편으로는 한 개념의 의미가 다른 개념들의 의미에서 파생하는 것처럼 보이는데, 아이에게 '단거리 경주'의 의미를 일종의 빨리 달리기라고 말하는 경우와 같다. 다른 한편으로는 개념의 의미가 세상 사물에 대한 관찰과 연결되는데, 아이가 어떤 사람이 단거리 달리기 하고 있는 것을 실제로 보는 경우와 같다. 한 개념의 의미는 일반적으로 다른 개념의 관점에서의 정의에 의해 주어지지 않는데, 성공적인 정확한 정의가 드물기 때문이다. 또한 마치 '개'의 개념과 개들의 집합을 동일시하는 것처럼, 일련의 예시에 의해 의미가 철저히 다루어지지도 않는다. 따라서 개념의 의미 이론에는 개념이 서로 그리고 세상 둘 다와 어떻게 관련되어 있는지에 대한 설명이 포함되어야 한다. 우리가 개념이 어떻게 언어 사용 능력의 기초가 되는지를 이해하기 위해서는 두 가지 측면이 모두 필요하다.

문제 해설
한 개념의 의미는 다른 개념의 의미에서 파생되기도 하고, 세상 사물에 대한 관찰과 연결되기도 한다는 내용이다. 따라서 빈칸에 들어갈 말로 가장 적절한 것은 ② '개념이 서로 그리고 세상 둘 다와 어떻게 관련되어 있는지에 대한 설명이 포함되어야'이다.
① 언어 사용자의 언어 능력과 인지 과정을 둘 다 반영해야
③ 개념의 실제적 적용뿐만 아니라 정서적 부분도 설명해야
④ 일반적인 용어뿐만 아니라 다양한 문화의 다양한 언어적 표현을 다뤄야
⑤ 개념의 내재적 모호성과 개인에 의한 해석의 역할을 다뤄야

구조 해설

- **Nor is meaning** exhausted by a set of examples, [**as if** one identified the concept of *dog* with a set of dogs].
  부정어인 Nor가 맨 앞으로 나오면서 조동사(is)와 주어(meaning)가 도치되었다. [ ]는 '마치 ~하는 것처럼'이라는 의미를 가진 접속사 as if가 이끄는 부사절이다.

어휘 및 어구
contribute to ~에 기여하다
particularly 특히
derive from ~에서 파생하다
sprint 단거리 경주, 전력 질주; 단거리 달리기를 하다
observation 관찰
definition 정의
rare 드문
exhaust 철저히 다루다
identify 동일시하다
underlie (~의) 기초가 되다

18
정답 | ③

소재 공동체 간의 의사소통

해석 서로를 잘 알고 있는 지역 공동체에서는 대개 화자와 청자가 겹치는 언어 습관의 지식을 활용하여 도덕적, 정치적 문제에 대해 대화하거나 논쟁할 수 있다. 실제적인 활동에서 정기적으로 서로 관련을 맺는 화자 공동체가 모두 같은 언어를 사용하지는 않거나, 같은 수준으로 유창하게 말하지 않는 경우에도 어느 정도까지는 이것이 여전히 사실일 수도 있다. 그러나 때로는 언어적 대화의 잠재적 당사자들이 서로에 대한 물리적 근접성 외에는 공유하는 것이 거의 없음을 발견하는 경우가 있다. (물리적 접촉이 그토록 강력한 문화적 의미를 가진다는 사실은 그것이 전 세계적으로 비언어적 의사소통의 중요한 요소임을 보여 준다.) 이러한 상황은 근본적으로 다른 언어 전통을 가지고 있으며, 이전에 서로 접촉한 적이 없는 공동체의 구성원들이 대면하고, 의사소통을 강요받을 때 발생한다. 이러한 강제된 접촉이 각 언어 공동체에 미치는 결과를 예측할 방법은 없지만, 이러한 새로운 공유 경험에서 피진어라는 새로운 형태의 언어를 포함하는 새로운 형태의 관례가 발전할 수도 있다.

문제 해설
서로를 잘 아는 지역 공동체나, 실제적인 활동에서 정기적으로 관련을 맺는 공동체는 겹치는 언어 습관을 통해 소통할 수 있지만, 근본적으로 다른 언어적 전통을 가지고 있으며, 서로 접촉한 적이 없는 공동체가 소통해야만 하는 상황에서는 새로운 소통 행위가 발전할 수도 있다는 내용의 글이다. ③은 물리적 접촉이 가지는 비언어적 의사소통의 중요성에 관한 내용이므로 글의 전체 흐름과 관계가 없다.

구조 해설

- Sometimes, however, [potential parties to a verbal exchange] find themselves [sharing **little more than** physical proximity to one another].

첫 번째 [ ]는 주어 역할을 하는 명사구이고, 두 번째 [ ]는 find의 목적격 보어 역할을 하고 있다. 「little more than ~」은 '~ 외에는 거의 없는, ~에 불과한'이라는 뜻이다.

19

정답 | ②

소재 과학적 과정의 자동화

해석 기계는 수십 년 동안 과학적 과정을 도와 왔다. 그들이 다음 단계로 나아가 우리가 유망한 새로운 발견과 기술을 자동으로 찾는 데 도움을 줄 수 있을까? (B) 만일 그렇다면 그것은 과학의 발전을 획기적으로 가속할 수 있다. 실제로 과학자들은 약물을 검사하고 게놈의 배열 순서를 밝히는 데 로봇이 구동하는 실험실 기기에 의존해 왔다. (A) 하지만 가설을 세우고, 실험을 설계하고, 결론을 도출하는 일은 여전히 인간이 책임지고 있다. 만약 기계가 가설 수립, 실험 설계 및 실행, 데이터 분석, 그리고 다음에 실행할 실험을 결정하는 전체 과학적 과정을 인간의 개입 없이 모두 책임질 수 있다면 어떨까? (C) 이 생각은 미래의 공상 과학 소설의 줄거리처럼 들릴 수도 있지만, 사실 그러한 공상과학 시나리오는 이미 실현된 적이 있다. 실제로 지난 2009년에 한 로봇 시스템이 인간의 지적 입력이 거의 없이 새로운 과학적 발견을 해냈다.

문제 해설
주어진 글에서 제기한, 기계에 의한 새로운 발견과 기술의 자동적 식별 가능성에 대한 답변의 성격을 띠고 있는 (B)가 이어진 후, (B)의 후반부에 언급된 실험실 기기를 구동하는 로봇의 역할과는 별도로 인간이 책임져야 하는 부분을 설명하는 (A)가 와야 한다. (A)의 후반부에 언급된 인간의 개입 없는 전체 과학적 과정의 자동화 가능성에 대해 그러한 발상이 공상 과학에만 존재하는 것이 아니라 실제로 실현된 적이 있다는 내용의 (C)가 마지막에 와야 한다. 따라서 주어진 글 다음에 이어질 글의 순서로 가장 적절한 것은 ② '(B)-(A)-(C)'이다.

(구조 해설)
- **What if** a machine could be responsible for the entire scientific process — [formulating a hypothesis], [designing and running the experiment], [analyzing data], and [deciding which experiment to run next] — all without human intervention?
「What if ~?」는 '~하면 어쩌지?, ~이라면 어떻게 될까?'라는 뜻이고, 네 개의 [ ]가 콤마와 and로 대등하게 연결되었다.

20

정답 | ③

소재 가치 판단에 따른 환경 문제 해결책 결정

해석 환경 문제를 해결하기 위해 선택하는 해결책은 사람과 환경을 우리가 어떻게 중시하냐에 달려 있다. (B) 예를 들어, 우리가 인구 증가가 문제라고 믿는다면, 인구 증가를 줄이기 위한 의식적인 결정은 우리가 사회로서 승인하고 실행하기로 선택하는 가치 판단을 반영한다. 또 다른 예로 도시의 작은 하천이 범람하는 경우를 생각해 보라. (하천의) 범람은 많은 지역 사회에서 경험하는 위험이다. (C) 강과 그것의 자연적 과정에 대한 연구를 통해 특정한 홍수 위험에 대한 여러 가지 잠재적 해결책을 찾을 수 있다. 우리는 홍수 위험을 크게 줄일 수 있는 해결책으로 하천을 콘크리트 박스 안에 넣는 방법을 선택할 수도 있다. 그렇지 않으면 우리의 도시 하천과, 주기적으로 범람하는 강에 인접한 평지인 그것의 범람원을 녹지대로 복원하는 방법을 선택할 수도 있다. (A) 이 선택은 범람으로 인한 피해를 줄이는 동시에 하천 환경을 이용하는 너구리, 여우, 비버, 사향쥐 등 다양한 동물, 강 근처에서 둥지를 틀고 먹이를 먹고 휴식을 취하는 텃새와 철새, 수계에 서식하는 다양한 물고기에게 서식지를 제공할 것이다. 또한 우리도 강과 상호 작용할 때 더 편안해질 것이다. 그래서 강변 공원이 그렇게 인기가 있는 것이다.

(문제 해설)
환경 문제의 해결을 위해 선택하는 해결책은 사람과 환경을 우리가 어떻게 중시하냐에 달려 있다는 내용의 주어진 글 다음에는, 이에 대한 예시로 인구 증가에 대해 우리가 선택하는 해결책은 우리의 가치 판단을 반영한다고 설명하는 (B)가 이어져야 한다. (B)의 뒷부분에서는 도시 하천 범람이라는 또 다른 예시가 시작되는데, 이에 대한 해결책으로 하천을 콘크리트 박스 안에 넣는 것과 하천과 범람원을 녹지대로 복원하는 것의 두 가지가 차례로 제시되는 (C)가 이어지고, 두 번째로 제시된 하천과 범람원을 녹지대로 복원하는 것의 장점에 대해 부연 설명하는 (A)가 그 뒤에 이어지는 것이 자연스럽다. 따라서 주어진 글 다음에 이어질 글의 순서로 가장 적절한 것은 ③ '(B)-(C)-(A)'이다.

(구조 해설)
- Alternatively, we may choose [to restore our urban streams and their floodplains, {the flat land ⟨adjacent to the river that periodically floods⟩}, as greenbelts].
  [ ]는 choose의 목적어 역할을 하는 to부정사구이고, 그 안의 { }는 their floodplains와 동격을 이루는 명사구이며, ⟨ ⟩는 the flat land를 수식하는 형용사구이다.

(어휘 및 어구)
habitat 서식지
migratory bird 철새
interact with ~과 상호 작용하다
conscious 의식적인
implement 실행하다
urban 도시의
hazard 위험
potential 잠재적인
remedy 해결책
significantly 크게, 상당히
alternatively 그렇지 않으면(둘째 대안을 소개할 때 씀)
restore 복원하다
floodplain 범람원
periodically 주기적으로

21
소재 고대 운동의 목적

정답 | ②

(해석) 성인의 운동이 현대적이라는 일반화는 다소 명백하다. 초기 농부들은 수렵 채집인보다 더 힘들게는 아니라고 하더라도 그만큼 힘들게 일해야 했고, 지난 수천 년 동안 농부들은 전투에 대비하기 위해 자주 스포츠를 통해 주로 운동을 했다. *The Iliad*와 같은 고대 문헌, 파라오 시대의 이집트 그림, 그리고 메소포타미아 조각품들은 레슬링, 단거리 달리기, 그리고 창 던지기와 같은 스포츠가 전사가 되려는 사람들의 체력을 유지하고 전투 기술을 연마하는 데 도움이 되었다는 것을 보여 준다. 하지만 고대 세계의 모든 운동이 전투와 관련된 것은 아니었다. 아테네의 훌륭한 철학 학교 중 하나에 다닐 만큼 부유했다면 여러분은 신체 교육의 일부로 운동을 권유받았을 것이다. 플라톤, 소크라테스, 그리고 키티온의 제논과 같은 철학자들은 가능한 최고의 삶을 살기 위해서 정신뿐만 아니라 신체도 연마해야 한다고 설파했다. 이러한 생각은 서양에만 국한된 것이 아니다. 공자를 비롯한 중국의 저명한 철학자들도 운동이 신체적, 정신적 건강에 똑같이 필수적이라고 가르치며 규칙적인 체조와 무술을 권장했다.

(문제 해설)
주어진 문장에서 모든 고대의 운동이 전투와 관련된 것은 아니라고 언급하고 있는 것으로 보아 주어진 문장이 들어갈 곳을 기준으로 앞쪽에는 고대에 운동이 전투와 관련되었다는 내용이, 뒤쪽에는 고대에 운동이 전투와 관련 없이 사용된 경우에 대한 내용이 나와야 한다. 따라서 주어진 문장이 들어가기에 가장 적절한 곳은 ②이다.

(구조 해설)
- Early farmers had to toil **as hard as if not** harder than hunter-gatherers, and [for the last few thousand years] farmers primarily exercised, often through sports, to prepare for fighting.
  '~만큼 …하게'의 의미를 가진 「as+부사+as ~」 구문과 '~은 아니라고 하더라도'의 의미를 가진 「if not ~」이 사용되었다.
  [ ]는 기간을 나타내는 전치사구이다.

(어휘 및 어구)
ancient 고대의
combat 전투
generalization 일반화
obvious 명백한, 분명한
primarily 주로
carving 조각품
testify (어떤 것이 사실임을) 보여 주다
sprinting 단거리 달리기
preach 설파하다
gymnastics 체조
martial art 무술

Mini Test

22

소재 현재와 과거를 통한 미래 예측하기

해석 미래가 실제로 존재하는지 여부와 (미래가) 존재한다면, 그것을 어디에서 찾을 수 있을지에 대한 상당한 논쟁이 있다. 시간 여행자들이 '미래가 생기기 전에' 그것을 실제로 경험할 수 있는 공상 과학 소설의 영역은 제쳐두고, 미래가 과거 및 현재와 연결되어 있기 때문에 어쨌든 그것을 알 수 있는 그러한 시각에 집중해 보자. 일부 학자들은 시간을 생물학자들이 생물체를 보는 것과 거의 같은 방식으로 본다. 생물학자들이 유기체의 유전 물질과 그것이 사는 환경(예를 들어, 그것의 영양 섭취)을 알고, 유기체의 발달을 좌우하는 법칙을 안다면, 그들은 성숙한 유기체가 어떻게 보이고 행동할지를 예측할 수 있다. 일부 학자들이 주장한 바에 따르면, <u>거의 같은 방식으로 미래는 과거와 현재에 있는 환경으로부터 어쨌든 '성장한다.'</u> 과거와 미래 사이에는 직접적이면서 대체로 물리적인 연결이 있다. 만약 현재와 과거를 올바르게 인식한다면, 그리고 발달이나 성장을 좌우하는 법칙 또한 알고 있다면, 그렇다면 미래가 어떨지 꽤 정확하게 예측할 수 있다.

문제 해설
주어진 문장은 In much the same way로 시작하면서 미래가 과거와 현재의 환경에서 성장한다는 내용이다. 여기서 거의 같은 방식이란 생물학자가 유기체의 행동을 예측하는 방식과 거의 같은 방식을 의미하므로, 주어진 문장이 들어가기에 가장 적절한 곳은 유기체의 행동 예측 방식에서 미래를 예측하는 방식으로 바뀌는 부분인 ④이다.

구조 해설
- There is considerable debate about [whether or not the future really exists] and, [{if it **does**}, where one could find it].
  두 개의 [ ]가 and로 연결되어 about의 목적어 역할을 한다. 두 번째 [ ] 안의 { }는 조건의 부사절이고, does는 exists를 의미한다.

어휘 및 어구
scholar 학자
considerable 상당한
debate 논쟁
biologist 생물학자
genetic 유전의
govern 좌우하다, 지배하다
predict 예측하다
forecast 예측하다
accurately 정확하게

23

소재 두려움에 대한 호소를 통한 행동 변화

해석 감정은 건강 관련 행동에서 강력한 역할을 하는데, 현재의 행동, 그리고 미래의 행동을 변화시키려는 노력 모두에 동기를 부여하고, 우리가 건강 관련 정보를 처리하는 방식을 바꾸며, 건강 관련 판단과 결정에 영향을 미친다. 건강 행동의 이론적 모델, 그리고 이러한 모델에 기반한 개입 노력이 지식, 신념, 태도, 자기 효능감을 강조하는 많은 세월이 지난 후, 감정은 결정적인 역할을 하는 것으로 점점 더 인식되고 있다. 그러나 행동 변화 개입에 감정적인 요소를 추가하려는 시도는 거의 전적으로 두려움에 초점을 맞추어 왔다. 행동 변화(예를 들어, 금연, 건강 식단)를 독려하는 공익 광고(PSAs)는 시청자에게 겁을 주어 더 건강한 생활 방식을 채택해 보려는 노력으로 무서운 사실과 이미지를 판에 박힌 듯 제시한다. 비록 공포 소구(恐怖訴求)가 기억에 아주 잘 남을 수 있다 하더라도, 그것이 또한 자기 효능감을 북돋아 결국 시청자가 그 위험한 결과를 피하기 위해 필요한 것은 무엇이든지 할 수 있다고 믿을 경우에만 그것(공포 소구)이 행동 변화를 촉진하는 데 효과가 있으며, 그렇지 않으면 시청자는 그 메시지를 그저 무시한다. 일반적인 시청자는 문제의 그 변화를 만들기 위해 이미 여러 번 시도해서 실패했기 때문에, 이것은 상당히 힘든 일이다.
→ 감정은 건강 관련 행동을 <u>조종하는 것</u>에서 중요한 역할을 해 왔지만, 두려움을 <u>불러일으키는 것</u>은 자기 효능감을 증가시키지 않고는 효과적이지 않을 수 있다.

문제 해설
감정은 동기 부여, 정보 처리 방식, 판단과 결정 등 건강 관련 행동에서 강력하고 결정적인 역할을 한다고 했다. 그리고 행동 변화 개입에 감정적인 요소를 추가하려는 시도는 거의 전적으로 두려움에 초점을 맞추어 왔지만, 두려움에 대한 호소는 자기 효능감을 증진해야만 행동 변화를 촉진하는 데 효과적이라고 했다. 따라서 요약문의 빈칸 (A), (B)에 들어갈 말로 가장 적절한 것은 ① '조종하는 것(steering) - 불러일으키는 것(evoking)'이다.
② 조종하는 것 - 상쇄하는 것
③ 영향을 미치는 것 - 정복하는 것
④ 억제하는 것 - 감소시키는 것
⑤ 억제하는 것 - 처리하는 것

구조 해설
- Emotions play powerful roles in health-related behaviors, [motivating both current behavior and efforts to change future behavior], [altering the way {we process health-related information}] and [shaping health-related judgments and decisions].
  세 개의 [ ]는 and로 연결되어 주절의 내용에 대해 부가적으로 설명하는 분사구문이다. 두 번째 [ ] 안의 { }는 the way를 수식하는 관계절이다.

(어휘 및 어구)

current 현재의

alter 바꾸다, 변화시키다

judgment 판단

theoretical 이론적인

intervention 개입

emphasize 강조하다

attitude 태도

crucial 결정적인, 중대한

component 요소

exclusively 전적으로, 독점적으로

scare 겁을 주다

memorable 기억에 남는

boost 북돋다, 증진하다

outcome 결과

tune out ~을 무시하다

substantial 상당한

(Tips)

fear appeal(공포 소구): 사람들에게 공포심이나 불안감을 자극하여 특정 행동을 유도하거나 변화시키려는 설득 기법을 말한다.

24~25

정답 | 24 ⑤ 25 ④

(소재) 창의력 개발을 위한 수평적 사고의 중요성

(해석) 교육은 항상 규칙과 논리적 사고를 강조해 왔다. 단계별 학습은 미리 정해진 방향으로 진행된다(수직적 사고). 연구자들은 길러 줄 수는 있으나 가르칠 수는 없는 신비로운 능력으로 창의력을 간주해 왔다. 심리학자들의 연구에 의하면 아이들의 타고난 창의력은 학교에 다닌 지 2년 정도 후 감소하는 경우가 많다. 심리학자 Edward de Bono에 따르면, 창의력은 뿌리 깊은 패턴을 재구성하는 것을 요구한다. 이전 패턴의 한계를 넘어서는 능력은, 타고난 것이든 개발된 것이든, 그 과정에 필수적이다. 수평적 사고가 필요한데, 그것의 근본 원칙은 어떤 특정한 관점도 많은 가능성 중 하나일 뿐이라는 것이다. 어떤 현상에 대한 설명은 정말 관점만큼이나 많다. 수평적 사고는 단순히 기존의 모델을 기반으로 하는 대신 대안적인 패턴을 추구한다. 수직적 사고는 점진적으로 증가하는 과정이고, 반면에 수평적 사고는 유의미한 도약을 허용한다. 최종 결론이 문제를 해결하는 한, 수평적 사고에 포함된 각각의 단계는 올바를 필요가 없다. 그 순간에 올바르게 보이는 것을 끊임없이 평가하고 받아들이는 대신에, 평가가 연기된다. 일단 특정 지점에 도달하면, 많은 경우 출발점으로 가는 논리적 경로를 되짚어 가는 것이 가능하다. 일단 그것이 이루어지면, 다양한 단계의 방향과 순서가 크게 중요할 것 같다(→ 할 것 같지는 않다). 수평적 사고는 일단 공사가 끝나면 허물어지는 비계를 가지고 아치형 천장이나 다리를 건설하는 것과 같다. 프로젝트의 각 단계 동안 다리의 여러 부분은 스스로 서 있을 필요가 없다. 그러나 일단 쐐기돌이 놓이면 전체 구조물은 틀림없이 스스로를 지탱하고 있을 것이다.

(문제 해설)

24 창의력은 뿌리 깊은 패턴을 재구성하는 것을 요구하고, 그 과정에 필수적인 것이 이전 패턴의 한계를 넘어서는 능력인데 이를 위해서는 수평적 사고가 필요하다는 것이 이 글의 중심 내용이다. 따라서 제목으로 가장 적절한 것은 ⑤ '수평적 사고: 유의미한 도약을 통해 창의성 열기'이다.

① 혼돈 받아들이기: 더 분명한 사고로 가는 길

② 창의적 문제 해결의 환상 드러내기

③ 업무를 위한 수평적 사고 기술 향상 방법

④ 수평에서 수직으로: 문제 해결 전략 발전시키기

25 수평적 사고는 수직적 사고와는 달리 최종 결론이 문제를 해결하는 한, 포함된 각각의 단계가 정확할 필요가 없다고 했다. 따라서 다양한 단계의 방향과 순서가 크게 중요하지는 않을 것이므로 (d)의 likely를 unlikely와 같은 낱말로 바꿔야 한다.

(구조 해설)

- Lateral thinking is needed, [the fundamental principle of which is {that any particular point of view is but one of many possibilities}].

  [ ]는 Lateral thinking에 대해 부가적으로 설명하고, 그 안의 { }는 주격 보어 역할을 하는 명사절이다.

- Lateral thinking is like building a vault or a bridge with scaffolding [that is torn down {once construction has been completed}].

  [ ]는 scaffolding을 수식하는 관계절이고, 그 안의 { }는 접속사 once가 이끄는 부사절이다.

(어휘 및 어구)

emphasize 강조하다

logical 논리적인

proceed 진행되다

predetermined 미리 정해진

vertical 수직적인, 수직의

foster 기르다

decline 감소하다, 감퇴하다

innate 타고난

integral 필수적인

fundamental 근본적인

phenomenon 현상
perspective 관점
alternative 대안적인
conclusion 결론
assessment 평가
postpone 연기하다
retrace 되짚어 가다, 역추적하다
keystone (아치 꼭대기의) 쐐기돌[이맛돌]

26~28

정답 | 26 ④ 27 ② 28 ③

소재 David와 Albert의 우정

해석 (A) 나의 아들 David는 그가 더 어렸을 때 스카우트에 참여해서 불을 피우는 법부터 우리 지역 사회의 사람들을 돕는 법까지 모든 것을 배웠다. 그가 자랑스럽다고 말하는 것은 절제된 표현일 수 있지만, 내 기억에 내 가슴을 자랑스러움으로 부풀어 오르게 했던 특별한 한 순간이 있었다. 우리 옆집에 Albert라는 연로한 이웃이 있었는데, 그는 몇 년 동안 홀로 지내고 있었다. 그의 아내는 세상을 떠났고, 그는 휠체어에 갇혀 지냈다.

(D) 어느 날 저녁, 끔찍한 뇌우가 있었다. 우리가 다음 날 아침에 일어났을 때, David는 Albert의 집이 특히 심한 타격을 받았다는 것을 알아차렸다. 나뭇가지들이 잔디를 가로질러 흩어져 있었고, 현관에 있던 물건들이 그의 앞 마당 잔디밭 사방에 흩어져 있었다. "엄마, Albert 씨를 돕기 위해 무언가 해야겠어요."라며 내가 팬케이크를 만드는 동안 David가 말했다. "아침은 그를 돕는 것을 마친 후까지 기다렸다가 먹을까 봐요." 그리고 그는 자신의 도구를 가지러 차고로 뛰어갔다.

(B) 나는 David의 의무감과 친절함에 미소를 지었다. 나는 David가 나뭇가지들을 모아 쌓고 잔디밭에 어지럽혀진 것을 치우는 모습을 창문으로 지켜보았다. 그리고 나서 그는 Albert의 문을 두드렸다. 나는 그가 무슨 말을 했는지 들을 수 없었지만, 그가 우리 집으로 달려 돌아와서 접시 두 개와 그를 기다리고 있던 팬케이크 더미를 집어 들고 Albert의 집으로 돌아가는 모습을 조용히 지켜보았다. 그 둘은 현관에 앉아 아침 식사를 나눠 먹으며 담소를 나누었다. 그 후, 그들은 좋은 친구가 되었다.

(C) 몇 년 후, Albert가 세상을 떠났다. David는 친구를 잃은 것으로 인해 비통해 하며 감정적으로 힘들어했다. 그러던 어느 날, Albert의 변호사가 우리를 찾아와 봉투를 건네주었다. 그 안에는 수표 형태의 거액이 들어 있었다. Albert가 David의 대학 첫해를 위해 돈을 저축해 두었던 것이었다! 그리고 편지가 있었는데, 이는 그가 David에게 쓴 마지막 편지였다. "네가 있기에 세상은 더 나은 곳이야. 계속해서 친절을 전파하고 선행을 계속하렴. 지금 우리에게는 너 같은 사람이 그 어느 때보다 필요해. 외로운 노인에게 친구가 되어줘서 고맙다."

문제 해설

26 자신의 아들 David를 소개하고 휠체어에 의지해서 홀로 지내던 이웃 Albert에 대해 말하는 내용의 (A) 다음에, 끔찍한 뇌우가 닥친 다음 날, David가 Albert의 집이 특히 심하게 타격을 받았음을 알아차리고 그를 돕기 위해 도구를 가지러 차고로 뛰어가는 내용의 (D)가 이어져야 한다. 다음으로 David가 Albert의 집을 치운 후 함께 팬케이크를 먹으며 담소를 나누고, 그날 이후로 그들이 좋은 친구가 되는 내용의 (B)가 이어져야 하고, Albert가 세상을 떠난 후 David에게 자신이 쓴 편지와 함께 대학 첫해를 위한 돈을 남기는 내용의 (C)가 이어지는 것이 자연스럽다. 따라서 주어진 글 (A)에 이어질 글의 순서로 가장 적절한 것은 ④ '(D)-(B)-(C)'이다.

27 (a), (c), (d), (e)는 모두 Albert를 가리키고 (b)는 David를 가리키므로, 가리키는 대상이 나머지 넷과 다른 것은 ②이다.

28 David는 집에 돌아와 팬케이크를 가지고 Albert의 집에 가서 아침 식사를 나눠 먹었으므로, 글에 관한 내용으로 적절하지 않은 것은 ③이다.

구조 해설

- **My son, David**, [participated in Scouts when he was younger] and [learned everything **from** {how to build a fire} to {how to help those in our community}].

  My son과 David는 동격 관계이고, 두 개의 [ ]가 and로 대등하게 연결되어 문장의 술부 역할을 하고 있다. 두 개의 { }는 「from ~ to ...」로 연결되었다.

어휘 및 어구

scout 스카우트
understatement 절제된 표현
swell 부풀어 오르다
pass away 세상을 떠나다
be confined to ~에 갇혀 지내다
sense of duty 의무감
gather 모으다
branch 나뭇가지
pile 쌓아 놓은 것, 더미
mess 어지럽혀진 것
lawn 잔디밭
knock 두드리다
stack 더미
chat 담소[이야기]를 나누다
awful 끔찍한
thunderstorm 뇌우
scatter 흩어지게 하다, 흩뿌리다
garage 차고

Week 8 Mini Test ②

본문 176~201쪽

01 ⑤	02 ②	03 ④	04 ⑤	05 ②	06 ③	07 ④
08 ⑤	09 ④	10 ④	11 ③	12 ④	13 ⑤	14 ④
15 ①	16 ②	17 ⑤	18 ④	19 ④	20 ②	21 ⑤
22 ⑤	23 ①	24 ④	25 ⑤	26 ④	27 ⑤	28 ⑤

01

정답 | ⑤

소재 프로젝트 논의를 위한 회의 참석 요청

해석 Ruth Allen 씨께

이 메시지가 귀하에게 도달했을 때 건강하시길 바랍니다. 제 이름은 Orville Rivera이고 최근에 팀에 합류하여 다양한 기업 전략 기획을 수행하고 있습니다. 현재 제가 맡고 있는 업무 중 하나는 품질 시스템 프로젝트를 이끄는 것입니다. 이 업무를 시작하면서, 저는 기존 시스템에 대한 정보를 수집하고 사용자의 관점에서 요구 사항을 이해하는 초기 단계에 있습니다. 귀하의 역할과 전문 지식을 고려할 때, 저는 귀하의 식견이 이번 노력의 성공을 보장하는 데 매우 귀중할 것이라고 믿습니다. 현재 업무 흐름, 이용자 기대, 그리고 귀하께서 가지고 계실 기타의 식견을 포함하여, 귀하와 프로젝트에 대해 더 자세히 논의할 기회를 주시면 대단히 감사하겠습니다. 저희가 이번 주 후반에 귀하가 편하신 시간에 회의 일정을 잡아도 될까요? 제 요청을 고려해 주셔서 감사합니다.
Ruth.
Orville Rivera 드림

문제 해설

품질 시스템 프로젝트를 이끌게 된 상황에서 Ruth Allen의 식견이 필요하여 그에게 프로젝트를 논의하기 위해 만나 주기를 요청하는 내용의 글이다. 따라서 글의 목적으로 가장 적절한 것은 ⑤이다.

구조 해설

- As I embark on this task, I'm in the initial stages of [gathering information about the existing systems] and [understanding the requirements from the users' perspective].
  두 개의 [ ]가 and로 연결되어 전치사 of의 목적어 역할을 한다.

어휘 및 어구

recently 최근에
corporate 기업의, 회사의
strategy 전략
responsibility 맡고 있는 업무, 책임
initial 초기의

requirement 요구 사항
perspective 관점
expertise 전문 지식
insight 식견, 통찰력
invaluable 매우 귀중한
endeavor 노력
appreciate 감사하다
workflow 업무 흐름

02

정답 | ②

소재 Sal의 암 수술

해석 수술할 날이 왔고, 우리는 아침 일찍 병원에 도착했다. Sal은 수술실로 옮겨졌고, 나는 기도하기 위해 예배당으로 내려갔다. 눈물이 얼굴을 타고 흘러내리고 있었다. 나는 전에 한 번도 기도해 본 적이 없는 것처럼 기도했다. 나는 내가 너무나 사랑하는 이 남자가 수술실에서 나와 이 위협적인 병에서 벗어나는 것을 볼 수 있기만을 바랐다. 나는 대기실로 다시 올라가 마치 영원과도 같았던 시간 동안 딸과 함께 그곳에 앉아 있었다. 의사가 마침내 대기실로 들어왔다. 내 눈에 보이는 것이라고는 그의 눈뿐이었다. 그는 수술이 잘 되었고, 자신들이 혹시 계속 남아 있을지도 모르는 암세포를 모두 제거했다고 말했다. 나는 다시 울기 시작했지만, 이번에는 내가 사랑하는 이 남자가 건강하게 미소를 지으면서, 그리고 삶이 우리에게 일어나기를 기다리고 있는 온갖 아름다운 것들을 기대하면서 다시 내 곁에 있을 것이라는 기쁨 때문이었다.

문제 해설

수술을 받기 위해 수술실로 들어간 Sal에 대한 걱정으로 눈물을 흘리며 기도했다고 했고, 수술이 잘 되었다는 의사의 말을 듣고 기쁨의 눈물을 흘렸다고 했으므로, I의 심경 변화로 가장 적절한 것은 ② '걱정하는 → 안도한'이다.

① 지루한 → 놀란
③ 의심하는 → 자신하는
④ 감사하는 → 겁에 질린
⑤ 매우 신이 나는 → 실망한

구조 해설

- I just wanted to see [this man {whom I loved so much}] [{come out of the surgery} and {be free of this threatening disease}].
  첫 번째 [ ]는 see의 목적어이고, 그 안의 { }는 this man을 수식하는 관계절이다. 두 번째 [ ]는 see의 목적격 보어이고, 그 안에 두 개의 { }가 and로 연결되어 있다.

어휘 및 어구

operation 수술

surgery 수술실, 수술
pray 기도하다
threatening 위협적인
cancer 암
in store (for) (~에게) 일어나기를 기다리고 있는

03

소재 과학 전문어를 독자의 눈높이에 맞게 다시 표현하기

해석 대중을 대상으로, 정부와 심지어 산업계를 대상으로 작성되는 글은 일반적으로 과학 자체가 아니라 과학의 응용에 초점을 맞추기 때문에 과학 전문어를 피하는 것은 보이는 것만큼 어렵지 않다. 평이한 언어로 과학의 응용을 설명하는 것은 거의 항상 가능하다. 그럼에도 불구하고, 과학자들은 때때로 과학을 평이한 언어로 바꾸어 말하면 과학이 '가치가 떨어지거나' '지나치게 단순화된다'고 불평한다. 그러나 과학 용어의 사용이 독자를 오해하게 한다거나, 더 심하게는, 말하고 있는 내용을 완전히 잘못 해석하게 할 뿐이라면, 그렇다면 그 결과는 혼란만 초래할 뿐이므로 그것을 사용하는 것은 말이 되지 않는다. 과학자는 그들(과학자)의 학문 분야 밖에 있는 사람들이 그들(과학자)이 어떤 전문적인 용어에 속한다고 생각하는 정확한 의미를 이해하기를 절대로 기대해서는 안 되는데, 심지어 '모형'과 같이 겉보기에 단순해 보이는 용어도 마찬가지이다. 독자에게 의미가 있도록 그 언어를 다시 표현하기 위해 모든 노력을 기울여야 한다. 이는 때때로 일부 연구자들이 할애할 수 있는 것보다 더 많은 시간과 노력을 필요로 하고, 과학과 사회 사이의 메신저이자 통역사로서 숙련된 소통자의 가치가 점점 더 커지고 있는 이유이다.

문제 해설
평이한 언어로 과학의 응용을 설명하는 것은 거의 항상 가능하며, 독자에게 의미가 있도록 그 언어를 다시 표현하기 위해 모든 노력을 기울여야 한다는 것이 주요 내용이므로, 필자가 주장하는 바로 가장 적절한 것은 ④이다.

구조 해설
- Scientists should never expect people outside their discipline [to understand the exact meaning {they ascribe to a specialised term—even an apparently simple one like 'model'}].
  [ ]는 expect의 목적격 보어이고, 그 안의 { }는 the exact meaning을 수식하는 관계절이다.

어휘 및 어구
avoid 피하다
article (신문이나 잡지의) 글, 기사, 논문
application 응용

plain 평이한
translation 바꾸어 말하기
devalue 가치를 떨어뜨리다, 평가절하하다
misinterpret 잘못 해석하다
confusion 혼란
discipline 학문 분야
ascribe ~ to ... ~이 …에 속한다고 생각하다, ~을 …에 부여하다
specialised 전문적인, 전문화된
apparently 겉보기에, 외관상
re-phrase 다시 표현하다
spare 할애하다
interpreter 통역사

04

소재 창의성을 중시하지 않는 기업 환경

해석 창의성은 탁월함을 추구하는 것과는 상당히 다르다. 탁월함에 대한 한계가 작동하기 시작할 때, 오직 창의성과 혁신만이 기업을 다른 궤도로 이끌 수 있다. 경쟁이 치열하고 불확실한 세상에서 창의적인 기업은 그렇지 않은 기업보다 거의 확실히 우위를 점할 것이다. 그럼에도 불구하고 창의성은 제도화해야 할 업무 또는 조직 윤리로서 기업 의제에서 덜 중요하다. 대부분의 조직에서는 정해진 관행을 따르는 직원의 눈에 보이는 정신 작용만이 중요하다. 직원의 정신은 블랙박스와 유사하다고 여겨진다("인간의 뇌 안에서 무슨 일이 일어나는지 결코 헤아릴 수 없고, 그럴 필요도 없다")! 대부분의 조직에서 그들의 주위 환경을 보는 시각은 이미 탐험된 비즈니스라는 캔버스를 넘어 아직 탐험되지 않은 화이트 스페이스로 향하는 일이 거의 없다("화이트 스페이스에서 돈을 잃으려면 돈이 많아야 한다")! 블랙박스를 통과하는 사고 과정을 살펴보지 못함으로써, 기업들은 자신들 앞으로 펼쳐져 있는 화이트 스페이스의 성장 기회를 놓친다.

문제 해설
직원의 창의성을 중시하지 않는 기업은 직원의 정신을 블랙박스로 보고, 그 속에서 일어나는 일에 무관심하여 정해진 업무 수행만을 중시하고 탐험되지 않은 영역으로 나아가는 일이 거의 없다는 내용이다. 따라서 밑줄 친 부분이 의미하는 바로 가장 적절한 것은 ⑤ '직원들 정신 내부의 창의적 사고 과정을 충분히 탐구하지 못함'이다.
① 작업장 안전에 대한 정해진 관행을 부주의하게 간과함
② 필수 자격을 충족하는 직원을 지속적으로 고용하지 않음
③ 개방되고 투명한 조직 환경을 억제함
④ 직원의 창의성을 효과적으로 평가하는 시스템을 제대로 갖추고 있지 않음

구조 해설

■ In a competitive and uncertain world, companies [that are creative] would most certainly have advantage over companies [that are not].

두 개의 [ ]는 각각 바로 앞의 companies를 수식하는 관계절이다. 두 번째 [ ]는 that are not creative를 의미한다.

어휘 및 어구

pursuit 추구
figure 중요하다
corporate 기업의
ethic 윤리
institutionalize 제도화하다
prescribed 정해진
relevant 중요한, 유의미한
white space 화이트 스페이스(새로운 기회나 시장에서 탐험되지 않는 영역)
have deep pockets 돈이 많다
dip into ~을 살펴보다

Tips

black box(블랙박스): 기능은 알지만 내부 구조는 알 수 없는 기계나 전자 장치

05 정답 | ②

소재 삶에 대한 해석과 반응

해석 삶은 아마도 5%(혹은 더 적게)가 우리에게 일어나는 일이고, 95%(혹은 더 많게)가 일어나고 있는 일을 우리가 어떻게 해석하고 반응하는가이다. 외부 사건에 의미를 부여하는 것은 다름 아닌 그것들에 대한 우리의 해석이다. 상황과 환경은 우리가 그것이 의미하는 바가 긍정적인지, 부정적인지, 혹은 중요하지 않은지를 결정하기 전까지는 중립적이다. 탁월함 범주에 속한 사람들은 다른 사람들의 제한하는 편견과 판단을 받아들이지 않는다. 그들은 자신만의 해석을 하고, 그에 따라 결정한다. 신발 사업의 확장 가능성을 살피기 위해 두 명의 마케팅 조사원을 먼 나라로 보낸 한 신발 공장에 관한 자주 언급되는 비유가 있다. 첫 번째 조사원은 상황이 절망적이며, 아무도 신발을 신고 있지 않아 시장이 없다는 답신을 보낸다. 다른 조사원은 아무도 신발을 가지고 있지 않기 때문에 이는 절호의 사업 기회라고 열정적으로 응답한다. 어떤 상황에서든 우리가 환경에 부여하는 의미가 우리가 무엇을 보고 어떻게 생각하고 행동하는지를 결정한다.

문제 해설

삶은 대부분 우리에게 일어나는 일을 우리가 어떻게 해석하고 반

응하는가에 달려 있고, 상황은 중립적이며, 우리가 부여하는 의미에 따라 달라진다는 내용이므로, 글의 요지로 가장 적절한 것은 ②이다.

구조 해설

■ Situations and circumstances are neutral [until we decide {what they mean whether positive, negative, or insignificant}].

[ ]는 시간의 부사절이며, 그 안의 { }는 decide의 목적어 역할을 하는 명사절이다.

어휘 및 어구

interpret 해석하다
respond 반응하다
external 외부의
circumstance (보통 복수로) 환경
neutral 중립적인
excellence 탁월함
bias 편견, 성향
scout 조사원, 정찰자
distant 먼
hopeless 절망적인
outstanding 뛰어난
opportunity 기회

06 정답 | ③

소재 향기와 음식 맛의 관계

해석 식품 과학자들 사이에서 점점 더 인기를 얻은 연구 분야 하나는 착각에 의한 향기 분야이다. 가끔 여러분은 실제로 주변에 없는 냄새를 상상할 때가 있을 수도 있다. 향기는 기억과 가장 강하게 연관된 감각이며, 요리하는 동안 이 관계를 솜씨 있게 처리하여 뇌가 그 음식의 맛이 어때야 하는지에 대한 자신의 인식을 재구성하도록 속일 수 있다. 햄은 소금 절임 된 고기로, 우리는 햄의 향기를 짠맛과 연관 짓도록 배워 왔다. 한 맛 실험에서, 음식 샘플 속 '햄 향기'의 존재는 한 그룹의 사람들이 그들의 음식이 더 짠맛이 난다고 확신하게 했다. 여러분은 스스로 이것을 다른 방법으로 사용할 수 있다. 디저트에 계피, 장미수, 그리고 바닐라와 같은 향기 있는 양념을 반복적으로 사용하면, 그러한 향기를 단맛과 연관 짓기 시작한다. 다음번에 옥수수 키르와 같은 디저트를 만들 때, 첨가하는 감미료의 양을 줄이고 '더 달콤한' 향기 있는 양념을 조금 더 추가해 보라. 여러분의 저녁 식사 동반자들은 아마도 디저트가 매우 달다고 느낄 것이다.

문제 해설

햄 향기가 음식을 더 짜게 느끼게 하고, 달콤한 향기 있는 양념을

사용해 감미료를 줄인 디저트를 매우 달게 느끼게 할 수 있는 것처럼, 음식의 향기는 기억과 강하게 연관되어 있으며, 요리 중에 이를 솜씨 있게 처리하여 음식의 맛을 재구성할 수 있다는 내용이다. 따라서 글의 주제로 가장 적절한 것은 ③ '음식 지각에서 향기와 기억의 관계를 솜씨 있게 처리하기'이다.

① 마케팅 분야에서 대중적인 맛과 향기의 사용
② 특정 맛에 대한 일부 개인들의(이 가진) 민감성 원인
④ 일상식의 영양가를 높이기 위해 적용되는 과학적 기법
⑤ 인공 향료로 더 맛있는 디저트를 만드는 데 사용되는 기술

구조 해설

- Aroma is the sense [that is most strongly associated with memory], and this relationship can be manipulated during cooking to trick our brain to reconstruct [what {it perceives} the food should taste like].

  첫 번째 [ ]는 the sense를 수식하는 관계절이고, 두 번째 [ ]는 reconstruct의 목적어 역할을 하는 명사절이며, { }는 삽입절이다.

어휘 및 어구

arena 분야, 무대
aroma 향기
memory 기억
manipulate 솜씨 있게 처리하다
reconstruct 재구성하다
perceive 지각하다, 인식하다
convince 확신시키다, 설득하다
aromatic 향기가 있는, 향기로운
spice 양념, 향신료
cinnamon 계피, 시나몬
rosewater 장미수, 로즈워터
vanilla 바닐라
sweetener 감미료
companion 동반자, 친구

07
정답 | ④

소재 대표성 편향과 과도한 낙관주의

해석 과도한 낙관주의는 부분적으로 '대표성 편향'에 의해 설명될 수도 있다. 대표성 편향은 더 나은 일반 통계 정보가 존재하더라도 사람들이 특정 '대표적인' 사례로부터 일반화하는 경험칙의 결과이다. 대표성 경험칙이 일상적인 추론에서 유용할 수도 있는 이유를 쉽게 알 수 있는데, 특정 정보가 일반 정보보다 더 정확하거나 더 유용할 때가 많고, 그것을 선호하는 것이 확률을 평가할

때 유용한 인지적 지름길일 수도 있기 때문이다. 그러나 이 가정이 실패할 때, 대표성 경험칙은 기이하고 잘못된 판단을 초래할 수 있다. 과도한 낙관주의는 대표성에서 비롯될 수도 있는데, 예를 들어, 사람들이 서투른 운전자가 연루된 특정 자동차 사고에 대한 지식을 통해 서투른 운전이 실제로 그런 것보다 사고와 더 높은 상관관계를 가진다고 추론할 때가 그러한 경우이다. 이 경우, 자신이 평균적인 운전자라고 믿더라도, 대표성 경험칙이 평균적인 운전자가 사고에 연루될 확률을 과소평가하게 만들었기 때문에 그들은 자신이 사고에 연루될 확률을 체계적으로 과소평가할 것이다.

문제 해설

우리가 접하는 특정 대표적인 사례로부터 일반화하는 대표성 경험칙은 일상적인 추론에서 유용한 인지적 지름길일 수도 있으나, 평균적인 운전자가 사고에 연루될 확률을 과소평가하는 과도한 낙관주의와 같은 잘못된 판단을 초래할 수 있다는 내용의 글이다. 따라서 글의 제목으로 가장 적절한 것은 ④ '사고의 실수: 어떻게 정신적 지름길이 과도한 낙관주의를 초래하는가'이다.

① 정신적 지름길이 우리가 빠른 결정을 내리는 데 도움을 주는 방법
② 진전을 향해 나아가기 위한 긍정적인 태도의 힘
③ 디딤돌로서의 장애물: 도전을 승리로 바꾸기
⑤ 여러 모자 쓰기: 다중 작업의 관리가 생산성을 높일 수 있는 방법

구조 해설

- It is easy [to see {why the representativeness heuristic might be useful in everyday reasoning}] ~.

  It은 형식상의 주어이고 [ ]가 내용상의 주어이다. 그 안의 { }는 see의 목적어 역할을 하는 명사절이다.

어휘 및 어구

overoptimism 과도한 낙관주의
bias 편향, 편견
generalize 일반화하다
representative 대표적인
statistical 통계의
represent (~)이다, (~에) 해당하다
cognitive 인지적인
shortcut 지름길
assess 평가하다
probability 확률
erroneous 잘못된
infer 추론하다
correlate 상관관계를 가지다
underestimate 과소평가하다

08

정답 | ⑤

소재 사이버 범죄를 경험한 인터넷 사용자의 비율

해석 위 도표는 2022년 선정된 국가에서 사이버 범죄를 경험한 인터넷 사용자의 비율을 보여 주며, 전 세계 평균에 대한 언급을 포함한다. 조사된 국가 중 인도는 사이버 범죄를 경험한 인터넷 사용자 비율을 가장 높게 응답했고, 그다음이 미국인데 인터넷 사용자 중 거의 절반이 그러한 사건을 겪었다. 반대로 일본은 인터넷 사용자 중 사이버 범죄를 경험한 비율을 가장 낮은 것으로 응답했고, 그리고 독일은 응답자의 4분의 1 넘게 사이버 범죄 활동을 경험했다고 인정해 두 번째로 낮은 비율을 가졌다. 호주는 선택된 국가들 중에서 사이버 범죄를 경험한 인터넷 사용자의 비율을 세 번째로 높게 기록했으며, 이는 전 세계 평균보다 1퍼센트포인트 높다. 전 세계 평균보다 높은 세 국가에서는 10명의 응답자 중 최소 4명이 사이버 범죄를 경험했다고 말했다. 사이버 범죄를 언급한 응답자의 비율이 전 세계 평균보다 낮은 모든 국가들에서는, 인터넷 사용자의 최소 3분의 1보다 많은 수가 어떤 형태로든 사이버 범죄를 경험했다.

문제 해설 사이버 범죄를 언급한 비율이 전 세계 평균보다 낮은 국가는 뉴질랜드, 프랑스, 영국, 독일, 일본인데, 독일(30%)과 일본(21%)은 3분의 1보다 적은 수가 사이버 범죄를 경험했다. 따라서 도표의 내용과 일치하지 않는 것은 ⑤이다.

구조 해설

- [In the three countries above the global average], at least four out of ten respondents said [that they experienced any cybercrime].

  첫 번째 [ ]는 전치사구이며, 두 번째 [ ]는 said의 목적어 역할을 하는 명사절이다.

어휘 및 어구

reference 언급, 참조
encounter 겪다; 경험
incident 사건
respondent 응답자
proportion 비율

09

정답 | ④

소재 Francis William Aston의 생애

해석 Francis William Aston은 영국의 화학자이자 물리학자로, 잉글랜드 Harborne에서 금속 상인의 아들로 태어났다. 그는 Birmingham University의 전신인 Mason College에서 교육을 받았고, 그곳에서 화학을 공부했다. 1898년부터 1900년까지 그는 P. F. Frankland의 지도하에 선광도에 대한 연구를 했다. 1900년에 Birmingham을 떠나 Wolverhampton의 한 맥주 공장에서 3년 동안 일했다. 이 기간 동안 그는 가정 실험실에서 과학 연구를 계속했으며, 그곳에서 엑스선 방전관을 위한 진공 생성에 대해 연구했다. 이 연구는 University of Birmingham의 J. H. Poynting의 주목을 받았고, 그는 Aston에게 자신과 함께 연구하자고 요청했다. 그는 1910년까지 Birmingham에 머물다가, 그 해에 Cambridge로 옮겨 J. J. Thomson의 연구 조수로 일했다. 그는 1920년에 Cambridge의 연구원이 되었고, 전쟁 기간 동안 Farnborough의 왕립 항공 연구소에서 보낸 시간을 제외하고는 그곳에서 평생을 보냈다. Aston의 주요 연구는 질량 분석기를 설계하고 사용하는 것에 관한 것으로, 이 연구로 인해 그는 1922년 노벨 화학상을 받았는데, 질량 분석기는 여러 가지 미해결 문제를 해결하는 데 사용되었고 새로운 원자 물리학의 기본 도구 중 하나가 되었다.

문제 해설 전쟁 시기에는 Farnborough에 있는 왕립 항공 연구소에서 시간을 보냈다고 했으므로, 글의 내용과 일치하지 않는 것은 ④이다.

구조 해설

- He was educated at [[Mason College], {the forerunner of Birmingham University}], [where he studied chemistry].

  첫 번째 [ ] 안에서 두 개의 { }는 서로 동격 관계이고, 두 번째 [ ]는 첫 번째 [ ]를 부수적으로 설명하는 관계절이다.

어휘 및 어구

merchant 상인
optical rotation 선광(旋光)(도) (광선이 광학 활성 물질을 통과할 때 편광면(偏光面)을 회전시키는 각도)
laboratory 실험실
vacuum 진공(*pl.* vacua)
discharge tube 방전관
research fellow 연구원
outstanding 미해결된, 뛰어난

10

정답 | ④

소재 신발 건조기 사용 설명서

해석 Caremaxy 신발 건조기

이 장치는 신발을 건조하여 습기를 제거하고, 활성 산소(오존) 기술을 사용하여 박테리아를 제거합니다.

작동법:

1. 신발을 홀더 위에 놓습니다.

2. 전원 버튼을 눌러 시작합니다. 필요하다면 타이머를 조정할 수 있으며, 최초 120분으로 설정되어 있습니다.
3. 건조 버튼을 누릅니다. 건조 과정이 시작되면서 건조등이 켜집니다. (건조) 과정이 끝나면 건조기가 자동으로 정지됩니다.

오존 기능 사용법:
– 오존 버튼을 누르면 기능이 작동합니다. 15분이 지나면 자동으로 꺼집니다.
– 필요한 경우 오존 버튼을 다시 눌러 오존 기능을 다시 시작합니다.
– 오존 기능이 켜져 있으면 기기가 열을 발생시키지 않는다는 것에 주의하세요. 열은 오존 기능이 꺼진 후에만 발생합니다.

안전 지침:
– 과도한 오존은 공기질 저하로 이어져 두통을 유발할 수 있습니다.
– 손상을 방지하기 위해 건조기를 물 가까이 두지 마십시오.

(문제)해설
오존 기능이 작동 중일 때 기기에서 열이 발생하지 않는다고 했으므로, 안내문의 내용과 일치하지 않는 것은 ④이다.

(구조)해설
■ Note [that {when the ozone function is on}, the unit will not generate heat].
[ ]는 Note의 목적어 역할을 하는 명사절이고, 그 안의 { }는 시간을 나타내는 부사절이다.

(어휘 및 어구)
moisture 습기
eliminate 제거하다
adjust 조정하다, 맞추다
initially 초기에, 처음에
generate 발생시키다
excessive 과도한

11
정답 | ③

(소재) 학교 도서 박람회 홍보
(해석)
            Woodlaw School 도서 박람회

Woodlaw School 도서 박람회는 모든 연령대의 학생들이 가정의 서재를 풍요롭게 하고 독서에 대한 사랑을 키울 수 있는 기회를 제공합니다.

행사 내용 안내:
– 도서 박람회는 3월 10일부터 3월 14일까지 Woodlaw 도서관에서 열립니다.
– 학생들은 지정된 수업 시간 동안 책을 탐색하고 구매할 수 있는 기회를 가집니다.

– 또한, 3월 12일 수요일 오후 4시 45분부터 7시 30분까지 가족의 밤이 열립니다.

결제 방법:
– 도서 박람회에서는 현금의 안전한 대안인 eWallet을 제공합니다.
– 간단히 무료 계정을 만들어 자금을 추가하고, 가족 및 친구들을 초대하여 기부하도록 할 수 있습니다.
– 모든 판매 금액과 미사용 기금은 전액 지역주민센터 어린이 도서관 건립 기금으로 기부됩니다.

더 많은 정보가 필요하시면 bookfair@woodlawschool.ac.us로 저희에게 이메일을 보내 주세요.

(문제)해설
3월 12일 수요일 오후 4시 45분부터 가족의 밤 행사가 예정되어 있다고 했으므로, 안내문의 내용과 일치하는 것은 ③이다.

(구조)해설
■ The Woodlaw School Book Fair provides an opportunity [for students of all ages to enrich their home libraries and cultivate a love for reading].
[ ]는 students of all ages를 의미상의 주어로 하는 to부정사구로 an opportunity의 구체적 내용을 설명하고 있다.

(어휘 및 어구)
fair 박람회
enrich 풍요롭게 하다, 부유하게 하다
cultivate 키우다, 기르다, 경작하다
designated 지정된
secure 안전한
alternative 대안
account 계정, 계좌
contribute 기부[기증]하다

12
정답 | ④

(소재) 정치 참여 모집의 사회경제적 지위 편향
(해석) 정치적 참여 요청은 소셜 미디어를 통해, 혹은 교회에서의 설교에서, 혹은 자신이 회원으로 있는 조직의 소식지에서, 혹은 친구, 이웃, 또는 동료로부터 직접적으로와 같이 다양한 출처에서 발생한다. 다른 사람들을 정치에 참여시키려 시도하는 사람들은 '합리적 탐사자'로 행동하기 때문에, 사람들이 참여하도록 요청을 받는 일반적인 과정은 개인의 정치적 목소리에 있는 계급 편향성을 개선하지 않는다. 가능한 한 효율적으로 결과를 얻고자 하기 때문에, 합리적 탐사자들은 동의할 뿐만 아니라 동의하자마자 사실상 참여할 가능성이 있는 사람들을 향해 자신들의 요청을 겨냥

한다. 흔히 그들은 과거에 적극적이었던 사람들을 대상으로 한다. 그 결과, 다른 사람들의 요청에 응하여 하게 되는 참여는 이전 활동가들의 불균형적인 몫을 가져오고(이전 활동가들의 비율이 균형에 맞지 않을 정도로 높아지게 하고), 이는 결국 정치 참여의 사회경제적 지위 편향성을 더 두드러지게 한다. 이러한 방식으로, 정치적 참여로의 일반적인 모집 과정은 단순히 정치적 참여의 사회경제적 구조화를 복제하는 것이 아니라, 실제로 그것을 증폭시킨다.

(문제 해설)

④ participation undertaken in response to requests from others가 that이 이끄는 절의 주어이고 주어의 핵심어인 participation이 단수형이므로, 복수형 동사 bring을 brings로 고쳐야 한다.

① an organization을 수식하는 관계절에서 which 이하가 필수 구성 요소를 갖춘 완전한 절이므로, 「전치사+관계사」의 구조인 of which를 쓰는 것은 어법상 적절하다.

② get의 목적격 보어 역할을 해야 하고, 목적어인 others가 행위의 주체가 아니라 대상이므로, involved는 어법상 적절하다.

③ are likely에 이어져 「not only ~ but also ...」 구문으로 to say yes와 대등하게 연결되어야 하므로, to부정사 형태인 to participate는 어법상 적절하다.

⑤ 단수 표현인 the socioeconomic structuring of political participation을 가리키므로, 대명사 it은 어법상 적절하다.

(구조 해설)

■ [Seeking to get results **as** efficiently **as possible**], rational prospectors aim their requests at those [who are likely not only to say yes but also, on assenting, to participate effectively].

첫 번째 [ ]는 이유를 나타내는 분사구문이고, rational prospectors를 의미상의 주어로 한다. 그 안에 '가능한 한 ~하게'라는 의미의 「as ~ as possible」 구문이 사용되었다. 두 번째 [ ]는 those를 수식하는 관계절이다.

(어휘 및 어구)

request 요청
take part 참여하다
politically 정치적으로
arise from ~에서 발생하다
source 출처
sermon 설교
attempt 시도하다
rational 합리적인
prospector 탐사자
bias 편향성

aim 겨냥하다
target 대상으로 하다
undertake (착수)하다, 떠맡다
in response to ~에 응하여
disproportionate 불균형적인
share 몫, 역할
previous 이전의
exaggerate 더 두드러지게 하다, 과장하다
recruitment 모집
structuring 구조(화)
amplify 증폭시키다

13
정답 | ⑤

(소재) 감정이 기억에 미치는 영향

(해석) 너무 많은 기억을 저장하는 것은 인출 과정을 늦출 수 있어서, 무관하다고 여겨지는 기억들은 조용히 지워진다. 이것이 여러분이 두 주 전 금요일에 아침으로 무엇을 먹었는지 아마도 기억할 수 없는 이유인데, 그런 정보는 유용할 가능성이 없기 때문이다. 물론 이것은 어떤 기억이 미래에 유용할 가능성이 있는지 여부를 우리가 어떻게 사전에 판단할 수 있는지에 관하여 의문을 제기한다. 한 가지 요인은 감정이다. 엄청난 스트레스를 받을 때 사람들은 흔히 그 사건에 대한 매우 상세한 기억을 갖게 될 것이고, 극단적인 경우 이것은 외상 후 스트레스 장애(PTSD)로 나타날 수 있다. 외상 후 스트레스 장애를 겪는 사람들은 외상적 사건(매우 충격적인 사건)의 기억에 압도되어 위협적인 상황에서 과도한 경계를 보일 수 있다. 비록 외상 후 스트레스 장애가 분명히 부적응적이지만, 그것은 진화적으로 유용한 과정의 극단적인 형태일 수도 있다. 예를 들어, 거의 죽을 뻔한 경험을 한 동물이 반드시 그 경험에서 배워서 유사한 상황에서 경계를 억누르게(→ 보이게) 하는 것은 분명히 좋은 설계일 것이다.

(문제 해설)

외상 후 스트레스 장애가 부적응적이지만 진화적으로 유용한 과정의 극단적인 형태일 수도 있다는 앞 문장과 연결되려면, 거의 죽을 뻔한 경험으로부터 배워서 유사한 상황이 왔을 때 생존을 위해 경계를 보여야 한다는 문맥이 되어야 한다. 따라서 ⑤의 suppress를 display와 같은 낱말로 바꿔야 한다.

(구조 해설)

■ This of course raises the question **as to** [how we can determine in advance {whether or not a memory is likely to be useful in the future}].

as to는 '~에 관하여'라는 뜻이고, [ ]는 as to의 목적어 역할을 한다. { }는 determine의 목적어 역할을 한다.

어휘 및 어구 (좌측)

in advance 사전에
extreme 극단적인
manifest itself 나타나다
overwhelm 압도하다
maladaptive 부적응적인
evolutionarily 진화적으로
ensure 반드시 ~하게 하다

vehicle 차량
sophisticated 정교한
aim at ~을 목표로 하다
detect 감지하다
aspect 측면
forward-looking 전방 주시의
alert 경고하다
rapid 빠른
take place 발생하다
act on ~에 따라 조치를 취하다, 행동하다
intervention 개입
distract 주의를 산만하게 하다
present 제공하다

14

정답 | ④

소재 조작하는 사람이 인식할 때 가치 있는 정보

해석 조작하는 사람의 인식 속에 있을 때 가장 큰 가치를 지니는 정보의 한 가지 예는 운전 상황에서 찾을 수 있다. 오늘날의 차량은 주행 환경의 다양한 측면을 감지하는 것을 목표로 하는 점점 더 정교해지는 센서 패키지와 함께 설계되어 있다. 예를 들어, 전방 주시 카메라와 전방 주시 레이더 시스템은 운전자 앞에 있는 차량과의 거리를 판단할 수 있다. 이 정보를 시간이 지남에 따라 측정하는 계산은 거리의 변화를 드러낸다. 이 정보는 주어진 차량 속도에 대한 거리의 변화가 충돌이 발생할 수 있음을 시사할 정도로 충분히 빠를 때 운전자에게 경고하는 데 사용될 수 있다. 그 정보는 중요하지만, 운전자가 그것에 따라 행동해야만 가치가 있다. (물론, 차량 자체가 운선자 개입 없이 그것에 따라 조치를 취하는 경우는 제외된다.) 여기서 핵심은 그 정보가 운전자에게 가치가 있으려면 운전자가 그 정보에 반드시 주의를 기울일 수 있어야 한다는 것이다. 예를 들어, 운전자가 전화에 의해 주의가 산만해지면, 차량이 제공하고 있는 중요한 정보를 처리하지 못할 수도 있다.

문제 해설
운전자와 앞에 있는 차량과의 거리와 그 변화 정보가 가치를 가지려면 운전자가 그 정보에 반드시 주의를 기울일 수 있어야 한다는 내용이므로, 빈칸에 들어갈 말로 가장 적절한 것은 ④ '인식'이다.
① 비판
② 상상
③ 양심
⑤ 인내

구조 해설
- [One example of information {that has greatest value when it is in the awareness of a human operator}] can be found in the context of driving.
  [ ]는 문장의 주어 역할을 하고 있고, { }는 information을 수식하는 관계절이다.

어휘 및 어구
operator (장비·기계를) 조작하는 사람

15

정답 | ①

소재 세포의 성장과 지속 사이의 균형

해석 생물학자인 Tyler Volk는 세포가 어떤 특정한 순간에도 항상 지속과 소멸의 경계에 있는 스스로 생성되는 역동적인 독립체라고 지적한다. 그것은 이 게임에서 우위를 유지하기 위해 자신의 신진대사를 이용함으로써 어떻게든 생존한다. 신진대사 노폐물이 배출되면 그 결과로 분자의 손실이 있다. (이를) 벌충하기 위해 세포는 또한 신진대사를 사용하여 새로운 분자를 성장시킨다. 그 교환이 적어도 동등하면 세포는 그것의 현재 형태로 지속될 수 있다. 손실되는 것보다 더 많은 분자가 생성이 된다면, 이는 소멸에 대한 보호를 더해 주는데, 순 성장이 발생하고 세포는 더 커진다. 그러나 더 큰 세포는 더 많은 영양분을 필요로 하고, 구가 커질수록 그것의 내부는 표면적보다 더욱 상당한 정도로 증가한다는 물리학의 기본 원칙에 직면하기 때문에 세포는 어느 정도까지만 성장할 수 있다. 세포의 경우, 이는 표면이 엄청나게 더 커진 내부를 지탱할 만큼 충분히 많은 영양분의 흐름을 유지하는 것을 더 어렵게 만든다. 그렇다면 세포는 무엇을 해야 할까? 그것은 유용한 크기 한계에 가까워질 때 반으로 분열하고 그 과정을 전부 다시 시작한다. 이는 성장과 지속 사이의 균형을 달성한다.

문제 해설
분자 손실을 보상하기 위해 세포는 새로운 분자를 성장시키는데, 손실되는 것보다 더 많은 분자가 생성이 될 경우 세포가 더 커지고 순 성장이 발생하지만, 더 큰 세포는 표면이 더 커진 내부를 지탱할 만큼 영양분의 흐름을 충분히 많이 유지하지 못하게 되므로, 세포는 지속을 위해 더 이상의 성장 대신 반으로 분열하고 그 과정을 전부 다시 시작한다는 내용이다. 따라서 빈칸에 들어갈 말로 가장 적절한 것은 ① '성장과 지속 사이의 균형을 달성한다'이다.
② 새로운 세포가 자신의 손상된 부분을 재성장시키는 것을 어렵

게 한다
③ 세포가 다른 의존할 영양분 원천을 찾게 한다
④ 세포가 무기한으로 더 커지도록 에너지 섭취를 증가시킨다
⑤ 세포가 서로 간에 분자를 교환할 수 있는 경로를 만든다

구조 해설

- A biologist, Tyler Volk, points out [that cells are self-generated dynamic entities {that at any given moment are always on the cusp between persisting and dying}].

  [ ]는 points out의 목적어 역할을 하는 명사절이고, { }는 self-generated dynamic entities를 수식하는 관계절이다.

어휘 및 어구

point out ~을 지적하다
self-generated 스스로 생성되는
dynamic 역동적인
persisting 지속
manage to 어떻게든 ~하다, 간신히 해내다
stay ahead 우위를 유지하다, 앞서 있다
expel 배출하다
molecule 분자
compensate 벌충하다, 보상하다
exchange 교환
net 순(純)
so much 어느 정도까지, 얼마만큼
nutrient 영양분
run up against ~에 직면하다, ~과 맞닥뜨리다
principle 원칙
physics 물리학
sphere 구
interior 내부
sustain 지탱하다, 유지하다

16
정답 | ②

소재 단기적인 반복 연습과 장기 기억

해석 한 연구에서 아이오와 주립 대학교 연구진은 사람들에게 단어 목록을 읽어 준 다음, 각 목록을 즉시, 15초의 반복 연습 후, 또는 반복 연습을 방해하는 매우 간단한 수학 문제를 15초간 푼 후에 다시 암송하도록 요청했다. 목록을 들은 직후에 바로 다시 생각해 낼 수 있었던 피험자들이 최고로 잘했다. 암송하기 전에 15초의 반복 연습 시간을 가진 피험자들은 2위를 차지했다. 수학 문제로 집중이 흐트러진 그룹은 최하위를 기록했다. 나중에 모두가 다 끝났다고 생각했을 때, 목록에서 기억해 낼 수 있는 단어를

모두 적으라는 돌발 퀴즈가 나와 모두를 놀라게 했다. 갑자기 최하위 그룹이 최고가 되었다. 단기적인 반복 연습은 순전히 단기적인 이득만 가져왔다. 정보를 붙잡은 다음 그것을 기억해 내기 위해 고군분투한 것은 수학 문제로 집중이 흐트러진 그룹이 정보를 단기 기억에서 장기 기억으로 옮기도록 도왔다. 더 많은, 즉각적인 반복 연습 기회를 가진 그룹은 돌발 퀴즈에서 거의 아무것도 기억해 내지 못했다. 반복은 고군분투보다 덜 중요하다는 것이 밝혀졌다.

문제 해설

단어 목록 기억 실험에서 반복 연습 시간을 가진 그룹보다 집중을 방해받은 그룹이 단어를 기억해 내려는 고군분투를 통해 단어에 대한 더 높은 장기 기억을 보였다는 내용의 글이므로, 빈칸에 들어갈 말로 가장 적절한 것은 ② '고군분투보다 덜 중요하다'이다.
① 암기에 필수적이다
③ 집중 연습 전에 요구된다
④ 조기 언어 발달에 도움이 된다
⑤ 복잡한 아이디어를 이해하는 데 효과적이지 않다

구조 해설

- [Struggling to hold on to information and then recall it] had **helped the group** [distracted by math problems] **transfer** the information from short-term to long-term memory.

  첫 번째 [ ]는 문장의 주어 역할을 하는 동명사구이다. 두 번째 [ ]는 the group을 수식하는 분사구이다. '~이 …하도록 돕다'의 의미를 가진 「help+목적어(the group distracted by math problems)+목적격 보어(transfer the information from short-term to long-term memory)」 구문이 사용되었다.

어휘 및 어구

recite 암송하다
distract 집중이 흐트러지게 하다
pop quiz 돌발 퀴즈
recall 기억해 내다, 기억하다
purely 순전히
struggle 고군분투하다; 고군분투
immediate 즉각적인

17
정답 | ⑤

소재 조직의 창발적 특성인 심리적 안전감

해석 심리적 안전감은 잘 실패하기 학문에서 강력한 역할을 한다. 그것은 사람들이 감당할 수 없는 상황에 처했을 때 도움을 요청하는 것을 허용하여 예방 가능한 실패를 없애는 데 도움이 된

다. 그것은 더 나쁜 결과를 피하기 위해 사람들이 오류를 보고하고, 이를 통해 오류를 발견하고 수정하도록 도와주며, 새로운 발견물을 만들어 내기 위해 신중한 방식으로 실험하는 것을 가능하게 해 준다. 여러분이 직장이나 학교에서 소속되었던 팀에 대해 생각해 보라. 이러한 집단은 심리적 안전감의 정도가 아마 다양했을 것이다. 아마 어떤 팀에서는 새로운 아이디어를 거리낌 없이 말하거나 팀 리더의 의견에 동의하지 않는 것이 완전히 편안하게 느껴졌을 수도 있다. 다른 팀에서는 억제하는 것이, 즉 여러분이 위험을 자초하기 전에 일어나는 일이나 다른 사람들의 행동과 말을 지켜보며 기다리는 것이 더 낫다고 느꼈을 수도 있다. 그러한 차이를 이제 심리적 안전감이라고 하는데, 내 연구를 통해 이것이 성격의 차이가 아니라 집단의 창발성이라는 사실을 발견했다. 이는 직장에서 의견을 거리낌 없이 말하는 것이 안전한지에 대한 인식은 외향적인 사람인지 내성적인 사람인지와는 무관함을 의미한다. 그 대신, 그것은 <u>주변 사람들이 여러분과 다른 사람들의 말과 행동에 어떻게 반응하는지</u>에 따라 형성된다.

(문제 해설)

심리적 안전감이 보장되어야 도움을 요청하고, 오류를 보고하며, 실패나 오류를 예방할 수 있도록 자신의 의견을 자유로이 말할 수 있는데, 이는 개인의 성격이 아니라 집단의 창발성에서 비롯된다는 내용의 글이므로, 빈칸에 들어갈 말로 가장 적절한 것은 ⑤ '주변 사람들이 여러분과 다른 사람들의 말과 행동에 어떻게 반응하는지'이다.
① 여러분이 이전에 비슷한 상황을 겪어본 적이 있는지
② 논의되고 있는 문제에 여러분이 얼마나 많은 관심이 있는지
③ 당면한 문제에 대해 여러분이 다른 사람들보다 얼마나 더 많이 알고 있는지
④ 가족 구성원들이 여러분의 생각을 소중하게 여기는지 아닌지

(구조 해설)

- It helps them report — and hence catch and correct — errors [to avoid worse outcomes], and it makes **it** possible [to experiment in thoughtful ways to generate new discoveries].
  첫 번째 [ ]는 목적을 나타내는 to부정사구이다. it은 형식상의 목적어이고 두 번째 [ ]가 내용상의 목적어이다.

(어휘 및 어구)

psychological 심리적인
be in over one's head 감당할 수 없는 상황에 처하다
eliminate 없애다
preventable 예방 가능한
avoid 피하다
outcome 결과
experiment 실험하다

hold back 억제하다
stick one's neck out 위험을 자초하다
perception 인식
extrovert 외향적인
introvert 내성적인

18
정답 | ④

(소재) 공공 공간이 이주민에게 미치는 영향

(해석) 우리는 사람들이 오래 살던 고향 지역을 자주 떠나는, 이동이 매우 잦은 세상에 살고 있다. 일부는 선택에 의해 그렇게 하지만, 다른 사람들은 정치적 또는 경제적 상황으로 인해 물리적, 문화적으로 꽤 다른 곳으로 이주해야 한다. 이러한 상황에서 사적 영역 외에도 공공 공간이 새로운 장소에 대한 소속감을 형성하는 데 역할을 한다. 바다 냄새, 익숙한 나무, 울타리의 디자인, 빛의 질, 축구장과 같은 공공 공간의 감각적 특성은 새로 도착한 사람들에게 '고향'과의 연결성을 만들어 내는 데 최고의 능력을 발휘한다. 겉보기에 일시적인 공공 공간의 이러한 특성은 친숙한 기억을 불러일으키거나, 상호 연결성을 만들어 내거나 과거와 현재 사이의 연속성을 형성함으로써 깊은 심리적 위안을 가져다줄 수 있다. (어떤 면에서 도시는 사적 공간으로 나누어진 거대한 공공 공간인 반면, 교외는 대부분 주거 지역으로 구성된 사적 영역의 집합체이다.) 공공 공간에서의 그리고 공공 공간에 대한 이러한 경험은 현지인과 새로 온 사람 사이에 공통의 또는 공유된 경험을 형성할 수도 있다.

(문제 해설)

공공 공간의 감각적 특성은 새로 이주한 사람들에게 소속감과 심리적 위안을 가져다준다는 내용의 글인데, ④는 도시와 교외 지역의 공간 구성 차이에 대한 설명이므로, 글의 전체 흐름과 관계가 없는 문장은 ④이다.

(구조 해설)

- [These seemingly ephemeral qualities of public space] can bring deep psychological comfort by [triggering a familiar memory], [creating an interconnectedness], or [forming a continuum between the past and the present].
  첫 번째 [ ]는 문장의 주어이고, 핵심어는 qualities이다. 두 번째, 세 번째, 네 번째 [ ]는 콤마와 or로 연결되어 by의 목적어 역할을 한다.

(어휘 및 어구)

regularly 자주
uproot (오래 살던 곳에서) 떠나게 만들다
locale 지역, 장소

political 정치적인

conditions 상황

relocate 이주하다

vastly 굉장히

aside from ~ 외에도

private 사적인

realm 영역

play a role 역할을 하다

sense of belonging 소속감

sensory 감각적인

fence 울타리

utmost 최고의

trigger 불러일으키다

continuum 연속성, 연속체

establish 형성하다

19
정답 | ④

소재 감각 자극을 표현하는 뇌

해석 뇌는 수십억 개의 뉴런을 모두 사용하여 모든 것을 표현하려고 하지 않으며, 서로 다른 뇌 부위가 서로 다른 종류의 감각 자극을 표현한다. (C) 예를 들어, 뇌 뒤쪽의 시각 피질에는 서로 다른 시각적 입력에 반응하는 뉴런이 있다. 발화 패턴이 입력의 구조와 공간적으로 일치하는 뉴런 그룹이 있는데, 선이 시각적 자극의 일부라는 사실을 나타내기 위해 뉴런 기둥이 함께 발화할 때가 이에 해당하는 예이다. (A) 이와 같이 뇌의 서로 다른 부위에는 서로 다른 종류의 시각, 후각(냄새), 미각, 청각, 촉각 자극이 제시될 때 발화하는 뉴런 그룹이 있다. 한 뉴런 그룹이 여러 뉴런 그룹의 입력에 반응할 수 있기 때문에 인간의 뇌는 제시된 자극을 단순히 표현만 하는 것보다 훨씬 더 많은 일을 할 수 있다. (B) 이는 입력 뉴런이 표현하는 것을 결합된 형태로 나타낼 수 있다. 예를 들어, 원숭이의 전두엽 피질에는 미각, 시각, 후각의 감각 양상이 함께 모여, 과일과 그 주요 속성을 나타낼 수 있는 부위가 있다. 따라서 뇌가 뛰어난 표상 장치라는 것은 분명하다.

문제 해설
뇌의 서로 다른 부위가 서로 다른 종류의 감각 자극을 표현한다는 내용의 주어진 글 다음에는, 이에 대한 예시로 뇌 뒤쪽의 시각 피질에는 감각 자극 중 시각적 입력에 반응하는 뉴런이 있다는 내용의 (C)가 이어져야 한다. 그다음으로, (C)의 사례로부터 뇌의 각 부위에서 서로 다른 종류의 자극이 제시될 때 서로 다른 뉴런 그룹이 발화된다는 것을 일반화하는 (A)가 이어져야 한다. (A)의 뒷부분에서는 한 뉴런 그룹이 여러 뉴런 그룹의 입력에 반응할 수 있기 때문에 제시된 자극을 단순히 표현만 하는 것보다 훨씬 더

많은 일을 할 수 있다는 내용이 나오므로, 그 뒤에는 입력 뉴런이 표현하는 것을 결합된 형태로 나타낼 수 있다는 내용을 예시와 함께 설명하는 (B)가 이어지는 것이 자연스럽다. 따라서 주어진 글 다음에 이어질 글의 순서로 가장 적절한 것은 ④ '(C)−(A)−(B)'이다.

구조 해설

- There are neuronal groups [whose firing patterns correspond spatially to the structure of the input]—for example, when a column of neurons fires together to represent the fact [that a line is part of the visual stimulus].

첫 번째 [ ]는 neuronal groups를 수식하는 관계절이고, 두 번째 [ ]는 the fact의 내용을 설명하는 동격절이다.

어휘 및 어구
neuron 뉴런, 신경 세포

region 부위

sensory 감각의

stimulus 자극(*pl*. stimuli)

fire 발화하다, 활성화하다

input 입력

combined 결합된

representation 표현, 나타냄

property 속성, 특성

device 장치

correspond to ~과 일치하다

spatially 공간적으로

column 기둥, 열

20
정답 | ②

소재 운동을 강조하는 민족주의적 움직임

해석 지난 몇 세기 동안 전문가들은 우리가 운동을 충분히 하고 있지 않다고 끊임없이 걱정해 왔다. (B) 민족주의는 이러한 불안감의 하나의 주요 원인이다. 군인으로서 싸울 수 있을 만큼 충분히 튼튼할 것을 고대 스파르타인이 요구받고 로마인이 권고받은 것과 꼭 마찬가지로, 애국심을 부추기는 지도자와 교육자들은 일반 시민들에게 군 복무를 준비하기 위해 스포츠와 다른 형태의 운동에 참여하도록 점점 더 권장했다. (A) 이런 움직임의 특히 영향력 있는 지지자는 '체조의 아버지'로 불리는 Friedrich Jahn이었다. 19세기 초 나폴레옹이 독일군에게 굴욕감을 주는 승리를 연이어 거둔 후 Jahn은 교육자들이 체조, 등산, 달리기 등을 통해 자국의 청소년들의 신체적, 도덕적 힘을 회복시켜야 할 책임이 있다고 주장했다. (C) 이후 미국에서도 제1, 2차 세계 대전에 입대

하거나 징집된 많은 남성들 사이의 부끄러운 체력 부족과 냉전이 시작될 무렵 학생들 사이의 한심한 체력 상태가 비슷한 우려를 불러일으켰다. 국가를 위해 체력을 선전하는 민족주의적 움직임이 중국과 다른 나라에서도 여전히 일어나고 있다.

문제 해설
주어진 글에서 지난 몇 세기 동안 우리가 운동을 충분히 하고 있지 않다는 데 대해 전문가들이 걱정해 왔다는 내용이 언급된 다음, (B)에서 이런 걱정의 주요 원인이 민족주의라고 설명하며 군 복무 대비를 위해 일반 시민이 다양한 운동을 하도록 권장했다는 말 다음에, 이런 움직임의 지지자였던 Friedrich Jahn의 주장을 언급하고 있는 (A)가 오는 것이 자연스러우며, 그 후 이와 같은 유사한 걱정을 했던 미국의 사례를 설명하는 (C)가 와야 글의 흐름이 자연스럽다. 따라서 주어진 글 다음에 이어질 글의 순서로 가장 적절한 것은 ② '(B)-(A)-(C)'이다.

구조 해설
- Later, similar worries in America were spurred [by the embarrassing lack of fitness among many men {who enlisted or were drafted for World Wars I and II}] and [by the poor state of fitness among schoolchildren at the start of the Cold War].

  두 개의 [ ]가 and로 대등하게 연결되어 were spurred에 이어지고, 첫 번째 [ ] 안의 { }는 many men을 수식하는 관계절이다.

어휘 및 어구
influential 영향력이 있는
humiliating 굴욕감을 주는
restore 회복시키다
physical 신체적인
moral 도덕적인
nationalism 민족주의
anxiety 불안감
urge 권고하다
flag-waving 애국심을 부추기는
participate 참여하다
military service 군 복무
draft 징집하다
drum up ~을 선전하다
for the sake of ~을 위해서
occur 일어나다, 발생하다

21
정답 | ⑤

소재 지구 기후 온난화로 인한 지역별 혜택과 피해

해석
많은 기후학자들은 강화된 온실 효과로 인해 지구 기후 온난화가 실제로 일어날 가능성이 있다고 인정한다. 많은 대규모 변화와 마찬가지로 승자와 패자가 있을 것이다. 만약 기후대가 더 높은 위도 지역으로 이동한다면, 이는 전체적으로 가장 가능성이 높은 것으로 보이는 결과인데, 일부 사하라 사막과 남부 러시아 지역은 강우량이 증가할 것이다. 그 지역들은 혜택을 받고 농업 생산량이 증가할 것이다. 반면에 남유럽과 미국의 곡창지대는 더 건조해질 수도 있다. 만약 온난화로 인해 증발 속도가 강수량 증가 속도를 초과한다면 토양은 더 메마를 것이다. 그러나 낮 기온은 거의 또는 전혀 변화 없이, 흐린 날씨의 증가로 인해 밤 기온 하강이 감소하는 방식으로 온난화를 경험할 수도 있다. 그 경우, 밤에 서리가 내리는 빈도가 더 줄어들고 토양이 다소 더 촉촉해져 농업에 도움이 될 것이다.

문제 해설
지구 기후 변화로 인해 혜택을 볼 수 있는 지역과 피해를 볼 수 있는 지역에 관한 내용이다. 주어진 문장은 역접의 의미를 나타내는 however와 함께 쓰여 낮 기온은 거의 또는 전혀 변화 없이, 흐린 날씨의 증가로 인해 밤 기온의 하강이 감소하는 방식으로 온난화를 경험할 수도 있다고 했는데, ⑤의 앞에서는 온난화로 인해 토양이 더 메마를 것이라 하였고, ⑤의 뒤에서는 토양이 더 촉촉해져 농업에 도움이 될 것이라 하면서 문장을 '그 경우'라는 말로 시작하고 있으므로 논리적 단절이 있음을 볼 수 있다. ⑤의 뒤에서 언급된 '그 경우'란 바로 주어진 문장에 나온 흐린 날씨의 증가로 인해 밤 기온의 하강이 감소하는 방식으로 온난화가 진행될 경우를 의미하므로, 주어진 문장이 들어가기에 가장 적절한 곳은 ⑤이다.

구조 해설
- [Were climate belts to shift toward higher latitudes], [which seems the most likely overall result], parts of the Sahara and southern Russia would receive increased rainfall.

  조건절인 첫 번째 [ ]에서 if가 생략되면서 동사(Were)와 주어(climate belts)가 도치되었다. 두 번째 [ ]는 첫 번째 [ ]를 부연 설명하는 관계절이다.

어휘 및 어구
reduction 감소
climatologist 기후학자
climatic 기후의
enhanced 강화된
greenhouse effect 온실 효과
evaporation 증발
exceed 초과하다
soil 토양
frost 서리

moist 촉촉한

22
정답 | ⑤

소재 소셜 미디어 플랫폼의 교육적 활용

해석 소셜 미디어 플랫폼은 학습 경험을 향상시키고, 학생 간의 상호 작용을 증가시키며, 참여를 촉진한다. 그러나 다른 교육적인 활동과 꼭 마찬가지로, 소셜 미디어로 무엇을 하는지의 이면에 있는 이유를 이해하라. 이러한 이유를 알면 교육 목표 달성에 가장 효과적인 디지털 도구를 선택하는 데 도움이 될 것이다. 교육 목표와 웹사이트 및 앱에서 접근 가능한 디지털 도구를 반드시 일치시켜라. 디지털 도구를 선택할 때, 불꽃놀이와 날아다니는 색깔(요란한 효과와 화려한 디자인)을 가진 앱에 현혹되지 않도록 자제해야 하는데, 그러한 매우 매력적인 앱이 반드시 학업 목적도, 학생들의 학습 필요도 충족시키는 것이 아닐 수도 있기 때문이다. 소셜 미디어 플랫폼을 선택하는 것은 목표 기술에 잘 맞는지에 근거하는 실용적인 결정이다. 예를 들어, 많은 소셜 미디어 플랫폼은 에세이 작성과 격식 있는 발표와 같은 기술에 잘 맞지 않는다. 그러나 이러한 플랫폼은 학생들이 사회적 상호 작용을 통해 주제에 대해 가상으로 의견을 제공할 수 있게 함으로써 교실 평가 기법과 같은 다른 교육 활동을 향상시킬 수 있다.

문제 해설
④ 뒤에서 소셜 미디어 플랫폼을 선택하는 것은 목표 기술에 잘 맞는지에 근거하는 실용적인 결정이라는 내용이 나온 이후에 ⑤ 뒤에서는 이러한 플랫폼에서 학생들이 가상으로 의견을 제공하게 한다는 내용이 이어져 논리적 단절이 일어난다. 많은 소셜 미디어가 에세이 작성이나 격식 있는 발표와 같은 목표 기술에 잘 맞지 않는다는 주어진 문장이 들어가면 ⑤ 다음의 내용이 역접 관계로 잘 이어져 단절이 해소된다. 따라서 주어진 문장이 들어가기에 가장 적절한 곳은 ⑤이다.

구조 해설
- [When selecting digital tools], refrain from being attracted to apps with fireworks and flying colors [because such highly attractive apps may **not** necessarily serve {your academic purposes} **nor** {your students' learning needs}].
  첫 번째 [ ]는 접속사 When이 명시된 분사구문이고, 두 번째 [ ]는 이유의 부사절이다. 그 안에서 두 개의 { }가 '~도 (아니고), …도 아닌'이라는 의미를 나타내는 「not ~ nor ….」에 의해 연결되어 serve의 목적어 역할을 한다.

어휘 및 어구
enhance 향상시키다
foster 촉진하다

engagement 참여
instructional goal 교육 목표
accessible 접근 가능한
refrain from ~을 자제하다
firework 불꽃놀이
purpose 목적
practical 실용적인
classroom assessment technique 교실 평가 기법(교사가 학생들의 학습을 평가하고 이해하는 데 도움을 주는 다양한 방법이나 도구)
virtual 가상의

23
정답 | ①

소재 명료한 글쓰기를 위한 방법

해석 아주 명료한 글쓰기를 하는 유일한 방법은 글과 그래픽을 작성할 때 여러분이 하는 선택에 독자가 어떻게 반응할지를 아는 것이다. 여러분은 어떤 문장 구조가 가장 쉽게 이해되는지, 자료를 섹션으로 어떻게 조직해야 (독자가) 가장 쉽게 따라올 수 있는지 등을 알아야 한다. 다음과 같은 몇몇 일반적인 규칙을 제공하는 것이 확실히 가능한데, 예를 들면 '능동태를 사용하기', '논문을 '서론', '연구 방법', '연구 결과' 및 '토의' 섹션으로 나누기', 그리고 '양이 비교되어야 할 때는 표 대신에 도표를 사용하기'이다. 원칙적으로 여러분은 그런 규칙의 긴 목록을 여러분의 컴퓨터 위에 테이프로 붙여 놓고 그것을 독자들에게 닿는 방법에 관한 권위 있는 목소리로 취급할 수도 있을 것이다. 그러나 긴 규칙 목록은 지루하다. 게다가 그것을 사용하는 것은 글쓰기를 기계적으로 만드는데, 좋은 글쓰기는 때로 언제 그 규칙을 따르는 대신에 그것들을 유연하게 적용할지를 아는 것을 수반한다. 더욱이, 규칙 목록을 사용하는 것은 이상하게도 간접적인데, 여러분이 명료한 텍스트를 만들어 낼 것이라고 들었던 규칙에 의존하는 것 대신에, 독자가 사고하는 방식을 이해하고 그 이해에 부합하게 글을 쓰는 것이 분명히 더 효과적일 것이다.

→ 명료한 글쓰기를 하기 위해서, 여러분은 여러분의 독자의 관점을 파악하고 일반적인 글쓰기 규칙을 융통성 있게 적용할 필요가 있다.

문제 해설
독자의 반응 방식을 생각하여 글쓰기의 일반적인 규칙을 목록화해서 활용할 수 있지만, 좋은 글쓰기는 기계적이지 않고 규칙을 유연하게 적용하며 독자의 사고방식에 부합하여 글을 쓴다는 내용의 글이다. 따라서 요약문의 빈칸 (A), (B)에 들어갈 말로 가장 적절한 것은 ① '독자(audience) – 융통성 있게(flexibly)'이다.
② 독자 – 일관성 있게
③ 편집자 – 체계적으로

④ 편집자 – 일관성 있게
⑤ 동료들 – 융통성 있게

구조 해설

- Furthermore, using a list of rules is oddly indirect: instead of [relying on rules {you've been told will produce clear text}], surely **it** would be more effective [to {understand how readers think}, and {write to that understanding}].

첫 번째 [ ]는 instead of의 목적어이다. 그 안의 { }는 rules를 수식하는 관계절인데, you've의 앞에 관계사(which나 that)가 생략되어 있다. it은 형식상의 주어이고 두 번째 [ ]가 내용상의 주어이다. 두 번째와 세 번째 { }는 and로 연결되어 to에 이어진다.

어휘 및 어구

crystal-clear 아주 명료한
compose 작성하다, 작곡하다
graphic 그래픽(사진, 그림, 도형, 그래프 등 다양한 시각적 형상)
active voice 능동태
figure 도표, 도형
quantity 양
in principle 원칙적으로
authority 권위
entail 수반하다
bend (사정에 따라) 유연하게 적용하다, 수정하다
oddly 이상하게, 기이하게
indirect 간접적인

24~25

정답 | 24 ④ 25 ⑤

소재 문화적 차이가 자원봉사자들에게 미친 영향

해석 1970년대 초, Botswana의 평화 봉사단 사무소는 '탈진한' 상태에 빠져, 임무에 실패하고, 배정된 마을을 떠나며, 주거를 제공하는 Tswana 주민에게 점점 더 적대감을 느끼는 것처럼 보이는 자원봉사자 수에 대해 걱정하고 있었다. 평화 봉사단은 Tswana 문화와 사회에 익숙한 미국인 인류학자 Hoyt Alverson에게 조언을 구했다. Alverson은 평화 봉사단 자원봉사자들이 겪고 있는 한 가지 주요 문제가 바로 비슷한 행동이 매우 다른 의미를 가지는 문제와 관련이 있음을 발견했다. 자원봉사자들은 Tswana 사람들이 결코 자신들을 혼자 내버려두려 하지 않는다고 불평했다. 그들이 떨어져서 혼자 앉아 잠시 개인 시간을 좀 가지려고 시도할 때마다, 한 명 이상의 Tswana 사람들이 재빨리 그들과 함께했다. 이것이 미국인들을 화나게 했다. 그들의 관점에서 모든 사람은 일정량의 사생활과 혼자만의 시간에 대한 권리가 있다. 그러나 Tswana 사람들에게 인간의 삶은 사회적 삶으로, 혼자 있고 싶어 하는 사람은 마녀와 제정신이 아닌 사람뿐이다. 이 젊은 미국인들이 그 둘 중 어느 쪽도 아닌 것처럼 보였기 때문에, 그들이 혼자 앉아 있는 것을 본 Tswana 사람들은 당연히 자신들이 환대에 실패했다고 생각하고, 자원봉사자들이 누군가 함께 있어 주는 것을 <u>거절할(→ 반길)</u>것이라고 여겼다. 여기서, 한 가지 행동, 즉 사람이 들판으로 걸어 나가 혼자 앉아 있는 것은 두 가지 매우 다른 의미를 지니고 있었다.

문제 해설

24 1970년대 초 Botswana에서 사생활을 중시해 혼자 있는 시간을 갖고 싶어 했던 미국인 자원봉사자들의 행동이 Tswana 사람들에게는 비정상적인 행동이었기 때문에 환대에 실패했다고 여겨 자원봉사자들과 계속 함께 있으려 했고, 결과적으로 이 문화적 차이가 갈등을 일으켰다는 내용의 글이다. 따라서 글의 제목으로 가장 적절한 것은 ④ '문화 간 오해 밝혀내기: 행동 대 의미'이다.
① 사회적 진화: 연결된 세상에서 변화에 적응하기
② 과제에 대처하고 관리하기 위한 창의적인 해결책
③ 자원봉사의 핵심: (자원봉사가) 꼭 필요한 곳에서 변화를 만드는 것
⑤ 문화 전체에서 보편적 가치관과 환대의 다양한 표현

25 미국인 자원봉사자에게 사생활이 중요한 반면, Tswana 사람들은 혼자 있는 것을 비정상적으로 생각했다고 했으므로, 혼자 개인 시간을 가지기 위해 앉아 있는 자원봉사자들을 본 Tswana 사람들은 자원봉사자가 혼자 있고 싶어 하지 않을 것이며, 함께 있어 주는 것을 반길 것이라고 여겼음을 추론할 수 있다. 따라서 (e)의 reject를 welcome과 같은 낱말로 바꿔야 한다.

구조 해설

- Because these young Americans did not seem to be either, the Tswana [who saw them sitting alone] naturally assumed [that there had been a breakdown in hospitality] and [that the volunteers would welcome some company].

첫 번째 [ ]는 the Tswana를 수식하는 관계절이고, 두 번째와 세 번째 [ ]는 and로 연결되어 assumed의 목적어 역할을 하는 명사절이다.

어휘 및 어구

burned out 탈진한
assignment 임무
hostile 적대적인
host 주인, 주최자
anthropologist 인류학자
entitled to ~에 대한 권리가 있는

privacy 사생활
witch 마녀
breakdown 실패
hospitality 환대, 접대
company 함께 있음, 동반자

26~28

정답 | 26 ④ 27 ⑤ 28 ⑤

소재 꿈을 실현한 어린 독수리 Edge

해석 (A) 숲 한가운데에 Edge라는 어린 독수리와 그의 쌍둥이 형제 Jo가 살고 있었다. Jo는 하늘 높이 날아오르며 비행의 자유를 경험할 수 있었지만, Edge는 손상된 날개를 가지고 태어나서 날 수 없었다. 이것은 그를 크게 낙담시킬 수도 있었지만, 그는 절대 희망을 잃지 않았고 언젠가 Jo와 함께 비행하는 자신의 꿈을 이룰 것이라고 항상 믿었다.

(D) 어느 날, Jo와 Edge는 먹이를 찾던 중 나이 많은 한 현자를 만났다. 지혜로 유명한 현자는 나무들 사이에 숨겨진 작은 오두막에 혼자 살았다. 자신의 쌍둥이 형제가 꿈을 이루기를 바라는 마음에 Jo는 현자가 Edge가 날 수 있도록 도와줄 수 있는지 간절히 물었다. 현자는 어린 독수리를 불쌍히 여겨 Edge에게 신비로운 약초에 대해 말해 주었다. 그 약초는 숲에서 가장 높은 산의 정상에서 자라며 푸른 빛을 발했다. 그는 만약 Edge가 그 신비한 약초를 찾아서 그것을 자기에게 가져올 수 있다면, 그는 그가 날 수 있게 그의 날개를 고쳐줄 수 있다고 말했다.

(B) 신비로운 약초를 찾는 것은 어렵고 위험한 일이었기에, Jo는 가지 않도록 그를 설득하려고 했다. 그러나 Edge는 약초를 찾기로 굳게 결심했다. 결국, Jo는 (그의 결심을) 이해하고 안전하고 성공적인 여정을 기원했다. Edge는 여정을 떠났는데, 그것은 그의 손상된 날개 때문에 특히 어려웠고, 어쩔 수 없이 그가 산을 걸어서 올라가게 만들었다. 어느 순간, 갑작스러운 산사태가 그가 따라가던 길을 쓸어 버려, 다시 처음부터 등반을 시작하게 만들었다. 그는 가는 길에 다른 장애물들을 많이 만났지만, 결코 포기하지 않았다. 마침내, 며칠 동안의 등반 끝에 그는 산 정상에 도착했고 신비로운 약초를 발견했다.

(C) Edge가 현자에게 돌아왔을 때, Edge의 날개를 고치기 위해 현자는 그 신비로운 약초를 사용하였다. Jo가 응원하는 가운데, Edge는 날개를 펼치고 처음으로 하늘로 날아올랐다. 그는 바람이 자신을 스쳐 지나가고 태양이 깃털에 비치는 것을 느꼈다. 그날 이후로 두 형제는 자주 함께 날며 광활한 하늘을 탐험했다. Edge는 희망과 결단력의 상징이 되었다. 그리고 비록 Edge는 자신의 꿈을 이루도록 도와준 현자를 절대 잊지 않았지만, 그는 또한 자신의 진정한 정신이 내면에서 비롯된다는 것을 알고 있었

고, 그것은 그에게 어떤 장애물이든 극복할 수 있는 힘을 주었다.

문제 해설

26 어린 독수리 Edge는 손상된 날개로 인해 날 수 없었지만, 희망을 잃지 않고 자신의 쌍둥이 형제인 Jo와 함께 비행하는 꿈을 이룰 수 있다고 믿었다는 내용의 (A) 다음에, 오두막에 사는 현자를 만나게 된 Jo는 Edge가 날 수 있도록 도와줄 수 있는지 간절히 묻자 Edge를 불쌍히 여긴 현자가 신비로운 약초를 가져오면 날개를 고쳐 주겠다고 제안하는 내용의 (D)가 이어져야 한다. 다음으로, Edge가 손상된 날개로 힘든 여정을 거쳐 마침내 신비로운 약초를 발견하는 내용의 (B)가 이어지고, 마지막으로 현자가 신비로운 약초를 사용하여 Edge의 날개를 고쳐 주고, Edge가 처음으로 하늘로 날아오르는 내용의 (C)가 이어지는 것이 자연스럽다. 따라서 주어진 글 (A) 다음에 이어질 글의 순서로 가장 적절한 것은 ④ '(D)-(B)-(C)'이다.

27 (a), (b), (c), (d)는 Edge를 가리키고, (e)는 나이 많은 현자를 가리키므로, 가리키는 대상이 나머지 넷과 다른 하나는 ⑤이다.

28 약초는 숲에서 가장 높은 산의 정상에서 자라며 푸른 빛을 발한다고 했으므로, 글에 관한 내용으로 적절하지 않은 것은 ⑤이다.

구조 해설

■ [When Edge returned to the sage], the sage used the mystical herb [to fix Edge's wing].

첫 번째 [ ]는 시간의 부사절이고, 두 번째 [ ]는 목적을 나타내는 to부정사구이다.

어휘 및 어구

freedom 자유
determined 굳게 결심한
journey 여정, 여행
wipe out ~을 쓸어 버리다, ~을 완전히 파괴하다
encounter 만나다, 마주치다
obstacle 장애물
feather 깃털
overcome 극복하다
hut 오두막
take pity on ~을 불쌍히 여기다

01 ②	02 ④	03 ②	04 ③	05 ⑤	06 ⑤	07 ②
08 ④	09 ⑤	10 ⑤	11 ④	12 ③	13 ④	14 ②
15 ①	16 ④	17 ②	18 ④	19 ②	20 ②	21 ④
22 ②	23 ②	24 ⑤	25 ⑤	26 ④	27 ④	28 ⑤

01

정답 | ②

소재 폭설로 인한 휴교 공지

해석 [2025년 1월 20일 월요일]
TBS 고등학교 학부모/보호자 및 학생 여러분께
Calbary에 40cm가 넘는 눈이 내린 폭설 기간 동안 인내심을 가져주셔서 감사합니다. 오늘 저녁, Calbary시는 눈보라 경보를 발령했고 주민들에게 집에 머물 것을 권장하고 있습니다. 저희 TBS 고등학교에서도 몇 가지 우려에 직면하고 있는데, 학생들이 안전하게 걸어서 등교하지 못할 수도 있고, 많은 통학 버스가 지연되거나, 경우에 따라 취소될 가능성도 있습니다. 이러한 모든 점을 고려하여, 본교는 1월 21일 화요일에 휴교하기로 결정했습니다. 학생들은 집에서 이전에 부과받은 과제를 하거나 이전에 다루었던 자료를 복습하며 시간을 보낼 수 있습니다. 여러분의 인내심과 양해에 감사드립니다.
TBS 고등학교 교장
Caroline R. Robbins 드림

문제 해설
폭설로 인하여 학생들이 안전하게 등교할 수 없을 수도 있고, 많은 통학 버스가 지연되거나 경우에 따라 취소될 가능성이 있는 등 여러 가지 우려에 직면해 있어 학교를 휴교하기로 결정했다고 말하고 있으므로, 글의 목적으로 가장 적절한 것은 ②이다.

구조 해설
■ This evening, the City of Calbary [has issued a snowstorm alert] and [is encouraging residents to stay home].
두 개의 [ ]는 and로 대등하게 연결되어 문장의 술어 역할을 하고 있다.

어휘 및 어구
heavy snowstorm 폭설
issue 발령하다, 공표하다
resident 주민
previously 이전에
assign 부과하다
review 복습하다

02

정답 | ④

소재 Santa Teresa 교회에서 생긴 일

해석 우리 가족이 Santa Teresa 교회에 도착했을 때, 나는 조용히 예배당의 아름다움을 만끽하려고 했다. 그러나 나는 한 무리의 사람들이 예배당 구석에서 큰 소리로 노래를 부르는 것을 발견했다. 이 사람들은 왜 내 여행을 망치고 있는 거지? 저들은 이곳이 어떤 곳인지 알기나 하는 걸까? 설상가상으로, 그들은 춤까지 추기 시작했다! 내가 예배당에서 보낸 그 짧은 시간은 그들의 사려 깊지 못한 행동으로 완전히 망쳐졌다. 갑자기, 무리 중 몇몇 여성들이 나에게 함께하자고 손짓하기 시작했다. 나는 뒷걸음질을 함으로써 거절하려고 했다. 그런데도 그들은 그저 계속해서 나를 향해 손을 뻗었다. 나는 내 딸을 내려다보았고, 그녀는 호기심 어린 눈으로 나를 다시 쳐다보았다. 나는 어깨를 으쓱하고 미소 지었다. 나도 모르는 새에, 우리는 얼굴에 커다란 미소를 지은 채 함께 춤을 추고 있었다. 내가 딸의 손을 꼭 잡고 경쾌한 리듬에 장단을 맞추기 위해 최선을 다하면서 내 마음은 기쁨으로 가득 찼다. 나는 이것이 여행에서 내가 가장 좋아하는 기억이 될 것임을 알았다.

문제 해설
'I'는 Santa Teresa 교회에서 조용히 예배당의 아름다움을 만끽하려다가 예배당 구석에서 큰 소리로 노래를 부르는 한 무리의 사람들 때문에 짜증이 났다가, 무리 중 몇몇 여성들의 권유로 딸과 함께 무리 속으로 들어가 춤을 추면서 자신의 마음이 기쁨으로 가득 차게 되었다는 내용이다. 따라서 'I'의 심경 변화로 가장 적절한 것은 ④ '짜증이 난 → 매우 기쁜'이다.
① 희망에 찬 → 스트레스 받는
② 부끄러운 → 느긋한
③ 지루한 → 두려운
⑤ 신이 난 → 실망한

구조 해설
■ [As I {squeezed my daughter's hand} and {did my best to keep time with the lively beat}], my heart was filled with joy.
[ ]는 '~하면서, ~할 때'라는 의미의 접속사 As가 이끄는 부사절이고, 그 안에 두 개의 { }가 and로 대등하게 연결되어 술어 역할을 하고 있다.

어휘 및 어구
soak up ~을 만끽하다, (분위기 등에) 흠뻑 젖다
chapel 예배당
to make matters worse 설상가상으로
motion 손짓하다
shrug 어깨를 으쓱하다
the next thing I knew 나도 모르는 새에, 어느 틈엔가
squeeze 꼭 잡다

keep time with ~에 장단[박자]을 맞추다
lively 경쾌한, 활기찬

03

소재 자신의 성과 뒤에 존재하는 조력자들

해석 좋든 싫든, 우리는 흔히 우리를 특정 편견의 희생자가 되게 하는 특이한 렌즈를 통해 세상을 본다. 나는 성공적인 사업체를 구축하고 자신들의 모든 성과를 자신들만의 재간 덕분으로 돌리는 많은 사업가들을 보아 왔다. 이 사람들은 자신들의 성취가 자기 외부의 다양한 사람들과 기관들의 도움 없이는 가능하지 못했을 수도 있다는 사실을 볼 수 있는 자기 성찰과 객관성이 부족한 경향이 있다. 우리 인생의 실패를 순전히 또는 주로 밖의 (외부의) 힘 때문으로만 여기고, 우리 인생의 성공을 순전히 또는 주로 안의 (내부의) 힘 때문으로만 여기는 것은 너무나 쉽다. 따라서 우리는 공로를 인정하고 공유해야만 한다는 것을 알아야 한다. Apple의 성공은 전적으로 Steve Jobs만의 것이 아니었고, Microsoft의 성공도 전적으로 Bill Gates만의 것이 아니었다. 산타에게는 그가 불가능한 일을 달성할 수 있게 해 준 수천 명의 작은 조력자들이 있다. 우리가 무엇이든 성취하려면 사람들(또는 요정들)로 구성된 팀과 거대한 네트워크가 필요하다.

문제 해설
우리가 무엇이든 성취하려면 조력자들의 공로를 인정하고 공유해야만 한다는 내용의 글이다. 따라서 필자가 주장하는 바로 가장 적절한 것은 ②이다.

구조 해설
- It's all too easy [to **see** {failures in our lives **as** purely or predominantly due to outside (external) forces}, and {successes in our lives **as** purely or predominantly due to inside (internal) forces}].

 It은 형식상의 주어이고 [ ]는 내용상의 주어이다. 두 개의 { }가 and로 대등하게 연결되어 '~을 …으로 여기다'라는 의미의 「see ~ as …」 구문을 이루고 있다.

어휘 및 어구
for better or worse 좋든 싫든[나쁘든]
singular 특이한, 색다른
fall victim to ~의 희생자가 되다
bias 편견
attribute ~ to … ~을 … 덕분으로 돌리다
self-reflection 자기 성찰
objectivity 객관성
accomplishment 성취
institution 기관

external 외부의
purely 순전히
predominantly 주로, 대부분
credit 공로
exclusively 전적으로, 오로지

04

소재 집단적 인식론

해석 우리는 시간을 들여 모든 정보를 사실 확인하지 않은 채 우리의 믿음을 공유하는 사람들을 믿는 경향이 있다. '정체성 정치'의 출현과 함께 이것은 가짜 뉴스라는 최근의 현상을 야기했다. 여러분의 집단의 관점과 일치하지 않는 것은 무엇이든 거부되고 가짜 뉴스라고 낙인찍힌다. 그것이 항상 허위 정보는 아니지만, 정치인과 언론 둘 다에 의해 가짜 뉴스라고 낙인찍히는 것은 다름 아닌 우리 자신의 것과 크게 다른 의견들인 경우가 흔하다. *Guardian*지의 Jonathan Freedland는 그것을 '집단적 인식론'이라고 불리는 새로운 유형의 인지적 편향이라고 일컬었는데, 이는 진실이 더 이상 사실이나 증거와 일치하지 않고 오히려 특정 주장이 자신이 속한 집단이나 사회 집단의 관점과 일치할 때를 말한다. '우리에게 적절하다' 또는 '우리에게 좋다'와 '진실'의 경계가 흐릿해졌다. 물론, 이것은 새로운 현상이 아니며, 우리는 인류 진화의 시작부터 우리 스스로를 집단으로 분류해 왔다. 다만 그것이 최근의 정치 풍토, 언론과 기술, 그리고 사실보다는 이야기와 이야기 속 묘사를 믿는 우리의 경향에 의해 불이 붙은 것이다. 모두가 좋은 이야기를 좋아한다.

문제 해설
특정 주장이 자신이 속한 집단이나 사회 집단의 관점과 일치할 때 그것이 진실이 된다는 내용의 글이므로, 밑줄 친 부분이 글에서 의미하는 바로 가장 적절한 것은 ③ '우리는 우리의 사회적 집단의 관점과 부합하는 이야기를 맹목적으로 따른다.'이다.
① 좋은 이야기는 사회 내에서 모두를 결속시키는 유일한 방법이다.
② 우리는 정보 과부하 때문에 가짜 뉴스의 세상에서 살고 있다.
④ 스토리텔링의 기술은 다른 사람들과 연결할 방법을 찾는 것을 포함한다.
⑤ 오래된 경험과 새로운 경험을 합치는 것은 우리가 매력적인 이야기를 만드는 것을 돕는다.

구조 해설
- It is not always falsified information, but [**it is** often {opinions ⟨that differ hugely from **our own**⟩} **that** are labelled fake news by politicians and the media alike].

 [ ]는 「it is ~ that …」 형식으로 { }가 강조된 구문인데, { }

안의 〈 〉는 opinions를 수식하는 관계절이다. our own은 our own opinion(s)의 뜻이다.

어휘 및 어구

emergence 출현
identity 정체성
give rise to ~을 야기하다, ~을 일으키다
phenomenon 현상
fake 가짜의
conform to ~과 일치하다
tribe (같은 직업·흥미·습관을 가진) 집단, 부족
reject 거부하다
be labelled (as) ~이라고 낙인찍히다
falsified 허위의
term (특정한 이름·용어로) 일컫다
cognitive 인지적인
bias 편향
correspond to ~과 일치하다, ~과 부합하다
assertion 주장
viewpoint 관점
belong to ~에 속하다
boundary 경계
sort 분류하다
evolution 진화
political climate 정치 풍토
tendency 경향
narrative 이야기 속 묘사

05
정답 | ⑤

소재 모호함에 대처하는 자세

해석 모호함은 대부분의 사람들에게 불편한 감정이다. 따라서 그것은 우리가 조급하게 그 원인을 잘못 생각하여 무언가 잘못되었다고 결론을 내리는 경향이 있는 감정이다. 하지만 사실, 모호함은 생산적이고 심지어 긍정적인 상태일 수 있다. 우리가 연구를 수행할 때, 우리는 우리 자신, 우리 학생들, 그리고 우리의 동료들이 모호함의 상태를 즐기도록 격려하려고 애를 쓴다. 그 논리는 답을 모를 때 진정한 지식이 생겨날 수 있으므로, 천천히 진행하고 신중하게 생각하며 무슨 일이 일어나고 있는지 알아내려는 과정을 즐기는 것이 최선이라는 것이다. 안타깝게도, 모호함에 대한 이러한 애정은 흔하지 않은 사고방식이다. 보통 우리는 빠르고 명확한 답을 원하며, 특히 스트레스를 받을 때 그렇다. 어떤 사람들에게는, 모호함에 대한 이러한 반감이 그들이 잘못된 믿음으로 전락하는 것의 원인이 된다. 성급하게 결론짓지 않고, 여러 가설을

염두에 두며, 새로운 정보와 가능성에 계속 열려 있는 능력은 잘못된 믿음에 빠지지 않는 비결이다. 우리는 확신과 자신감을 동경하고 추구하는 경향이 있다. 하지만 우리가 모호함의 상태를 동경하고 즐기는 법을 배운다면 우리에게 더 나을 것이다.

문제 해설
모호함은 불편한 감정이지만 모호함의 상태를 즐기면서 새로운 정보와 가능성에 계속 열려 있으면 잘못된 믿음에 빠지지 않는다는 내용의 글이므로, 글의 요지로 가장 적절한 것은 ⑤이다.

구조 해설
- The logic is [that {when we don't know the answer}, real knowledge can arise, {so **it** is best 〈to go slowly, think carefully, and enjoy the process of trying to find out what is going on〉}].
  [ ]는 주격 보어 역할을 하는 명사절이며, [ ] 안의 첫 번째 { }는 시간의 부사절이다. 두 번째 { } 안에 있는 it은 형식상의 주어이고, 〈 〉는 내용상의 주어이다.

어휘 및 어구

ambiguity 모호함
as such 따라서
misattribute (~의) 원인을 잘못 생각하다
mindset 사고방식
contribute to ~의 원인이 되다
misbelief 잘못된 믿음
multiple 여러, 다수의
hypothesis 가설(pl. hypotheses)
admire 동경하다
conviction 확신, 신념

06
정답 | ⑤

소재 인공 지능에 대한 투자

해석 기업, 비영리 단체, 또는 정부의 리더들은 인공 지능에 투자할 때, 기계 학습 전문가를 고용하거나 도구에 비용을 내는 데 대부분의 관심을 기울인다. 하지만 이는 매우 중요한 기회를 놓치는 것이다. 조직이 인공 지능으로부터 얻을 수 있는 최대의 효과를 얻기 위해서는 모든 팀 구성원들이 이 기술을 더 잘 이해하도록 돕는 데에도 투자해야 한다. 기계 학습을 이해하는 것은 직원이 자신의 업무에서 잠재적으로 적용할 수 있는 부분을 발견할 가능성을 더 크게 만들 수 있다. 기계 학습의 가장 유망한 용도 중 많은 부분은 단조로울 것인데, 이것이 기술이 가장 유용할 수 있는, 즉 사람들이 기계보다 뛰어난 여러 작업에 집중할 수 있도록 그들에게 시간을 절약해 줄 수 있는 부분이다. 기계 학습에 대한

이해가 더 높은 비서라면 일정 관리 소프트웨어가 시간이 지남에 따라 발생하는 패턴으로부터 더 명시적으로 학습하여, 상사가 어떤 팀원과 비정상적으로 오랫동안 만나지 않았을 때 자신에게 이를 상기시켜 주도록 하면 좋겠다고 제안할 수도 있을 것이다. 패턴을 학습하는 일정 관리 도구는 상사가 팀을 관리하는 것을 돕는 것과 같은, 업무에 있어서 인간에게 특화된 것을 위한 더 많은 시간을 비서에게 줄 수도 있을 것이다.

(문제 해설)

기업, 비영리 단체, 정부가 인공 지능에 투자할 때 기계 학습 전문가를 고용하거나 도구에 비용을 내는 데 관심을 기울이지만, 모든 팀 구성원들이 이 기술을 더 잘 이해하도록 돕는 데에도 투자하면 이들이 자신의 업무에서 잠재적으로 적용할 수 있는 부분을 발견할 가능성을 더 크게 만들 수 있다는 내용의 글이므로, 글의 주제로 가장 적절한 것은 ⑤ '모든 직원의 인공 지능 이해 능력 향상에 대한 투자의 필요성'이다.
① 인공 지능 기반 관리 도구 도입의 이점
② 유능한 기계 학습 전문가를 고용하는 것의 어려움
③ 인공 지능 시대에 인간 직원을 유지해야 하는 이유
④ 인공 지능 도입에 대한 경영진의 거리낌을 극복하는 방법

(구조 해설)

■ [An executive assistant {who has a better understanding of machine learning}] might suggest [that calendar software **learn** more explicitly from patterns that develop over time, {reminding them when their boss has not met with a team member for an unusually long time}].

첫 번째 [ ]는 문장의 주어이고, 그 안의 { }는 An executive assistant를 수식하는 관계절이다. 두 번째 [ ]는 suggest의 목적어 역할을 하는 명사절인데, 그 안에 '당위'의 뜻이 포함되어 있어서 「(should+) 동사원형」이 술어동사로 사용되었다. 그 안의 { }는 calendar software의 부수적인 동작을 나타내는 분사구문이다.

(어휘 및 어구)

nonprofit 비영리 단체
machine learning 기계 학습
spot 발견하다
application 적용, 응용
promising 유망한
executive assistant 비서, 보좌관
calendar 일정표, 달력
explicitly 명시적으로, 명확하게
specialty 특화된 것, 전문 분야

07

정답 | ②

소재 디지털 시대의 라디오의 가치

해석 라디오가 '마음의 극장'이라는 것은 유용한 상투적인 표현인데, (업계) 종사자들과 라디오 교육자들에 의해 똑같이 반복되는 말이다. 말, 음악, 그리고 음향 효과를 사용하여 이야기를 전달함으로써, 라디오는 상상력을 사로잡아 각각의 개별 청취자에게 고유하게 경험되는 일종의 서사를 만들어 내는 아이디어와 이미지를 전달할 수 있다. 시각적인 단서의 생략을 통해, 그리고 작업의 개방성을 수용함으로써, 라디오 스토리텔링은 개인적인 유대를 형성하고, 소리로 그림을 그리며, 실제로 다른 상황에서는 불가능할 장면을 만들 수 있는 능력을 갖추고 있다. 그러한 이야기들이 전달되는 제작, 유통, 그리고 소비 문화와 기술은 근본적으로 변화했지만 창의적이고 흥미진진한 스토리텔링을 위한 매체로서의 오디오의 능력은 디지털 시대에도 줄어들지 않았다. 실제로 디지털 시대에 라디오의 소리 효과에 의한 서사의 가능성이 여러 면에서 확장되었는데, 이는 연구 기술자, 전문가 및 열정적인 사람들이 새로운 제작 과정과 플랫폼, 그리고 상호 작용을 하는 기회의 특성과 더불어 라디오 스토리텔링을 라디오 전문가의 영역 밖으로 가져갈 기회도 탐구하기 때문이다.

(문제 해설)

라디오는 상상력을 사로잡아 개별 청취자에게 고유하게 경험되는 서사를 만들어 내는 능력을 갖추고 있는데, 이러한 창의적이고 흥미진진한 라디오 스토리텔링의 능력은 디지털 시대에도 줄어들지 않았고 오히려 라디오의 소리 효과에 의한 서사의 가능성이 디지털 시대를 맞아 여러 면에서 확장되었다는 내용의 글이다. 따라서 글의 제목으로 가장 적절한 것은 ② '디지털 시대에서 라디오 스토리텔링의 지속적인 가치'이다.
① 라디오 스토리텔링이 글로 쓰인 스토리텔링과 다른 점
③ 변화하는 미디어 분야에서 라디오가 직면한 어려움
④ 엔터테인먼트를 넘어: 라디오가 문화 형성에 있어 미친 지속적인 영향
⑤ 신호에서 스트리밍까지: 라디오 기술의 진화 탐구

(구조 해설)

■ Through the omission of visual cues and by embracing the openness of the work, radio storytelling has the capacity [to {make personal connections}, {paint pictures with sound}, and indeed {create scenes ⟨that would be impossible in another context⟩}].

[ ]는 the capacity의 구체적 내용을 설명하는 to부정사구이다. 그 안의 세 개의 { }는 콤마와 and로 연결되어 to에 이어진다. ⟨ ⟩는 scenes를 수식하는 관계절이다.

practitioner (전문업) 종사자

theatre 극장

engage 사로잡다, 끌어들이다

narrative 서사

uniquely 고유하게

cue 단서

capacity 능력

connection 유대, 관계

compelling 흥미진진한

undiminished 줄어들지 않은

distribution 유통, 배포

mediate 전달하다

enthusiast 열정적인 사람, 애호가

parameter (보통 복수로) 특성, 요소

realm 영역

08
정답 | ④

소재 소셜 미디어가 민주주의에 미치는 영향에 대한 인식

해석 위 도표는 선정된 6개국에서 소셜 미디어가 민주주의에 미치는 긍정적인 영향에 대한 성인들의 견해를 보여 준다. 도표의 6개국 전부에서 소셜 미디어가 민주주의에 유익했다고 말하는 나이가 더 어린 성인(18~39세)의 비율이 나이가 더 많은 성인(40세 이상)의 비율보다 더 높다. 선정된 6개국 중 두 연령 집단 간 퍼센트포인트 차이는 폴란드에서 가장 크고, 브라질에서 가장 작다. 일본과 남아프리카 공화국 모두에서 3분의 2가 넘는 18~39세 성인이 소셜 미디어가 민주주의에 유익했다고 말한다. 영국에서는 나이가 더 많은 연령 집단의 절반이 넘는 성인이 소셜 미디어가 자국의 민주주의에 유익했다고 말한다. 6개국 중 미국은 소셜 미디어가 자국의 민주주의에 긍정적이었다고 말하는 40세 이상 성인의 비율이 30퍼센트 미만으로 가장 낮다.

문제 해설

영국에서는 두 성인 집단 중 나이가 더 많은 40세 이상의 집단의 45퍼센트가 소셜 미디어가 자국의 민주주의에 유익했다고 말했으므로, 도표의 내용과 일치하지 않는 것은 ④이다.

구조 해설

- Among the six countries, the U.S. shows [the lowest percentage of adults ages 40 and older {who say social media has been positive for democracy in their country}, at below 30 percent].
  [ ]는 shows의 목적어이고, 그 안의 { }는 adults ages 40 and older를 수식하는 관계절이다.

democracy 민주주의

positive 긍정적인

benefit (~에) 유익하다

09
정답 | ⑤

소재 탐험가 Barbara Hillary

해석 Barbara Hillary는 1931년 New York City에서 태어나 Harlem에서 자랐다. 그녀의 집안은 가난했지만, 그녀에게 독서를 권장했다. 그녀는 New York City의 New School에서 노인학을 전공했고, 졸업 후 간호사가 되어 은퇴하기 전까지 55년간 일했다. 67세에 Hillary는 폐암 진단을 받았는데, 그것은 제거 수술이 필요했다. 그 수술로 인해 호흡 능력이 25% 감소했지만, 그녀는 이 일이 자신을 멈추게 하도록 내버려두지 않았다. 그녀는 캐나다를 방문해 북극곰 사진을 찍은 후 북극에 관심이 생겼다. 2007년 봄, 75세의 나이에 그녀는 북극에 스키를 타고 간 최고령 여성이자 최초의 아프리카계 미국인 여성이 되었다. 2011년에는 79세의 나이로 남극에 도착하여 양쪽 극에 모두 도달한 최초의 아프리카계 미국인 여성이 되었다.

문제 해설

2011년에 남극에 도착하여 양쪽 극에 모두 도달한 최초의 아프리카계 미국인 여성이 되었다고 했으므로, 글의 내용과 일치하지 않는 것은 ⑤이다.

구조 해설

- At age 67, Hillary was diagnosed with lung cancer, [which required surgery to remove].
  [ ]는 lung cancer에 대해 부가적으로 설명하는 관계절이다.

major in ~을 전공하다

retire 은퇴하다

be diagnosed with ~이라는 진단을 받다

surgery 수술

reduction 감소

10
정답 | ⑤

소재 Greenfields 말 캠프

해석 2025 Greenfields 말 캠프

잊을 수 없는 경험을 위해 저희와 함께하세요! 이 캠프는 6~10세 어린이를 대상으로 하며, 말과 평생 인연을 맺을 수 있는 안전하고 재미있는 환경을 제공합니다.

날짜
– 1기: 7월 7일~11일
– 2기: 7월 14일~18일

활동
캠프 기간 동안 참가자는
– 우리의 실내 및 실외 시설에서 말을 타는 법을 배울 것입니다.
– 말을 안전하게 다루고, 먹이를 주며, 돌보는 법을 배울 것입니다.
– 말에 대해 배우는 데 중점을 둔 다양한 게임과 활동을 즐길 것입니다.

참고 사항
– 캠프는 오전 9시부터 오후 2시까지 운영됩니다. 추가 요금 지불 시 오후 5시까지 선택 활동을 이용할 수 있습니다.
– 음료는 제공되지만, 참가자는 점심 도시락을 가져와야 합니다.
– 기별 비용은 250달러이며, 별도로 50달러의 등록비가 있습니다.
– 결석일에 대한 환불은 하지 않습니다.

805-222-3990으로 전화하거나 greenhorseclub@abc.com 으로 이메일을 보내 자리를 예약하세요!

(문제 해설)
기별 비용은 250달러이며, 별도로 50달러의 등록비가 있다고 했으므로, 안내문의 내용과 일치하지 않는 것은 ⑤이다.

(구조 해설)
▪ During camp, participants will enjoy a variety of games and activities [focused on learning about horses].
[ ]는 a variety of games and activities를 수식하는 분사구이다.

(어휘 및 어구)
unforgettable 잊을 수 없는
facility 시설
handle 다루다
available 이용할 수 있는

11
정답 | ④

(소재) 반려동물 자동 급식기
(해석) Smart 반려동물 자동 급식기 사용 설명서

이 반려동물 자동 급식기는 여러분이 특정 시간에 반려동물에게 먹이를 주는 것을 돕도록 설계되었습니다. 이것은 실내 가정 전용입니다.

사료 통 채우기
1. 사료 통을 분리하려면 그것을 누르거나 비틀지 말고 위로 밀어

올립니다.
2. 통에 반려동물 사료를 채우고 사료 통을 다시 제자리에 밀어 넣습니다.

먹이 주는 시간 관리하기
1. 설정 버튼을 누릅니다. 화살표 키를 사용해 원하는 급식 시간을 선택합니다.
2. 설정을 다시 눌러 저장합니다.
3. 단계를 반복하여 먹이 주는 시간을 추가로 설정합니다. 하루에 최대 4번까지 먹이 주는 시간을 설정할 수 있습니다.

개인 음성 메시지 녹음하기
1. 녹음 버튼을 누르고 그 상태를 유지합니다.
2. 녹음 포트에 대고 말하여 메시지를 녹음합니다. 개인 음성 메시지는 최대 10초까지 가능합니다.
3. 녹음 버튼에서 손을 뗍니다.

주의 사항
• 젖은 사료를 사용하지 마세요.
• 사료 트레이와 사료 통은 분리하여 손으로 세척할 수 있지만, 식기세척기로 세척하면 안 됩니다.

(문제 해설)
개인 음성 메시지는 최대 10초까지 녹음할 수 있다고 했으므로, 안내문의 내용과 일치하는 것은 ④이다.

(구조 해설)
▪ [Slide the food compartment upwards to remove it], [without pressing or twisting it].
첫 번째 [ ]는 명령문이고, 두 번째 [ ]에서는 동명사 pressing과 twisting이 or로 대등하게 연결되어 전치사 without 뒤에 이어진다.

(어휘 및 어구)
automatic 자동의
feeder 급식기
specific 특정한
arrow 화살표
desired 원하는
precaution 주의 사항, 예방 조치
dishwasher 식기세척기

12
정답 | ③

(소재) 동물의 기질 측정
(해석) 가축의 경우, 연구자들은 일반적으로 실험용 또는 상업용 농장의 동물 무리에서 기질 특성을 측정하고 있다. 더 통제된 실험용의 실험실 환경에 비해, 이러한 환경은 기질 특성을 측정하고 이

해하는 데 기회와 제약을 모두 제공한다. 농장은 농장 및 그 밖의 장소에 있는 다른 동물과의 혈통상의 연관성이 자주 알려지는 많은 동물의 공급원이다. 어떤 특정 농장 내에서든 동물들은 상당히 표준화된 사육 과정을 경험했을 수도 있다. 그러나 한 농장은 여러 면에서 다른 농장과 다른데, 예컨대 사람과의 상호 작용의 질, 제공되는 영양의 수준, 그리고 무엇보다도 그 동물이 속한 사회적 무리이며, 그 모든 것이 기질에 영향을 미칠 수 있다. 예를 들어, 돼지 무리가 다른 것과 섞여 있을 때의 개별 돼지의 공격성을 평가하기 위해서는, 모든 동물이 통제된 체중 범위 내에서, 같은 크기의 우리 안에, 표준화된 수의 낯선 동물과 섞여 있는 실험을 사용해야 한다. 그러한 표준화는 일부 농장에서만 달성할 수 있다. 이러한 요소 모두는 농장을 통계적 복제본으로 취급할 수 없으며, 각 농장을 고유한 것으로 간주해야 한다는 것을 의미한다.

(문제 해설)
③ 앞의 절과 뒤의 절을 연결하는 접속사가 없으므로 대명사 them을 관계사 which로 고쳐야 한다.
① 전치사 to 뒤에 사용된 동명사구를 이끄는 동명사 measuring은 어법상 적절하다.
② 관계절의 주어의 핵심어구는 pedigree links이므로 복수형 동사 are는 어법상 적절하다.
④ 목적을 나타내는 to부정사구를 이끄는 to assess는 어법상 적절하다.
⑤ 동사 mean의 목적어 역할을 할 명사절이 필요하므로 명사절을 이끄는 접속사 that은 어법상 적절하다.

(구조 해설)
- For instance, [to assess aggressiveness in individual pigs {when groups of pigs are mixed with others}], a test must be used [in which all animals are mixed with a standardized number of unfamiliar animals, within a controlled weight range, and into pens of the same size].
첫 번째 [ ]는 목적을 나타내는 to부정사구이고, 그 안의 { }는 시간을 나타내는 부사절이다. 두 번째 [ ]는 a test를 수식하는 관계절이다.

(어휘 및 어구)
livestock 가축
measure 측정하다
commercial 상업용의
laboratory 실험실
constraint 제약
trait 특성
standardized 표준화된
rearing 사육

respect 측면
nutrition 영양
aggressiveness 공격성
pen 우리
statistical 통계적인

13
정답 | ④

(소재) 생존을 위한 생물의 정보 저장과 사용

(해석) 정보를 저장하고 상기하는 것은 생존에 매우 중요해서 자연은 이를 수행하는 수천 가지 기술을 진화시켰다. 예를 들어, 파리지옥은 수를 셀 수 있다. 덫이 단 한 방울의 빗방울로 인해 갑자기 닫히지 않도록 확실히 하기 위해, 그것은 일정한 짧은 시간 내에 덫이 세 번 접촉된 후에 필요한 에너지를 소비하여 덫을 닫는다. 따라서, 그것은 접촉 횟수를 셀 수 있어야 하고, 아울러 접촉 사이의 시간 간격도 저장할 수 있어야 한다. 우리는 실제로 이것이 어떻게 이루어지는지 알고 있다. 덫이 접촉되면 그 식물은 약간의 칼슘을 방출하는데, 그것은 서서히, 몇 초에 걸쳐, 소멸된다. 그 이후의 접촉은 더 많은 칼슘을 축적하고(→ 방출하고), 특정 임계치에 도달하면 덫이 툭 닫힌다. 칼슘이 끊임없이 소멸되기 때문에, 임계치에 도달하려면 연속적인 접촉이 특정 시간 간격 이내에 이루어져야 한다. 이 능력은 분명히 DNA 어딘가에 암호화되어 있지만, 다시 말하는데, 거의 인식하지 않은 채 정보에 의거하여 행동하고 있는 것이다.

(문제 해설)
정보를 저장하고 상기하는 것은 생존에 매우 중요하기 때문에 자연은 이를 위한 다양한 기술을 진화시켰는데, 예를 들어, 파리지옥은 접촉에 따라 칼슘을 방출하며 세 번의 접촉으로 칼슘 방출이 임계치에 이르면 덫을 닫는 능력을 가지고 있다는 내용의 글이다. 따라서 ④의 accumulate를 release와 같은 낱말로 바꿔야 한다.

(구조 해설)
- Storing and retrieving information is **so** key to survival **that** nature has evolved thousands of techniques to **do so**.
「so ~ that ...」 구문이 사용되어 '매우 ~해서 …하다'라는 의미를 나타낸다. do so는 store and retrieve information 정도의 의미라고 이해할 수 있다.

(어휘 및 어구)
retrieve 상기하다, 회수하다
evolve 진화하다
interval 간격
release 방출하다
subsequent 그 이후의

snap shut 툭 닫히다
constantly 끊임없이
sequential 연속적인, 순차적인
encoded 암호화된
cognition 인식, 인지

14

정답 | ②

소재 귀납적 추론의 한계

해석 과학 철학자 Hans Reichenbach 및 기타 철학자들은 까마귀의 색깔을 사용하여, 가설 검증을 기반으로 한 추론이 어떤 상황이 항상 발생한다는 것을 확실하게 증명할 수 없는 이유를 설명한다. '모든 까마귀는 검다'라는 가설을 생각해 보자. 비록 100만 마리의 검은 까마귀를 보았고 다른 색의 까마귀를 본 적이 없다고 해도 다음 까마귀가 검은색일 것이라는 보장은 없다. 누군가 자신이 모든 까마귀를 다 보았다고 생각하더라도, 그 사람이 잘못 생각하고 있는 것일 수도 있다. 더구나, 모든 미래 또는 과거의 까마귀를 다 볼 수는 없다. 따라서 아무리 많은 까마귀가 연구되었다 하더라도, 그 연구가 모두 모든 까마귀는 검다는 가설과 일치한다 해도, 모든 까마귀가 검다는 것을 확신할 수 있는, 즉 증명할 수 있는 방법은 없다. 내가 지금까지 보고 들은 많은 까마귀를 고려하면, 나는 까마귀의 대다수가 검은색이라고 확신하지만, 모든 까마귀가 검은색이라고 확실히 아는 것은 아니다. 이것이 귀납적 추론의 본질적 한계다.

문제 해설
과학 철학자들은 '모든 까마귀는 검다'는 가설을 통해 가설 검증을 기반으로 한 추론이 어떤 상황이 항상 발생한다는 것을 확실하게 증명할 수 없음을 설명한다는 내용의 글로, 아무리 많은 검은 까마귀를 보았어도 다음 까마귀가 검을 것이라고 보장할 수 없으며, 과거와 미래의 모든 까마귀를 확인할 수 없기 때문에 모든 까마귀가 검다는 것을 확실하게 보장할 수는 없다고 했으므로, 빈칸에 들어갈 말로 가장 적절한 것은 ② '한계'이다.
① 유연성
③ 포괄성
④ 실용성
⑤ 의도성

구조 해설

- Philosopher of science Hans Reichenbach and others use the colour of crows [to illustrate {why inference based upon hypothesis tests cannot demonstrate with certainty ⟨that a circumstance always occurs⟩}].
[ ]는 목적의 의미를 나타내는 to부정사구이고, 그 안에서 { } 는 illustrate의 목적어 역할을 하는 명사절이며, ⟨ ⟩는

demonstrate의 목적어 역할을 하는 명사절이다.

어휘 및 어구
illustrate 설명하다, 예증하다
inference 추론, 추리
hypothesis 가설, 가정
demonstrate 증명하다, 논증하다
circumstance (보통 복수로) 상황, 환경
occur 발생하다, 일어나다
guarantee 보장
consistent 일치하는, 일관된
for a fact 확실히, 사실로서
inherent 본질적인, 타고난
inductive reasoning 귀납적 추론

15

정답 | ①

소재 반려동물과의 진정한 의사소통

해석 우리의 반려동물은 야생 동물과 매우 다르다. 수천 년간의 가축화를 거치면서 그것들은 야생의 조상과는 거의 알아볼 수 없을 정도로 변했다. 하지만 완전히 변한 것은 아니다. 적어도 개와 고양이의 경우 우리는 그것들과 함께 앉아서, 보고, 교감할 수 있으며, 그것들도 우리가 보고 교감하는 것에 기뻐한다. 우리 반려동물과의 진정한 의사소통은 그것들에게 인간의 말('앉아', '누워', '그걸 내려놔')을 가르치는 것과 관련된 것이 아니며, 진정한 의사소통은 그것들 안에 있는 동물의 본성을 관찰하는 것과 관련되어 있다. 만약 개를 '앉도록' 훈련시키는 어떤 의도도 기억하지 못한다면, 그것과 시간을 보낼 때 무엇을 보게 될까? 늑대 무리와 시간을 보낸다면 보게 될 것과 매우 비슷한 것을 보게 될 것이다. 짜증을 표현하거나(아마도 그냥 일어나서 방의 다른 장소로 이동하는 것), 사회적 상호 작용에 대한 욕구를 표현하는 것(여러분의 손을 핥거나 여러분 옆에 눕는 것)과 같은 행동들이다. 단어들이나 문장들이 아니라, 이러한 개념들이 동물들의 진정한 어휘이며, 동물들과 더 많은 시간을 보낼수록, 이 진정한 어휘를 더 많이 알아차릴 것이다.

문제 해설
개와 고양이 같은 반려동물과의 진정한 의사소통은 인간의 말을 가르치는 것이 아니라, 그들의 자연스러운 행동과 감정을 관찰하고 상호 작용하는 것이라고 했으므로, 빈칸에 들어갈 말로 가장 적절한 것은 ① '그것들 안에 있는 동물의 본성을 관찰하는 것'이다.
② 그것들을 인간으로 인식하는 것
③ 그것들의 야생성을 복원하고 시험하는 것
④ 그것들의 지적 잠재력을 극대화하는 것
⑤ 그것들이 비언어적 신호를 이해하게 만드는 것

■ ~ and **the more** you spend time with your animals, **the more** of this true vocabulary you will notice.

「the + 비교급(~), the + 비교급(...)」 구문이 사용되어 '~할수록 더욱 …하다'의 의미를 나타낸다.

어휘 및 어구

domestication 가축화
beyond recognition 알아볼 수 없을 정도로
ancestor 조상
communication 의사소통
intention 목적, 의도
annoyance 짜증
lick 핥다

16

정답 | ④

소재 공포 반응의 세대 간 전이

해석 Dias와 Ressler는 생쥐 수컷에게 벚꽃나무 향기를 두려워하도록 훈련시키는 생쥐 실험을 수행했다. 벚꽃 향기를 접하게 할 때마다 그들은 생쥐들에게 전기 충격을 가했다. 3일 후, 그들은 그 수컷 생쥐들을 암컷 생쥐들과 교배시켰다. 그 결과로 생긴 새끼 생쥐들은 성체가 될 때까지 전기 충격을 받거나 벚꽃 향기를 접하지 않았다. 그들이 마침내 새끼 생쥐들에게 벚꽃 향기를 접하게 하자, 새끼 생쥐들은 몸을 떨고 다른 불안 증상을 보이기 시작했다. 또한, 이 새끼 생쥐들은 공기 중의 미묘한 벚꽃의 자취도 감지할 수 있었으며, 이것이 뇌 구조를 바꾸어 디지털 방식으로 관찰했을 때 특정 영역이 활성화되었다. 이러한 똑같은 행동이 전기 충격을 받지 않았음에도 불구하고 그 생쥐의 다음 세대에서도 관찰되었다. 이를 통해 연구자들은 우리가 인간으로서 어떻게 <u>충격적인 상황과 사건에 대한 우리 조상들의 정서적 반응을 가지고 있</u>는지 이해할 수 있었다.

문제 해설

생쥐에게 벚꽃 향기를 두려워하도록 훈련시키는 실험을 진행한 후 태어난 새끼 생쥐들은 전기 충격이나 벚꽃 향기를 경험하지 않았음에도 성체가 되었을 때 벚꽃 향기에 불안 증상을 보였는데, 이러한 행동은 전기 충격을 받지 않은 다음 세대에서도 계속 관찰되었다고 했으므로, 빈칸에 들어갈 말로 가장 적절한 것은 ④ '충격적인 상황과 사건에 대한 우리 조상들의 정서적 반응을 가지고'이다.

① 다른 동물들과 비교적 제한된 진화적 유사성을 공유하고
② 특정 질병을 해결하기 위해 특화된 치료법을 고안하고
③ 우리 유전자 내의 놀라운 능력을 드러냄으로써 진화하고 적응하고

⑤ 유전이 아닌 환경적 영향을 통해 정서적 특성을 발달시키고

구조 해설

■ [When they eventually introduced the offspring to the scent of cherry blossoms], the pups started [shaking] and [displaying other symptoms of anxiety].

첫 번째 [ ]는 시간의 의미를 나타내는 부사절이고, 두 번째와 세 번째 [ ]는 and로 연결되어 started의 목적어 역할을 한다.

어휘 및 어구

conduct (업무 등을) 수행하다, 처리하다
experiment 실험
cherry blossom 벚꽃
mate 교배하다
resultant (앞에 언급한) 그 결과로 생긴
offspring 새끼, 자식
subtle 미묘한

17

정답 | ②

소재 대상 위치에 대한 정보

해석 많은 경우, (목표로 하는) 대상의 위치에 대한 정보는 우리의 장기 기억에 저장된다. 그런 정보는 대상과 보는 사람의 특정 위치와의 연관성보다는 대상과 특정 랜드마크와의 연관성과 관련지어질 가능성이 더 높다. 예를 들어, 열쇠의 빈번한 위치는 현관 근처의 서랍장 위일 수도 있다. 그러나 보는 사람은 아마도 서랍장과 관련된 습관적인 위치를 가지고 있지 않기 때문에, 열쇠를 얻는 것은 그 서랍장과 관련된 자신의 위치뿐만 아니라 서랍장과 관련된 열쇠의 위치도 아는 것을 필요로 한다. 물론, 어떤 경우에는 사람이 특정 공간 상황에서 랜드마크와 관련된 습관적인 위치를 실제로 가지고 있으며, 그 위치가 대상물과 직접적으로 연관되어 있다. 예를 들어, 부엌에서 요리사는 요리용 화덕과 관련된 습관적인 위치를 가지고 있을 수도 있으며, 조미료는 그 요리용 화덕과 관련된 특정 위치에 있을 수도 있다. 이 상황에서는 <u>자기 중심적 코드화로도 족할 것이며</u>, 요리용 화덕을 통한 조미료와 자신과의 관계에 대한 추론은 필요하지 않을 것이다.

문제 해설

대상 위치에 대한 정보는 주로 랜드마크와 연관되어 장기 기억에 저장되며, 그 예로 열쇠는 현관 근처 서랍장 위에 있을 수 있지만, 사람은 서랍장에 대한 습관적인 위치를 가지고 있지 않다고 하였다. 그러나 부엌과 같은 특정 공간 상황에서는 요리사가 화덕의 위치를 습관적으로 알고 있어 조미료와 같은 대상물과 화덕과 같은 랜드마크 사이의 관계를 별도로 추론할 필요가 없다. 즉, 개인의 위치와 랜드마크의 위치를 기준으로 삼지 않아도 직접적으로 대상물을 찾을 수 있다는 내용이므로, 빈칸에 들어갈 말로 가장

적절한 것은 ② '자기 중심적 코드화로도 족할 것이며'이다.
① 보는 사람의 위치가 변경될 수도 있으며
③ 특정 랜드마크의 위치가 주어지며
④ 습관적인 위치가 단기 기억에 저장되며
⑤ 공간 배치를 아는 것으로는 불충분할 수도 있으며

구조 해설

- ~, so obtaining the keys involves [knowing the position of the keys relative to the chest **as well as** one's own position relative to that chest].
  [ ]는 동사 involves의 목적어로 쓰인 동명사구로 그 안의 「~ as well as ...」은 '…뿐만 아니라 ~도'라는 의미이다.

어휘 및 어구

location 위치
long-term memory 장기 기억
involve 관련시키다, 필요로 하다
association 연관성
chest of drawers 서랍장, 옷장
habitual 습관적인
relative to ~에 관련된
spatial 공간의, 공간적인
inference 추론
ego-centered 자기 중심적인
sufficient (~하기에) 족한, 충분한

18

정답 | ④

소재 불황 속 마케팅 투자와 기업 생존 전략

해석 힘든 시기와 변화하는 시장 상황은 기업의 성과와 생존에 심각하게 영향을 미칠 수 있다. 일부 기업들은 힘든 시기를 자신들의 사업을 강화할 기회로 간주한다. 따라서 그들은 마케팅 활동, 연구 및 개발, 그리고 신제품 개발에 공격적으로 투자함으로써 자신들의 경쟁업체에 대한 우위를 확보하려고 노력하는데, 반면에 일부 기업들은 마케팅 지출을 줄이고 불황이 지나가기를 기다린다. 어려운 시기에 불황에 대한 기업들의 가장 빈번한 대응은 경제가 회복될 때까지 생존하기 위해 마케팅 투자에 대한 경비를 삭감하는 것이다. 광고와 커뮤니케이션 같은 마케팅 비용을 줄이는 것은 단기적으로 기업에 안도감을 줄 수도 있다. (불황 동안 소셜 미디어를 광고에 활용하면 고객의 독립성을 향상시켜 그들이 제품 구매에 자유 의지를 행사하게 할 수 있다.) 그러나 장기적으로 기업들이 자신들의 고객 기반과 시장 점유율을 잃을 위험을 무릅쓸 수도 있다.

문제 해설

힘든 시기에 일부 기업은 투자를 늘려 경쟁 우위를 노리고, 일부

는 마케팅 지출을 줄여 불황이 지나가기를 기다리지만, 이는 장기적으로는 고객 기반과 시장 점유율을 잃을 위험이 있다는 내용의 글이다. 불황 동안 소셜 미디어를 광고에 활용하면 고객의 독립성을 향상시켜 그들이 제품 구매에 자유 의지를 행사할 수 있게 할 수 있다는 내용인 ④는 글의 전체 흐름과 관계가 없다.

구조 해설

- Some enterprises **view** tough times **as** an opportunity [to strengthen their businesses].
  「view ~ as ...」는 '~을 …으로 간주하다'라는 의미이다. [ ]는 an opportunity를 수식하는 to부정사구이다.

어휘 및 어구

severely 심각하게
survival 생존
enterprise 기업
strengthen 강화하다
aggressively 공격적으로
expense 비용
free will 자유 의지
risk (~의) 위험을 무릅쓰다, (~을) 각오해야 할 짓을 하다
market share 시장 점유율

19

정답 | ②

소재 자율 주행 차량과 프로그래밍

해석 사고를 당한 자율 주행차의 경우, 우리는 다음 현상을 다루게 되는데, 즉 사고 직전의 상황에서 더 이상 결정을 내릴 수 없다는 것이다. (B) 자율 주행차의 행동에 대한 결정은 그 프로그래밍에 대한 결정이 내려졌을 때 이루어졌다. 이것은 적절한 법적 규제를 만드는 것과 제조업체부터 개별 프로그래머에 이르기까지 그것의 이행 둘 다를 포함하는 아주 긴 과정이 될 수 있다. 현재 특정 상황에 특정 도덕 이론을 적용하도록 기계를 프로그래밍하려는 시도 외에도 인간의 판단을 최대한 모방하는 것을 목표로 하는 시도도 있다. (A) 그러나 이것이 자율 주행 차량이 '도덕적 주체'로의 지위를 획득하는 것으로 이어지지는 않을 것이다. 그것의 행동은 진정한 의사 결정의 결과라는 의미에서의 행동으로 간주되지 않을 것이다. 자율 주행 차량은 단지 그것의 소프트웨어에 프로그래밍된 규칙을 이행할 뿐이다. (C) 이는 자가 학습 인공 지능의 형태들이 사용될 때도 마찬가지이다. 여기에서도 인간이 훈련의 예시들을 선택하고 각 경우에 올바른 답이 무엇인지를 결정할 것이다. 그들은 프로그램이 무엇을 '학습해야' 하고 언제 충분히 '학습했는지' 결정한다.

문제 해설

자율 주행차는 사고 직전에 더 이상 결정을 내릴 수 없다는 내용의

Mini Test

주어진 글 다음에 자율 주행차의 행동 결정이 프로그래밍 단계에서 이루어졌으며, 이는 적절한 법적 규제를 만드는 것과 제조업체 및 개별 프로그래머까지 규제의 이행을 포함하는 긴 과정이 될 수 있고, 심지어 인간의 판단을 최대한 모방하려는 시도도 있었다는 내용의 (B)가 오고, 이러한 시도에도 자율 주행 차량이 '도덕적 주체'로 간주되지는 않을 것이라는 내용의 (A)가 이어져야 한다. 그런 다음 (A)에서 설명한 자율 주행 차량의 제한점처럼 자가 학습 인공지능 형태를 사용하는 경우에도 마찬가지라고 설명하는 내용의 (C)로 이어지는 것이 자연스럽다. 따라서 주어진 글 다음에 이어질 글의 순서로 가장 적절한 것은 ② '(B)-(A)-(C)'이다.

(구조 해설)

■ The decision [about the behavior of an autonomous car] was made [when a decision was made about its programming].
첫 번째 [ ]는 The decision을 수식하는 전치사구이고, 두 번째 [ ]는 시간을 나타내는 부사절이다.

(어휘 및 어구)

vehicle 차량
status 지위
moral agent 도덕적 주체
genuine 진정한
implement 이행하다
lengthy 아주 긴
legal 법률의
regulation 규제
attempt 시도; 시도하다

20
정답 | ②

(소재) 다문화주의와 상호 문화주의

(해석) 다문화주의는 사회가 독특한 문화적, 종교적 집단으로 구성되거나 적어도 (그러한 집단을) 공정하게 허용하고 포함해야 한다고 주장하는 이데올로기이다. 일부 국가들은 이민자 집단의 문화적 정체성을 보존하는 것을 목표로 한 공식적인 다문화주의 정책을 가지고 있다. (B) 이러한 상황에서 다문화주의는 하나의 문화가 우위를 차지하지 않고, 독특한 문화적, 종교적 집단에 공정성을 베푸는 사회를 옹호한다. 그러나 이 용어는 지배적인 토착 문화와 나란히 존재하는 소수 이민자 문화로 구성된 사회를 묘사하는 데 더 자주 사용된다. 흔히 다문화주의는 '상호 문화주의'라는 용어와 상호 호환적으로 사용된다. (A) 어떤 용어가 사용되는지는 사람들이 어떤 언어를 사용하는지에 따라 다르다. 예를 들어, 영어를 사용하는 유럽 연구자들은 보통 다문화라는 용어를 사용하는 반면, 비영어권 연구자들은 상호 문화라는 용어를 사용한다. 또

한 다문화는 그 구성원이 서로 다른 민족적, 종교적 집단 출신인 사회의 성격을 묘사하는 반면, 상호 문화는 그들의 상호 작용, 협상 및 과정을 묘사한다는 주장도 있다. (C) 또 다른 견해는 상호 문화는 문화적으로 다른 두 개 집단의 사람들을 지칭하고, 다문화는 문화적으로 다른 두 개가 넘는 집단의 사람들을 지칭한다는 것이다. 따라서 다수의 문화를 언급할 때는 다문화라는 용어가 적합하다.

(문제 해설)

다문화주의는 사회가 독특한 문화적, 종교적 집단을 공정하게 허용하고 포용해야 한다고 주장하는 이데올로기라는 주어진 글 다음에 다문화주의는 특정 문화가 지배하지 않고, 독특한 문화적, 종교적 집단에 공정성을 베푸는 사회를 옹호하며, 다문화주의와 상호 문화주의 용어가 혼용된다는 점을 언급하는 내용의 (B)가 오고, 두 용어 중 어떤 용어가 사용되는지는 언어에 따라 다르다는 점을 설명하는 (A)로 이어진 다음, (A)의 후반부에서 다문화와 상호 문화의 차이를 설명하는 내용에 이어, 다문화와 상호 문화에 대한 또 다른 견해를 언급한 내용의 (C)로 이어지는 흐름이 자연스럽다. 따라서 주어진 글 다음에 이어질 글의 순서로 가장 적절한 것은 ② '(B)-(A)-(C)'이다.

(구조 해설)

■ However, the term is more commonly used [to describe a society consisting of minority immigrant cultures {existing alongside a predominant, indigenous culture}].
[ ]는 목적을 나타내는 to부정사구이며, 그 안의 { }는 minority immigrant cultures를 수식하는 분사구이다.

(어휘 및 어구)

multiculturalism 다문화주의
ideology 이데올로기, 이념
advocate 주장하다, 옹호하다
fairness 공정성, 공정함
preserve 보존하다
immigrant 이민자
intercultural 상호 문화의, 문화 간의
negotiation 협상
minority 소수의
interchangeably 상호 호환적으로
multiple 다수의

21
정답 | ④

(소재) 과학에서의 과거에 대한 현재의 우월감

(해석) 과학의 역사를 살펴보면 과학이라고 불리는 이것은 무엇인지에 관한 질문의 답은 과학이라고 불리는 단일한 것이 없다는

것을 시사한다. 과학은 확실히 통합되고 계속 이어지는 믿음의 집합체가 아니다. 그것은 단 하나의 과학적 방법으로 정확히 담아낼 수도 없다. 만약 우리가 과학이 무엇인지 이해하고 싶다면, 즉 현재와 과거 모두에서 지식이 생산되어 온 다양한 방식, 그리고 서로 다른 민족과 문화가 자연 세계에 대해 가지고 있는 믿음을 지니게 된 방식을 이해하고 싶다면, 우리는 그저 역사를 살펴볼 수밖에 없다. 이것을 성공적으로 수행하려면, 과학의 역사에서 흔히 특히 만연한, 과거에 대한 현대의 그 우월감을 극복해야 한다. 분명한 것은 우리의 현재 지식에 비추어 볼 때, 과거 사람들이 알고 있다고 생각했던 것의 대부분이 틀렸다는 것이다. 하지만 같은 비관적인 귀납적 추정에 의해 우리가 현재 알고 있다고 생각하는 것의 대부분도 미래의 기준으로는 틀린 것으로 판명될 것이라는 점은 기억할 가치가 있다. 과학의 역사는 과거의 승자와 패자에게 상벌을 주는 게임이 되어서는 안 된다.

(문제 해설)

과학은 통합되고 계속 이어지는 믿음의 집합체가 아니며 과학의 역사를 살펴보는 과정을 통해 과학이 무엇인지를 이해할 수 있다는 글이다. ④ 바로 앞 문장의 내용은 과학이 무엇인지 이해하기 위해서는 과거에 대한 현대의 우월감을 극복해야 한다는 것이다. ④ 바로 뒤 문장은 우리가 현재 알고 있다고 생각하는 것의 대부분도 미래의 기준으로는 틀린 것으로 판명될 것이라는 내용이다. 주어진 문장은 우리의 현재 지식에 비추어 볼 때 과거 사람들이 알고 있다고 생각했던 것의 대부분은 틀렸다는 내용으로 ④에 주어진 문장이 들어가야 내용상 단절이 해소된다. 따라서 주어진 문장이 들어가기에 가장 적절한 곳은 ④이다.

(구조 해설)

- [Looking at science's history] suggests [that {the answer to the question, ⟨what is this thing called science⟩}, is {that there is no single thing called science}].

첫 번째 [ ]는 문장의 주어 역할을 하는 동명사구이다. 두 번째 [ ]는 suggests의 목적어 역할을 하는 명사절이다. 첫 번째 { }는 명사절의 주어인데, 그 안의 ⟨ ⟩는 the question과 동격 관계이다. 두 번째 { }는 주격 보어로 사용된 명사절이다.

(어휘 및 어구)

current 현재의
continuous 계속 이어지는, 지속적인
capture 정확히 담아내다
overcome 극복하다
modern 현대의
particularly 특히
prevalent 만연한, 널리 퍼진
pessimistic 비관적인

induction 귀납적 추정

22
정답 | ②

(소재) 저작권 보호의 자격

(해석) 일반적으로 작품은 '독창적인' 경우, 즉 작가나 예술가에 의한 창의적인 의사 결정의 표시를 포함하는 경우에 저작권 보호를 받을 자격이 있다. 이름의 알파벳순 목록과 같이 가공되지 않았거나 흔한 것은 저작권 보호를 받을 자격이 없을 것이다. 하지만 이름의 알파벳순 목록을 가사로 사용한 노래 멜로디는 저작권 보호를 받을 자격이 있을 것이다. 미국 저작권법을 인용하자면, 저작권 보호의 자격을 얻으려면 작품은 '유형의 표현 매체에 고정'되어야 한다. 이것은 내가 사람들이 있는 데서 일어나 허공에 시를 낭송하면 저작권 보호를 받지 못한다는 것을 의미한다. 하지만 만약에 내가 허공에 시를 낭송하는 소리를 녹음하거나 키보드로 컴퓨터 하드 드라이브에 시를 입력하면 그것은 즉시 보호를 받는다. 내 컴퓨터 하드 드라이브는 '유형의 표현 매체'이다. 필름, 사진 용지, 콘크리트, 천, 그리고 조각가 Richard Serra가 자신의 조각품을 만드는 데 사용하는 거대한 강철 평판도 또한 마찬가지이다.

(문제 해설)

주어진 문장은 작품이 저작권 보호의 자격을 얻기 위해서는, 미국 저작권을 인용하자면, '유형의 표현 매체에 고정'되어야 한다는 내용이다. ② 앞 문장에는 이름의 알파벳순 목록을 가사로 사용한 노래 멜로디는 저작권 보호의 자격을 얻는다고 하여 알파벳이 독창적으로 활용되었을 때 저작권 보호에 관한 내용이 있고 ② 바로 뒤 문장은 This(이것은)로 시작하여 허공에 시를 낭송하는 것과 같이 유형의 매체에 고정되어 있지 않으면 저작권 보호를 받지 못한다는 내용을 설명한다. 따라서 유형의 매체에 고정되는 것에 관해 논하는 주어진 문장이 ②에 들어가면 내용상 단절이 해소되므로, 주어진 문장이 들어가기에 가장 적절한 곳은 ②이다.

(구조 해설)

- **So are** [film, photographic paper, concrete, cloth, and the huge slabs of steel {that the sculptor Richard Serra uses to make his sculptures}].

'~도 또한 마찬가지다'라는 의미를 나타내는 「so+are+주어」가 사용되었으며, [ ]가 주어이다. { }는 the huge slabs of steel을 수식하는 관계절이다.

(어휘 및 어구)

medium 매체
quote 인용하다
copyright 저작권
qualify for ~에 대한 자격을 얻다

original 독창적인
marker 표시
raw 가공되지 않은
lyrics 가사
steel 강철
sculpture 조각품

23

정답 | ②

소재 얼굴 표정을 통한 타인의 감정 인지

해석 여러분은 어떤 사람들이 얼굴 표정을 통해 타인의 감정을 잘 인지하는 한 이유가 그들이 보통 자기 얼굴에 감정을 분명히 드러내기 때문이라고 생각할 수도 있을 것이다. 하지만 놀랍게도, 이러한 생각과 관련된 연구 결과는 엇갈린다. 이용 가능한 증거를 검토한 결과, 감정을 읽는 것과 표현하는 것은 둘 다 서로 관련이 있지만, 개인이 자신의 감정을 다른 사람에게 '전달'하려고 노력하고' 있을 때만 그런 것으로 나타난다. 예컨대 갑자기 멋진 일이 일어날 때의 기쁨의 표현과 같이 이러한 표현이 저절로 발생할 때, 그런 경우에는 다른 사람의 얼굴 표정을 인식할 수 있는 것과 그러한 단서를 스스로 명확하게 표시하는 것은 서로 관련이 '없다'. 다시 말해, 자신의 감정을 솔직하고 쉽게 표현하는 사람들이 타인의 얼굴 표정을 인지하는 데 있어서 반드시 정확하지는 않다. 정확성은 자신의 얼굴 표정에 자신의 감정을 보여 주는 것에 관한 사람들의 의도적인 집중과 관련이 있는 것으로 보인다. 아마도 그렇게 함으로써 그들은 타인의 표정의 본질에 대한 더 큰 통찰을 얻는 듯하며, 이것이 그들이 타인의 기저에 있는 감정을 더 정확하게 인지하는 데 도움을 줄 것이다.

→ 연구 결과는 자신의 감정을 얼굴 표정을 통해 명확하게 전달하는 우리의 경향이 반드시 타인의 감정을 정확하게 읽는 것을 보장하지 않는데, 이는 (감정) 읽기의 정확성이 우리의 의도적인 감정 표현과 관련이 있는 것 같기 때문이라는 점을 보여 준다.

문제 해설
자기 얼굴에 감정을 분명히 드러내는 것과 타인의 얼굴 표정을 통해 타인의 감정을 인지하는 능력은 반드시 관련이 있지는 않으며, 타인의 감정을 인지하는 능력의 정확성은 자신의 감정을 얼굴 표정을 통해 의도적으로 보여 주는 것과 관련이 있는 것으로 보인다는 내용의 글이므로, 요약문의 빈칸 (A), (B)에 들어갈 말로 가장 적절한 것은 ② '보장하지(guarantee) - 의도적인(deliberate)'이다.
① 보장하지 - 사적인
③ 방지하지 - 무심코 한
④ 발생시키지 - 빈번한
⑤ 방지하지 - 과장된

구조 해설

■ A review of available evidence indicates [that {both reading and expressing emotions} are related but {only when individuals are *trying* to communicate their feelings to others}].

[ ]는 indicates의 목적어 역할을 하는 명사절이다. 첫 번째 { }는 명사절의 주어이고, 두 번째 { }는 시간의 부사절이다.

어휘 및 어구
recognize 인지하다, 인식하다
typically 보통, 일반적으로
mixed 엇갈리는, 뒤섞인
review 검토
indicate 나타내다, 시사하다
spontaneously 저절로, 자연스럽게
cue 단서, 신호
accuracy 정확성
be tied to ~과 관련이 있다
intentional 의도적인
insight 통찰(력), 이해
underlying 기저에 있는, 근본적인

24~25

정답 | 24 ⑤ 25 ⑤

소재 포유류의 몸집과 수명

해석 포유류를 보면 일반적으로 동물이 클수록 수명이 더 길다. 이것은 진화론적으로 이치에 맞는다. 작은 동물은 포식자에게 더 취약하고, 늙어서 죽기 오래 전에 잡아먹힐 것이라면 긴 수명을 갖는 것이 의미가 없을 것이다. 그러나 크기와 수명의 관계에 대한 더 근본적인 이유는 크기가 대사율과 관련이 있기 때문인데, 그것은 대략 동물이 기능을 하는 데 필요한 에너지를 마련하기 위해 먹이 형태의 연료를 연소하는 속도이다. 작은 포유류는 크기에 비해 표면적이 더 넓어서 열을 더 쉽게 잃는다. 이를 보상하기 위해 그것들은 더 많은 열을 발생시켜야 하는데, 이는 더 높은 대사율을 유지하고 체중에 비해 더 많이 먹어야 한다는 것을 의미한다. 이는 동물이 시간당 연소하는 총칼로리 수는 동물의 질량보다 더 느리게 증가함을 의미한다. 몸집이 10배 큰 동물은 시간당 4~5배 많은 칼로리만 연소한다. 따라서 몸무게에 비해 더 작은 동물은 더 큰 동물보다 칼로리를 덜(→ 더) 연소한다. 동물의 칼로리 연소 속도와 질량 사이의 관계는 Max Kleiber의 이름을 딴 Kleiber의 법칙이라고 이름 붙여졌는데, 그는 1930년대에 동물의 대사율이 질량의 3/4제곱에 비례한다는 사실을 밝혀냈다. 정확한 제곱은 논란의 여지가 있으며 일부는 포유류의 경우 2/3제곱이 데이터에 더 잘 맞는다는 것을 보여 준다.

문제 해설

24 동물의 크기가 작으면 표면적이 넓어 먹이를 더 많이 섭취해야 하는 반면, 동물의 몸집이 크면 그렇지 않다는 내용의 글이다. 따라서 글의 제목으로 가장 적절한 것은 ⑤ '동물의 수명에 있어서 크기는 중요하다'이다.
① 소형 포유류의 생존 전략
② 크기 변이: 유전적 다양성의 증거
③ 수명: 성공적인 적응의 지표
④ 다양한 음식 종류, 다양한 칼로리 양

25 동물의 몸집과 수명에 관한 글로 시간당 연소하는 총칼로리 수는 동물의 질량보다 덜 느리게 증가한다고 했다. 즉, 몸집이 클수록 연소하는 칼로리 수는 작은 배율로 증가하고, 몸집이 작을수록 연소하는 칼로리 수는 큰 배율로 증가한다는 의미이므로, (e)의 fewer를 more와 같은 낱말로 바꿔야 한다.

구조 해설

■ [The relationship **between** how fast an animal burns calories **and** its mass] is named Kleiber's law after Max Kleiber, [who showed in the 1930s {that an animal's metabolic rate scales to the 3/4 power of its mass}].

첫 번째 [ ]는 문장의 주어이고, 핵심어는 relationship이며, 'A와 B 사이의'라는 의미를 가진 「between A and B」의 구조를 사용하였다. 두 번째 [ ]는 Max Kleiber에 대해 부가적으로 설명하는 관계절이다. { }는 showed의 목적어 역할을 하는 명사절이다.

어휘 및 어구

mammal 포유류
life span 수명
vulnerable 취약한
predator 포식자
fundamental 근본적인
metabolic rate 대사율
surface area 표면적
compensate 보상하다
generate 발생시키다
mass 질량

26~28

정답 | 26 ④ 27 ④ 28 ⑤

소재 Larry Jordan과 그의 남동생 Michael Jordan

해석 (A) Larry의 남동생은 충족시켜야 할 것이 많았다. Larry는 가족 중 스타 농구 선수였다. 비록 5피트 8인치밖에 되지 않았지만, Larry는 놀랄 만한 능력을 갖추고 있었다. 대학 스카우트들은 입을 떡 벌리게 만드는 그의 경기력에 놀랐다. Larry의 남동생은 한 살 어렸고, 훨씬 더 키가 작았으며, 그 정도로 잘하지는 못했다. Larry는 그들의 끝없는 호스 게임(농구 슈팅 게임)과 일대일 게임에서 항상 이겼고, 그것은 영웅을 숭배하는 그의 남동생에게 괜찮았다. 하지만 그것은 또한 그 아이가 자신의 경기 실력을 향상시키기 위해 자기 자신을 더 강하게 밀어붙이도록 했다.

(D) 그 마른 소년이 2학년이 되기 바로 직전에, 코치는 그에게 농구 캠프에서 형과 함께하도록 권했다. 그는 Larry의 DNA가 그의 남동생에게 틀림없이 있을 것으로 생각했다. 코치는 실망했다. 그는 그 아이의 다듬어지지 않은 속도와 발달 중인 기술에 감탄했지만, 그는 자신의 나머지 가족들처럼 여전히 키가 작을 운명이라고 판단했다. 코치가 대표팀 명단을 발표했을 때, 6피트 혹은 그보다 더 큰 그 아이의 친구들은 모두 선발되었다. 그러나 Larry의 동생은 잘렸다.

(B) 그는 망연자실했지만, 이 실패로 인해 그는 그 다음 해까지 더 열심히 노력하였으며, 그런 다음 그는 Larry가 코트에서 주인공이 되는 동안 벤치를 지키는 특권을 얻었다. 그러고 나서 Larry의 남동생은 자신의 2학년과 3학년 사이에 기적적으로 5인치가 자랐다. 그는 이어서 McDonald 고등학교의 전 미국 대표선수가 되었고, 그의 대학팀을 NCAA 전국 챔피언십으로, 그의 시카고 불스를 6개의 NBA 타이틀로, 그리고 미국 올림픽 팀을 금메달로 이끌었다. 그는 어떤 다른 선수보다 더 많은 개인 기록을 세웠다. Larry의 남동생은 지금 13억 1천만 달러의 가치가 있었다.

(C) 우리는 때때로 Larry Jordan의 남동생도 코트 안팎에서 자기 몫의 실수를 했다는 것을 잊는다. Michael이 CBS 인터뷰에서 다음과 같이 고백한 것은 유명하다. "저는 제 경력에서 9천 개가 넘는 슛을 놓쳤습니다. 저는 거의 300번의 경기에서 졌습니다. 26번이나 저는 경기의 승부를 결정짓는 슛을 하도록 맡겨졌지만 놓쳤습니다." 그는 또한 키로 따져보면 자신이 가족 내에서 두 번째로 잘하는 선수라고 으레 말한다. 그는 "여러분이 '에어 조던'이라고 할 때, 저는 2위이고, 그가 1위입니다."라고 말한다.

문제 해설

26 가족 중 스타 농구 선수였던 Larry는 그의 남동생의 숭배를 받았고, Larry의 남동생은 자신의 경기를 개선하기 위해 더 강하게 자신을 밀어붙였다는 내용의 글 (A) 다음으로 Larry의 DNA가 그의 남동생에게도 있을 것으로 생각한 코치가 Larry의 남동생을 농구 캠프로 불렀으나 키가 자라지 않을 것으로 판단하여 대표팀에 그를 선발하지 않았다는 내용의 (D)가 이어지고, 이로 인해 Larry의 남동생이 망연자실했지만, 더 열심히 노력하여 엄청난 가치가 있는 선수로 성장했다는 내용의 (B)가 이어진 다음, 엄청난 가치가 있는 선수였지만, 그도 역시 자기 몫의 실수를 했으며, 자신은 가족 중에서 두 번째로 가장 훌륭한 선수라고 말한

Larry의 남동생 Michael의 말을 소개한 (C)로 이어지는 것이 가장 적절하다.

27 (d)는 Larry를 가리키고, 나머지 넷은 Larry의 동생을 가리킨다.

28 코치가 대표팀 명단을 발표했을 때, 6피트 혹은 그보다 더 큰 그 아이의 친구들은 모두 선발되었지만, Larry의 동생은 잘렸다고 했으므로, 글에 관한 내용으로 적절하지 않은 것은 ⑤이다.

구조 해설

- He [admired the kid's raw speed and developing skills] but [figured {that he was destined to remain small like the rest of his family}].

  두 개의 [ ]가 but으로 연결되어 문장의 술어를 이루고 있다. 두 번째 [ ] 안의 { }는 figured의 목적어 역할을 하는 명사절이다.

- When the coach announced the varsity roster, all of the kid's friends [who were six feet or taller] were chosen.

  [ ]는 all of the kid's friends를 수식하는 관계절이다.

어휘 및 어구

live up to (기대, 요구 등) ~을 충족시키다, ~에 부응하다
miraculous 놀랄 만한
scout 스카우트, 신인을 발굴하는 사람
jaw-dropping (놀라서) 입을 떡 벌리게 만드는
hero-worshiping 영웅을 숭배하는
devastated 망연자실한
privilege 특권
junior 3학년(의)
cool (돈의 액수가 많음을 강조하여) 거금의
confess 고백하다
destined to *do* ~할 운명인

고2~N수, 수능 집중

구분	수능 입문 >	기출/연습 >	연계 + 연계 보완	>	고난도 >	모의고사

국어
- 윤혜정의 개념/패턴의 나비효과
- 윤혜정의 기출의 나비효과
- 수능특강 문학 연계 기출
- 수능특강 사용설명서
- 하루 3개 1등급 국어독서
- FINAL 실전모의고사

영어
- 기본서 수능 빌드업
- 수능특강 Light
- 강의노트 수능개념
- 수능 기출의 미래
- 수능연계교재의 VOCA 1800
- 수능연계 기출 Vaccine VOCA 2200
- 수능 영어 간접연계 서치라이트
- 수능완성 사용설명서
- 하루 6개 1등급 영어독해
- 수능연계완성 3주 특강
- 만점마무리 봉투모의고사 시즌1
- 만점마무리 봉투모의고사 시즌2

수학
- 수능 감(感)잡기
- 수능 기출의 미래 미니모의고사

한국사 사회 / 과학
- 수능 스타트
- 수능특강Q 미니모의고사

수능 연계교재
- 감수 수능특강 | 감수 수능완성

eBook 전용
- 수능완성R 모의고사 | 수능 등급을 올리는 변별 문항 공략

- 박봄의 사회·문화 표 분석의 패턴
- 만점마무리 봉투모의고사 고난도 Hyper
- 수능 직전보강 클리어 봉투모의고사

구분	시리즈명	특징	난이도	영역
수능 입문	윤혜정의 개념/패턴의 나비효과	윤혜정 선생님과 함께하는 수능 국어 개념/패턴 학습		국어
	수능 빌드업	개념부터 문항까지 한 권으로 시작하는 수능 특화 기본서		국/수/영
	수능 스타트	2028학년도 수능 예시 문항 분석과 문항 연습		사/과
	수능 감(感) 잡기	동일 소재·유형의 내신과 수능 문항 비교로 수능 입문		국/수/영
	수능특강 Light	수능 연계교재 학습 전 가볍게 시작하는 수능 도전		영어
	수능개념	EBS*i* 대표 강사들과 함께하는 수능 개념 다지기		전 영역
기출/연습	윤혜정의 기출의 나비효과	윤혜정 선생님과 함께하는 까다로운 국어 기출 완전 정복		국어
	수능 기출의 미래	올해 수능에 딱 필요한 문제만 선별한 기출문제집		전 영역
	수능 기출의 미래 미니모의고사	부담 없는 실전 훈련을 위한 기출 미니모의고사		국/수/영
	수능특강Q 미니모의고사	매일 15분 연계교재 우수문항 풀이 미니모의고사		국/수/영/사/과
	수능완성R 모의고사	과년도 수능 연계교재 수능완성 실전편 수록		수학
연계 + 연계 보완	수능특강	최신 수능 경향과 기출 유형을 반영한 종합 개념 학습		전 영역
	수능특강 사용설명서	수능 연계교재 수능특강의 국어·영어 지문 분석		국/영
	수능특강 문학 연계 기출	수능특강 수록 작품과 연관된 기출문제 학습		국어
	수능완성	유형·테마 학습 후 실전 모의고사로 문항 연습		전 영역
	수능완성 사용설명서	수능 연계교재 수능완성의 국어·영어 지문 분석		국/영
	수능 영어 간접연계 서치라이트	출제 가능성이 높은 핵심 간접연계 대비		영어
	수능연계교재의 VOCA 1800	수능특강과 수능완성의 필수 중요 어휘 1800개 수록		영어
	수능연계 기출 Vaccine VOCA 2200	수능 - EBS 연계와 평가원 최다 빈출 어휘 선별 수록		영어
고난도	하루 N개 1등급 국어독서/영어독해	매일 꾸준한 기출문제 학습으로 완성하는 1등급 실력		국/영
	수능연계완성 3주 특강	단기간에 끝내는 수능 1등급 변별 문항 대비		국/수/영
	박봄의 사회·문화 표 분석의 패턴	박봄 선생님과 사회·문화 표 분석 문항의 패턴 연습		사회탐구
	수능 등급을 올리는 변별 문항 공략	EBS*i* 선생님이 직접 선별한 고변별 문항 연습		수/영
모의고사	FINAL 실전모의고사	EBS 모의고사 중 최다 분량 최다 과목 모의고사		전 영역
	만점마무리 봉투모의고사 시즌1/시즌2	실제 시험지 형태와 OMR 카드로 실전 연습 모의고사		전 영역
	만점마무리 봉투모의고사 고난도 Hyper	고난도 문항까지 국·수·영 논스톱 훈련 모의고사		국·수·영
	수능 직전보강 클리어 봉투모의고사	수능 직전 성적을 끌어올리는 마지막 모의고사		국/수/영/사/과

memo

memo

memo

첫 수업부터 실전이다!

취업률 1위
전국여자대학교
4년제 포함 5년연속

(2017~2021,4년제 포함 대학알리미
졸업생 1000명 이상 2000명 미만)

반값 등록금 수준의
장학금 지급

· 1인당 평균 325만원
· 전체 장학금 지급 금액
  약 132억원
  (2023년 대학알리미 기준)

편리한 교통

· 1호선 수원역 스쿨버스 상시 운행
· 수인선 오목천(수원여대)역 개통
· 광역 스쿨버스 운행(사당,부평,잠실 등)
· 5개 광역버스와 21개 시내버스 운행

Come Fly with SWWU

홍보대사
박하은 학생

간호교육인증평가 5년 인증 / 보건의료정보관리교육인증평가 / 사회공헌대상 선정 / 대학기본역량진단 일반재정지원대학 / 기관평가

2026학년도 신입학 모집

입학처 https://entr.swwu.ac.kr/swc.do
입학상담 및 문의 031-290-8025,8028

수시1차	2025.09.08(월) ~ 09.30(화)
수시2차	2025.11.07(금) ~ 11.21(금)
정 시	2025.12.29(월) ~ 2026.01.14(수)

수 원 여 자 대 학
SUWON WOMEN'S UNIVERS

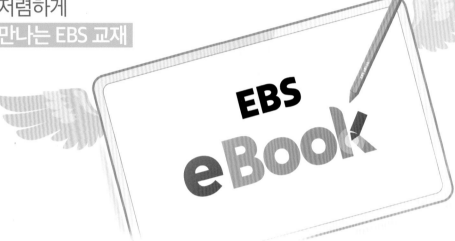

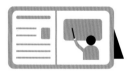